Alan M. Turing

The Essential Turing

Seminal Writings in Computing, Logic, Philosophy,
Artificial Intelligence, and Artificial Life
plus The Secrets of Enigma

Edited by **B. Jack Copeland**

CLARENDON PRESS · OXFORD

OXFORD
UNIVERSITY PRESS

Great Clarendon Street, Oxford OX2 6DP

Oxford University Press is a department of the University of Oxford.
It furthers the University's objective of excellence in research, scholarship,
and education by publishing worldwide in

Oxford New York

Auckland Bangkok Buenos Aires Cape Town Chennai
Dar es Salaam Delhi Hong Kong Istanbul Karachi Kolkata
Kuala Lumpur Madrid Melbourne Mexico City Mumbai Nairobi
São Paulo Shanghai Taipei Tokyo Toronto

Oxford is a registered trade mark of Oxford University Press
in the UK and in certain other countries

Published in the United States
by Oxford University Press Inc., New York

© In this volume the Estate of Alan Turing 2004

Supplementary Material © the several contributors 2004

The moral rights of the author have been asserted

Database right Oxford University Press (maker)

First published 2004

All rights reserved. No part of this publication may be reproduced,
stored in a retrieval system, or transmitted, in any form or by any means,
without the prior permission in writing of Oxford University Press,
or as expressly permitted by law, or under terms agreed with the appropriate
reprographics rights organization. Enquiries concerning reproduction
outside the scope of the above should be sent to the Rights Department,
Oxford University Press, at the address above

You must not circulate this book in any other binding or cover
and you must impose this same condition on any acquirer

British Library Cataloguing in Publication Data

Data available

Library of Congress Cataloging in Publication Data

Data available

ISBN 0-19-825079-7
ISBN 0-19-825080-0 (pbk.)

10 9 8 7 6 5 4 3 2 1

Typeset by Kolam Information Services Pvt. Ltd, Pondicherry, India
Printed in Great Britain
on acid-free paper by Biddles Ltd, King's Lynn, Norfolk.

Acknowledgements

Work on this book began in 2000 at the Dibner Institute for the History of Science and Technology, Massachusetts Institute of Technology, and was completed at the University of Canterbury, New Zealand. I am grateful to both these institutions for aid, and to the following for scholarly assistance: John Andreae, Friedrich Bauer, Frank Carter, Alonzo Church Jnr, David Clayden, Bob Doran, Ralph Erskine, Harry Fensom, Jack Good, John Harper, Geoff Hayes, Peter Hilton, Harry Huskey, Eric Jacobson, Elizabeth Mahon, Philip Marks, Elisabeth Norcliffe, Rolf Noskwith, Gualtiero Piccinini, Andrés Sicard, Wilfried Sieg, Frode Weierud, Maurice Wilkes, Mike Woodger, and especially Diane Proudfoot. This book would not have existed without the support of Turing's literary executor, P. N. Furbank, and that of Peter Momtchiloff at Oxford University Press.

B.J.C.

Contents

Alan Turing 1912–1954   1
Jack Copeland

Computable Numbers: A Guide   5
Jack Copeland

1. On Computable Numbers, with an Application to the Entscheidungsproblem (*1936*)   58

2. On Computable Numbers: Corrections and Critiques   91
   Alan Turing, Emil Post, and Donald W. Davies

3. Systems of Logic Based on Ordinals (*1938*), including excerpts from Turing's correspondence, 1936–1938   125

4. Letters on Logic to Max Newman (*c.1940*)   205

Enigma   217
Jack Copeland

5. History of Hut 8 to December 1941 (*1945*), featuring an excerpt from Turing's 'Treatise on the Enigma'   265
   Patrick Mahon

6. Bombe and Spider (*1940*)   313

7. Letter to Winston Churchill (*1941*)   336

8. Memorandum to OP-20-G on Naval Enigma (*c.1941*)   341

Artificial Intelligence   353
Jack Copeland

9. Lecture on the Automatic Computing Engine (*1947*)   362

10. Intelligent Machinery (*1948*)   395

11. Computing Machinery and Intelligence (*1950*)   433

12. Intelligent Machinery, A Heretical Theory (*c.1951*)   465

13. Can Digital Computers Think? (*1951*)   476

14. Can Automatic Calculating Machines Be Said to Think? (*1952*)   487
    Alan Turing, Richard Braithwaite, Geoffrey Jefferson, and Max Newman

Artificial Life   507
Jack Copeland

15. The Chemical Basis of Morphogenesis (*1952*)   519

16. Chess (*1953*)   562

17. Solvable and Unsolvable Problems (*1954*)   576

Index   597

Alan Turing 1912–1954

Jack Copeland

Alan Mathison Turing was born on 23 June 1912 in London[1]; he died on 7 June 1954 at his home in Wilmslow, Cheshire. Turing contributed to logic, mathematics, biology, philosophy, cryptanalysis, and formatively to the areas later known as computer science, cognitive science, Artificial Intelligence, and Artificial Life.

Educated at Sherborne School in Dorset, Turing went up to King's College, Cambridge, in October 1931 to read Mathematics. He graduated in 1934, and in March 1935 was elected a Fellow of King's, at the age of only 22. In 1936 he published his most important theoretical work, 'On Computable Numbers, with an Application to the Entscheidungsproblem [Decision Problem]' (Chapter 1, with corrections in Chapter 2). This article described the abstract digital computing machine—now referred to simply as the universal Turing machine—on which the modern computer is based. Turing's fundamental idea of a universal stored-programme computing machine was promoted in the United States by John von Neumann and in England by Max Newman. By the end of 1945 several groups, including Turing's own in London, were devising plans for an electronic stored-programme universal digital computer—a Turing machine in hardware.

In 1936 Turing left Cambridge for the United States in order to continue his research at Princeton University. There in 1938 he completed a Ph.D. entitled 'Systems of Logic Based on Ordinals', subsequently published under the same title (Chapter 3, with further exposition in Chapter 4). Now a classic, this work addresses the implications of Gödel's famous incompleteness result. Turing gave a new analysis of mathematical reasoning, and continued the study, begun in 'On Computable Numbers', of uncomputable problems—problems that are 'too hard' to be solved by a computing machine (even one with unlimited time and memory).

Turing returned to his Fellowship at King's in the summer of 1938. At the outbreak of war with Germany in September 1939 he moved to Bletchley Park, the wartime headquarters of the Government Code and Cypher School (GC & CS). Turing's brilliant work at Bletchley Park had far-reaching consequences.

---

[1] At 2 Warrington Crescent, London W9, where now there is a commemorative plaque.

'I won't say that what Turing did made us win the war, but I daresay we might have lost it without him', said another leading Bletchley cryptanalyst.[2] Turing broke Naval Enigma—a decisive factor in the Battle of the Atlantic—and was the principal designer of the 'bombe', a high-speed codebreaking machine. The ingenious bombes produced a flood of high-grade intelligence from Enigma. It is estimated that the work done by Turing and his colleagues at GC & CS shortened the war in Europe by at least two years.[3] Turing's contribution to the Allied victory was a state secret and the only official recognition he received, the Order of the British Empire, was in the circumstances derisory. The full story of Turing's involvement with Enigma is told for the first time in this volume, the material that forms Chapters 5, 6, and 8 having been classified until recently.

In 1945, the war over, Turing was recruited to the National Physical Laboratory (NPL) in London, his brief to design and develop an electronic digital computer—a concrete form of the universal Turing machine. His design (for the Automatic Computing Engine or ACE) was more advanced than anything else then under consideration on either side of the Atlantic. While waiting for the engineers to build the ACE, Turing and his group pioneered the science of computer programming, writing a library of sophisticated mathematical programmes for the planned machine.

Turing founded the field now called 'Artificial Intelligence' (AI) and was a leading early exponent of the theory that the human brain is in effect a digital computer. In February 1947 he delivered the earliest known public lecture to mention computer intelligence ('Lecture on the Automatic Computing Engine' (Chapter 9)). His technical report 'Intelligent Machinery' (Chapter 10), written for the NPL in 1948, was effectively the first manifesto of AI. Two years later, in his now famous article 'Computing Machinery and Intelligence' (Chapter 11), Turing proposed (what subsequently came to be called) the Turing test as a criterion for whether machines can think. *The Essential Turing* collects together for the first time the series of five papers that Turing devoted exclusively to Artificial Intelligence (Chapters 10, 11, 12, 13, 16). Also included is a discussion of AI by Turing, Newman, and others (Chapter 14).

In the end, the NPL's engineers lost the race to build the world's first working electronic stored-programme digital computer—an honour that went to the Computing Machine Laboratory at the University of Manchester in June 1948. The concept of the universal Turing machine was a fundamental influence on the Manchester computer project, via Newman, the project's instigator. Later in

---

[2] Jack Good in an interview with Pamela McCorduck, on p. 53 of her *Machines Who Think* (New York: W. H. Freeman, 1979).

[3] This estimate is given by Sir Harry Hinsley, official historian of the British Secret Service, writing on p. 12 of his and Alan Stripp's edited volume *Codebreakers: The Inside Story of Bletchley Park* (Oxford: Oxford University Press, 1993).

1948, at Newman's invitation, Turing took up the deputy directorship of the Computing Machine Laboratory (there was no Director). Turing spent the rest of his short career at Manchester University. He was elected a Fellow of the Royal Society of London in March 1951 (a high honour) and in May 1953 was appointed to a specially created Readership in the Theory of Computing at Manchester.

It was at Manchester, in March 1952, that he was prosecuted for homosexual activity, then a crime in Britain, and sentenced to a period of twelve months' hormone 'therapy'—the shabbiest of treatment from the country he had helped save, but which he seems to have borne with amused fortitude.

Towards the end of his life Turing pioneered the area now known as Artificial Life. His 1952 article 'The Chemical Basis of Morphogenesis' (Chapter 15) describes some of his research on the development of pattern and form in living organisms. This research dominated his final years, but he nevertheless found time to publish in 1953 his classic article on computer chess (Chapter 16) and in 1954 'Solvable and Unsolvable Problems' (Chapter 17), which harks back to 'On Computable Numbers'. From 1951 he used the Computing Machine Laboratory's Ferranti Mark I (the first commercially produced electronic stored-programme computer) to model aspects of biological growth, and in the midst of this groundbreaking work he died.

Turing's was a far-sighted genius and much of the material in this book is of even greater relevance today than in his lifetime. His research had remarkable breadth and the chapters range over a diverse collection of topics—mathematical logic and the foundations of mathematics, computer design, mechanical methods in mathematics, cryptanalysis and chess, the nature of intelligence and mind, and the mechanisms of biological growth. The chapters are united by the overarching theme of Turing's work, his enquiry into (as Newman put it) 'the extent and the limitations of mechanistic explanations'.[4]

Biographies of Turing

Gottfried, T., *Alan Turing: The Architect of the Computer Age* (Danbury, Conn.: Franklin Watts, 1996).
Hodges, A., *Alan Turing: The Enigma* (London: Burnett, 1983).
Newman, M. H. A., 'Alan Mathison Turing, 1912–1954', *Biographical Memoirs of Fellows of the Royal Society*, 1 (1955), 253–63.
Turing, S., *Alan M. Turing* (Cambridge: W. Heffer, 1959).

---

[4] M. H. A. Newman, 'Alan Mathison Turing, 1912–1954', *Biographical Memoirs of Fellows of the Royal Society*, 1 (1955), 253–63 (256).

Computable Numbers: A Guide

Jack Copeland

Part I The Computer
1. Turing Machines   *6*
2. Standard Descriptions and Description Numbers   *10*
3. Subroutines   *12*
4. The Universal Computing Machine   *15*
5. Turing, von Neumann, and the Computer   *21*
6. Turing and Babbage   *27*
7. Origins of the Term 'Computer Programme'   *30*

Part II Computability and Uncomputability
8. Circular and Circle-Free Machines   *32*
9. Computable and Uncomputable Sequences   *33*
10. Computable and Uncomputable Numbers   *36*
11. The Satisfactoriness Problem   *36*
12. The Printing and Halting Problems   *39*
13. The Church-Turing Thesis   *40*
14. The *Entscheidungsproblem*   *45*

'On Computable Numbers, with an Application to the Entscheidungsproblem' appeared in the *Proceedings of the London Mathematical Society* in 1936.[1] This,

---

[1] *Proceedings of the London Mathematical Society*, 42 (1936–7), 230–65. The publication date of 'On Computable Numbers' is sometimes cited, incorrectly, as 1937. The article was published in two parts, both parts appearing in 1936. The break between the two parts occurred, rather inelegantly, in the middle of Section 5, at the bottom of p. 240 (p. 67 in the present volume). Pages 230–40 appeared in part 3 of volume 42, issued on 30 Nov. 1936, and the remainder of the article appeared in part 4, issued on 23 Dec. 1936. This information is given on the title pages of parts 3 and 4 of volume 42, which show the contents of each part and their dates of issue. (I am grateful to Robert Soare for sending me these pages. See R. I. Soare, 'Computability and Recursion', *Bulletin of Symbolic Logic*, 2 (1996), 284–321.)

The article was published bearing the information 'Received 28 May, 1936.—Read 12 November, 1936.' However, Turing was in the United States on 12 November, having left England in September 1936 for what was to be a stay of almost two years (see the introductions to Chapters 3 and 4). Although papers were read at the meetings of the London Mathematical Society, many of those published in the *Proceedings* were 'taken as read', the author not necessarily being present at the meeting in question. Mysteriously, the minutes of the meeting held on 18 June 1936 list 'On Computable Numbers, with an Application to the Entscheidungsproblem' as one of 22 papers taken as read at that meeting. The minutes of an Annual General Meeting held

Turing's second publication,[2] contains his most significant work. Here he pioneered the theory of computation, introducing the famous abstract computing machines soon dubbed 'Turing machines' by the American logician Alonzo Church.[3] 'On Computable Numbers' is regarded as the founding publication of the modern science of computing. It contributed vital ideas to the development, in the 1940s, of the electronic stored-programme digital computer. 'On Computable Numbers' is the birthplace of the fundamental principle of the modern computer, the idea of controlling the machine's operations by means of a programme of coded instructions stored in the computer's memory.

In addition Turing charted areas of mathematics lying beyond the scope of the Turing machine. He proved that not all precisely stated mathematical problems can be solved by computing machines. One such is the *Entscheidungsproblem* or 'decision problem'. This work—together with contemporaneous work by Church[4]—initiated the important branch of mathematical logic that investigates and codifies problems 'too hard' to be solvable by Turing machine.

In this one article, Turing ushered in both the modern computer and the mathematical study of the uncomputable.

Part I The Computer

1. Turing Machines

A Turing machine consists of a scanner and a limitless memory-tape that moves back and forth past the scanner. The tape is divided into squares. Each square may be blank or may bear a single symbol—'0' or '1', for example, or some other symbol taken from a finite alphabet. The scanner is able to examine only one square of tape at a time (the 'scanned square').

The scanner contains mechanisms that enable it to *erase* the symbol on the scanned square, to *print* a symbol on the scanned square, and to *move* the tape to the left or right, one square at a time.

In addition to the operations just mentioned, the scanner is able to alter what Turing calls its '*m*-configuration'. In modern Turing-machine jargon it is usual to

---

on 12 Nov. 1936 contain no reference to the paper. (I am grateful to Janet Foster, Archives Consultant to the London Mathematical Society, for information.)

[2] The first was 'Equivalence of Left and Right Almost Periodicity', *Journal of the London Mathematical Society*, 10 (1935), 284–5.

[3] Church introduced the term 'Turing machine' in a review of Turing's paper in the *Journal of Symbolic Logic*, 2 (1937), 42–3.

[4] A. Church, 'An Unsolvable Problem of Elementary Number Theory', *American Journal of Mathematics*, 58 (1936), 345–63, and 'A Note on the Entscheidungsproblem', *Journal of Symbolic Logic*, 1 (1936), 40–1.

Computable Numbers: A Guide | 7

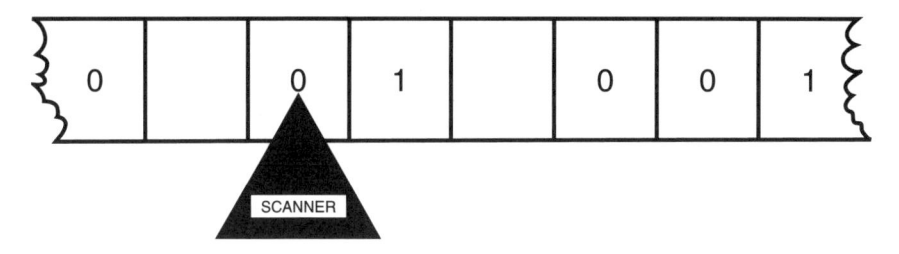

| 0 | | 0 | 1 | | 0 | 0 | 1 |

SCANNER

use the term 'state' in place of '*m*-configuration'. A device within the scanner is capable of adopting a number of different states (*m*-configurations), and the scanner is able to alter the state of this device whenever necessary. The device may be conceptualized as consisting of a dial with a (finite) number of positions, labelled 'a', 'b', 'c', etc. Each of these positions counts as an *m*-configuration or state, and changing the *m*-configuration or state amounts to shifting the dial's pointer from one labelled position to another. This device functions as a simple memory. As Turing says, 'by altering its *m*-configuration the machine can effectively remember some of the symbols which it has "seen" (scanned) previously' (p. 59). For example, a dial with two positions can be used to keep a record of which binary digit, 0 or 1, is present on the square that the scanner has just vacated. (If a square might also be blank, then a dial with three positions is required.)

The operations just described—erase, print, move, and change state—are the *basic* (or *atomic*) operations of the Turing machine. Complexity of operation is achieved by chaining together large numbers of these simple basic actions. Commercially available computers are hard-wired to perform basic operations considerably more sophisticated than those of a Turing machine—add, multiply, decrement, store-at-address, branch, and so forth. The precise list of basic operations varies from manufacturer to manufacturer. It is a remarkable fact, however, that despite the austere simplicity of Turing's machines, they are capable of computing anything that any computer on the market can compute. Indeed, because they are abstract machines, with unlimited memory, they are capable of computations that no actual computer could perform in practice.

Example of a Turing machine

The following simple example is from Section 3 of 'On Computable Numbers' (p. 61). The once-fashionable Gothic symbols that Turing used in setting out the example—and also elsewhere in 'On Computable Numbers'—are not employed in this guide. I also avoid typographical conventions used by Turing that seem likely to hinder understanding (for example, his special symbol 'ə', which he used to mark the beginning of the tape, is here replaced by '!').

The machine in Turing's example—call it **M**—starts work with a blank tape. The tape is endless. The problem is to set up the machine so that if the scanner is positioned over any square of the tape and the machine set in motion, the scanner will print alternating binary digits on the tape, 0 1 0 1 0 1 ..., working to the right from its starting place, and leaving a blank square in between each digit:

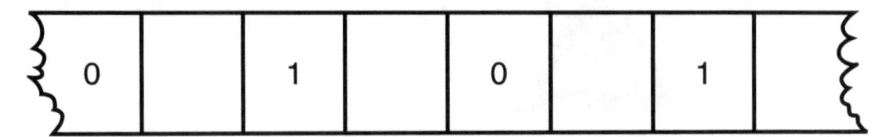

| 0 | | 1 | | 0 | | 1 |

In order to do its work, M makes use of four states or *m*-configurations. These are labelled 'a', 'b', 'c', and 'd'. (Turing employed less familiar characters.) M is in state **a** when it starts work.

The operations that M is to perform can be set out by means of a table with four columns (Table 1). 'R' abbreviates the instruction 'reposition the scanner one square to the right'. This is achieved by moving the tape one square to the left. 'L' abbreviates 'reposition the scanner one square to the left', 'P[0]' abbreviates 'print 0 on the scanned square', and likewise 'P[1]'. Thus the top line of Table 1 reads: if you are in state **a** and the square you are scanning is blank, then print 0 on the scanned square, move the scanner one square to the right, and go into state **b**.

A machine acting in accordance with this table of instructions—or *programme*—toils endlessly on, printing the desired sequence of digits while leaving alternate squares blank.

Turing does not explain how it is to be brought about that the machine acts in accordance with the instructions. There is no need. Turing's machines are abstractions and it is not necessary to propose any specific mechanism for causing the machine to act in accordance with the instructions. However, for purposes of visualization, one might imagine the scanner to be accompanied by a bank of switches and plugs resembling an old-fashioned telephone switchboard. Arranging the plugs and setting the switches in a certain way causes the machine to act in accordance with the instructions in Table 1. Other ways of setting up the 'switchboard' cause the machine to act in accordance with other tables of instructions. In fact, the earliest electronic digital computers, the British Colossus (1943) and the American ENIAC (1945), were programmed in very much this way. Such machines are described as 'programme-controlled', in order to distinguish them from the modern 'stored-programme' computer.

Table 1

State	Scanned Square	Operations	Next State
a	blank	P[0], R	b
b	blank	R	c
c	blank	P[1], R	d
d	blank	R	a

As everyone who can operate a personal computer knows, the way to set up a stored-programme machine to perform some desired task is to open the appropriate programme of instructions stored in the computer's memory. The stored-programme concept originates with Turing's *universal* computing machine, described in detail in Section 4 of this guide. By inserting different programmes into the memory of the universal machine, the machine is made to carry out different computations. Turing's 1945 technical report 'Proposed Electronic Calculator' was the first relatively complete specification of an electronic stored-programme digital computer (see Chapter 9).

E-squares and F-squares

After describing M and a second example of a computing machine, involving the start-of-tape marker '!' (p. 62), Turing introduces a convention which he makes use of later in the article (p. 63). Since the tape is the machine's general-purpose storage medium—serving not only as the vehicle for data storage, input, and output, but also as 'scratchpad' for use during the computation—it is useful to divide up the tape in some way, so that the squares used as scratchpad are distinguished from those used for the various other functions just mentioned.

Turing's convention is that every alternate square of the tape serves as scratchpad. These he calls the 'E-squares', saying that the 'symbols on E-squares will be liable to erasure' (p. 63). The remaining squares he calls 'F-squares'. ('E' and 'F' perhaps stand for 'erasable' and 'fixed'.)

In the example just given, the 'F-squares' of M's tape are the squares bearing the desired sequence of binary digits, 0 1 0 1 0 1... In between each pair of adjacent F-squares lies a blank E-square. The computation in this example is so simple that the E-squares are never used. More complex computations make much use of E-squares.

Turing mentions one important use of E-squares at this point (p. 63): any F-square can be 'marked' by writing some special symbol, e.g. '*', on the E-square immediately to its right. By this means, the scanner is able to find its way back to a particular string of binary digits—a particular item of data, say. The scanner locates the first digit of the string by finding the marker '*'.

Adjacent blank squares

Another useful convention, also introduced on p. 63, is to the effect that the tape must never contain a run of non-blank squares followed by two or more adjacent blank squares that are themselves followed by one or more non-blank squares. The value of this convention is that it gives the machine an easy way of finding the last non-blank square. As soon as the machine finds two adjacent blank squares, it knows that it has passed beyond the region of tape that has been written on and has entered the region of blank squares stretching away endlessly.

The start-of-tape marker

Turing usually considers tapes that are endless in one direction only. For purposes of visualization, these tapes may all be thought of as being endless to the right. By convention, each of the first two squares of the tape bears the symbol '!', mentioned previously. These 'signposts' are never erased. The scanner searches for the signposts when required to find the beginning of the tape.

2. Standard Descriptions and Description Numbers

In the final analysis, a computer programme is simply a (long) stream, or row, of characters. Combinations of characters encode the instructions. In Section 5 of 'On Computable Numbers' Turing explains how an instruction table is to be converted into a row of letters, which he calls a 'standard description'. He then explains how a standard description can be converted into a single number. He calls these 'description numbers'.

Each line of an instruction table can be re-expressed as a single 'word' of the form $q_i S_j S_k M q_l$. q_i is the state shown in the left-hand column of the table. S_j is the symbol on the scanned square (a blank is counted as a type of symbol). S_k is the symbol that is to be printed on the scanned square. M is the direction of movement (if any) of the scanner, left or right. q_l is the next state. For example, the first line of Table 1 can be written: a-0Rb (using '-' to represent a blank). The third line is: c-1Rd.

The second line of the table, which does not require the contents of the scanned square (a blank) to be changed, is written: b--Rc. That is to say we imagine, for the purposes of this new notation, that the operations column of the instruction table contains the redundant instruction P[-]. This device is employed whenever an instruction calls for no change to the contents of the scanned square, as in the following example:

State	Scanned Square	Operations	Next State
d	x	L	c

It is imagined that the operations column contains the redundant instruction P[x], enabling the line to be expressed: dxxLc.

Sometimes a line may contain no instruction to move. For example:

State	Scanned Square	Operations	Next State
d	*	P[1]	c

The absence of a move is indicated by including 'N' in the instruction-word: d*1Nc.

Sometimes a line may contain an instruction to erase the symbol on the scanned square. This is denoted by the presence of 'E' in the 'operations' column:

State	Scanned Square	Operations	Next State
m	*	E, R	n

Turing notes that E is equivalent to P[-]. The corresponding instruction-word is therefore m*-Rn.

Any table of instructions can be rewritten in the form of a stream of instruction-words separated by semicolons.[5] Corresponding to Table 1 is the stream:

a-0Rb; b--Rc; c-1Rd; d--Ra;

This stream can be converted into a stream consisting uniformly of the letters A, C, D, L, R, and N (and the semicolon). Turing calls this a *standard description* of the machine in question. The process of conversion is done in such a way that the individual instructions can be retrieved from the standard description.

The standard description is obtained as follows. First, '-' is replaced by 'D', '0' by 'DC', and '1' by 'DCC'. (In general, if we envisage an ordering of all the printable symbols, the *n*th symbol in the ordering is replaced by a 'D' followed by *n* repetitions of 'C'.) This produces:

aDDCRb; bDDRc; cDDCCRd; dDDRa;

Next, the lower case state-symbols are replaced by letters. 'a' is replaced by 'DA', 'b' by 'DAA', 'c' by 'DAAA', and so on. An obvious advantage of the new notation is that there is no limit to the number of states that can be named in this way.

The standard description corresponding to Table 1 is:

DADDCRDAA; DAADDRDAAA; DAAADDCCRDAAAA; DAAAADDRDA;

Notice that occurrences of 'D' serve to mark out the different segments or regions of each instruction-word. For example, to determine which symbol an instruction-word says to print, find the third 'D' to the right from the beginning of the word, and count the number of occurrences of 'C' between it and the next D to the right.

The standard description can be converted into a number, called a *description number*. Again, the process of conversion is carried out in such a way that the individual instructions can be retrieved from the description number. A standard description is converted into a description number by means of replacing each 'A' by '1', 'C' by '2', 'D' by '3', 'L' by '4', 'R' by '5', 'N' by '6', and ';' by 7. In the case of the above example this produces:

31332531173113353111731113322531111731111335317.[6]

---

[5] There is a subtle issue concerning the placement of the semicolons. See Davies's 'Corrections to Turing's Universal Computing Machine', Sections 3, 7, 10.

[6] Properly speaking, the description number is not the string '3133253117311335311173111332253 1111731111335317', but is the number denoted by this string of numerals.

Occurrences of '7' mark out the individual instruction-words, and occurrences of '3' mark out the different regions of the instruction-words. For example: to find out which symbol the third instruction-word says to print, find the second '7' (starting from the left), then the third '3' to the right of that '7', and count the number of occurrences of '2' between that '3' and the next '3' to the right. To find out the exit state specified by the third instruction-word, find the last '3' in that word and count the number of occurrences of '1' between it and the next '7' to the right.

Notice that *different* standard descriptions can describe the behaviour of one and the same machine. For example, interchanging the first and second lines of Table 1 does not in any way affect the behaviour of the machine operating in accordance with the table, but a different standard description—and therefore a different description number—will ensue if the table is modified in this way.

This process of converting a table of instructions into a standard description or a description number is analogous to the process of *compiling* a computer programme into 'machine code'. Programmers generally prefer to work in so-called high-level languages, such as Pascal, Prolog, and C. Programmes written in a high-level language are, like Table 1, reasonably easy for a trained human being to follow. Before a programme can be executed, the instructions must be translated, or compiled, into the form required by the computer (machine code).

The importance of standard descriptions and description numbers is explained in what follows.

3. Subroutines

Subroutines are programmes that are used as components of other programmes. A subroutine may itself have subroutines as components. Programmers usually have access to a 'library' of commonly used subroutines—the programmer takes ready-made subroutines 'off the shelf' whenever necessary.

Turing's term for a subroutine was 'subsidiary table'. He emphasized the importance of subroutines in a lecture given in 1947 concerning the Automatic Computing Engine or ACE, the electronic stored-programme computer that he began designing in 1945 (see Chapter 9 and the introduction to Chapter 10):

Probably the most important idea involved in instruction tables is that of standard *subsidiary tables*. Certain processes are used repeatedly in all sorts of different connections, and we wish to use the same instructions...every time...We have only to think out how [a process] is to be done once, and forget then how it is done.[7]

In 'On Computable Numbers'—effectively the first programming manual of the computer age—Turing introduced a library of subroutines for Turing machines (in Sections 4 and 7), saying (p. 63):

---

[7] The quotation is from p. 389 below.

There are certain types of process used by nearly all machines, and these, in some machines, are used in many connections. These processes include copying down sequences of symbols, comparing sequences, erasing all symbols of a given form, etc.

Some examples of subroutines are:

cpe(A, B, x, y) (p. 66):

'cpe' may be read 'compare for equality'. This subroutine compares the string of symbols marked with an *x* to the string of symbols marked with a *y*. The subroutine places the machine in state *B* if the two strings are the same, and in state *A* if they are different. Note: throughout these examples, '*A*' and '*B*' are variables representing any states; '*x*' and '*y*' are variables representing any symbols.

f(A, B, x) (p. 63):

'f' stands for 'find'. This subroutine finds the leftmost occurrence of *x*. f(A, B, x) moves the scanner left until the start of the tape is encountered. Then the scanner is moved to the right, looking for the first *x*. As soon as an *x* is found, the subroutine places the machine in state *A*, leaving the scanner resting on the *x*. If no *x* is found anywhere on the portion of tape that has so far been written on, the subroutine places the machine in state *B*, leaving the scanner resting on a blank square to the right of the used portion of the tape.

e(A, B, x) (p. 64):

'e' stands for 'erase'. The subroutine e(A, B, x) contains the subroutine f(A, B, x). e(A, B, x) finds the leftmost occurrence of symbol *x* and erases it, placing the machine in state *A* and leaving the scanner resting on the square that has just been erased. If no *x* is found the subroutine places the machine in state *B*, leaving the scanner resting on a blank square to the right of the used portion of the tape.

The subroutine f(A, B, x)

It is a useful exercise to construct f(A, B, x) explicitly, i.e. in the form of a table of instructions. Suppose we wish the machine to enter the subroutine f(A, B, x) when placed in state **n**, say. Then the table of instructions is as shown in Table 2. (Remember that by the convention mentioned earlier, if ever the scanner encounters two adjacent blank squares, it has passed beyond the region of tape that has been written on and has entered the region of blank squares stretching away to the right.)

As Turing explains, f(A, B, x) is in effect built out of two further subroutines, which he writes $f_1(A, B, x)$ and $f_2(A, B, x)$. The three rows of Table 2 with an 'm' in the first column form the subroutine $f_1(A, B, x)$, and the three rows with 'o' in the first column form $f_2(A, B, x)$.

Skeleton tables

For ease of defining subroutines Turing introduces an abbreviated form of instruction table, in which one is allowed to write expressions referring to

Table 2

State	Scanned Square	Operations	Next State	Comments
n	does not contain !	L	n	Search for the first square.
n	!	L	m	Found right-hand member of the pair '!!'; move left to first square of tape; go into state **m**. (Notice that x might be '!'.)
m	x	none	A	Found x; go into state A; subroutine ends.
m	neither x nor blank	R	m	Keep moving right looking for x or a blank.
m	blank	R	o	Blank square encountered; go into state **o** and examine next square to the right.
o	x	none	A	Found x; go into state A; subroutine ends.
o	neither x nor blank	R	m	Found a blank followed by a non-blank square but no x; switch to state **m** and keep looking for x.
o	blank	R	B	Two adjacent blank squares encountered; go into state B; subroutine ends.

Table 3

$$f(A, B, x) \begin{cases} \text{not !} & L & f(A, B, x) \\ ! & L & f_1(A, B, x) \end{cases}$$

$$f_1(A, B, x) \begin{cases} x & & A \\ \text{neither } x \text{ nor blank} & R & f_1(A, B, x) \\ \text{blank} & R & f_2(A, B, x) \end{cases}$$

$$f_2(A, B, x) \begin{cases} x & & A \\ \text{neither } x \text{ nor blank} & R & f_1(A, B, x) \\ \text{blank} & R & B \end{cases}$$

subroutines in the first and fourth columns (the state columns). Turing calls these abbreviated tables 'skeleton tables' (p. 63). For example, the skeleton table corresponding to Table 2 is as in Table 3.

Turing's notation for subroutines is explained further in the appendix to this guide ('Subroutines and *m*-functions').

4. The Universal Computing Machine

In Section 7 of 'On Computable Numbers' Turing introduces his 'universal computing machine', now known simply as the universal Turing machine. The universal Turing machine is the stored-programme digital computer in abstract conceptual form.

The universal computing machine has a single, fixed table of instructions (which we may imagine to have been set into the machine, once and for all, by way of the switchboard-like arrangement mentioned earlier). Operating in accordance with this table of instructions, the universal machine is able to carry out *any* task for which an instruction table can be written. The trick is to put an instruction table—programme—for carrying out the desired task onto the tape of the universal machine.

The instructions are placed on the tape in the form of a standard description—i.e. in the form of a string of letters that encodes the instruction table. The universal machine reads the instructions and carries them out on its tape.

The universal Turing machine and the modern computer

Turing's greatest contributions to the development of the modern computer were:

- The idea of controlling the function of a computing machine by storing a programme of symbolically encoded instructions in the machine's memory.
- His demonstration (in Section 7 of 'On Computable Numbers') that, by this means, a *single* machine of *fixed structure* is able to carry out every computation that can be carried out by any Turing machine whatsoever, i.e. is universal.

Turing's teacher and friend Max Newman has testified that Turing's interest in building a stored-programme computing machine dates from the time of 'On Computable Numbers'. In a tape-recorded interview Newman stated, 'Turing himself, right from the start, said it would be interesting to try and make such a machine'.[8] (It was Newman who, in a lecture on the foundations of mathematics and logic given in Cambridge in 1935, launched Turing on the research that led to the universal Turing machine; see the introduction to Chapter 4.[9]) In his obituary of Turing, Newman wrote:

The description that [Turing] gave of a 'universal' computing machine was entirely theoretical in purpose, but Turing's strong interest in all kinds of practical experiment

---

[8] Newman in interview with Christopher Evans ('The Pioneers of Computing: An Oral History of Computing', London, Science Museum).
[9] Ibid.

made him even then interested in the possibility of actually constructing a machine on these lines.[10]

Turing later described the connection between the universal computing machine and the stored-programme digital computer in the following way (Chapter 9, pp. 378 and 383):

Some years ago I was researching on what might now be described as an investigation of the theoretical possibilities and limitations of digital computing machines. I considered a type of machine which had a central mechanism, and an infinite memory which was contained on an infinite tape... It can be shown that a single special machine of that type can be made to do the work of all... The special machine may be called the universal machine; it works in the following quite simple manner. When we have decided what machine we wish to imitate we punch a description of it on the tape of the universal machine. This description explains what the machine would do in every configuration in which it might find itself. The universal machine has only to keep looking at this description in order to find out what it should do at each stage. Thus the complexity of the machine to be imitated is concentrated in the tape and does not appear in the universal machine proper in any way... [D]igital computing machines such as the ACE... are in fact practical versions of the universal machine. There is a certain central pool of electronic equipment, and a large memory. When any particular problem has to be handled the appropriate instructions for the computing process involved are stored in the memory of the ACE and it is then 'set up' for carrying out that process.

Turing's idea of a universal stored-programme computing machine was promulgated in the USA by von Neumann and in the UK by Newman, the two mathematicians who, along with Turing himself, were by and large responsible for placing Turing's abstract universal machine into the hands of electronic engineers.

By 1946 several groups in both countries had embarked on creating a universal Turing machine in hardware. The race to get the first electronic stored-programme computer up and running was won by Manchester University where, in Newman's Computing Machine Laboratory, the 'Manchester Baby' ran its first programme on 21 June 1948. Soon after, Turing designed the input/output facilities and the programming system of an expanded machine known as the Manchester Mark I.[11] (There is more information about the Manchester computer in the introductions to Chapters 4, 9, and 10, and in 'Artificial Life'.) A small pilot version of Turing's Automatic Computing Engine first ran in 1950, at the National Physical Laboratory in London (see the introductions to Chapters 9 and 10).

[10] 'Dr. A. M. Turing', *The Times*, 16 June 1954, p. 10.

[11] F. C. Williams described some of Turing's contributions to the Manchester machine in a letter written in 1972 to Brian Randell (parts of which are quoted in B. Randell, 'On Alan Turing and the Origins of Digital Computers', in B. Meltzer and D. Michie (eds.), *Machine Intelligence 7* (Edinburgh: Edinburgh University Press, 1972)); see the introduction to Chapter 9 below. A digital facsimile of Turing's *Programmers' Handbook for Manchester Electronic Computer* (University of Manchester Computing Machine Laboratory, 1950) is in The Turing Archive for the History of Computing <www.AlanTuring.net/programmers_handbook>.

By 1951 electronic stored-programme computers had begun to arrive in the market place. The first model to go on sale was the Ferranti Mark I, the production version of the Manchester Mark I (built by the Manchester firm Ferranti Ltd.). Nine of the Ferranti machines were sold, in Britain, Canada, the Netherlands, and Italy, the first being installed at Manchester University in February 1951.[12] In the United States the first UNIVAC (built by the Eckert-Mauchly Computer Corporation) was installed later the same year. The LEO computer also made its debut in 1951. LEO was a commercial version of the prototype EDSAC machine, which at Cambridge University in 1949 had become the second stored-programme electronic computer to function.[13] 1953 saw the IBM 701, the company's first mass-produced stored-programme electronic computer. A new era had begun.

How the universal machine works

The details of Turing's universal machine, given on pp. 69–72, are moderately complicated. However, the basic principles of the universal machine are, as Turing says, simple.

Let us consider the Turing machine **M** whose instructions are set out in Table 1. (Recall that **M**'s scanner is positioned initially over any square of **M**'s endless tape, the tape being completely blank.) If a standard description of **M** is placed on the universal machine's tape, the universal machine will *simulate* or *mimic* the actions of **M**, and will produce, on specially marked squares of its tape, the output sequence that **M** produces, namely:

$$0\ 1\ 0\ 1\ 0\ 1\ 0\ 1\ 0\ 1 \ldots$$

The universal machine does this by reading the instructions that the standard description contains and carrying them out on its own tape.

In order to start work, the universal machine requires on its tape not only the standard description but also a record of **M**'s intial state (**a**) and the symbol that **M** is initially scanning (a blank). The universal machine's own tape is initially blank except for this record and **M**'s standard description (and some ancillary punctuation symbols mentioned below). As the simulation of **M** progresses, the universal machine prints a record on its tape of:

- the symbols that **M** prints
- the position of **M**'s scanner at each step of the computation
- the symbol 'in' the scanner
- **M**'s state at each step of the computation.

---

[12] S. Lavington, 'Computer Development at Manchester University', in N. Metropolis, J. Howlett, and G. C. Rota (eds.), *A History of Computing in the Twentieth Century* (New York: Academic Press, 1980).

[13] See M. V. Wilkes, *Memoirs of a Computer Pioneer* (Cambridge, Mass.: MIT Press, 1985).

When the universal machine is started up, it reads from its tape M's initial state and initial symbol, and then searches through M's standard description for the instruction beginning: 'when in state a and scanning a blank...' The relevant instruction from Table 1 is:

        a   blank   P[0], R   b

The universal machine accordingly prints '0'. It then creates a record on its tape of M's new state, b, and the new position of M's scanner (i.e. immediately to the right of the '0' that has just been printed on M's otherwise blank tape). Next, the universal machine searches through the standard description for the instruction beginning 'when in state b and scanning a blank...'. And so on.

How does the universal machine do its record-keeping? After M executes its first instruction, the relevant portion of M's tape would look like this—using 'b' both to record M's state and to indicate the position of the scanner. All the other squares of M's tape to the left and right are blank.

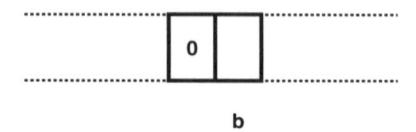

The universal machine keeps a record of this state of affairs by employing three squares of tape (pp. 62, 68):

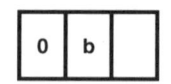

The symbol 'b' has the double function of recording M's state and indicating the position of M's scanner. The square immediately to the right of the state-symbol displays the symbol 'in' M's scanner (a blank).

What does the universal machine's tape look like before the computation starts? The standard description corresponding to Table 1 is:

    DADDCRDAA; DAADDRDAAA; DAAADDCCRDAAAA; DAAAADDRDA;

The operator places this programme on the universal machine's tape, writing only on F-squares and beginning on the second F-square of the tape. The first F-square and the first E-square are marked with the start-of-tape symbol '!'. The E-squares (shaded in the diagram) remain blank (except for the first).

On the F-square following the final semicolon of the programme, the operator writes the end-of-programme symbol '::'. On the next F-square to the right of this symbol, the operator places a record of M's initial state, a, and leaves the following F-square blank in order to indicate that **M** is initially scanning a blank. The next F-square to the right is then marked with the punctuation symbol ':'. This completes the setting-up of the tape:

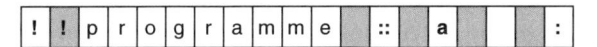

| ! | ! | p | r | o | g | r | a | m | m | e | | :: | | a | | | : |

What does the universal machine's tape look like as the computation progresses? In response to the first instruction in the standard description, the universal machine creates the record '0b-:' (in describing the tape, '-' will be used to represent a blank) on the next four F-squares to the right of the first ':'. Depicting only the portion of tape to the right of the end-of-programme marker '::' (and ignoring any symbols which the universal machine may have written on the E-squares in the course of dealing with the first instruction), the tape now looks like this:

| :: | | a | | | : | | 0 | | b | | | : |

Next the universal machine searches for the instruction beginning 'when in state **b** and scanning a blank...'. The relevant instruction from Table 1 is

    b   blank   R   c

This instruction would put **M** into the condition:

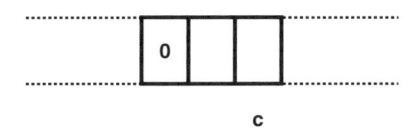

| | 0 | | |

       c

So the universal machine creates the record '0-c-:' on its tape:

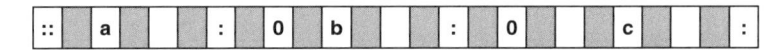

| :: | | a | | | : | | 0 | | b | | | : | | 0 | | | c | | | : |

Each pair of punctuation marks frames a representation (on the F-squares) of **M**'s tape extending from the square that was in the scanner at start-up to the furthest square to the right to have been scanned at that stage of the computation.

The next instruction is:

    c   blank   P[1], R   d

This causes the universal machine to create the record '0-1d-:'. (The diagram represents a single continuous strip of tape.)

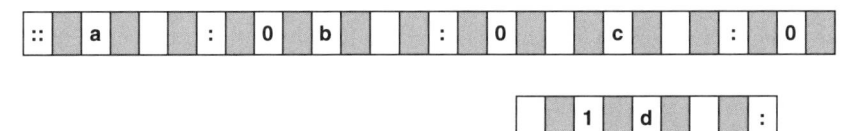

| :: | | a | | | : | | 0 | | b | | | : | | 0 | | | c | | | : | | 0 |
| | | | | | | | | 1 | | d | | | : |

And so on. Record by record, the outputs produced by the instructions in Table 1 appear on the universal machine's tape.

Turing also introduces a variation on this method of record-keeping, whereby the universal machine additionally prints on the tape a second record of the binary digits printed by M. The universal machine does this by printing *in front of* each record shown in the above diagram a record of any digit newly printed by M (plus an extra colon):

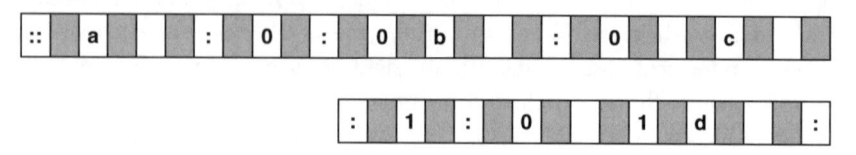

| :: | a | | : | 0 | : | 0 | b | | : | 0 | | c | |

| : | 1 | : | 0 | | 1 | d | | : |

These single digits bookended by colons form a representation of what has been printed by M on the F-squares of its tape.

Notice that the record-keeping scheme employed so far requires the universal machine to be able to print each type of symbol that the machine being simulated is able to print. In the case of M this requirement is modest, since M prints only '0', '1', and the blank. However, if the universal machine is to be able to simulate each of the infinitely many Turing machines, then this record-keeping scheme requires that the universal machine have the capacity to print an endless variety of types of discrete symbol. This can be avoided by allowing the universal machine to keep its record of M's tape in the same notation that is used in forming standard descriptions, namely with 'D' replacing the blank, 'DC' replacing '0', 'DCC' replacing '1', 'DA' replacing 'a', 'DAA' replacing 'b', and so on.

The universal machine's tape then looks like this (to the right of the end-of-programme symbol '::' and not including the second record of digits printed by M):

| :: | D | A | D | : | D | C | D | A | A | D | : | D | C | D | etc |

In this elegant notation of Turing's, 'D' serves to indicate the start of each new term on the universal machine's tape. The letters 'A' and 'C' serve to distinguish terms representing M's states from terms representing symbols on M's tape.

The E-squares and the instruction table

The universal machine uses the E-squares of its tape to mark up each instruction in the standard description. This facilitates the copying that the universal machine must do in order to produce its records of M's activity. For example, the machine temporarily marks the portion of the current instruction specifying M's next state with 'y' and subsequently the material marked 'y' is copied to the appropriate place in the record that is being created. The universal machine's records of M's tape are also temporarily marked in various ways.

In Section 7 Turing introduces various subroutines for placing and erasing markers on the E-squares. He sets out the table of instructions for the universal machine in terms of these subroutines. The table contains the detailed instructions for carrying out the record-keeping described above.

In Section 2.4 of Chapter 2 Turing's sometime colleague Donald Davies gives an introduction to these subroutines and to Turing's detailed table of instructions for the universal machine (and additionally corrects some errors in Turing's own formulation).

5. Turing, von Neumann, and the Computer

In the years immediately following the Second World War, the Hungarian-American logician and mathematician John von Neumann—one of the most important and influential figures of twentieth-century mathematics—made the concept of the stored-programme digital computer widely known, through his writings and his charismatic public addresses. In the secondary literature, von Neumann is often said to have himself invented the stored-programme computer. This is an unfortunate myth.

From 1933 von Neumann was on the faculty of the prestigious Institute for Advanced Study at Princeton University. He and Turing became well acquainted while Turing was studying at Princeton from 1936 to 1938 (see the introduction to Chapter 3). In 1938 von Neumann offered Turing a position as his assistant, which Turing declined. (Turing wrote to his mother on 17 May 1938: 'I had an offer of a job here as von Neumann's assistant at $1500 a year but decided not to take it.'[14] His father had advised him to find a job in America,[15] but on 12 April of the same year Turing had written: 'I have just been to see the Dean [Luther Eisenhart] and ask him about possible jobs over here; mostly for Daddy's information, as I think it unlikely I shall take one unless you are actually at war before July. He didn't know of one at present, but said he would bear it all in mind.')

It was during Turing's time at Princeton that von Neumann became familiar with the ideas in 'On Computable Numbers'. He was to become intrigued with Turing's concept of a universal computing machine.[16] It is clear that von

---

[14] Turing's letters to his mother are among the Turing Papers in the Modern Archive Centre, King's College Library, Cambridge (catalogue reference K 1).
[15] S. Turing, *Alan M. Turing* (Cambridge: Heffer, 1959), 55.
[16] 'I know that von Neumann was influenced by Turing ... during his Princeton stay before the war,' said von Neumann's friend and colleague Stanislaw Ulam (in an interview with Christopher Evans in 1976; 'The Pioneers of Computing: An Oral History of Computing', Science Museum, London). When Ulam and von Neumann were touring in Europe during the summer of 1938, von Neumann devised a mathematical game involving Turing-machine-like descriptions of numbers (Ulam reported by W. Aspray on pp. 178, 313 of his *John von Neumann and the Origins of Modern Computing* (Cambridge, Mass.: MIT Press, 1990)). The word

Neumann held Turing's work in the highest regard.[17] One measure of his esteem is that the only names to receive mention in his pioneering volume *The Computer and the Brain* are those of Turing and the renowned originator of information theory, Claude Shannon.[18]

The Los Alamos physicist Stanley Frankel—responsible with von Neumann and others for mechanizing the large-scale calculations involved in the design of the atomic and hydrogen bombs—has recorded von Neumann's view of the importance of 'On Computable Numbers':

> I know that in or about 1943 or '44 von Neumann was well aware of the fundamental importance of Turing's paper of 1936 'On computable numbers...', which describes in principle the 'Universal Computer' of which every modern computer (perhaps not ENIAC as first completed but certainly all later ones) is a realization. Von Neumann introduced me to that paper and at his urging I studied it with care. Many people have acclaimed von Neumann as the 'father of the computer' (in a modern sense of the term) but I am sure that he would never have made that mistake himself. He might well be called the midwife, perhaps, but he firmly emphasized to me, and to others I am sure, that the fundamental conception is owing to Turing—insofar as not anticipated by Babbage, Lovelace, and others. In my view von Neumann's essential role was in making the world aware of these fundamental concepts introduced by Turing and of the development work carried out in the Moore school and elsewhere.[19]

In 1944 von Neumann joined the ENIAC group, led by Presper Eckert and John Mauchly at the Moore School of Electrical Engineering (part of the University of Pennsylvania).[20] At this time von Neumann was involved in the Manhattan Project at Los Alamos, where roomfuls of clerks armed with desk calculating machines were struggling to carry out the massive calculations required by the physicists. Hearing about the Moore School's planned computer during a chance encounter on a railway station (with Herman Goldstine), von Neumann immediately saw to it that he was appointed as consultant to the project.[21] ENIAC—under construction since 1943—was, as previously mentioned, a programme-controlled (i.e. not stored-programme) computer: programming consisted of

'intrigued' is used in this connection by von Neumann's colleague Herman Goldstine on p. 275 of his *The Computer from Pascal to von Neumann* (Princeton: Princeton University Press, 1972).)

[17] Turing's universal machine was crucial to von Neumann's construction of a self-reproducing automaton; see the chapter 'Artificial Life', below.

[18] J. von Neumann, *The Computer and the Brain* (New Haven: Yale University Press, 1958).

[19] Letter from Frankel to Brain Randell, 1972 (first published in B. Randell, 'On Alan Turing and the Origins of Digital Computers', in Meltzer and Michie (eds.), *Machine Intelligence 7*. I am grateful to Randell for giving me a copy of this letter.

[20] John Mauchly recalled that 7 September 1944 'was the first day that von Neumann had security clearance to see the ENIAC and talk with Eckert and me' (J. Mauchly, 'Amending the ENIAC Story', *Datamation*, 25/11 (1979), 217–20 (217)). Goldstine (*The Computer from Pascal to von Neumann*, 185) suggests that the date of von Neumann's first visit may have been a month earlier: 'I probably took von Neumann for a first visit to the ENIAC on or about 7 August'.

[21] Goldstine, *The Computer from Pascal to von Neumann*, 182.

rerouting cables and setting switches. Moreover, the ENIAC was designed with only one very specific type of task in mind, the calculation of trajectories of artillery shells. Von Neumann brought his knowledge of 'On Computable Numbers' to the practical arena of the Moore School. Thanks to Turing's abstract logical work, von Neumann knew that by making use of coded instructions stored in memory, a single machine of fixed structure could in principle carry out *any* task for which an instruction table can be written.

Von Neumann gave his engineers 'On Computable Numbers' to read when, in 1946, he established his own project to build a stored-programme computer at the Institute for Advanced Study.[22] Julian Bigelow, von Neumann's chief engineer, recollected:

The person who really...pushed the whole field ahead was von Neumann, because he understood logically what [the stored-programme concept] meant in a deeper way than anybody else...The reason he understood it is because, among other things, he understood a good deal of the mathematical logic which was implied by the idea, due to the work of A. M. Turing...in 1936–1937. ...Turing's [universal] machine does not sound much like a modern computer today, but nevertheless it was. It was the germinal idea...So...[von Neumann] saw...that [ENIAC] was just the first step, and that great improvement would come.[23]

Von Neumann repeatedly emphasized the fundamental importance of 'On Computable Numbers' in lectures and in correspondence. In 1946 von Neumann wrote to the mathematician Norbert Wiener of 'the great positive contribution of Turing', Turing's mathematical demonstration that 'one, definite mechanism can be "universal"'.[24] In 1948, in a lecture entitled 'The General and Logical Theory of Automata', von Neumann said:

The English logician, Turing, about twelve years ago attacked the following problem. He wanted to give a general definition of what is meant by a computing automaton...Turing carried out a careful analysis of what mathematical processes can be effected by automata of this type...He...also introduce[d] and analyse[d] the concept of a 'universal automaton'...An automaton is 'universal' if any sequence that can be produced by any automaton at all can also be solved by this particular automaton. It will, of course, require in general a different instruction for this purpose. *The Main Result of the Turing Theory.* We might expect a priori that this is impossible. How can there be an automaton which is

---

[22] Letter from Julian Bigelow to Copeland (12 Apr. 2002). See also Aspray, *John von Neumann*, 178.

[23] Bigelow in a tape-recorded interview made in 1971 by the Smithsonian Institution and released in 2002. I am grateful to Bigelow for sending me a transcript of excerpts from the interview.

[24] The letter, dated 29 Nov. 1946, is in the von Neumann Archive at the Library of Congress, Washington, DC. In the letter von Neumann also remarked that Turing had 'demonstrated in absolute...generality that anything and everything Brouwerian can be done by an appropriate mechanism' (a Turing machine). He made a related remark in a lecture: 'It has been pointed out by A. M. Turing [in "On Computable Numbers"]...that effectively constructive logics, that is, intuitionistic logics, can be best studied in terms of automata' ('Probabilistic Logics and the Synthesis of Reliable Organisms from Unreliable Components', in vol. v of von Neumann's *Collected Works*, ed. A. H. Taub (Oxford: Pergamon Press, 1963), 329).

at least as effective as any conceivable automaton, including, for example, one of twice its size and complexity? Turing, nevertheless, proved that this is possible.[25]

The following year, in a lecture delivered at the University of Illinois entitled 'Rigorous Theories of Control and Information', von Neumann said:

> The importance of Turing's research is just this: that if you construct an automaton right, then any additional requirements about the automaton can be handled by sufficiently elaborate instructions. This is only true if [the automaton] is sufficiently complicated, if it has reached a certain minimal level of complexity. In other words... there is a very definite finite point where an automaton of this complexity can, when given suitable instructions, do anything that can be done by automata at all.[26]

Von Neumann placed Turing's abstract 'universal automaton' into the hands of American engineers. Yet many books on the history of computing in the United States make no mention of Turing. No doubt this is in part explained by the absence of any explicit reference to Turing's work in the series of technical reports in which von Neumann, with various co-authors, set out a logical design for an electronic stored-programme digital computer.[27] Nevertheless there is evidence in these documents of von Neumann's knowledge of 'On Computable Numbers'. For example, in the report entitled 'Preliminary Discussion of the Logical Design of an Electronic Computing Instrument' (1946), von Neumann and his co-authors, Burks and Goldstine—both former members of the ENIAC group, who had joined von Neumann at the Institute for Advanced Study—wrote the following:

> 3.0. *First Remarks on the Control and Code*: It is easy to see by formal-logical methods, that there exist codes that are in abstracto adequate to control and cause the execution of any sequence of operations which are individually available in the machine and which are, in their entirety, conceivable by the problem planner. The really decisive considerations from the present point of view, in selecting a code, are more of a practical nature: Simplicity of the equipment demanded by the code, and the clarity of its application to the actually important problems together with the speed of its handling of those problems.[28]

Burks has confirmed that the first sentence of this passage is a reference to Turing's universal computing machine.[29]

[25] The text of 'The General and Logical Theory of Automata' is in vol. v of von Neumann, *Collected Works*; see pp. 313–14.
[26] The text of 'Rigorous Theories of Control and Information' is printed in J. von Neumann, *Theory of Self-Reproducing Automata*, ed. A. W. Burks (Urbana: University of Illinois Press, 1966); see p. 50.
[27] The first papers in the series were the 'First Draft of a Report on the EDVAC' (1945, von Neumann; see n. 31), and 'Preliminary Discussion of the Logical Design of an Electronic Computing Instrument' (1946, Burks, Goldstine, von Neumann; see n. 28).
[28] A. W. Burks, H. H. Goldstine, and J. von Neumann, 'Preliminary Discussion of the Logical Design of an Electronic Computing Instrument', 28 June 1946, Institute for Advanced Study, Princeton University, Section 3.1 (p. 37); reprinted in vol. v of von Neumann, *Collected Works*.
[29] Letter from Burks to Copeland (22 Apr. 1998). See also Goldstine, *The Computer from Pascal to von Neumann*, 258.

The situation in 1945–1946

The passage just quoted is an excellent summary of the situation at that time. In 'On Computable Numbers' Turing had shown *in abstracto* that, by means of instructions expressed in the programming code of standard descriptions, a single machine of fixed structure is able to carry out any task that a 'problem planner' is able to analyse into effective steps. By 1945, considerations *in abstracto* had given way to the practical problem of devising an equivalent programming code that could be implemented efficiently by means of thermionic valves (vacuum tubes).

A machine-level programming code in effect specifies the basic operations that are available in the machine. In the case of Turing's universal machine these are move left one square, scan one symbol, write one symbol, and so on. These operations are altogether too laborious to form the basis of efficient electronic computation. A *practical* programming code should not only be universal, in the sense of being adequate in principle for the programming of any task that can be carried out by a Turing machine, but must in addition:

- employ basic operations that can be realized simply, reliably, and efficiently by electronic means;
- enable the 'actually important problems' to be solved on the machine as rapidly as the electronic hardware permits;
- be as easy as possible for the human 'problem planner' to work with.

The challenge of designing a practical code, and the underlying mechanism required for its implementation, was tackled in different ways by Turing and the several American groups.

Events at the Moore School

The 'Preliminary Discussion of the Logical Design of an Electronic Computing Instrument' was not intended for formal publication and no attempt was made to indicate those places where reference was being made to the work of others. (Von Neumann's biographer Norman Macrae remarked: 'Johnny borrowed (we must not say plagiarized) anything from anybody.'[30] The situation was the same in the case of von Neumann's 1945 paper 'First Draft of a Report on the EDVAC'.[31] This described the Moore School group's proposed stored-programme computer, the EDVAC. The 'First Draft' was distributed (by Goldstine and a Moore School administrator) before references had been added—and indeed without consideration of whether the names of Eckert and Mauchly

---

[30] N. Macrae, *John von Neumann* (New York: Pantheon Books, 1992), 23.
[31] J. von Neumann, 'First Draft of a Report on the EDVAC', Moore School of Electrical Engineering, University of Pennsylvania, 1945; reprinted in full in N. Stern, *From ENIAC to UNIVAC: An Appraisal of the Eckert-Mauchly Computers* (Bedford, Mass.: Digital Press, 1981).

should appear alongside von Neumann's as co-authors.[32] Eckert and Mauchly were outraged, knowing that von Neumann would be given credit for everything in the report—their ideas as well as his own. There was a storm of controversy and von Neumann left the Moore School group to establish his own computer project at Princeton. Harry Huskey, a member of the Moore School group from the spring of 1944, emphasizes that the 'First Draft' should have contained acknowledgement of the considerable extent to which the design of the proposed EDVAC was the work of other members of the group, especially Eckert.[33]

In 1944, before von Neumann came to the Moore School, Eckert and Mauchly had rediscovered the idea of using a single memory for data and programme.[34] (They were far, however, from rediscovering Turing's concept of a universal machine.) Even before the ENIAC was completed, Eckert and Mauchly were thinking about a successor machine, the EDVAC, in which the ENIAC's most glaring deficiencies would be remedied. Paramount among these, of course, was the crude wire'n'plugs method of setting up the machine for each new task. Yet if pluggable connections were not to be used, how was the machine to be controlled without a sacrifice in speed? If the computation were controlled by means of existing, relatively slow, technology—e.g. an electro-mechanical punched-card reader feeding instructions to the machine—then the high-speed electronic hardware would spend much of its time idle, awaiting the next instruction. Eckert explained to Huskey his idea of using a mercury 'delay line':

Eckert described a mercury delay line to me, a five foot pipe filled with mercury which could be used to store a train of acoustic pulses... [O]ne recirculating mercury line would store more than 30 [32 bit binary] numbers... My first question to Eckert: thinking about the pluggable connections to control the ENIAC, 'How do you control the operations?' 'Instructions are stored in the mercury lines just like numbers,' he said. Of course! Once he said it, it was so obvious, and the *only* way that instructions could come available at rates comparable to the data rates. That was the *stored program computer*.[35]

[32] See N. Stern, 'John von Neumann's Influence on Electronic Digital Computing, 1944–1946', *Annals of the History of Computing*, 2 (1980), 349–62.

[33] Huskey in interview with Copeland (Feb. 1998). (Huskey was offered the directorship of the EDVAC project in 1946 but other commitments prevented him from accepting.)

[34] Mauchly, 'Amending the ENIAC Story'; J. P. Eckert, 'The ENIAC', in Metropolis, Howlett, and Rota, *A History of Computing in the Twentieth Century*; letter from Burks to Copeland (16 Aug. 2003): 'before von Neumann came' to the Moore School, Eckert and Mauchly were 'saying that they would build a mercury memory large enough to store the program for a problem as well as the arithmetic data'. Burks points out that von Neumann was however the first of the Moore School group to note the possibility, implicit in the stored-programme concept, of allowing the computer to modify the addresses of selected instructions in a programme while it runs (A. W. Burks, 'From ENIAC to the Stored-Program Computer: Two Revolutions in Computers', in Metropolis, Howlett, and Rota, *A History of Computing in the Twentieth Century*, 340–1). Turing employed a more general form of the idea of instruction modification in his 1945 technical report 'Proposed Electronic Calculator' (in order to carry out conditional branching), and the idea of instruction modification lay at the foundation of his theory of machine learning (see Chapter 9).

[35] H. D. Huskey, 'The Early Days', *Annals of the History of Computing*, 13 (1991), 290–306 (292–3). The date of the conversation was 'perhaps the spring of 1945' (letter from Huskey to Copeland (5 Aug. 2003)).

Following his first visit to the ENIAC in 1944, von Neumann went regularly to the Moore School for meetings with Eckert, Mauchly, Burks, Goldstine, and others.[36] Goldstine reports that 'these meetings were scenes of greatest intellectual activity' and that 'Eckert was delighted that von Neumann was so keenly interested' in the idea of the high-speed delay line memory. It was, says Goldstine, 'fortunate that just as this idea emerged von Neumann should have appeared on the scene'.[37]

Eckert had produced the means to make the abstract universal computing machine of 'On Computable Numbers' concrete! Von Neumann threw himself at the key problem of devising a practical code. In 1945, Eckert and Mauchly reported that von Neumann 'has contributed to many discussions on the logical controls of the EDVAC, has prepared certain instruction codes, and has tested these proposed systems by writing out the coded instructions for specific problems'.[38] Burks summarized matters:

Pres [Eckert] and John [Mauchly] invented the circulating mercury delay line store, with enough capacity to store program information as well as data. Von Neumann created the first modern order code and worked out the logical design of an electronic computer to execute it.[39]

Von Neumann's embryonic programming code appeared in May 1945 in the 'First Draft of a Report on the EDVAC'.

So it was that von Neumann became the first to outline a 'practical version of the universal machine' (the quoted phrase is Turing's; see p. 16). The 'First Draft' contained little engineering detail, however, in particular concerning electronics. Turing's own practical version of the universal machine followed later the same year. His 'Proposed Electronic Calculator' set out a detailed programming code—very different from von Neumann's—together with a detailed design for the underlying hardware of the machine (see Chapter 9).

6. Turing and Babbage

Charles Babbage, Lucasian Professor of Mathematics at the University of Cambridge from 1828 to 1839, was one of the first to appreciate the enormous potential of computing machinery. In about 1820, Babbage proposed an

---

[36] Goldstine, *The Computer from Pascal to von Neumann*, 186.
[37] Ibid.
[38] J. P. Eckert and J. W. Mauchly, 'Automatic High Speed Computing: A Progress Report on the EDVAC', Moore School of Electrical Engineering, University of Pennsylvania (Sept. 1945), Section 1; this section of the report is reproduced on pp. 184–6 of L. R. Johnson, *System Structure in Data, Programs, and Computers* (Englewood Cliffs, NJ: Prentice-Hall, 1970).
[39] Burks, 'From ENIAC to the Stored-Program Computer: Two Revolutions in Computers', 312.

'Engine' for the automatic production of mathematical tables (such as logarithm tables, tide tables, and astronomical tables).[40] He called it the 'Difference Engine'. This was the age of the steam engine, and Babbage's Engine was to consist of more accurately machined forms of components found in railway locomotives and the like—brass gear wheels, rods, ratchets, pinions, and so forth.

Decimal numbers were represented by the positions of ten-toothed metal wheels mounted in columns. Babbage exhibited a small working model of the Engine in 1822. He never built the full-scale machine that he had designed, but did complete several parts of it. The largest of these—roughly 10 per cent of the planned machine—is on display in the London Science Museum. Babbage used it to calculate various mathematical tables. In 1990 his 'Difference Engine No. 2' was finally built from the original design and this is also on display at the London Science Museum—a glorious machine of gleaming brass.

In 1843 the Swedes Georg and Edvard Scheutz (father and son) built a simplified version of the Difference Engine. After making a prototype they built two commercial models. One was sold to an observatory in Albany, New York, and the other to the Registrar-General's office in London, where it calculated and printed actuarial tables.

Babbage also proposed the 'Analytical Engine', considerably more ambitious than the Difference Engine.[41] Had it been completed, the Analytical Engine would have been an all-purpose mechanical digital computer. A large model of the Analytical Engine was under construction at the time of Babbage's death in 1871, but a full-scale version was never built.

The Analytical Engine was to have a memory, or 'store' as Babbage called it, and a central processing unit, or 'mill'. The behaviour of the Analytical Engine would have been controlled by a programme of instructions contained on punched cards, connected together by ribbons (an idea Babbage adopted from the Jacquard weaving loom). The Analytical Engine would have been able to select from alternative actions on the basis of outcomes of previous actions—a facility now called 'conditional branching'.

Babbage's long-time collaborator was Ada, Countess of Lovelace (daughter of the poet Byron), after whom the modern programming language ADA is named. Her vision of the potential of computing machines was in some respects perhaps more far-reaching even than Babbage's own. Lovelace envisaged computing that

---

[40] C. Babbage, *Passages from the Life of a Philosopher*, vol. xi of *The Works of Charles Babbage*, ed. M. Campbell-Kelly (London: William Pickering, 1989); see also B. Randell (ed.), *The Origins of Digital Computers: Selected Papers* (Berlin: Springer-Verlag, 3rd edn. 1982), ch. 1.

[41] See Babbage, *Passages from the Life of a Philosopher*; A. A. Lovelace and L. F. Menabrea, 'Sketch of the Analytical Engine Invented by Charles Babbage, Esq.' (1843), in B. V. Bowden (ed.), *Faster than Thought* (London: Pitman, 1953); Randell, *The Origins of Digital Computers: Selected Papers*, ch. 2; A. Bromley, 'Charles Babbage's Analytical Engine, 1838', *Annals of the History of Computing*, 4 (1982), 196–217.

went beyond pure number-crunching, suggesting that the Analytical Engine might compose elaborate pieces of music.[42]

Babbage's idea of a general-purpose calculating engine was well known to some of the modern pioneers of automatic calculation. In 1936 Vannevar Bush, inventor of the Differential Analyser (an analogue computer), spoke in a lecture of the possibility of machinery that 'would be a close approach to Babbage's large conception'.[43] The following year Howard Aiken, who was soon to build the digital—but not stored-programme and not electronic—Harvard Automatic Sequence Controlled Calculator, wrote:

> Hollerith... returned to the punched card first employed in calculating machinery by Babbage and with it laid the groundwork for the development of... machines as manufactured by the International Business Machines Company, until today many of the things Babbage wished to accomplish are being done daily in the accounting offices of industrial enterprises all over the world.[44]

Babbage's ideas were remembered in Britain also, and his proposed computing machinery was on occasion a topic of lively mealtime discussion at Bletchley Park, the wartime headquarters of the Government Code and Cypher School and birthplace of the electronic digital computer (see 'Enigma' and the introductions to Chapters 4 and 9).[45]

It is not known when Turing first learned of Babbage's ideas.[46] There is certainly no trace of Babbage's influence to be found in 'On Computable Numbers'. Much later, Turing generously wrote (Chapter 11, p. 446):

> The idea of a digital computer is an old one. Charles Babbage... planned such a machine, called the Analytical Engine, but it was never completed. Although Babbage had all the essential ideas, his machine was not at that time such a very attractive prospect.

Babbage had emphasized the generality of the Analytical Engine, claiming that 'the conditions which enable a finite machine to make calculations of unlimited extent are fulfilled in the Analytical Engine'.[47] Turing states (Chapter 11, p. 455) that the Analytical Engine was *universal*—a judgement possible only from the vantage point of 'On Computable Numbers'. The Analytical Engine was not, however, a stored-programme computer. The programme resided externally on

---

[42] Lovelace and Menabrea, 'Sketch of the Analytical Engine', 365.

[43] V. Bush, 'Instrumental Analysis', *Bulletin of the American Mathematical Society*, 42 (1936), 649–69 (654) (the text of Bush's 1936 Josiah Willard Gibbs Lecture).

[44] H. Aiken, 'Proposed Automatic Calculating Machine' (1937), in Randell, *The Origins of Digital Computers: Selected Papers*, 196.

[45] Thomas H. Flowers in interview with Copeland (July 1996).

[46] Dennis Babbage, chief cryptanalyst in Hut 6, the section at Bletchley Park responsible for Army, Airforce, and Railway Enigma, is sometimes said to have been a descendant of Charles Babbage. This was not in fact so. (Dennis Babbage in interview with Ralph Erskine.)

[47] Babbage, *Passages from the Life of a Philosopher*, 97.

punched cards, and as each card entered the Engine, the instruction marked on that card would be obeyed.

Someone might wonder what difference there is between the Analytical Engine and the universal Turing machine in that respect. After all, Babbage's cards strung together with ribbon would in effect form a tape upon which the programme is marked. The difference is that in the universal Turing machine, but not the Analytical Engine, there is no fundamental distinction between programme and data. It is the absence of such a distinction that marks off a stored-programme computer from a programme-controlled computer. As Gandy put the point, Turing's 'universal machine is a stored-program machine [in that], unlike Babbage's all-purpose machine, the mechanisms used in reading a program are *of the same kind* as those used in executing it'.[48]

7. Origins of the Term 'Computer Programme'

As previously mentioned, Turing's tables of instructions for Turing machines are examples of what are now called computer programmes. When he turned to designing an electronic computer in 1945 (the ACE), Turing continued to use his term 'instruction table' where a modern writer would use 'programme' or 'program'.[49] Later material finds Turing referring to the actual process of writing instruction tables for the electronic computer as 'programming' but still using 'instruction table' to refer to the programme itself (see Chapter 9, pp. 388, 390–91).[50]

In an essay published in 1950 Turing explained the emerging terminology to the layman (Chapter 11, p. 445): 'Constructing instruction tables is usually described as "programming". To "programme a machine to carry out the operation A" means to put the appropriate instruction table into the machine so that it will do A.'

Turing seems to have inherited the term 'programming' from the milieu of punched-card plug-board calculators. (These calculators were electro-mechanical, not electronic. Electro-mechanical equipment was based on the *relay*—a small electrically driven mechanical switch. Relays operated much more slowly than the thermionic valves (vacuum tubes) on which the first electronic computers were based; valves owe their speed to the fact that they

---

[48] R. Gandy, 'The Confluence of Ideas in 1936', in R. Herken (ed.), *The Universal Turing Machine: A Half-Century Survey* (Oxford: Oxford University Press, 1998), 90. Emphasis added.

[49] 'Program' is the original English spelling, in conformity with 'anagram', 'diagram', etc. The spelling 'programme' was introduced into Britain from France in approximately 1800 (*Oxford English Dictionary*). The earlier spelling persisted in the United States. Turing's spelling is followed in this volume (except in quotations from other authors and in the section by Davies).

[50] See also 'The Turing-Wilkinson Lecture Series on the Automatic Computing Engine' (ed. Copeland), in K. Furukawa, D. Michie, and S. Muggleton (eds.), *Machine Intelligence 15* (Oxford: Oxford University Press, 1999).

have no moving parts save a beam of electrons—hence the term 'electronic'.) Plug-board calculators were set up to perform a desired sequence of arithmetical operations by means of plugging wires into appropriate sockets in a board resembling a telephone switchboard. Data was fed into the calculator from punched cards, and a card-punching device or printer recorded the results of the calculation. An early example of a punched-card machine was constructed in the USA by Herman Hollerith for use in processing statistical data gathered in the 1890 census. By the mid-twentieth century most of the world's computing was being done by punched-card calculators. Gradually the technology was displaced by the electronic computer.

When Turing joined the National Physical Laboratory in 1945 there was a large room filled with punched-card calulating equipment. David Clayden, one of the engineers who built the ACE, describes the punched-card equipment and the terminology in use at that time:

When I started at NPL in 1947 there was a well established punched card department, mainly Hollerith. The workhorse of punched card equipment is the 'Reproducer', which has a broadside card reader and a broadside card punch. By taking a stack of cards from the punch and putting them into the reader, it is possible to do iterative calculations. All functions are controlled by a plugboard on which there are two sets of 12 × 80 sockets, one for the reader and one for the punch. In addition there is a relay store [i.e. memory]. The plugboard can be connected in many ways (using short plugleads) in order to perform many functions, including addition, subtraction, and multiplication. The plugboards are removable. NPL had a stack of them and called them 'programme' boards.[51]

Turing's own preference for 'instruction table' over the noun 'programme' was not shared by all his colleagues at the NPL. Mike Woodger, Turing's assistant from 1946, says: ' "Programme" of course was an ordinary English word meaning a planned sequence of events. We adopted it naturally for any instruction table that would give rise to a desired sequence of events.'[52] The noun 'programme' was in use in its modern sense from the earliest days of the ACE project. A report (probably written by Turing's immediate superior, Womersley) describing work done by Turing and his assistants during 1946 stated: 'It is intended to prepare the instructions to the machine [the ACE] on Hollerith cards, and it is proposed to maintain a library of these cards with programmes for standard operations.'[53] By the early 1950s specially printed ruled sheets used at the NPL for writing out programmes bore the printed heading 'ACE Pilot Model Programme'.[54]

---

[51] Letter from Clayden to Copeland (3 Oct. 2000).
[52] Letter from Woodger to Copeland (6 Oct. 2000).
[53] 'Draft Report of the Executive Committee for the Year 1946', National Physical Laboratory, paper E.910, section Ma. 1, anon., but probably by Womersley (NPL Library; a digital facsimile is in The Turing Archive for the History of Computing <www.AlanTuring.net/annual_report_1946>).
[54] J. G. Hayes, 'The Place of Pilot Programming', MS, 2000.

A document written by Woodger in 1947 used the single 'm' spelling: 'A Program for Version H'.[55] Woodger recalls: 'We used both spellings carelessly for some years until Goodwin (Superintendent of Mathematics Division from 1951) laid down the rule that the "American" spelling should be used.'[56] It is possible that the single 'm' spelling first came to the NPL via the American engineer Huskey, who spent 1947 with the ACE group. Huskey was responsible for 'Version H', a scaled-down form of Turing's design for the ACE (see Chapter 10).

Like Turing, Eckert and Mauchly, the chief architects of ENIAC, probably inherited the terms 'programming' and 'program' from the plug-board calculator. In 1942, while setting out the idea of a high-speed electronic calculator, Mauchly used the term 'programming device' (which he sometimes shortened to 'program device') to refer to a mechanism whose function was to determine how and when the various component units of a calculator shall perform.[57] In the summer of 1946 the Moore School organized a series of influential lectures entitled 'Theory and Techniques for Design of Electronic Digital Computers'. In the course of these, Eckert used the term 'programming' in a similar sense when describing the new idea of storing instructions in high-speed memory: 'We ... feed those pieces of information which relate to programming from the memory.'[58] Also in 1946, Burks, Goldstine, and von Neumann (all ex-members of the Moore School group) were using the verb-form 'program the machine', and were speaking of 'program orders' being stored in memory.[59] The modern nominalized form appears not to have been adopted in the USA until a little later. Huskey says, 'I am pretty certain that no one had written a "program" by the time I left Philadelphia in June 1946.'[60]

Part II Computability and Uncomputability

8. Circular and Circle-Free Machines

Turing calls the binary digits '0' and '1' symbols 'of the first kind'. Any symbols that a computing machine is able to print apart from the binary digits—such as

[55] M. Woodger, 'A Program for Version H', handwritten MS, 1947 (in the Woodger Papers, National Museum of Science and Industry, Kensington, London (catalogue reference N30/37)).

[56] Letter from Woodger to Copeland (6 Oct. 2000).

[57] J. W. Mauchly, 'The Use of High Speed Vacuum Tube Devices for Calculating' (1942), in Randell, *The Origins of Digital Computers: Selected Papers.*

[58] J. P. Eckert, 'A Preview of a Digital Computing Machine' (15 July 1946), in M. Campbell-Kelly and M. R. Williams (eds.), *The Moore School Lectures* (Cambridge, Mass.: MIT Press, 1985), 114.

[59] Sections 1.2, 5.3 of Burks, Goldstine, and von Neumann, 'Preliminary Discussion of the Logical Design of an Electronic Computing Instrument' (von Neumann, *Collected Works*, vol. v, 15, 43).

[60] Letter from Huskey to Copeland (3 Feb. 2002).

'2', '∗', 'x', and blank—Turing calls 'symbols of the second kind' (p. 60). He also uses the term 'figures' for symbols of the first kind.

A computing machine is said by Turing to be *circular* if it never prints more than a finite number of symbols of the first kind. A computing machine that will print an infinite number of symbols of the first kind is said to be *circle-free* (p. 60). For example, a machine operating in accordance with Table 1 is circle-free. (The terms 'circular' and 'circle-free' were perhaps poor choices in this connection, and the terminology has not been followed by others.)

A simple example of a circular machine is one set up to perform a single calculation whose result is an integer. Once the machine has printed the result (in binary notation), it prints nothing more.

A circular machine's scanner need not come to a halt. The scanner may continue moving along the tape, printing nothing further. Or, after printing a finite number of binary digits, a circular machine may work on forever, printing only symbols of the second kind.

Many real-life computing systems are circle-free, for example automated teller machine networks, air traffic control systems, and nuclear reactor control systems. Such systems never terminate by design and, barring hardware failures, power outages, and the like, would continue producing binary digits forever.

In Section 8 of 'On Computable Numbers' Turing makes use of the circular/circle-free distinction in order to formulate a mathematical problem that cannot be solved by computing machines.

9. Computable and Uncomputable Sequences

The sequence of binary digits printed by a given computing machine on the F-squares of its tape, starting with a blank tape, is called the *sequence computed by the machine*. Where the given machine is circular, the sequence computed by the machine is finite. The sequence computed by a circle-free machine is infinite.

A sequence of binary digits is said to be a *computable* sequence if it is the sequence computed by some circle-free computing machine. For example, the infinite sequence 010101 . . . is a computable sequence.

Notice that although the finite sequence 010, for example, is the sequence computed by some machine, this sequence is *not* a *computable* sequence, according to Turing's definition. This is because, being finite, 010 is not the sequence computed by any circle-free machine. According to Turing's definition, no finite sequence is a computable sequence. Modern writers usually define 'computable' in such a way that every finite sequence is a computable sequence, since each of them can be computed (e.g. by means of an instruction table that simply prints the desired sequence). Turing, however, was not much interested in finite sequences.

The focus of Turing's discussion is his discovery that not every infinite sequence of binary digits is a computable sequence. That this is so is shown by what mathematicians call a *diagonal* argument.

The diagonal argument

Imagine that all the computable sequences are listed one under another. (The order in which they are listed does not matter.) The list stretches away to infinity both downwards and to the right. The top left-hand corner might look like this:

0110010101100010010110100100011101...
0101110100111000111111111111110111...
1101000001101101010000011001000011...
  .
  .
  .

Let's say that this list was drawn up in the following way (by an infinite deity, perhaps). The first sequence on the list is the sequence computed by the machine with a description number that is smaller than any description number of any other circle-free machine. The second sequence on the list is the one computed by the circle-free machine with the *next smallest* description number, and so on. Every computable sequence appears somewhere on this list. (Some will in fact be listed twice, since sometimes different description numbers correspond to the same sequence.)

To prove that not all infinite binary sequences are computable, it is enough to describe one that does not appear on this list. To this end, consider the infinite binary sequence formed by moving diagonally down and across the list, starting at the top left:

01100...
0**1**011...
11**0**10...
  .
  .
  .

The twist is to transform this sequence into a different one by switching each '0' lying on the diagonal to '1' and each '1' to '0'. So the first digit of this new sequence is formed by switching the first digit of the first sequence on the list (producing 1); the second digit of the sequence is formed by switching the second digit of the second sequence on the list (producing 0); the third digit is formed by switching the third digit of the third sequence on the list (producing 1); and so on. Turing calls this sequence 'β' (p. 72).

A moment's reflection shows that β cannot itself be one of the listed sequences, since it has been constructed in such a way that it differs from each of these. It differs from the first sequence on the list at the first digit. It differs from the second sequence on the list at the second digit. And so on. Therefore, since every computable sequence appears somewhere on this list, β is not among the computable sequences.

Why the computable sequences are listable

A sceptic might challenge this reasoning, saying: 'Perhaps the computable sequences *cannot* be listed. In assuming that the computable sequences can be listed, one, two, three, and so on, you are assuming in effect that each computable sequence can be paired off with an integer (no two sequences being paired with the same integer). But what if the computable sequences cannot be paired off like this with the integers? Suppose that there are just *too many* computable sequences for this to be possible.' If this challenge were successful, it would pull the rug out from under the diagonal argument.

The response to the challenge is this. Each circle-free Turing machine produces just one computable sequence. So there cannot be more computable sequences than there are circle-free Turing machines. But there certainly cannot be more circle-free Turing machines than there are integers. This is because every Turing machine has a description number, which *is* an integer, and this number is not shared by any other Turing machine.

This reasoning shows that each computable sequence can be paired off with an integer, one sequence per integer. As Turing puts this, the computable sequences are 'enumerable' (p. 68).

The totality of infinite binary sequences, however, is *non-enumerable*. Not all the sequences can be paired off with integers in such a way that no integer is allocated more than one sequence. This is because, once *every* integer has had an infinite binary sequence allocated to it, one can 'diagonalize' in the above way and produce an extra sequence.

Starting with a blank tape

Incidentally, notice the significance, in Turing's definition of *sequence computed by the machine*, of the qualification 'starting with a blank tape'. If the computing machine were allowed to make use of a tape that had already had an infinite sequence of figures printed on it by some means, then the concept of a computable sequence would be trivialized. Every infinite binary sequence would become computable, simply because any sequence of digits whatever—e.g. β—could be present on the tape before the computing machine starts printing.

The following trivial programme causes a machine to run along the tape printing the figures that are already there!

a	1	P[1], R	a
a	0	P[0], R	a
a	-	P[-], R	a

(The third line is required to deal with blank E-squares, if any.)

10. Computable and Uncomputable Numbers

Prefacing a binary sequence by '0' produces a real number expressed in the form of a binary decimal. For example, prefacing the binary sequence 010101... produces 0.010101... (the binary form of the ordinary decimal 0.363636...). If B is the sequence of binary digits printed by a given computing machine, then 0.B is called the *number computed by the machine*.

Where the given machine is circular, the number computed by the machine is always a rational number. A circle-free machine may compute an irrational number (π, for example).

A number computed by a circle-free machine is said to be a *computable number*. Turing also allows that any number 'that differs by an integer' from a number computed by a circle-free machine is a computable number (p. 61). So if B is the infinite sequence of binary digits printed by some circle-free machine, then the number computed by the machine, 0.B, is a computable number, as are all the numbers that differ from 0.B by an integer: 1.B, 10.B, etc.

In Section 10 of 'On Computable Numbers', Turing gives examples of large classes of numbers that are computable. In particular, he proves that the important numbers π and e are computable.

Not all real numbers are computable, however. This follows immediately from the above proof that not all infinite binary sequences are computable. If S is an infinite binary sequence that is uncomputable, then 0.S is an uncomputable number.

11. The Satisfactoriness Problem

In Section 8 of 'On Computable Numbers' Turing describes two mathematical problems that cannot be solved by computing machines. The first will be referred to as the *satisfactoriness problem*.

Satisfactory descriptions and numbers

A standard description is said to be *satisfactory* if the machine it describes is circle-free. (Turing's choice of terminology might be considered awkward, since there need be nothing at all unsatisfactory, in the usual sense of the word, about a circular machine.)

A number is satisfactory if it is a description number of a circle-free machine. A number is unsatisfactory if either it is a description number of a circular machine, or it is not a description number at all.

The satisfactoriness problem is this: decide, of any arbitrarily selected standard description—or, equivalently, any arbitrarily selected description number—whether or not it is satisfactory. The decision must be arrived at in a finite number of steps.

The diagonal argument revisited

Turing approaches the satisfactoriness problem by reconsidering the above proof that not every binary sequence is computable.

Imagine someone objecting to the diagonal argument: 'Look, there must be something wrong with your argument, because β evidently *is* computable. In the course of the argument, you have in effect given instructions for computing each digit of β, in terms of counting out digits and switching the relevant ones. Let me try to describe how a Turing machine could compute β. I'll call this Turing machine BETA. BETA is similar to the universal machine in that it is able to simulate the activity of any Turing machine that one wishes. First, BETA simulates the circle-free machine with the smallest description number. BETA keeps up the simulation just as far as is necessary in order to discover the first digit of the sequence computed by this machine. BETA then switches this digit, producing the first digit of β. Next, BETA simulates the circle-free machine with the next smallest description number, keeping up the simulation until it finds the second digit of the sequence computed by this machine. And so on.'

The objector continues: 'I can make my description of BETA specific. BETA uses only the E-squares of its tape to do its simulations, erasing its rough work each time it begins a new simulation. It prints out the digits of β on successive F-squares. I need to take account of the restriction that, in order for it to be said that β is the sequence computed by BETA, BETA must produce the digits of β *starting from a blank tape*. What BETA will do first of all, starting from a blank tape, is find the smallest description number that corresponds to a circle-free machine. It does this by checking through the integers, one by one, starting at 1. As BETA generates the integers one by one, it checks each to see whether it is a description number. If the integer is not a description number, then BETA moves on to the next. If the integer is a description number, then BETA checks whether the number is satisfactory. Once BETA finds the first integer to describe a circle-free machine, it uses the instructions contained in the description number in order to simulate the machine. This is how BETA finds the first digit of β. Then BETA continues its search through the integers, until it finds the next smallest description number that is satisfactory. This enables BETA to calculate the second digit of β. And so on.'

Turing tackles this objection head on, proving that no computing machine can possibly do what BETA is supposed to do. It suffices for this proof to consider a slightly simplified version of BETA, which Turing calls 𝓝. 𝓝 is just like BETA except that 𝓝 does not switch the digits of the list's 'diagonal' sequence. 𝓝 is supposed to write out (on the F-squares) the successive digits not of β but of the 'diagonal' sequence itself: the sequence whose first digit is the first digit of the first sequence on the list, whose second digit is the second digit of the second sequence on the list, and so on. Turing calls this sequence β'. If no computing machine can compute β', then there is no such computing machine as BETA—because if there were, a machine that computes β' could be obtained from it, simply by deleting the instructions to switch each of the digits of the diagonal sequence.

What happens when 𝓝 meets itself?

Turing asks: *what happens when, as 𝓝 searches through the integers one by one, it encounters a number describing 𝓝 itself?* Call this description number K.

𝓝 must first check whether K is a description number. Having ascertained that it is, 𝓝 must test whether K is satisfactory. Since 𝓝 is supposed to be computing the endless binary sequence β', 𝓝 itself must be circle-free. So 𝓝 must pronounce K to be satisfactory.

In order to find the next digit of β', 𝓝 must now simulate the behaviour of the machine described by K. Since 𝓝 is that machine, 𝓝 must simulate its *own* behaviour, starting with its very first action. There is nothing wrong with the idea of a machine starting to simulate its own previous behaviour (just as a person might act out some episode from their own past). 𝓝 first simulates (on its E-squares) the series of actions that it performed up to and including writing down the first digit of β', then its actions up to and including writing down the second digit of β', and so on.

Eventually, however, 𝓝's simulation of its own past reaches the point where 𝓝 began to simulate the behaviour of the machine described by K. What must 𝓝 do now? 𝓝 must simulate the series of actions that it performed when simulating the series of actions that culminated in its writing down the first digit of β', and then simulate the series of actions that it performed when simulating the series of actions that culminated in its writing down the second digit of β', and so on! 𝓝 is doomed to relive its past forever.

From the point when it began simulating itself, 𝓝 writes only on the E-squares of its tape and never adds another digit to the sequence on its F-squares. Therefore, 𝓝 does *not* compute β'. 𝓝 computes some finite number of digits of β' and then sticks.

The problem lies with the glib assumption that 𝓝 and BETA are able to determine whether each description number is satisfactory.

No computing machine can solve the satisfactoriness problem

Since, as has just been shown, no computing machine can possibly do what 𝓗 was introduced to do, one of the various tasks that 𝓗 is supposed to carry out must be impossible for a computing machine. But all the things that 𝓗 is supposed to do *apart* from checking for satisfactoriness—decide whether a number is a description number, extract instructions from a description number, simulate a machine that follows those instructions, and so on—are demonstrably things that can be done by the universal machine.

By a process of elimination, then, the task that it is impossible for a computing machine to carry out must be that of determining whether each description number is satisfactory or not.

12. The Printing and Halting Problems

The printing problem

Some Turing-machine programmes print '0' at some stage in their computation; all the remaining programmes never print '0'. Consider the problem of deciding, given any arbitrarily selected programme, into which of these two categories it falls. This is an example of the printing problem.

The printing problem (p. 73) is the problem of determining whether or not the machine described by any arbitrarily selected standard description (or, equivalently, any arbitrarily selected description number) ever prints a certain symbol ('0', for example). Turing proves that if the printing problem were solvable by some computing machine, then the satisfactoriness problem would be too. Therefore neither is.

The halting problem

Another example of a problem that cannot be solved by computing machines, and a close relative of the printing problem, is the *halting problem*. This is the problem of determining whether or not the machine described by any arbitrarily selected standard description eventually halts—i.e. ceases moving altogether—when started on a given tape (e.g. a blank tape).

The machine shown in Table 1 is rather obviously one of those that never halt—but in other cases it is not at all obvious from a machine's table whether or not it halts. Simply watching the machine run (or a simulation of the machine) is of little help, for what can be concluded if after a week or a year the machine has not halted? If the machine does eventually halt, a watching human—or Turing machine—will sooner or later find this out; but in the case of a machine that has not yet halted, there is no systematic method for deciding whether or not it is going to halt.

The halting problem was so named (and, it appears, first stated) by Martin Davis.[61] The proposition that the halting problem cannot be solved by computing machine is known as the 'halting theorem'.[62] (It is often said that Turing stated and proved the halting theorem in 'On Computable Numbers', but strictly this is not true.)

13. The Church–Turing Thesis

Human computers

When Turing wrote 'On Computable Numbers', a computer was not a machine at all, but a human being. A computer—sometimes also spelt 'computor'—was a mathematical assistant who calculated by rote, in accordance with a systematic method. The method was supplied by an overseer prior to the calculation. Many thousands of human computers were employed in business, government, and research establishments, doing some of the sorts of calculating work that nowadays is performed by electronic computers. Like filing clerks, computers might have little detailed knowledge of the end to which their work was directed.

The term 'computing machine' was used to refer to small calculating machines that mechanized elements of the human computer's work. These were somewhat like today's non-programmable hand-calculators: they were not automatic, and each step—each addition, division, and so on—was initiated manually by the human operator. A computing machine was in effect a homunculus, calculating more quickly than an unassisted human computer, but doing nothing that could not in principle be done by a human clerk working by rote. For a complex calculation, several dozen human computers might be required, each equipped with a desk-top computing machine.

In the late 1940s and early 1950s, with the advent of electronic computing machines, the phrase 'computing machine' gave way gradually to 'computer'. During the brief period in which the old and new meanings of 'computer' coexisted, the prefix 'electronic' or 'digital' would usually be used in order to distinguish machine from human. As Turing stated, the new electronic machines were 'intended to carry out any definite rule of thumb process which could have been done by a human operator working in a disciplined but unintelligent manner'.[63] Main-frames, laptops, pocket calculators, palm-pilots—all carry out

---

[61] See M. Davis, *Computability and Unsolvability* (New York: McGraw-Hill, 1958), 70. Davis thinks it likely that he first used the term 'halting problem' in a series of lectures that he gave at the Control Systems Laboratory at the University of Illinois in 1952 (letter from Davis to Copeland, 12 Dec. 2001).

[62] It is interesting that if one lifts the restriction that the determination must be carried out in a *finite* number of steps, then Turing machines *are* able to solve the halting and printing problems, and moreover in a finite time. See B. J. Copeland, 'Super Turing-Machines', *Complexity*, 4 (1998), 30–2, and 'Accelerating Turing Machines', *Minds and Machines*, 12 (2002), 281–301.

[63] Turing's *Programmers' Handbook for Manchester Electronic Computer*, 1.

work that a human rote-worker could do, if he or she worked long enough, and had a plentiful enough supply of paper and pencils.

It must be borne in mind when reading 'On Computable Numbers' that Turing there used the word 'computer' in this now archaic sense. Thus he says, for example, 'Computing is normally done by writing certain symbols on paper' (p. 75) and 'The behaviour of the computer at any moment is determined by the symbols which he is observing, and his "state of mind"' (p. 75).

The Turing machine is an idealization of the human computer (p. 59): 'We may compare a man in the process of computing a real number to a machine which is only capable of a finite number of conditions... called "*m*-configurations". The machine is supplied with a "tape" ...' Wittgenstein put the point in a striking way: 'Turing's "Machines". These machines are *humans* who calculate.'[64]

In the primary sense, a computable number is a real number that can be calculated by a human computer—or in other words, a real number that a human being can calculate by means of a systematic method. When Turing asserts that 'the "computable" numbers include all numbers which would naturally be regarded as computable' (p. 74), he means that each number that is computable in this primary sense is also computable in the technical sense defined in Section 2 of 'On Computable Numbers' (see Section 10 of this introduction).

The thesis

Turing's thesis, that

> the 'computable' numbers include all numbers which would naturally be regarded as computable,

is now known as the *Church–Turing thesis*.

Some other ways of expressing the thesis are:

1. The universal Turing machine can perform any calculation that any human computer can carry out.
2. Any systematic method can be carried out by the universal Turing machine.

The Church–Turing thesis is sometimes heard in the strengthened form:

> Anything that can be made completely precise can be programmed for a universal digital computer.

However, this strengthened form of the thesis is false.[65] The printing, halting, and satisfactoriness problems are completely precise, but of course cannot be programmed for a universal computing machine.

---

[64] L. Wittgenstein, *Remarks on the Philosophy of Psychology*, vol. i (Oxford: Blackwell, 1980), § 1096.
[65] As Martin Davis emphasized long ago in his *Computability and Unsolvability*, p. vii.

Systematic methods

A systematic method—sometimes also called an *effective* method and a *mechanical* method—is any mathematical method of which all the following are true:

- the method can, in practice or in principle, be carried out by a human computer working with paper and pencil;
- the method can be given to the human computer in the form of a *finite* number of instructions;
- the method demands neither insight nor ingenuity on the part of the human being carrying it out;
- the method will definitely work if carried out without error;
- the method produces the desired result in a finite number of steps; or, if the desired result is some *infinite* sequence of symbols (e.g. the decimal expansion of π), then the method produces each individual symbol in the sequence in some finite number of steps.

The term 'systematic' and its synonyms 'effective' and 'mechanical' are terms of art in mathematics and logic. They do not carry their everyday meanings. For example: if some type of machine were able to solve the satisfactoriness problem, the method it used would not be systematic or mechanical in *this* sense. (Turing is sometimes said to have proved that *no machine* can solve the satisfactoriness problem. This is not so. He demonstrates only that his idealized human computers—Turing machines—cannot solve the satisfactoriness problem. This does not in itself rule out the possibility that some other type of machine might be able to solve the problem.[66])

Turing sometimes used the expression *rule of thumb* in place of 'systematic'. If this expression is employed, the Church–Turing thesis becomes (Chapter 10, p. 414):

LCMs can do anything that could be described as 'rule of thumb' or 'purely mechanical'.

'LCM' stands for 'logical computing machine', a term that Turing seems to have preferred to the (then current) 'Turing machine'.

Section 9 of 'On Computable Numbers' contains a bouquet of arguments for Turing's thesis. The arguments are persuasive, but do not offer the certainty of mathematical proof. As Turing says wryly of a related thesis in Chapter 17 (p. 588): 'The statement is...one which one does not attempt to prove. Propaganda is more appropriate to it than proof.'

Additional arguments and other forms of evidence for the thesis amassed. These, too, left matters short of absolute certainty. Nevertheless, before long it was, as Turing put it, 'agreed amongst logicians' that his proposal gives the

---

[66] See R. Gandy, 'Church's Thesis and Principles for Mechanisms', in J. Barwise, H. J. Keisler, and K. Kunen (eds.), *The Kleene Symposium* (Amsterdam: North-Holland, 1980).

'correct accurate rendering' of talk about systematic methods (Chapter 10, p. 414).[67] There have, however, been occasional dissenting voices over the years (for example, Kalmár and Péter).[68]

The converse of the thesis

The *converse* of the Church–Turing thesis is:

> Any number, or binary sequence, that can be computed by the universal Turing machine can be calculated by means of a systematic method.

This is self-evidently true—the instruction table on the universal machine's tape is itself a specification of a systematic method for calculating the number or sequence in question. In principle, a human being equipped with paper and pencil could work through the instructions in the table and write out the digits of the number, or sequence, without at any time exercising ingenuity or insight ('in principle' because we have to assume that the human does not throw in the towel from boredom, die of old age, or use up every sheet of paper in the universe).

Application of the thesis

The concept of a systematic method is an informal one. Attempts—such as the above—to explain what counts as a systematic method are not rigorous, since the requirement that the method demand neither insight nor ingenuity is left unexplicated.

One of the most significant achievements of 'On Computable Numbers'—and this was a large step in the development of the mathematical theory of computation—was to propose a rigorously defined expression with which the informal expression 'by means of a systematic method' might be replaced. The rigorously defined expression is, of course, 'by means of a Turing machine'.

The importance of Turing's proposal is this. If the proposal is correct—i.e. if the Church–Turing thesis is true—then talk about the existence or non-existence of systematic methods can be replaced throughout mathematics and logic by talk about the existence or non-existence of Turing-machine programmes. For instance, one can establish that there is no systematic method at all for doing such-and-such a thing by proving that no Turing machine can do the thing in question. This is precisely Turing's strategy with the *Entscheidungsproblem*, as explained in the next section.

---

[67] There is a survey of the evidence in chapters 12 and 13 of S. C. Kleene, *Introduction to Metamathematics* (Amsterdam: North-Holland, 1952).
[68] L. Kalmár, 'An Argument against the Plausibility of Church's Thesis', R. Péter, 'Rekursivität und Konstruktivität'; both in A. Heyting (ed.), *Constructivity in Mathematics* (Amsterdam: North-Holland, 1959).

Church's contribution

In 1935, on the other side of the Atlantic, Church had independently proposed a different way of replacing talk about systematic methods with formally precise language (in a lecture given in April of that year and published in 1936).[69] Turing learned of Church's work in the spring of 1936, just as 'On Computable Numbers' was nearing completion (see the introduction to Chapter 4).

Where Turing spoke of numbers and sequences, Church spoke of mathematical *functions*. (x^2 and $x + y$ are examples of mathematical functions. 4 is said to be the *value of the function* x^2 for $x = 2$.) Corresponding to each computable sequence S is a computable function fx (and vice versa). The value of fx for $x = 1$ is the first digit of S, for $x = 2$, the second digit of S, and so on. In 'On Computable Numbers' Turing said (p. 58): 'Although the subject of this paper is ostensibly the computable *numbers*, it is almost equally easy to define and investigate computable functions... I have chosen the computable numbers for explicit treatment as involving the least cumbrous technique.'

Church's analysis was in terms of his and Stephen Kleene's concept of a *lambda-definable* function. A function of positive integers is said to be lambda-definable if the values of the function can be calculated by a process of repeated substitution.

Thus we have alongside Turing's thesis

Church's thesis: every function of positive integers whose values can be calculated by a systematic method is lambda-definable.

Although Turing's and Church's approaches are different, they are nevertheless equivalent, in the sense that every lambda-definable function is computable by the universal machine and every function (or sequence) computable by the universal machine is lambda-definable.[70] Turing proved this in the Appendix to 'On Computable Numbers' (added in August 1936).

The name 'Church–Turing thesis', now standard, seems to have been introduced by Kleene, with a flourish of bias in favour of his mentor Church: 'So Turing's and Church's theses are equivalent. We shall usually refer to them both as *Church's thesis*, or in connection with that one of its... versions which deals with "Turing machines" as *the Church-Turing thesis*.'[71]

Although Turing's and Church's theses are equivalent in the logical sense, there is nevertheless good reason to prefer Turing's formulation. As Turing wrote in 1937: 'The identification of "effectively calculable" functions with computable

---

[69] Church, 'An Unsolvable Problem of Elementary Number Theory'.

[70] Equivalent, that is, if the computable functions are restricted to functions of positive integers. Turing's concerns were rather more general than Church's, in that whereas Church considered only functions of positive integers, Turing described his work as encompassing 'computable functions of an integral variable or a real or computable variable, computable predicates, and so forth' (p. 58, below). Turing intended to pursue the theory of computable functions of a real variable in a subsequent paper, but in fact did not do so.

[71] S. C. Kleene, *Mathematical Logic* (New York: Wiley, 1967), 232.

functions is possibly more convincing than an identification with the λ-definable [lambda-definable] or general recursive functions.'[72] Church acknowledged the point:

> As a matter of fact, there is... equivalence of three different notions: computability by a Turing machine, general recursiveness in the sense of Herbrand–Gödel–Kleene, and λ-definability in the sense of Kleene and [myself]. Of these, the first has the advantage of making the identification with effectiveness in the ordinary (not explicitly defined) sense evident immediately... The second and third have the advantage of suitability for embodiment in a system of symbolic logic.[73]

The great Kurt Gödel, it seems, was unpersuaded by Church's thesis until he saw Turing's formulation. Kleene wrote:

> According to a November 29, 1935, letter from Church to me, Gödel 'regarded as thoroughly unsatisfactory' Church's proposal to use λ-definability as a definition of effective calculability... It seems that only after Turing's formulation appeared did Gödel accept Church's thesis.[74]

Hao Wang reports Gödel as saying: 'We had not perceived the sharp concept of mechanical procedures sharply before Turing, who brought us to the right perspective.'[75]

Gödel described Turing's analysis of computability as 'most satisfactory' and 'correct... beyond any doubt'.[76] He also said: 'the great importance of... Turing's computability... seems to me... largely due to the fact that with this concept one has for the first time succeeded in giving an absolute definition of an interesting epistemological notion.'[77]

14. The *Entscheidungsproblem*

In Section 11 of 'On Computable Numbers', Turing turns to the *Entscheidungsproblem*, or *decision problem*. Church gave the following definition of the *Entscheidungsproblem*:

> By the Entscheidungsproblem of a system of symbolic logic is here understood the problem to find an effective method by which, given any expression Q in the notation of the system, it can be determined whether or not Q is provable in the system.[78]

---

[72] Turing, 'Computability and λ-Definability', *Journal of Symbolic Logic*, 2 (1937), 153–63 (153).
[73] Church's review of 'On Computable Numbers' in *Journal of Symbolic Logic*, 43.
[74] S. C. Kleene, 'Origins of Recursive Function Theory', *Annals of the History of Computing*, 3 (1981), 52–67 (59, 61).
[75] H. Wang, *From Mathematics to Philosophy* (New York: Humanities Press, 1974), 85.
[76] K. Gödel, *Collected Works*, ed. S. Feferman et al., vol. iii (Oxford: Oxford University Press, 1995), 304, 168.
[77] Ibid., vol. ii. (Oxford: Oxford University Press, 1990), 150.
[78] Church, 'A Note on the Entscheidungsproblem', 41.

The decision problem was brought to the fore of mathematics by the German mathematician David Hilbert (who in a lecture given in Paris in 1900 set the agenda for much of twentieth-century mathematics). In 1928 Hilbert described the decision problem as 'the main problem of mathematical logic', saying that 'the discovery of a general decision procedure is a very difficult problem which is as yet unsolved', and that the 'solution of the decision problem is of fundamental importance'.[79]

The Hilbert programme

Hilbert and his followers held that mathematicians should seek to express mathematics in the form of a complete, consistent, decidable formal system—a system expressing 'the whole thought content of mathematics in a uniform way'.[80] Hilbert drew an analogy between such a system and 'a court of arbitration, a supreme tribunal to decide fundamental questions—on a concrete basis on which everyone can agree and where every statement can be controlled'.[81] Such a system would banish ignorance from mathematics: given any mathematical statement, one would be able to tell whether the statement is true or false by determining whether or not it is provable in the system. As Hilbert famously declared in his Paris lecture: 'in mathematics there is no *ignorabimus*' (there is no *we shall not know*).[82]

It is important that the system expressing the 'whole thought content of mathematics' be *consistent*. An inconsistent system—a system containing contradictions—is worthless, since *any* statement whatsoever, true or false, can be derived from a contradiction by simple logical steps.[83] So in an inconsistent

---

[79] D. Hilbert and W. Ackermann, *Grundzüge der Theoretischen Logik* [Principles of Mathematical Logic] (Berlin: Julius Springer, 1928), 73, 77.

[80] D. Hilbert, 'The Foundations of Mathematics' (English translation of a lecture given in Hamburg in 1927, entitled 'Die Grundlagen der Mathematik'), in J. van Heijenoort (ed.), *From Frege to Gödel: A Source Book in Mathematical Logic, 1879–1931* (Cambridge, Mass.: Harvard University Press, 1967), 475.

[81] D. Hilbert, 'Über das Unendliche' [On the Infinite], *Mathematische Annalen*, 95 (1926), 161–90 (180); English translation by E. Putnam and G. Massey in R. L. Epstein and W. A. Carnielli, *Computability: Computable Functions, Logic, and the Foundations of Mathematics* (2nd edn. Belmont, Calif.: Wadsworth, 2000).

[82] D. Hilbert, 'Mathematical Problems: Lecture Delivered before the International Congress of Mathematicians at Paris in 1900', *Bulletin of the American Mathematical Society*, 8 (1902), 437–79 (445).

[83] To prove an arbitrary statement from a contradiction P & *not* P, one may use the following rules of inference (see further pp. 49–52, below):

(a) $not\ P \vdash not\,(P\ \&\ X)$
(b) $P\ \&\ not\,(P\ \&\ X) \vdash not\ X$.

Rule (*a*) says: from the statement that it is not the case that P, it can be inferred that not *both* P and X are the case—i.e. inferred that one at least of P and X is not the case—where X is any statement that you please. Rule (*b*) says: given that P is the case and that not *both* P and X are the case, it can be inferred that X is not the case. Via (*a*), the contradiction 'P & *not* P' leads to '*not* $(P\ \&\ X)$'; and since the contradiction also offers us P, we may then move, via (*b*), to '*not* X'. So we have deduced an arbitrary statement, '*not* X', from the contradiction. (To deduce simply X, replace X in (*a*) and (*b*) by '*not* X', and at the last step use the rule saying that two negations 'cancel out': $not\ not\ X \vdash X$.)

system, absurdities such as $0 = 1$ and $6 \neq 6$ are provable. An inconsistent system would indeed contain all true mathematical statements—would be *complete*, in other words—but would in addition also contain all false mathematical statements!

Hilbert's requirement that the system expressing the whole content of mathematics be *decidable* amounts to this: there must be a systematic method for telling, of each mathematical statement, whether or not the statement is provable in the system. If the system is to banish ignorance totally from mathematics then it must be decidable. Only then could we be confident of always being able to tell whether or not any given statement is provable. An undecidable system might sometimes leave us in ignorance.

The project of expressing mathematics in the form of a complete, consistent, decidable formal system became known as 'proof theory' and as the 'Hilbert programme'. In 1928, in a lecture delivered in the Italian city of Bologna, Hilbert said:

In a series of presentations in the course of the last years I have... embarked upon a new way of dealing with fundamental questions. With this new foundation of mathematics, which one can conveniently call proof theory, I believe the fundamental questions in mathematics are finally eliminated, by making every mathematical statement a concretely demonstrable and strictly derivable formula...

[I]n mathematics there is no *ignorabimus*, rather we are always able to answer meaningful questions; and it is established, as Aristotle perhaps anticipated, that our reason involves no mysterious arts of any kind: rather it proceeds according to formulable rules that are completely definite—and are as well the guarantee of the absolute objectivity of its judgement.[84]

Unfortunately for the Hilbert programme, however, it was soon to become clear that most interesting mathematical systems are, if consistent, *incomplete* and *undecidable*.

In 1931, Gödel showed that Hilbert's ideal is impossible to satisfy, even in the case of simple arithmetic.[85] He proved that the formal system of arithmetic set out by Whitehead and Russell in their seminal *Principia Mathematica*[86] is, if consistent, incomplete. That is to say: if the system is consistent, there are true

---

[84] D. Hilbert, 'Probleme der Grundlegung der Mathematik' [Problems Concerning the Foundation of Mathematics], *Mathematische Annalen*, 102 (1930), 1–9 (3, 9). Translation by Elisabeth Norcliffe.

[85] K. Gödel, 'Über formal unentscheidbare Sätze der Principia Mathematica und verwandter Systeme I.' [On Formally Undecidable Propositions of Principia Mathematica and Related Systems I], *Monatshefte für Mathematik und Physik*, 38 (1931), 173–98. English translation in M. Davis (ed.), *The Undecidable: Basic Papers on Undecidable Propositions, Unsolvable Problems and Computable Functions* (New York: Raven, 1965), 5–38.

[86] A. N. Whitehead and B. Russell, *Principia Mathematica*, vols. i–iii (Cambridge: Cambridge University Press, 1910–13).

statements of arithmetic that are not provable in the system—the formal system fails to capture the 'whole thought content' of arithmetic. This is known as Gödel's *first incompleteness theorem*.

Gödel later generalized this result, pointing out that 'due to A. M. Turing's work, a precise and unquestionably adequate definition of the general concept of formal system can now be given', with the consequence that incompleteness can 'be proved rigorously for *every* consistent formal system containing a certain amount of finitary number theory'.[87] The definition made possible by Turing's work is this (in Gödel's words): 'A formal system can simply be defined to be any mechanical procedure for producing formulas, called provable formulas.'[88]

In his incompleteness theorem, Gödel had shown that no matter how hard mathematicians might try to construct the all-encompassing formal system envisaged by Hilbert, the product of their labours would, if consistent, inevitably be incomplete. As Hermann Weyl—one of Hilbert's greatest pupils—observed, this was nothing less than 'a catastrophe' for the Hilbert programme.[89]

Decidability

Gödel's theorem left the question of decidability open. As Newman summarized matters:

The Hilbert decision-programme of the 1920's and 30's had for its objective the discovery of a general process... for deciding... truth or falsehood... A first blow was dealt at the prospects of finding this new philosopher's stone by Gödel's incompleteness theorem (1931), which made it clear that truth or falsehood of A could not be equated to provability of A or not-A in any finitely based logic, chosen once for all; but there still remained in principle the possibility of finding a mechanical process for deciding whether A, or not-A, or neither, was formally provable in a given system.[90]

The question of decidability was tackled head on by Turing and, independently, by Church.

On p. 84 of 'On Computable Numbers' Turing pointed out—by way of a preliminary—a fact that Hilbertians appear to have overlooked: if a system is complete then it follows that it is also decidable. Bernays, Hilbert's close collaborator, had said: 'One observes that [the] requirement of deductive completeness

---

[87] Gödel, 'Postscriptum', in Davis, *The Undecidable: Basic Papers on Undecidable Propositions, Unsolvable Problems and Computable Functions*, 71–3 (71); the Postscriptum, dated 1964, is to Gödel's 1934 paper 'On Undecidable Propositions of Formal Mathematical Systems' (ibid. 41–71).

[88] Ibid. 72.

[89] H. Weyl, 'David Hilbert and his Mathematical Work', *Bulletin of the American Mathematical Society*, 50 (1944), 612–54 (644).

[90] M. H. A. Newman, 'Alan Mathison Turing, 1912–1954', *Biographical Memoirs of Fellows of the Royal Society*, 1 (1955), 253–63 (256).

does not go as far as the requirement of decidability.'[91] Turing's simple argument on p. 84 shows that there is no conceptual room for the distinction that Bernays is claiming.

Nevertheless, the crucial question was still open: given that in fact simple arithmetic is (if consistent) *in*complete, is it or is it not decidable? Turing and Church both showed that no consistent formal system of arithmetic is decidable. They showed this by proving that not even the *functional calculus*—the weaker, purely logical system presupposed by any formal system of arithmetic—is decidable. The Hilbertian dream of a completely mechanized mathematics now lay in total ruin.

A tutorial on first-order predicate calculus

What Turing called the functional calculus (and Church, following Hilbert, the *engere Funktionenkalkül*) is today known as *first-order predicate calculus* (FOPC). FOPC is a formalization of deductive logical reasoning.

There are various different but equivalent ways of formulating FOPC. One formulation presents FOPC as consisting of about a dozen formal rules of inference. (This formulation, which is more accessible than the Hilbert–Ackermann formulation mentioned by Turing on p. 84, is due to Gerhard Gentzen.[92])

The following are examples of formal rules of inference. The symbol '⊢' indicates that the statement following it can be concluded from the statements (or statement) displayed to its left, the premises.

(i) X, if X then $Y \vdash Y$
(ii) X and $Y \vdash X$
(iii) X, $Y \vdash X$ and Y

So if, for example, 'X' represents 'It is sunny' and 'Y' represents 'We will go for a picnic', (i) says:

'We will go for a picnic' can be concluded from the premisses 'It is sunny' and 'If it is sunny then we will go for a picnic'.

(ii) says:

'It is sunny' can be concluded from the conjunctive premiss 'It is sunny and we will go for a picnic'.

Turing uses the symbol '→' to abbreviate 'if then' and the symbol '&' to abbreviate 'and'. Using this notation, (i)–(iii) are written:

---

[91] P. Bernays, 'Die Philosophie der Mathematik und die Hilbertsche Beweistheorie' [The Philosophy of Mathematics and Hilbert's Proof Theory], *Blätter für Deutsche Philosophie*, 4 (1930/1931), 326–67. See also H. Wang, *Reflections on Kurt Gödel* (Cambridge, Mass.: MIT Press, 1987), 87–8.
[92] G. Gentzen, 'Investigations into Logical Deduction' (1934), in *The Collected Papers of Gerhard Gentzen*, ed. M. E. Szabo (Amsterdam: North-Holland, 1969).

(i) $X, X \rightarrow Y \vdash Y$
(ii) $X \& Y \vdash X$
(iii) $X, Y \vdash X \& Y$

Some more rules of the formal calculus are as follows. *a* represents any object, *F* represents any property:

(iv) *a* has property $F \vdash$ there is an object that has property F
(v) each object has property $F \vdash a$ has property F

In Turing's notation, in which '*a* has property F' is abbreviated '$F(a)$', these are written:

(iv) $F(a) \vdash (\exists x) F(x)$
(v) $(x) F(x) \vdash F(a)$

'$(\exists x)$' is read: 'there is an object (call it x) which ...'. So '$(\exists x) F(x)$' says 'there is an object, call it x, which has property F'. '(x)' is read: 'each object, x, is such that ...'. So '$(x) F(x)$' says 'each object, x, is such that x has property F'.

Set out in full, FOPC contains not only rules like (i)–(v) but also several rules leading from statements containing '$\vdash$' to other statements containing '$\vdash$'. One such rule is the so-called 'cut rule', used in moving from lines (2) and (3) to (4) in the proof below.

Turing calls '$(\exists x)$' and '(x)' *quantors*; the modern term is *quantifiers*. A symbol, such as 'F', that denotes a property is called a *predicate*. Symbols denoting relationships, for example '$<$' (less than) and '$=$' (identity), are also classed as predicates. The symbol 'x' is called a *variable*.

(FOPC is *first-order* in the sense that the quantifiers of the calculus always involve variables that refer to individual objects. In *second-order* predicate calculus, on the other hand, the quantifiers can contain predicates, as in '$(\exists F)$'. The following are examples of second-order quantification: 'Jules and Jim have some properties in common,' 'Each relationship that holds between *a* and *b* also holds between *c* and *d*.')

Using the dozen or so basic rules of FOPC, more complicated rules of inference can be proved as *theorems* ('provable formulas') of FOPC. For example:

Theorem $(x)(G(x) \rightarrow H(x)), G(a) \vdash (\exists x) H(x)$

This theorem says: 'There is an object that has property H' can be concluded from the premises 'Each object that has property G also has property H' and '*a* has property G'.

The proof of the theorem is as follows:

(1) $(x)(G(x) \rightarrow H(x)) \vdash G(a) \rightarrow H(a)$ (rule (v))
(2) $G(a), (G(a) \rightarrow H(a)) \vdash H(a)$ (rule (i))
(3) $H(a) \vdash (\exists x) H(x)$ (rule (iv))
(4) $G(a), (G(a) \rightarrow H(a)) \vdash (\exists x) H(x)$ (from (2) and (3) by the cut rule)

(5) $(x)(G(x) \to H(x)), G(a) \vdash (\exists x)H(x)$

(from (1) and (4) by the cut rule)

The cut rule (or rule of transitivity) says in effect that whatever can be concluded from a statement Y (possibly in conjunction with additional premises P) can be concluded from any premiss(es) from which Y can be concluded (together with the additional premisses P, if any). For example, if $Y \vdash Z$ and $X \vdash Y$, then $X \vdash Z$. In the transition from (1) and (4) to (5), the additional premiss $G(a)$ in (4) is gathered up and placed among the premisses of (5).

So far we have seen how to prove further inference rules in FOPC. Often logicians are interested in proving not inference rules but single statements unbroken by commas and '$\vdash$'. An example is the complex statement

$$\textit{not } (F(a) \ \& \ \textit{not } (\exists x)F(x)),$$

which says 'It is not the case that *both* $F(a)$ and the denial of $(\exists x)F(x)$ are true'; or in other words, you are not going to find $F(a)$ true without finding $(\exists x)F(x)$ true.

To say that a single statement, as opposed to an inference rule, is provable in FOPC is simply to say that the result of *prefixing* that statement by '$\vdash$' can be derived by using the rules of the calculus. Think of a '$\vdash$' with no statements on its left as indicating that the statement on its right is to be concluded as a matter of 'pure logic'—no premises are required.

For example, the theorem

$$\vdash \textit{not } (F(a) \ \& \ \textit{not } (\exists x)F(x))$$

can be derived using rule (iv) and the following new rule.[93]

$$\frac{X \vdash Y}{\vdash \textit{not } (X \ \& \ \textit{not } Y)}$$

This rule is read:

If Y can be concluded from X, then it can be concluded that not *both* X and the denial of Y are true.

Much of mathematics and science can be formulated within the framework of FOPC. For example, a formal system of arithmetic can be constructed by adding a number of arithmetical axioms to FOPC. The axioms consist of very basic arithmetical statements, such as:

$$(x)(x + 0 = x)$$

and

$$(x)(y)(Sx = Sy \to x = y),$$

---

[93] In Gentzen's system this rule can itself be derived from the basic rules. It should be mentioned that in the full system it is permissible to write any finite number of statements (including zero) on the right hand side of '$\vdash$'.

where 'S' means 'the successor of'—the successor of 1 is 2, and so on. (In these axioms the range of the variables 'x' and 'y' is restricted to numbers.) Other arithmetical statements can be derived from these axioms by means of the rules of FOPC. For example, rule (v) tells us that the statement

$$1 + 0 = 1$$

can be concluded from the first of the above axioms.

If FOPC is undecidable then it follows that arithmetic is undecidable. Indeed, if FOPC is undecidable, then so are very many important mathematical systems. To find decidable logics one must search among systems that are in a certain sense *weaker* than FOPC. One example of a decidable logic is the system that results if all the quantifier rules—rules such as (iv) and (v)—are elided from FOPC. This system is known as the *propositional calculus*.

The proof of the undecidability of FOPC

Turing and Church showed that there is no systematic method by which, given any formula Q in the notation of FOPC, it can be determined whether or not Q is provable in the system (i.e. whether or not ⊢ Q). To put this another way, Church and Turing showed that the *Entscheidungsproblem* is *unsolvable* in the case of FOPC.

Both published this result in 1936.[94] Church's demonstration of undecidability proceeded via his lambda calculus and his thesis that to each effective method there corresponds a lambda-definable function. There is general agreement that Turing was correct in his view, mentioned above (p. 45), that his own way of showing undecidability is 'more convincing' than Church's.

Turing's method makes use of his proof that no computing machine can solve the printing problem. He showed that if a Turing machine could tell, of any given statement, whether or not the statement is provable in FOPC, then a Turing machine could tell, of any given Turing machine, whether or not it ever prints '0'. Since, as he had already established, no Turing machine can do the latter, it follows that no Turing machine can do the former. The final step of the argument is to apply Turing's thesis: if no Turing machine can perform the task in question, then there is no systematic method for performing it.

[94] In a lecture given in April 1935—the text of which was printed the following year as 'An Unsolvable Problem of Elementary Number Theory' (a short 'Preliminary report' dated 22 Mar. 1935 having appeared in the *Bulletin of the American Mathematical Society* (41 (1935), 332–3))—Church proved the undecidability of a system that includes FOPC as a part. This system is known as *Principia Mathematica*, or PM, after the treatise in which it was first set out (see n. 86). PM is obtained by adding mathematical axioms to FOPC. Church established the conditional result that if PM is *omega-consistent*, then PM is undecidable. Omega-consistency (first defined by Gödel) is a stronger property than consistency, in the sense that a consistent system is not necessarily omega-consistent. As explained above, a system is consistent when there is no statement *S* such that both *S* and *not-S* are provable in the system. A system is omega-consistent when there is no predicate *F* of integers such that all the following are provable in the system: $(\exists x)F(x)$, *not-F*(1), *not-F*(2), *not-F*(3), and so on, for every integer. In his later paper 'A Note on the Entscheidungsproblem' (completed in April 1936) Church improved on this earlier result, showing unconditionally that FOPC is undecidable.

In detail, Turing's demonstration contains the following steps.

1. Turing shows how to construct, for any computing machine **m**, a complicated statement of FOPC that says 'at some point, machine **m** prints 0'. He calls this formula 'Un(**m**)'. (The letters 'Un' probably come from 'undecidable' or the German equivalent 'unentscheidbare'.)
2. Turing proves the following:
   (*a*) If Un(**m**) is provable in FOPC, then at some point **m** prints 0.
   (*b*) If at some point **m** prints 0, then Un(**m**) is provable in FOPC.
3. Imagine a computing machine which, when given any statement Q in the notation of FOPC, is able to determine (in some finite number of steps) whether or not Q is provable in FOPC. Let's call this machine HILBERT'S DREAM. 2(*a*) and 2(*b*) tell us that HILBERT'S DREAM would solve the printing problem. Because if the machine were to indicate that Un(**m**) is provable then, in view of 2(*a*), it would in effect be indicating that **m** does print 0; and if the machine were to indicate that the statement Un(**m**) is not provable then, in view of 2(*b*), it would in effect be indicating that **m** does not print 0. Since no computing machine can solve the printing problem, it follows that HILBERT'S DREAM is a figment. *No computing machine is able to determine in some finite number of steps, of each statement Q, whether or not Q is provable in FOPC.*
4. If there were a systematic method by which, given any statement Q, it can be determined whether or not Q is provable in FOPC, then it would follow, by Turing's thesis, that there is such a computing machine as HILBERT'S DREAM. Therefore there is no such systematic method.

The significance of undecidability

Poor news though the unsolvability of the *Entscheidungsproblem* was for the Hilbert school, it was very welcome news in other quarters, for a reason that Hilbert's illustrious pupil von Neumann had given in 1927:

> If undecidability were to fail then mathematics, in today's sense, would cease to exist; its place would be taken by a completely mechanical rule, with the aid of which any man would be able to decide, of any given statement, whether the statement can be proven or not.[95]

As the Cambridge mathematician G. H. Hardy said in a lecture in 1928: 'if there were...a mechanical set of rules for the solution of all mathematical problems...our activities as mathematicians would come to an end.'[96]

---

[95] J. von Neumann, 'Zur Hilbertschen Beweistheorie' [On Hilbert's Proof Theory], *Mathematische Zeitschrift*, 26 (1927), 1–46 (12); reprinted in vol. i of von Neumann's *Collected Works*, ed. A. H. Taub (Oxford: Pergamon Press, 1961).

[96] G. H. Hardy, 'Mathematical Proof', *Mind*, 38 (1929), 1–25 (16) (the text of Hardy's 1928 Rouse Ball Lecture).

Further reading

Barwise, J., and Etchemendy, J., *Turing's World: An Introduction to Computability Theory* (Stanford, Calif.: CSLI, 1993). (Includes software for building and displaying Turing machines.)

Boolos, G. S., and Jeffrey, R. C., *Computability and Logic* (Cambridge: Cambridge University Press, 2nd edn. 1980).

Copeland, B. J., 'Colossus and the Dawning of the Computer Age', in R. Erskine and M. Smith (eds.), *Action This Day* (London: Bantam, 2001).

Epstein, R. L., and Carnielli, W. A., *Computability: Computable Functions, Logic, and the Foundations of Mathematics* (Belmont, Calif.: Wadsworth, 2nd edn. 2000).

Hopcroft, J. E., and Ullman, J. D., *Introduction to Automata Theory, Languages, and Computation* (Reading, Mass.: Addison-Wesley, 1979).

Minsky, M. L., *Computation: Finite and Infinite Machines* (Englewood Cliffs, NJ: Prentice-Hall, 1967).

Sieg, W., 'Hilbert's Programs: 1917–1922', *Bulletin of Symbolic Logic*, 5 (1999), 1–44.

Sipser, M., *Introduction to the Theory of Computation* (Boston: PWS, 1997).

Appendix

Subroutines and M-Functions[97]

Section 3 of this guide gave a brief introduction to the concept of a *skeleton table*, where names of subroutines are employed in place of letters referring to states of the machine. This appendix explains the associated idea of an *m-function*, introduced by Turing on p. 63. *m*-functions are subroutines with *parameters*—values that are plugged into the subroutine before it is used.

The example of the 'find' subroutine f makes this idea clear. The subroutine f(A, B, x) is defined in Section 3 (Tables 2 and 3). Recall that f(A, B, x) finds the leftmost x on the tape and places the machine in A, leaving the scanner resting on the x; or if no x is found, places the machine in B and leaves the scanner resting on a blank square to the right of the used portion of the tape. 'A', 'B', and 'x' are the parameters of the subroutine. Parameter 'x' may be replaced by any symbol (of the Turing machine in question). Parameters 'A' and 'B' may be replaced by names of states of the machine. Alternatively, Turing permits 'A' and 'B' (one or both) to be replaced by a name of a *subroutine*. For example, replacing 'A' by the subroutine name '$e_1(C)$' produces:

$$f(e_1(C), B, x)$$

This says: find the leftmost x, let the scanner rest on it, and go into subroutine $e_1(C)$; or, if there is no x, go into B (leaving the scanner resting on a blank square to the right of the used portion of the tape).

The subroutine $e_1(C)$ simply erases the scanned square and places the machine in C, leaving the scanner resting on the square that has just been erased. ('C' is another parameter of the same type as 'A' and 'B'.) Thus the subroutine $f(e_1(C), B, x)$ finds

[97] By Andrés Sicard and Jack Copeland.

the leftmost occurrence of the symbol *x* and erases it, placing the machine in *C* and leaving the scanner resting on the square that has just been erased (or if no *x* is found, leaves the scanner resting on a blank square to the right of the used portion of the tape and places the machine in *B*). Since in this case nothing turns on the choice of letter, the name of the subroutine may also be written '**f**(**e**$_1$(*A*), *B*, *x*)'.

The subroutine **f**(**e**$_1$(*A*), *B*, *x*) is one and the same as the subroutine **e**(*A*, *B*, *x*) (Section 3). The new notation exhibits the structure of the subroutine.

More examples of *m*-functions are given below. While the use of *m*-functions is not strictly necessary for the description of any Turing machine, *m*-functions are very useful in describing large or complex Turing machines. This is because of the possibilities they offer for generalization, reusability, simplification, and modularization. Generalization is achieved because tasks of a similar nature can be done by a single *m*-function, and modularization because a complex task can be divided into several simpler *m*-functions. Simplification is obtained because the language of *m*-functions submerges some of the detail of the language of instruction-words—i.e. words of the form $q_i S_j S_k M q_l$—so producing transparent descriptions of Turing machines. Reusability arises simply because we can employ the same *m*-function in different Turing machines.

Although it is difficult (if not impossible) to indicate the exact role that Turing's concept of an *m*-function played in the development of today's programming languages, it is worth emphasizing that some characteristics of *m*-functions are present in the subroutines of almost all modern languages. Full use was made of the idea of parametrized subroutines by Turing and his group at the National Physical Laboratory as they pioneered the science of computer programming during 1946. A contemporary report (by Huskey) outlining Turing's approach to programming said the following:

The fact that repetition of subroutines require[s] large numbers of orders has led to the abbreviated code methods whereby not only standard orders are used but special words containing parameters are converted into orders by an interpretation table. The general idea is that these describe the entries to subroutines, the values of certain parameters in the subroutine, how many times the subroutine is to be used, and where to go after the subroutine is finished.[98]

Rather than give a formal definition of an *m*-function we present a series of illustrative examples.

First, some preliminaries. An *alphabet* **A** is some set of symbols, for example {-, 0, 1, 2}, and a *word* of alphabet **A** is a finite sequence of non-blank symbols of **A**. The blank symbol, represented '-', is used to separate different words on the tape and is part of the alphabet, but never occurs within words. The following examples all assume that, at the start of operation, there is a single word *w* of the alphabet on an otherwise blank tape, with the scanner positioned over any symbol of *w*. The symbols of *w* are written on adjacent squares, using both E-squares and F-squares, and *w* is surrounded by blanks (some of the examples require there to be at least one blank in front of *w* and at least three following *w*).

---

[98] H. D. Huskey, untitled typescript, National Physical Laboratory, n.d. but c. Mar. 1947 (in the Woodger Papers, National Museum of Science and Industry, Kensington, London (catalogue reference M12/105); a digital facsimile is in The Turing Archive for the History of Computing <www.AlanTuring.net/huskey_1947>).

Let **M** be a Turing machine with alphabet **A** = {-, 0, 1, 2}. The following instructions result in **M** printing the symbol '1' at the end of *w*, replacing the first blank to the right of *w*:

$$q_100Rq_1, \ q_111Rq_1, \ q_122Rq_1, \ q_1\text{-}1Nq_2$$

The first three instructions move the scanner past the symbols '0', '1', and '2', and once the scanner arrives at the first blank square to the right of *w*, the fourth instruction prints '1' (leaving **M** in state q_2).

If the symbols '3', '4', ..., '9' are added to the alphabet, so **A** = {-, 0, 1, ..., 9}, then the necessary instructions for printing '1' at the end of *w* are lengthier:

$$q_100Rq_1, \ q_111Rq_1, \ \ldots, \ q_199Rq_1, \ q_1\text{-}1Nq_2$$

The *m*-function **add**(S, α) defined by Table 4 carries out the task of printing one symbol 'α' at the end of any word *w* of *any* alphabet (assuming as before that the machine starts operating with the scanner positioned over one or another symbol of *w* and that *w* is surrounded by blanks).

Table 4 is the skeleton table for the *m*-function **add**(S, α). (Skeleton tables are like tables of instructions but with some parameters to be replaced by concrete values.) Table 4 has two parameters, 'α' and 'S'. The second parameter 'S' is to be replaced by the state or *m*-function into which the machine is to go once **add**(S, α) completes its operation, and the first parameter 'α' is to be replaced by whatever symbol it is that we wish to be printed at the end of the word.

Both sets of instruction-words shown above can now be replaced by a simple call to the *m*-function **add**(S, α), where S = q_2 and α = 1.

If instead of adding '1' at the end of a word from alphabet **A** = {-, 0, 1, ..., 9}, we wanted to add a pair of symbols '5' and '4', then the instruction-words would be:

$$q_100Rq_1, \ q_111Rq_1, \ \ldots, \ q_199Rq_1, \ q_1\text{-}5Rq_2, \ q_2\text{-}4Nq_3$$

These instruction-words can be replaced by the *m*-function **add**(**add**(q_3, 4), 5). This *m*-function finds the end of the word and writes '5', going into *m*-function **add**(q_3, 4), which writes '4' and ends in state q_3.

Another example: suppose that '5' and '4' are to be printed as just described, and then each occurrence of the symbol '3' is to be replaced by '4'. The *m*-function **add**(**add**(**change**(q_n, 3, 4), 4), 5) carries out the required task, where the *m*-function **change**(S, α, β) is defined by Table 5. The *m*-function **change**$_1$(S, α, β) is a subroutine inside the *m*-function **change**(S, α, β).

m-functions can employ *internal variables*. Although internal variables are not strictly necessary, they simplify an *m*-function's description. Internal variables are not parameters of the *m*-function—we do not need to replace them with concrete values before the *m*-function is used. In the following example, the internal variable 'δ' refers to whatever symbol is present on the scanned square when the machine enters the *m*-function **repeat**$_1$(S).

Suppose we wish to print a repetition of the first symbol of *w* at the end of *w*. This can be achieved by the *m*-function **repeat**(S) defined by Table 5. (The *m*-function **add**(S, δ) is as given by Table 4.)

Every *m*-function has the form: **name**($S_1, S_2, \ldots, S_n, α_1, α_2, \ldots, α_m$), where $S_1, S_2, \ldots, S_n$ refer either to states or to *m*-functions, and $α_1, α_2, \ldots, α_m$ denote symbols. Each *m*-function is a Turing machine with parameters. To convert an *m*-function's

Table 4

State	Scanned Square	Operations	Next State
add(S, α)	not -	R	**add**(S, α)
add(S, α)	-	P[α]	S

Table 5

State	Scanned Square	Operations	Next State
change(S, α, β)	not -	L	**change**(S, α, β)
change(S, α, β)	-	R	**change**$_1$(S, α, β)
change$_1$(S, α, β)	α	P[β], R	**change**$_1$(S, α, β)
change$_1$(S, α, β)	not α	R	**change**$_1$(S, α, β)
change$_1$(S, α, β)	-	L	S

Table 6

State	Scanned Square	Operations	Next State
repeat(S)	not -	L	**repeat**(S)
repeat(S)	-	R	**repeat**$_1$(S)
repeat$_1$(S)	δ		**add**(S, δ)

skeleton table to a Turing-machine instruction table, where each row is an instruction-word of the form q$_i$S$_j$S$_k$Mq$_1$, it is necessary to know the context in which the *m*-function is to be used, namely, the underlying Turing machine's alphabet and states. It is necessary to know the alphabet because of the use in skeleton tables of expressions such as 'does not contain !', 'not α', 'neither α nor -', 'any'. Knowledge of the underlying machine's states is necessary to ensure that the *m*-function begins and ends in the correct state.

The economy effected by *m*-functions is illustrated by the fact that if the *m*-functions are eliminated from Turing's description of his universal machine, nearly 4,000 instruction-words are required in their place.[99]

---

[99] A. Sicard, 'Máquinas de Turing dinámicas: historia y desarrollo de una idea' [Dynamic Turing Machines: Story and Development of an Idea], appendix 3 (Master's thesis, Universidad EAFIT, 1998); 'Máquina universal de Turing: algunas indicaciones para su construcción' [The Universal Turing Machine: Some Directions for its Construction], *Revista Universidad EAFIT*, vol. 108 (1998), pp. 61–106.

CHAPTER 1

On Computable Numbers, with an Application to the Entscheidungsproblem (*1936*)

Alan Turing

The "computable" numbers may be described briefly as the real numbers whose expressions as a decimal are calculable by finite means. Although the subject of this paper is ostensibly the computable *numbers*, it is almost equally easy to define and investigate computable functions of an integral variable or a real or computable variable, computable predicates, and so forth. The fundamental problems involved are, however, the same in each case, and I have chosen the computable numbers for explicit treatment as involving the least cumbrous technique. I hope shortly to give an account of the relations of the computable numbers, functions, and so forth to one another. This will include a development of the theory of functions of a real variable expressed in terms of computable numbers. According to my definition, a number is computable if its decimal can be written down by a machine.

In §§ 9, 10 I give some arguments with the intention of showing that the computable numbers include all numbers which could naturally be regarded as computable. In particular, I show that certain large classes of numbers are computable. They include, for instance, the real parts of all algebraic numbers, the real parts of the zeros of the Bessel functions, the numbers π, e, etc. The computable numbers do not, however, include all definable numbers, and an example is given of a definable number which is not computable.

Although the class of computable numbers is so great, and in many ways similar to the class of real numbers, it is nevertheless enumerable. In § 8 I examine certain arguments which would seem to prove the contrary. By the correct application of one of these arguments, conclusions are reached which are

[Received 28 May, 1936.—Read 12 November, 1936.]
This article first appeared in *Proceedings of the London Mathematical Society*, Series 2, 42 (1936–7). It is reprinted with the permission of the London Mathematical Society and the Estate of Alan Turing.

superficially similar to those of Gödel.[1] These results have valuable applications. In particular, it is shown (§ 11) that the Hilbertian Entscheidungsproblem can have no solution.

In a recent paper Alonzo Church has introduced an idea of "effective calculability", which is equivalent to my "computability", but is very differently defined.[2] Church also reaches similar conclusions about the Entscheidungsproblem.[3] The proof of equivalence between "computability" and "effective calculability" is outlined in an appendix to the present paper.

1. Computing machines

We have said that the computable numbers are those whose decimals are calculable by finite means. This requires rather more explicit definition. No real attempt will be made to justify the definitions given until we reach § 9. For the present I shall only say that the justification lies in the fact that the human memory is necessarily limited.

We may compare a man in the process of computing a real number to a machine which is only capable of a finite number of conditions $q_1, q_2, \ldots, q_R$ which will be called "m-configurations". The machine is supplied with a "tape" (the analogue of paper) running through it, and divided into sections (called "squares") each capable of bearing a "symbol". At any moment there is just one square, say the r-th, bearing the symbol $\mathfrak{S}(r)$ which is "in the machine". We may call this square the "scanned square". The symbol on the scanned square may be called the "scanned symbol". The "scanned symbol" is the only one of which the machine is, so to speak, "directly aware". However, by altering its m-configuration the machine can effectively remember some of the symbols which it has "seen" (scanned) previously. The possible behaviour of the machine at any moment is determined by the m-configuration q_n and the scanned symbol $\mathfrak{S}(r)$. This pair $q_n, \mathfrak{S}(r)$ will be called the "configuration": thus the configuration determines the possible behaviour of the machine. In some of the configurations in which the scanned square is blank (*i.e.* bears no symbol) the machine writes down a new symbol on the scanned square: in other configurations it erases the scanned symbol. The machine may also change the square which is being scanned, but only by shifting it one place to right or left. In addition to any of these operations the m-configuration may be changed. Some of the symbols written down will form the sequence of figures which is the decimal of the real number which is being

---

[1] Gödel, "Über formal unentscheidbare Sätze der Principia Mathematica und verwandter Systeme, I", *Monatshefte Math. Phys.*, 38 (1931), 173–198.

[2] Alonzo Church, "An unsolvable problem of elementary number theory", *American J. of Math.*, 58 (1936), 345–363.

[3] Alonzo Church, "A note on the Entscheidungsproblem", *J. of Symbolic Logic*, 1 (1936), 40–41.

computed. The others are just rough notes to "assist the memory". It will only be these rough notes which will be liable to erasure.

It is my contention that these operations include all those which are used in the computation of a number. The defence of this contention will be easier when the theory of the machines is familiar to the reader. In the next section I therefore proceed with the development of the theory and assume that it is understood what is meant by "machine", "tape", "scanned", etc.

2. Definitions

Automatic machines

If at each stage the motion of a machine (in the sense of § 1) is *completely* determined by the configuration, we shall call the machine an "automatic machine" (or *a*-machine).

For some purposes we might use machines (choice machines or *c*-machines) whose motion is only partially determined by the configuration (hence the use of the word "possible" in § 1). When such a machine reaches one of these ambiguous configurations, it cannot go on until some arbitrary choice has been made by an external operator. This would be the case if we were using machines to deal with axiomatic systems. In this paper I deal only with automatic machines, and will therefore often omit the prefix *a*-.

Computing machines

If an *a*-machine prints two kinds of symbols, of which the first kind (called figures) consists entirely of 0 and 1 (the others being called symbols of the second kind), then the machine will be called a computing machine. If the machine is supplied with a blank tape and set in motion, starting from the correct initial *m*-configuration, the subsequence of the symbols printed by it which are of the first kind will be called the *sequence computed by the machine*. The real number whose expression as a binary decimal is obtained by prefacing this sequence by a decimal point is called the *number computed by the machine*.

At any stage of the motion of the machine, the number of the scanned square, the complete sequence of all symbols on the tape, and the *m*-configuration will be said to describe the *complete configuration* at that stage. The changes of the machine and tape between successive complete configurations will be called the *moves* of the machine.

Circular and circle-free machines

If a computing machine never writes down more than a finite number of symbols of the first kind, it will be called *circular*. Otherwise it is said to be *circle-free*.

A machine will be circular if it reaches a configuration from which there is no possible move, or if it goes on moving, and possibly printing symbols of the second kind, but cannot print any more symbols of the first kind. The significance of the term "circular" will be explained in § 8.

Computable sequences and numbers

A sequence is said to be computable if it can be computed by a circle-free machine. A number is computable if it differs by an integer from the number computed by a circle-free machine.

We shall avoid confusion by speaking more often of computable sequences than of computable numbers.

3. Examples of computing machines

I. A machine can be constructed to compute the sequence 010101... . The machine is to have the four m-configurations "b", "c", "f", "e" and is capable of printing "0" and "1". The behaviour of the machine is described in the following table in which "R" means "the machine moves so that it scans the square immediately on the right of the one it was scanning previously". Similarly for "L". "E" means "the scanned symbol is erased" and "P" stands for "prints". This table (and all succeeding tables of the same kind) is to be understood to mean that for a configuration described in the first two columns the operations in the third column are carried out successively, and the machine then goes over into the m-configuration described in the last column. When the second column is left blank, it is understood that the behaviour of the third and fourth columns applies for any symbol and for no symbol. The machine starts in the m-configuration b with a blank tape.

Configuration		Behaviour	
m-config.	symbol	operations	final m-config.
b	None	P0, R	c
c	None	R	e
e	None	P1, R	f
f	None	R	b

If (contrary to the description in § 1) we allow the letters L, R to appear more than once in the operations column we can simplify the table considerably.

m-config.	symbol	operations	final m-config.
b	None	P0	b
	0	R, R, P1	b
	1	R, R, P0	b

62 | Alan Turing

II. As a slightly more difficult example we can construct a machine to compute the sequence 001011011101111011111... . The machine is to be capable of five *m*-configurations, viz. "o", "q", "p", "f", "b" and of printing "ə", "x", "0", "1". The first three symbols on the tape will be "əə0"; the other figures follow on alternate squares. On the intermediate squares we never print anything but "x". These letters serve to "keep the place" for us and are erased when we have finished with them. We also arrange that in the sequence of figures on alternate squares there shall be no blanks.

m-config.	symbol	operations	final *m*-config.
b		Pə, R, Pə, R, P0, R, R, P0, L, L	o
o	1	R, Px, L, L, L	o
	0		q
q	Any (0 or 1)	R, R	q
	None	P1, L	p
p	x	E, R	q
	ə	R	f
	None	L, L	p
f	Any	R, R	f
	None	P0, L, L	o

To illustrate the working of this machine a table is given below of the first few complete configurations. These complete configurations are described by writing down the sequence of symbols which are on the tape, with the *m*-configuration written below the scanned symbol. The successive complete configurations are separated by colons.

```
              : ə ə 0    0 : ə ə 0    0 : ə ə 0    0 : ə ə 0    0      : ə ə 0   0   1 :
       b          o           q               q              q                      p
  ə ə 0   0   1 : ə ə 0    0   1 : ə ə 0    0   1 : ə ə 0    0   1 :
            p              p                 f                   f
  ə ə 0   0   1 : ə ə 0    0   1    : ə ə 0   0   1    0 :
              f                  f                       o
  ə ə 0   0   1 x 0 : ....
              o
```

This table could also be written in the form

$$\mathfrak{b} \; : \mathfrak{o}\, \mathfrak{o}\, 0\, 0 \quad : \mathfrak{o}\, \mathfrak{o}\, q\, 0\, 0 \quad : \ldots, \tag{C}$$

in which a space has been made on the left of the scanned symbol and the *m*-configuration written in this space. This form is less easy to follow, but we shall make use of it later for theoretical purposes.

The convention of writing the figures only on alternate squares is very useful: I shall always make use of it. I shall call the one sequence of alternate squares *F*-squares and the other sequence *E*-squares. The symbols on *E*-squares will be liable to erasure. The symbols on *F*-squares form a continuous sequence. There are no blanks until the end is reached. There is no need to have more than one *E*-square between each pair of *F*-squares: an apparent need of more *E*-squares can be satisfied by having a sufficiently rich variety of symbols capable of being printed on *E*-squares. If a symbol β is on an *F*-square *S* and a symbol α is on the *E*-square next on the right of *S*, then *S* and β will be said to be *marked* with α. The process of printing this α will be called marking β (or *S*) with α.

4. Abbreviated tables

There are certain types of process used by nearly all machines, and these, in some machines, are used in many connections. These processes include copying down sequences of symbols, comparing sequences, erasing all symbols of a given form, etc. Where such processes are concerned we can abbreviate the tables for the *m*-configurations considerably by the use of "skeleton tables". In skeleton tables there appear capital German letters and small Greek letters. These are of the nature of "variables". By replacing each capital German letter throughout by an *m*-configuration and each small Greek letter by a symbol, we obtain the table for an *m*-configuration.

The skeleton tables are to be regarded as nothing but abbreviations: they are not essential. So long as the reader understands how to obtain the complete tables from the skeleton tables, there is no need to give any exact definitions in this connection.

Let us consider an example:

m-config.	Symbol	Behaviour	Final *m*-config.	
$\mathfrak{f}(\mathfrak{C}, \mathfrak{B}, \alpha)$	$\begin{cases} \ni \\ \text{not } \ni \end{cases}$	L L	$\mathfrak{f}_1(\mathfrak{C}, \mathfrak{B}, \alpha)$ $\mathfrak{f}(\mathfrak{C}, \mathfrak{B}, \alpha)$	From the *m*-configuration $\mathfrak{f}(\mathfrak{C}, \mathfrak{B}, \alpha)$ the machine finds the symbol of form α which is farthest to the left (the "first α") and the *m*-configuration then becomes $\mathfrak{C}$. If there is no α then the *m*-configuration becomes $\mathfrak{B}$.
$\mathfrak{f}_1(\mathfrak{C}, \mathfrak{B}, \alpha)$	$\begin{cases} \alpha \\ \text{not } \alpha \\ \text{None} \end{cases}$	 R R	$\mathfrak{C}$ $\mathfrak{f}_1(\mathfrak{C}, \mathfrak{B}, \alpha)$ $\mathfrak{f}_2(\mathfrak{C}, \mathfrak{B}, \alpha)$	
$\mathfrak{f}_2(\mathfrak{C}, \mathfrak{B}, \alpha)$	$\begin{cases} \alpha \\ \text{not } \alpha \\ \text{None} \end{cases}$	 R R	$\mathfrak{C}$ $\mathfrak{f}_1(\mathfrak{C}, \mathfrak{B}, \alpha)$ $\mathfrak{B}$	

If we were to replace $\mathfrak{C}$ throughout by q (say), $\mathfrak{B}$ by r, and α by *x*, we should have a complete table for the *m*-configuration $\mathfrak{f}(q, r, x)$. $\mathfrak{f}$ is called an "*m*-configuration function" or "*m*-function".

The only expressions which are admissible for substitution in an *m*-function are the *m*-configurations and symbols of the machine. These have to be enumerated more or less explicitly: they may include expressions such as $\mathfrak{p}(e, x)$; indeed they must if there are any *m*-functions used at all. If we did not insist on this explicit enumeration, but simply stated that the machine had certain *m*-configurations (enumerated) and all *m*-configurations obtainable by substitution of *m*-configurations in certain *m*-functions, we should usually get an infinity of *m*-configurations; e.g., we might say that the machine was to have the *m*-configuration q and all *m*-configurations obtainable by substituting an *m*-configuration for $\mathfrak{C}$ in $\mathfrak{p}(\mathfrak{C})$. Then it would have q, $\mathfrak{p}(q)$, $\mathfrak{p}(\mathfrak{p}(q))$, $\mathfrak{p}(\mathfrak{p}(\mathfrak{p}(q)))$, ... as *m*-configurations.

Our interpretation rule then is this. We are given the names of the *m*-configurations of the machine, mostly expressed in terms of *m*-functions. We are also given skeleton tables. All we want is the complete table for the *m*-configurations of the machine. This is obtained by repeated substitution in the skeleton tables.

Further examples

(In the explanations the symbol "→" is used to signify "the machine goes into the *m*-configuration....")

$\mathfrak{e}(\mathfrak{C}, \mathfrak{B}, \alpha)$	$\mathfrak{f}(\mathfrak{e}_1(\mathfrak{C}, \mathfrak{B}, \alpha), \mathfrak{B}, \alpha)$	From $\mathfrak{e}(\mathfrak{C}, \mathfrak{B}, \alpha)$ the first α is erased
$\mathfrak{e}_1(\mathfrak{C}, \mathfrak{B}, \alpha)$ E	$\mathfrak{C}$	and $\to \mathfrak{C}$. If there is no $\alpha \to \mathfrak{B}$.
$\mathfrak{e}(\mathfrak{B}, \alpha)$	$\mathfrak{e}(\mathfrak{e}(\mathfrak{B}, \alpha), \mathfrak{B}, \alpha)$	From $\mathfrak{e}(\mathfrak{B}, \alpha)$ all letters α are erased
		and $\to \mathfrak{B}$.

The last example seems somewhat more difficult to interpret than most. Let us suppose that in the list of *m*-configurations of some machine there appears $\mathfrak{e}(\mathfrak{b}, x)$ ($=$ q, say). The table is

| | $\mathfrak{e}(\mathfrak{b}, x)$ | | $\mathfrak{e}(\mathfrak{e}(\mathfrak{b}, x), \mathfrak{b}, x)$ |
| or | q | | $\mathfrak{e}(q, \mathfrak{b}, x)$. |

Or, in greater detail:

q		$\mathfrak{e}(q, \mathfrak{b}, x)$
$\mathfrak{e}(q, \mathfrak{b}, x)$		$\mathfrak{f}(\mathfrak{e}_1(q, \mathfrak{b}, x), \mathfrak{b}, x)$
$\mathfrak{e}_1(q, \mathfrak{b}, x)$	E	q.

In this we could replace $\mathfrak{e}_1(q, \mathfrak{b}, x)$ by q′ and then give the table for $\mathfrak{f}$ (with the right substitutions) and eventually reach a table in which no *m*-functions appeared.

$\mathfrak{pe}(\mathfrak{C}, \beta)$		$\mathfrak{f}(\mathfrak{pe}_1(\mathfrak{C}, \beta), \mathfrak{C}, \partial)$	From $\mathfrak{pe}(\mathfrak{C}, \beta)$ the machine prints β at the end of the sequence of symbols and $\to \mathfrak{C}$.
$\mathfrak{pe}_1(\mathfrak{C}, \beta) \begin{cases} \text{Any} & R, R \\ \text{None} & P\beta \end{cases}$		$\mathfrak{pe}_1(\mathfrak{C}, \beta)$ $\mathfrak{C}$	
$\mathfrak{l}(\mathfrak{C})$	L	$\mathfrak{C}$	From $\mathfrak{f}'(\mathfrak{C}, \mathfrak{B}, \alpha)$ it does the same as for $\mathfrak{f}(\mathfrak{C}, \mathfrak{B}, \alpha)$ but moves to the left before $\to \mathfrak{C}$.
$\mathfrak{r}(\mathfrak{C})$	R	$\mathfrak{C}$	
$\mathfrak{f}'(\mathfrak{C}, \mathfrak{B}, \alpha)$		$\mathfrak{f}(\mathfrak{l}(\mathfrak{C}), \mathfrak{B}, \alpha)$	
$\mathfrak{f}''(\mathfrak{C}, \mathfrak{B}, \alpha)$		$\mathfrak{f}(\mathfrak{r}(\mathfrak{C}), \mathfrak{B}, \alpha)$	
$\mathfrak{c}(\mathfrak{C}, \mathfrak{B}, \alpha)$		$\mathfrak{f}'(\mathfrak{c}_1(\mathfrak{C}), \mathfrak{B}, \alpha)$	$\mathfrak{c}(\mathfrak{C}, \mathfrak{B}, \alpha)$. The machine writes at the end the first symbol marked α and $\to \mathfrak{C}$.
$\mathfrak{c}_1(\mathfrak{C})$	β	$\mathfrak{pe}(\mathfrak{C}, \beta)$	

The last line stands for the totality of lines obtainable from it by replacing β by any symbol which may occur on the tape of the machine concerned.

$\mathfrak{ce}(\mathfrak{C}, \mathfrak{B}, \alpha)$	$\mathfrak{c}(\mathfrak{e}(\mathfrak{C}, \mathfrak{B}, \alpha), \mathfrak{B}, \alpha)$	$\mathfrak{ce}(\mathfrak{B}, \alpha)$. The machine copies down in order at the end all symbols marked α and erases the letters α; $\to \mathfrak{B}$.
$\mathfrak{ce}(\mathfrak{B}, \alpha)$	$\mathfrak{ce}(\mathfrak{ce}(\mathfrak{B}, \alpha), \mathfrak{B}, \alpha)$	
$\mathfrak{re}(\mathfrak{C}, \mathfrak{B}, \alpha, \beta)$	$\mathfrak{f}(\mathfrak{re}_1(\mathfrak{C}, \mathfrak{B}, \alpha, \beta), \mathfrak{B}, \alpha)$	$\mathfrak{re}(\mathfrak{C}, \mathfrak{B}, \alpha, \beta)$. The machine replaces the first α by β and $\to \mathfrak{C} \to \mathfrak{B}$ if there is no α.
$\mathfrak{re}_1(\mathfrak{C}, \mathfrak{B}, \alpha, \beta)$ $E, P\beta$	$\mathfrak{C}$	
$\mathfrak{re}(\mathfrak{B}, \alpha, \beta)$	$\mathfrak{re}(\mathfrak{re}(\mathfrak{B}, \alpha, \beta), \mathfrak{B}, \alpha, \beta)$	$\mathfrak{re}(\mathfrak{B}, \alpha, \beta)$. The machine replaces all letters α by β; $\to \mathfrak{B}$.
$\mathfrak{cr}(\mathfrak{C}, \mathfrak{B}, \alpha)$	$\mathfrak{c}(\mathfrak{re}(\mathfrak{C}, \mathfrak{B}, \alpha, a), \mathfrak{B}, \alpha)$	$\mathfrak{cr}(\mathfrak{B}, \alpha)$ differs from $\mathfrak{ce}(\mathfrak{B}, \alpha)$ only in that the letters α are not erased. The m-configuration $\mathfrak{cr}(\mathfrak{B}, \alpha)$ is taken up when no letters "a" are on the tape.
$\mathfrak{cr}(\mathfrak{B}, \alpha)$	$\mathfrak{cr}(\mathfrak{cr}(\mathfrak{B}, \alpha), \mathfrak{re}(\mathfrak{B}, a, \alpha), \alpha)$	
$\mathfrak{cp}(\mathfrak{C}, \mathfrak{A}, \mathfrak{E}, \alpha, \beta)$		$\mathfrak{f}'(\mathfrak{cp}_1(\mathfrak{C}, \mathfrak{A}, \beta), \mathfrak{f}(\mathfrak{A}, \mathfrak{E}, \beta), \alpha)$
$\mathfrak{cp}_1(\mathfrak{C}, \mathfrak{A}, \beta)$	γ	$\mathfrak{f}'(\mathfrak{cp}_2(\mathfrak{C}, \mathfrak{A}, \gamma), \mathfrak{A}, \beta)$
$\mathfrak{cp}_2(\mathfrak{C}, \mathfrak{A}, \gamma)$	$\begin{cases} \gamma \\ \text{not } \gamma \end{cases}$	$\mathfrak{C}$ $\mathfrak{A}$.

The first symbol marked α and the first marked β are compared. If there is neither α nor β, → 𝕰. If there are both and the symbols are alike, → ℭ. Otherwise → 𝔄.

cpe(ℭ, 𝔄, 𝔈, α, β)   cp(e(e(ℭ, ℭ, β), ℭ, α), 𝔄, 𝔈, α, β)

cpe(ℭ, 𝔄, 𝔈, α, β) differs from cp(ℭ, 𝔄, 𝔈, α, β) in that in the case when there is similarity the first α and β are erased.

cpe(𝔄, 𝔈, α, β)   cpe(cpe(𝔄, 𝔈, α, β), 𝔄, 𝔈, α, β).

cpe(𝔄, 𝔈, α, β). The sequence of symbols marked α is compared with the sequence marked β. → 𝔈 if they are similar. Otherwise → 𝔄. Some of the symbols α and β are erased.

q(ℭ)	{ Any	R	q(ℭ)	q(ℭ, α). The machine finds the last symbol of form α. → ℭ.
	None	R	q₁(ℭ)	
q₁(ℭ)	{ Any	R	q(ℭ)	
	None		ℭ	
q(ℭ, α)			q(q₁(ℭ, α))	
q₁(ℭ, α)	{ α		ℭ	
	Not α	L	q₁(ℭ, α)	
pe₂(ℭ, α, β)			pe(pe(ℭ, β), α)	pe₂(ℭ, α, β). The machine prints α β at the end.
ce₂(𝔅, α, β)			ce(ce(𝔅, β), α)	ce₃(𝔅, α, β, γ). The machine copies down at the end first the symbols marked α, then those marked β, and finally those marked γ; it erases the symbols α, β, γ.
ce₃(𝔅, α, β, γ)			ce(ce₂(𝔅, β, γ), α)	
e(ℭ)	{ ə	R	e₁(ℭ)	From e(ℭ) the marks are erased from all marked symbols. → ℭ.
	Not ə	L	e(ℭ)	
e₁(ℭ)	{ Any	R, E, R	e₁(ℭ)	
	None		ℭ	

5. Enumeration of computable sequences

A computable sequence γ is determined by a description of a machine which computes γ. Thus the sequence 001011011101111... is determined by the table on p. [62], and, in fact, any computable sequence is capable of being described in terms of such a table.

It will be useful to put these tables into a kind of standard form. In the first place let us suppose that the table is given in the same form as the first table, for example, I on p. [61]. That is to say, that the entry in the operations column is always of one of the forms $E : E$, $R : E$, $L : P\alpha : P\alpha$, $R : P\alpha$, $L : R : L$: or no entry at all. The table can always be put into this form by introducing more m-configurations. Now let us give numbers to the m-configurations, calling them $q_1, \ldots, q_R$, as in § 1. The initial m-configuration is always to be called q_1. We also give numbers to the symbols $S_1, \ldots, S_m$ and, in particular, blank $= S_0$, $0 = S_1$, $1 = S_2$. The lines of the table are now of form

m-config.	Symbol	Operations	Final m-config.	
q_i	S_j	PS_k, L	q_m	(N_1)
q_i	S_j	PS_k, R	q_m	(N_2)
q_i	S_j	PS_k	q_m	(N_3)

Lines such as

| q_i | S_j | E, R | q_m |

are to be written as

| q_i | S_j | PS_0, R | q_m |

and lines such as

| q_i | S_j | R | q_m |

to be written as

| q_i | S_j | PS_j, R | q_m |

In this way we reduce each line of the table to a line of one of the forms (N_1), (N_2), (N_3).

From each line of form (N_1) let us form an expression $q_i S_j S_k L q_m$; from each line of form (N_2) we form an expression $q_i S_j S_k R q_m$; and from each line of form (N_3) we form an expression $q_i S_j S_k N q_m$.

Let us write down all expressions so formed from the table for the machine and separate them by semi-colons. In this way we obtain a complete description of the machine. In this description we shall replace q_i by the letter "D" followed by the letter "A" repeated i times, and S_j by "D" followed by "C" repeated j times. This new description of the machine may be called the *standard description* (S.D). It is made up entirely from the letters "A", "C", "D", "L", "R", "N", and from ";".

If finally we replace "A" by "1", "C" by "2", "D" by "3", "L" by "4", "R" by "5", "N" by "6", and ";" by "7" we shall have a description of the machine in the form of an arabic numeral. The integer represented by this numeral may be called a *description number* (D.N) of the machine. The D.N determine the S.D and the structure of the machine uniquely. The machine whose D.N is n may be described as $\mathcal{M}(n)$.

To each computable sequence there corresponds at least one description number, while to no description number does there correspond more than one computable sequence. The computable sequences and numbers are therefore enumerable.

Let us find a description number for the machine I of § 3. When we rename the m-configurations its table becomes:

q_1	S_0	PS_1, R	q_2
q_2	S_0	PS_0, R	q_3
q_3	S_0	PS_2, R	q_4
q_4	S_0	PS_0, R	q_1

Other tables could be obtained by adding irrelevant lines such as

q_1	S_1	PS_1, R	q_2

Our first standard form would be

$$q_1 S_0 S_1 R q_2; \quad q_2 S_0 S_0 R q_3; \quad q_3 S_0 S_2 R q_4; \quad q_4 S_0 S_0 R q_1;$$

The standard description is

DADDCRDAA ;DAADDRDAAA ;DAAADDCCRDAAAA ;DAAAADDRDA ;

A description number is

31332531173113353111731113322531111731111335317

and so is

3133253117311335311173111332253111173111133531731323253117

A number which is a description number of a circle-free machine will be called a *satisfactory* number. In § 8 it is shown that there can be no general process for determining whether a given number is satisfactory or not.

6. The universal computing machine

It is possible to invent a single machine which can be used to compute any computable sequence. If this machine $\mathcal{U}$ is supplied with a tape on the beginning of which is written the S.D of some computing machine $\mathcal{M}$, then $\mathcal{U}$ will compute the same sequence as $\mathcal{M}$. In this section I explain in outline the behaviour of the machine. The next section is devoted to giving the complete table for $\mathcal{U}$.

Let us first suppose that we have a machine $\mathcal{M}'$ which will write down on the F-squares the successive complete configurations of $\mathcal{M}$. These might be expressed in the same form as on p. [62], using the second description, (C), with all symbols on one line. Or, better, we could transform this description (as in § 5) by replacing each *m*-configuration by "*D*" followed by "*A*" repeated the appropriate number of times, and by replacing each symbol by "*D*" followed by "*C*" repeated the appropriate number of times. The numbers of letters "*A*" and "*C*" are to agree with the numbers chosen in § 5, so that, in particular, "0" is replaced by "*DC*", "1" by "*DCC*", and the blanks by "*D*". These substitutions are to be made after the complete configurations have been put together, as in (C). Difficulties arise if we do the substitution first. In each complete configuration the blanks would all have to be replaced by "*D*", so that the complete configuration would not be expressed as a finite sequence of symbols.

If in the description of the machine II of § 3 we replace "ɒ" by "*DAA*", "ə" by "*DCCC*", "q" by "*DAAA*", then the sequence (C) becomes:

$$DA : DCCCDCCCDAADCDDC : DCCCDCCCDAAADCDDC : \ldots \quad (C_1)$$

(This is the sequence of symbols on *F*-squares.)

It is not difficult to see that if $\mathcal{M}$ can be constructed, then so can $\mathcal{M}'$. The manner of operation of $\mathcal{M}'$ could be made to depend on having the rules of operation (*i.e.* the S.D) of $\mathcal{M}$ written somewhere within itself (*i.e.* within $\mathcal{M}'$); each step could be carried out by referring to these rules. We have only to regard the rules as being capable of being taken out and exchanged for others and we have something very akin to the universal machine.

One thing is lacking: at present the machine $\mathcal{M}'$ prints no figures. We may correct this by printing between each successive pair of complete configurations the figures which appear in the new configuration but not in the old. Then (C$_1$) becomes

$$DDA : 0 : 0 : DCCCDCCCDAADCDDC : DCCC \ldots \quad (C_2)$$

It is not altogether obvious that the *E*-squares leave enough room for the necessary "rough work", but this is, in fact, the case.

The sequences of letters between the colons in expressions such as (C$_1$) may be used as standard descriptions of the complete configurations. When the letters are replaced by figures, as in § 5, we shall have a numerical description of the complete configuration, which may be called its description number.

7. Detailed description of the universal machine

A table is given below of the behaviour of this universal machine. The *m*-configurations of which the machine is capable are all those occurring in the first and last columns of the table, together with all those which occur when we write out the unabbreviated tables of those which appear in the table in the form of *m*-functions. E.g., e(ɑnŧ) appears in the table and is an *m*-function. Its unabbreviated table is (see p. [66])

$$e(\mathfrak{a}\mathfrak{n}\mathfrak{f}) \begin{cases} \mathfrak{d} & R & e_1(\mathfrak{a}\mathfrak{n}\mathfrak{f}) \\ \text{not } \mathfrak{d} & L & e(\mathfrak{a}\mathfrak{n}\mathfrak{f}) \end{cases}$$

$$e_1(\mathfrak{a}\mathfrak{n}\mathfrak{f}) \begin{cases} \text{Any} & R, E, R & e_1(\mathfrak{a}\mathfrak{n}\mathfrak{f}) \\ \text{None} & & \mathfrak{a}\mathfrak{n}\mathfrak{f} \end{cases}$$

Consequently $e_1(\mathfrak{a}\mathfrak{n}\mathfrak{f})$ is an *m*-configuration of $\mathcal{U}$.

When $\mathcal{U}$ is ready to start work the tape running through it bears on it the symbol ə on an *F*-square and again ə on the next *E*-square; after this, on *F*-squares only, comes the S.D of the machine followed by a double colon "::" (a single symbol, on an *F*-square). The S.D consists of a number of instructions, separated by semi-colons.

Each instruction consists of five consecutive parts

(i) "*D*" followed by a sequence of letters "*A*". This describes the relevant *m*-configuration.
(ii) "*D*" followed by a sequence of letters "*C*". This describes the scanned symbol.
(iii) "*D*" followed by another sequence of letters "*C*". This describes the symbol into which the scanned symbol is to be changed.
(iv) "*L*", "*R*", or "*N*", describing whether the machine is to move to left, right, or not at all.
(v) "*D*" followed by a sequence of letters "*A*". This describes the final *m*-configuration.

The machine $\mathcal{U}$ is to be capable of printing "*A*", "*C*", "*D*", "0", "1", "*u*", "*v*", "*w*", "*x*", "*y*", "*z*". The S.D is formed from ";", "*A*", "*C*", "*D*", "*L*", "*R*", "*N*".

Subsidiary skeleton table

$$\text{con}(\mathfrak{C}, \alpha) \begin{cases} \text{Not } A & R, R \\ A & L, P\alpha, R \end{cases} \quad \begin{matrix} \text{con}(\mathfrak{C}, \alpha) \\ \text{con}_1(\mathfrak{C}, \alpha) \end{matrix}$$

$$\text{con}_1(\mathfrak{C}, \alpha) \begin{cases} A & R, P\alpha, R \\ D & R, P\alpha, R \end{cases} \quad \begin{matrix} \text{con}_1(\mathfrak{C}, \alpha) \\ \text{con}_2(\mathfrak{C}, \alpha) \end{matrix}$$

$$\text{con}_2(\mathfrak{C}, \alpha) \begin{cases} C & R, P\alpha, R \\ \text{Not } C & R, R \end{cases} \quad \begin{matrix} \text{con}_2(\mathfrak{C}, \alpha) \\ \mathfrak{C} \end{matrix}$$

$\text{con}(\mathfrak{C}, \alpha)$. Starting from an *F*-square, *S* say, the sequence *C* of symbols describing a configuration closest on the right of *S* is marked out with letters $\alpha. \rightarrow \mathfrak{C}$.

$\text{con}(\mathfrak{C},)$. In the final configuration the machine is scanning the square which is four squares to the right of the last square of *C*. *C* is left unmarked.

The table for $\mathcal{U}$

b		f($\mathfrak{b}_1, \mathfrak{b}_1,$ ::)	b. The machine prints : *DA* on
$\mathfrak{b}_1$	R, R, P :, R, R, PD, R, R, PA	anf	the *F*-squares after :: $\rightarrow$ anf.
anf		g(anf$_1$, :)	anf. The machine marks the
anf$_1$		con(fom, y)	configuration in the last complete configuration with $y. \rightarrow$ fom.

fom	$\begin{cases} ; \\ z \\ \text{not } z \text{ nor } ; \end{cases}$	R, Pz, L L, L L	con(fmp, x) fom fom	fom. The machine finds the last semi-colon not marked with z. It marks this semi-colon with z and the configuration following it with x.
fmp			cpe(e(fom, x, y), sim, x, y)	fmp. The machine compares the sequences marked x and y. It erases all letters x and y. → sim if they are alike. Otherwise → fom.

anf. Taking the long view, the last instruction relevant to the last configuration is found. It can be recognised afterwards as the instruction following the last semi-colon marked z. → sim.

sim sim₁			f'(sim₁, sim₁, z) con(sim₂,)	sim. The machine marks out the instructions. That part of the instructions which refers to operations to be carried out is marked with u, and the final m-configuration with y. The letters z are erased.
sim₂	$\begin{cases} A \\ \text{not } A \end{cases}$	R, Pu, R, R, R	sim₃ sim₂	
sim₃	$\begin{cases} \text{not } A \\ A \end{cases}$	L, Py L, Py, R, R, R	e(mf, z) sim₃	
mf			g(mf, :)	mf. The last complete configuration is marked out into four sections. The configuration is left unmarked. The symbol directly preceding it is marked with x. The remainder of the complete configuration is divided into two parts, of which the first is marked with v and the last with w. A colon is printed after the whole. → sh.
mf₁	$\begin{cases} \text{not } A \\ A \end{cases}$	R, R L, L, L, L	mf₁ mf₂	
mf₂	$\begin{cases} C \\ : \\ D \end{cases}$	R, Px, L, L, L R, Px, L, L, L	mf₂ mf₄ mf₃	
mf₃	$\begin{cases} \text{not : } \\ : \end{cases}$	R, Pv, L, L, L	mf₃ mf₄	
mf₄			con(l(l(mf₅)),)	
mf₅	$\begin{cases} \text{Any} \\ \text{None} \end{cases}$	R, Pw, R P:	mf₅ sh	
sh sh₁		L, L, L	f(sh₁, inst, u) sh₂	sh. The instructions (marked u) are examined. If it is found that they involve "Print 0" or "Print 1", then 0 : or 1 : is printed at the end.
sh₂	$\begin{cases} D \\ \text{not } D \end{cases}$	R, R, R, R	sh₂ inst	
sh₃	$\begin{cases} C \\ \text{not } C \end{cases}$	R, R	sh₄ inst	

$\mathfrak{sh}_4$	$\begin{cases} C \\ \text{not } C \end{cases}$	R, R	$\mathfrak{sh}_5$ $\mathfrak{pe}_2(\mathfrak{inst}, 0, :)$
$\mathfrak{sh}_5$	$\begin{cases} C \\ \text{not } C \end{cases}$		$\mathfrak{inst}$ $\mathfrak{pe}_2(\mathfrak{inst}, 1, :)$

$\mathfrak{inst}$			$\mathfrak{g}(\mathfrak{l}(\mathfrak{inst}_1), u)$
$\mathfrak{inst}_1$	α	R, E	$\mathfrak{inst}_1(\alpha)$
$\mathfrak{inst}_1(L)$			$\mathfrak{ce}_5(\mathfrak{ov}, v, y, x, u, w)$
$\mathfrak{inst}_1(R)$			$\mathfrak{ce}_5(\mathfrak{ov}, v, x, u, y, w)$
$\mathfrak{inst}_1(N)$			$\mathfrak{ec}_5(\mathfrak{ov}, v, x, y, u, w)$
$\mathfrak{ov}$			$\mathfrak{e}(\mathfrak{anf})$

$\mathfrak{inst}$. The next complete configuration is written down, carrying out the marked instructions. The letters u, v, w, x, y are erased. →$\mathfrak{anf}$.

8. Application of the diagonal process

It may be thought that arguments which prove that the real numbers are not enumerable would also prove that the computable numbers and sequences cannot be enumerable.[4] It might, for instance, be thought that the limit of a sequence of computable numbers must be computable. This is clearly only true if the sequence of computable numbers is defined by some rule.

Or we might apply the diagonal process. "If the computable sequences are enumerable, let α_n be the n-th computable sequence, and let $\phi_n(m)$ be the m-th figure in α_n. Let β be the sequence with $1 - \phi_n(n)$ as its n-th figure. Since β is computable, there exists a number K such that $1 - \phi_n(n) = \phi_K(n)$ all n. Putting $n = K$, we have $1 = 2\phi_K(K)$, i.e. 1 is even. This is impossible. The computable sequences are therefore not enumerable."

The fallacy in this argument lies in the assumption that β is computable. It would be true if we could enumerate the computable sequences by finite means, but the problem of enumerating computable sequences is equivalent to the problem of finding out whether a given number is the D.N of a circle-free machine, and we have no general process for doing this in a finite number of steps. In fact, by applying the diagonal process argument correctly, we can show that there cannot be any such general process.

The simplest and most direct proof of this is by showing that, if this general process exists, then there is a machine which computes β. This proof, although perfectly sound, has the disadvantage that it may leave the reader with a feeling that "there must be something wrong". The proof which I shall give has not this disadvantage, and gives a certain insight into the significance of the idea "circle-free". It depends not on constructing β, but on constructing β', whose n-th figure is $\phi_n(n)$.

Let us suppose that there is such a process; that is to say, that we can invent a machine $\mathcal{D}$ which, when supplied with the S.D of any computing machine $\mathcal{M}$

---
[4] Cf. Hobson, *Theory of functions of a real variable* (2nd ed., 1921), 87, 88.

will test this S.D and if $\mathcal{M}$ is circular will mark the S.D with the symbol "u" and if it is circle-free will mark it with "s". By combining the machines $\mathcal{D}$ and $\mathcal{U}$ we could construct a machine $\mathcal{H}$ to compute the sequence β'. The machine $\mathcal{D}$ may require a tape. We may suppose that it uses the E-squares beyond all symbols on F-squares, and that when it has reached its verdict all the rough work done by $\mathcal{D}$ is erased.

The machine $\mathcal{H}$ has its motion divided into sections. In the first $N-1$ sections, among other things, the integers 1, 2, ..., $N-1$ have been written down and tested by the machine $\mathcal{D}$. A certain number, say $R(N-1)$, of them have been found to be the D.N's of circle-free machines. In the N-th section the machine $\mathcal{D}$ tests the number N. If N is satisfactory, *i.e.*, if it is the D.N of a circle-free machine, then $R(N) = 1 + R(N-1)$ and the first $R(N)$ figures of the sequence of which a D.N is N are calculated. The $R(N)$-th figure of this sequence is written down as one of the figures of the sequence β' computed by $\mathcal{H}$. If N is not satisfactory, then $R(N) = R(N-1)$ and the machine goes on to the $(N+1)$-th section of its motion.

From the construction of $\mathcal{H}$ we can see that $\mathcal{H}$ is circle-free. Each section of the motion of $\mathcal{H}$ comes to an end after a finite number of steps. For, by our assumption about $\mathcal{D}$, the decision as to whether N is satisfactory is reached in a finite number of steps. If N is not satisfactory, then the N-th section is finished. If N is satisfactory, this means that the machine $\mathcal{M}(N)$ whose D.N is N is circle-free, and therefore its $R(N)$-th figure can be calculated in a finite number of steps. When this figure has been calculated and written down as the $R(N)$-th figure of β', the N-th section is finished. Hence $\mathcal{H}$ is circle-free.

Now let K be the D.N of $\mathcal{H}$. What does $\mathcal{H}$ do in the K-th section of its motion? It must test whether K is satisfactory, giving a verdict "s" or "u". Since K is the D.N of $\mathcal{H}$ and since $\mathcal{H}$ is circle-free, the verdict cannot be "u". On the other hand the verdict cannot be "s". For if it were, then in the K-th section of its motion $\mathcal{H}$ would be bound to compute the first $R(K-1) + 1 = R(K)$ figures of the sequence computed by the machine with K as its D.N and to write down the $R(K)$-th as a figure of the sequence computed by $\mathcal{H}$. The computation of the first $R(K) - 1$ figures would be carried out all right, but the instructions for calculating the $R(K)$-th would amount to "calculate the first $R(K)$ figures computed by $\mathcal{H}$ and write down the $R(K)$-th". This $R(K)$-th figure would never be found. *I.e.*, $\mathcal{H}$ is circular, contrary both to what we have found in the last paragraph and to the verdict "s". Thus both verdicts are impossible and we conclude that there can be no machine $\mathcal{D}$.

We can show further that *there can be no machine $\mathcal{E}$ which, when supplied with the S.D of an arbitrary machine $\mathcal{M}$, will determine whether $\mathcal{M}$ ever prints a given symbol* (0 say).

We will first show that, if there is a machine $\mathcal{E}$, then there is a general process for determining whether a given machine $\mathcal{M}$ prints 0 infinitely often. Let $\mathcal{M}_1$ be a machine which prints the same sequence as $\mathcal{M}$, except that in the position where the first 0 printed by $\mathcal{M}$ stands, $\mathcal{M}_1$ prints $\bar{0}$. $\mathcal{M}_2$ is to have the first two symbols 0 replaced by $\bar{0}$, and so on. Thus, if $\mathcal{M}$ were to print

$$A B A 0 1 A A B 0 0 1 0 A B \ldots,$$

then $\mathcal{M}_1$ would print

$$A B A \bar{0} 1 A A B 0 0 1 0 A B \ldots$$

and $\mathcal{M}_2$ would print

$$A B A \bar{0} 1 A A B \bar{0} 0 1 0 A B \ldots.$$

Now let $\mathcal{F}$ be a machine which, when supplied with the S.D of $\mathcal{M}$, will write down successively the S.D of $\mathcal{M}$, of $\mathcal{M}_1$, of $\mathcal{M}_2$, ... (there is such a machine). We combine $\mathcal{F}$ with $\mathcal{E}$ and obtain a new machine, $\mathcal{G}$. In the motion of $\mathcal{G}$ first $\mathcal{F}$ is used to write down the S.D of $\mathcal{M}$, and then $\mathcal{E}$ tests it, : 0 : is written if it is found that $\mathcal{M}$ never prints 0; then $\mathcal{F}$ writes the S.D of $\mathcal{M}_1$, and this is tested, : 0 : being printed if and only if $\mathcal{M}_1$ never prints 0, and so on. Now let us test $\mathcal{G}$ with $\mathcal{E}$. If it is found that $\mathcal{G}$ never prints 0, then $\mathcal{M}$ prints 0 infinitely often; if $\mathcal{G}$ prints 0 sometimes, then $\mathcal{M}$ does not print 0 infinitely often.

Similarly there is a general process for determining whether $\mathcal{M}$ prints 1 infinitely often. By a combination of these processes we have a process for determining whether $\mathcal{M}$ prints an infinity of figures, *i.e.* we have a process for determining whether $\mathcal{M}$ is circle-free. There can therefore be no machine $\mathcal{E}$.

The expression "there is a general process for determining..." has been used throughout this section as equivalent to "there is a machine which will determine ...". This usage can be justified if and only if we can justify our definition of "computable". For each of these "general process" problems can be expressed as a problem concerning a general process for determining whether a given integer n has a property $G(n)$ [*e.g.* $G(n)$ might mean "n is satisfactory" or "n is the Gödel representation of a provable formula"], and this is equivalent to computing a number whose n-th figure is 1 if $G(n)$ is true and 0 if it is false.

9. The extent of the computable numbers

No attempt has yet been made to show that the "computable" numbers include all numbers which would naturally be regarded as computable. All arguments which can be given are bound to be, fundamentally, appeals to intuition, and for this reason rather unsatisfactory mathematically. The real question at issue is "What are the possible processes which can be carried out in computing a number?"

The arguments which I shall use are of three kinds.

(*a*) A direct appeal to intuition.
(*b*) A proof of the equivalence of two definitions (in case the new definition has a greater intuitive appeal).
(*c*) Giving examples of large classes of numbers which are computable.

Once it is granted that computable numbers are all "computable", several other propositions of the same character follow. In particular, it follows that, if there is a general process for determining whether a formula of the Hilbert function calculus is provable, then the determination can be carried out by a machine.

I. [Type (*a*)]. This argument is only an elaboration of the ideas of § 1.

Computing is normally done by writing certain symbols on paper. We may suppose this paper is divided into squares like a child's arithmetic book. In elementary arithmetic the two-dimensional character of the paper is sometimes used. But such a use is always avoidable, and I think that it will be agreed that the two-dimensional character of paper is no essential of computation. I assume then that the computation is carried out on one-dimensional paper, *i.e.* on a tape divided into squares. I shall also suppose that the number of symbols which may be printed is finite. If we were to allow an infinity of symbols, then there would be symbols differing to an arbitrarily small extent.[5] The effect of this restriction of the number of symbols is not very serious. It is always possible to use sequences of symbols in the place of single symbols. Thus an Arabic numeral such as 17 or 999999999999999 is normally treated as a single symbol. Similarly in any European language words are treated as single symbols (Chinese, however, attempts to have an enumerable infinity of symbols). The differences from our point of view between the single and compound symbols is that the compound symbols, if they are too lengthy, cannot be observed at one glance. This is in accordance with experience. We cannot tell at a glance whether 9999999999999999 and 999999999999999 are the same.

The behaviour of the computer at any moment is determined by the symbols which he is observing, and his "state of mind" at that moment. We may suppose that there is a bound *B* to the number of symbols or squares which the computer can observe at one moment. If he wishes to observe more, he must use successive observations. We will also suppose that the number of states of mind which need be taken into account is finite. The reasons for this are of the same character as

---

[5] If we regard a symbol as literally printed on a square we may suppose that the square is $0 \leqslant x \leqslant 1, 0 \leqslant y \leqslant 1$. The symbol is defined as a set of points in this square, viz. the set occupied by printer's ink. If these sets are restricted to be measurable, we can define the "distance" between two symbols as the cost of transforming one symbol into the other if the cost of moving unit area of printer's ink unit distance is unity, and there is an infinite supply of ink at $x = 2, y = 0$. With this topology the symbols form a conditionally compact space.

those which restrict the number of symbols. If we admitted an infinity of states of mind, some of them will be "arbitrarily close" and will be confused. Again, the restriction is not one which seriously affects computation, since the use of more complicated states of mind can be avoided by writing more symbols on the tape.

Let us imagine the operations performed by the computer to be split up into "simple operations" which are so elementary that it is not easy to imagine them further divided. Every such operation consists of some change of the physical system consisting of the computer and his tape. We know the state of the system if we know the sequence of symbols on the tape, which of these are observed by the computer (possibly with a special order), and the state of mind of the computer. We may suppose that in a simple operation not more than one symbol is altered. Any other changes can be split up into simple changes of this kind. The situation in regard to the squares whose symbols may be altered in this way is the same as in regard to the observed squares. We may, therefore, without loss of generality, assume that the squares whose symbols are changed are always "observed" squares.

Besides these changes of symbols, the simple operations must include changes of distribution of observed squares. The new observed squares must be immediately recognisable by the computer. I think it is reasonable to suppose that they can only be squares whose distance from the closest of the immediately previously observed squares does not exceed a certain fixed amount. Let us say that each of the new observed squares is within L squares of an immediately previously observed square.

In connection with "immediate recognisability", it may be thought that there are other kinds of square which are immediately recognisable. In particular, squares marked by special symbols might be taken as immediately recognisable. Now if these squares are marked only by single symbols there can be only a finite number of them, and we should not upset our theory by adjoining these marked squares to the observed squares. If, on the other hand, they are marked by a sequence of symbols, we cannot regard the process of recognition as a simple process. This is a fundamental point and should be illustrated. In most mathematical papers the equations and theorems are numbered. Normally the numbers do not go beyond (say) 1000. It is, therefore, possible to recognise a theorem at a glance by its number. But if the paper was very long, we might reach Theorem 157767733443477; then, further on in the paper, we might find "... hence (applying Theorem 157767733443477) we have ...". In order to make sure which was the relevant theorem we should have to compare the two numbers figure by figure, possibly ticking the figures off in pencil to make sure of their not being counted twice. If in spite of this it is still thought that there are other "immediately recognisable" squares, it does not upset my contention so long as these squares can be found by some process of which my type of machine is capable. This idea is developed in III below.

The simple operations must therefore include:

(*a*) Changes of the symbol on one of the observed squares.
(*b*) Changes of one of the squares observed to another square within L squares of one of the previously observed squares.

It may be that some of these changes necessarily involve a change of state of mind. The most general single operation must therefore be taken to be one of the following:

(A) A possible change (*a*) of symbol together with a possible change of state of mind.
(B) A possible change (*b*) of observed squares, together with a possible change of state of mind.

The operation actually performed is determined, as has been suggested on p. [75], by the state of mind of the computer and the observed symbols. In particular, they determine the state of mind of the computer after the operation is carried out.

We may now construct a machine to do the work of this computer. To each state of mind of the computer corresponds an "*m*-configuration" of the machine. The machine scans B squares corresponding to the B squares observed by the computer. In any move the machine can change a symbol on a scanned square or can change any one of the scanned squares to another square distant not more than L squares from one of the other scanned squares. The move which is done, and the succeeding configuration, are determined by the scanned symbol and the *m*-configuration. The machines just described do not differ very essentially from computing machines as defined in § 2, and corresponding to any machine of this type a computing machine can be constructed to compute the same sequence, that is to say the sequence computed by the computer.

II. [Type (*b*)].

If the notation of the Hilbert functional calculus[6] is modified so as to be systematic, and so as to involve only a finite number of symbols, it becomes possible to construct an automatic[7] machine $\mathcal{K}$, which will find all the provable formulae of the calculus.[8]

---

[6] The expression "the functional calculus" is used throughout to mean the *restricted* Hilbert functional calculus.

[7] It is most natural to construct first a choice machine (§2) to do this. But it is then easy to construct the required automatic machine. We can suppose that the choices are always choices between two possibilities 0 and 1. Each proof will then be determined by a sequence of choices $i_1, i_2, \ldots, i_n$ ($i_1 = 0$ or 1, $i_2 = 0$ or 1, ..., $i_n = 0$ or 1), and hence the number $2^n + i_1 2^{n-1} + i_2 2^{n-2} + \ldots + i_n$ completely determines the proof. The automatic machine carries out successively proof 1, proof 2, proof 3, ...

[8] The author has found a description of such a machine.

Now let α be a sequence, and let us denote by $G_\alpha(x)$ the proposition "The x-th figure of α is 1", so that $-G_\alpha(x)$ means "The x-th figure of α is 0".[9] Suppose further that we can find a set of properties which define the sequence α and which can be expressed in terms of $G_\alpha(x)$ and of the propositional functions $N(x)$ meaning "x is a non-negative integer" and $F(x, y)$ meaning "$y = x + 1$". When we join all these formulae together conjunctively, we shall have a formula, $\mathfrak{A}$ say, which defines α. The terms of $\mathfrak{A}$ must include the necessary parts of the Peano axioms, viz.,

$$(\exists u)N(u) \ \& \ (x)(N(x) \to (\exists y)F(x,y)) \ \& \ (F(x,y) \to N(y)),$$

which we will abbreviate to P.

When we say "$\mathfrak{A}$ defines α", we mean that $-\mathfrak{A}$ is not a provable formula, and also that, for each n, one of the following formulae (A_n) or (B_n) is provable.[10]

$$\mathfrak{A} \ \& \ F^{(n)} \to G_\alpha(u^{(n)}), \qquad (A_n)$$

$$\mathfrak{A} \ \& \ F^{(n)} \to (-G_\alpha(u^{(n)})), \qquad (B_n),$$

where $F^{(n)}$ stands for $F(u, u') \ \& \ F(u', u'') \ \& \ \ldots \ F(u^{(n-1)}, u^{(n)})$.

I say that α is then a computable sequence: a machine $\mathcal{K}_\alpha$ to compute α can be obtained by a fairly simple modification of $\mathcal{K}$.

We divide the motion of $\mathcal{K}_\alpha$ into sections. The n-th section is devoted to finding the n-th figure of α. After the $(n-1)$-th section is finished a double colon :: is printed after all the symbols, and the succeeding work is done wholly on the squares to the right of this double colon. The first step is to write the letter "A" followed by the formula (A_n) and then "B" followed by (B_n). The machine $\mathcal{K}_\alpha$ then starts to do the work of $\mathcal{K}$, but whenever a provable formula is found, this formula is compared with (A_n) and with (B_n). If it is the same formula as (A_n), then the figure "1" is printed, and the n-th section is finished. If it is (B_n), then "0" is printed and the section is finished. If it is different from both, then the work of $\mathcal{K}$ is continued from the point at which it had been abandoned. Sooner or later one of the formulae (A_n) or (B_n) is reached; this follows from our hypotheses about α and $\mathfrak{A}$, and the known nature of $\mathcal{K}$. Hence the n-th section will eventually be finished. $\mathcal{K}_\alpha$ is circle-free; α is computable.

It can also be shown that the numbers α definable in this way by the use of axioms include all the computable numbers. This is done by describing computing machines in terms of the function calculus.

It must be remembered that we have attached rather a special meaning to the phrase "$\mathfrak{A}$ defines α". The computable numbers do not include all (in the ordinary sense) definable numbers. Let δ be a sequence whose n-th figure is

---

[9] The negation sign is written before an expression and not over it.
[10] A sequence of r primes is denoted by $^{(r)}$.

1 or 0 according as n is or is not satisfactory. It is an immediate consequence of the theorem of § 8 that δ is not computable. It is (so far as we know at present) possible that any assigned number of figures of δ can be calculated, but not by a uniform process. When sufficiently many figures of δ have been calculated, an essentially new method is necessary in order to obtain more figures.

III. This may be regarded as a modification of I or as a corollary of II.

We suppose, as in I, that the computation is carried out on a tape; but we avoid introducing the "state of mind" by considering a more physical and definite counterpart of it. It is always possible for the computer to break off from his work, to go away and forget all about it, and later to come back and go on with it. If he does this he must leave a note of instructions (written in some standard form) explaining how the work is to be continued. This note is the counterpart of the "state of mind". We will suppose that the computer works in such a desultory manner that he never does more than one step at a sitting. The note of instructions must enable him to carry out one step and write the next note. Thus the state of progress of the computation at any stage is completely determined by the note of instructions and the symbols on the tape. That is, the state of the system may be described by a single expression (sequence of symbols), consisting of the symbols on the tape followed by Δ (which we suppose not to appear elsewhere) and then by the note of instructions. This expression may be called the "state formula". We know that the state formula at any given stage is determined by the state formula before the last step was made, and we assume that the relation of these two formulae is expressible in the functional calculus. In other words, we assume that there is an axiom 𝔄 which expresses the rules governing the behaviour of the computer, in terms of the relation of the state formula at any stage to the state formula at the preceding stage. If this is so, we can construct a machine to write down the successive state formulae, and hence to compute the required number.

10. Examples of large classes of numbers which are computable

It will be useful to begin with definitions of a computable function of an integral variable and of a computable variable, etc. There are many equivalent ways of defining a computable function of an integral variable. The simplest is, possibly, as follows. If γ is a computable sequence in which 0 appears infinitely[11] often, and n is an integer, then let us define $\xi(\gamma, n)$ to be the number of figures

---

[11] If 𝔐 computes γ, then the problem whether 𝔐 prints 0 infinitely often is of the same character as the problem whether 𝔐 is circle-free.

1 between the n-th and the $(n+1)$-th figure 0 in γ. Then $\phi(n)$ is computable if, for all n and some γ, $\phi(n) = \xi(\gamma, n)$. An equivalent definition is this. Let $H(x, y)$ mean $\phi(x) = y$. Then, if we can find a contradiction-free axiom $\mathfrak{A}_\phi$, such that $\mathfrak{A}_\phi \to P$, and if for each integer n there exists an integer N, such that

$$\mathfrak{A}_\phi \ \& \ F^{(N)} \to H(u^{(n)}, u^{(\phi(n))}),$$

and such that, if $m \neq \phi(n)$, then, for some N',

$$\mathfrak{A}_\phi \ \& \ F^{(N')} \to \left(-H(u^{(n)}, u^{(m)})\right),$$

then ϕ may be said to be a computable function.

We cannot define general computable functions of a real variable, since there is no general method of describing a real number, but we can define a computable function of a computable variable. If n is satisfactory, let γ_n be the number computed by $\mathcal{M}(n)$, and let

$$\alpha_n = \tan\left(\pi\left(\gamma_n - \tfrac{1}{2}\right)\right),$$

unless $\gamma_n = 0$ or $\gamma_n = 1$, in either of which cases $\alpha_n = 0$. Then, as n runs through the satisfactory numbers, α_n runs through the computable numbers.[12] Now let $\phi(n)$ be a computable function which can be shown to be such that for any satisfactory argument its value is satisfactory.[13] Then the function f, defined by $f(\alpha_n) = \alpha_{\phi(n)}$, is a computable function and all computable functions of a computable variable are expressible in this form.

Similar definitions may be given of computable functions of several variables, computable-valued functions of an integral variable, etc.

I shall enunciate a number of theorems about computability, but I shall prove only (ii) and a theorem similar to (iii).

(i) A computable function of a computable function of an integral or computable variable is computable.

(ii) Any function of an integral variable defined recursively in terms of computable functions is computable. I.e. if $\phi(m, n)$ is computable, and r is some integer, then $\eta(n)$ is computable, where

$$\eta(0) = r,$$
$$\eta(n) = \phi(n, \eta(n-1)).$$

(iii) If $\phi(m, n)$ is a computable function of two integral variables, then $\phi(n, n)$ is a computable function of n.

---

[12] A function α_n may be defined in many other ways so as to run through the computable numbers.
[13] Although it is not possible to find a general process for determining whether a given number is satisfactory, it is often possible to show that certain classes of numbers are satisfactory.

(iv) If $\phi(n)$ is a computable function whose value is always 0 or 1, then the sequence whose n-th figure is $\phi(n)$ is computable.

Dedekind's theorem does not hold in the ordinary form if we replace "real" throughout by "computable". But it holds in the following form:

(v) If $G(\alpha)$ is a propositional function of the computable numbers and

(a) $(\exists \alpha)(\exists \beta)\{G(\alpha) \ \& \ (-G(\beta))\}$,
(b) $G(\alpha) \ \& \ (-G(\beta)) \rightarrow (\alpha < \beta)$,

and there is a general process for determining the truth value of $G(\alpha)$, then there is a computable number ξ such that

$$G(\alpha) \rightarrow \alpha \leq \xi,$$
$$-G(\alpha) \rightarrow \alpha \geq \xi.$$

In other words, the theorem holds for any section of the computables such that there is a general process for determining to which class a given number belongs.

Owing to this restriction of Dedekind's theorem, we cannot say that a computable bounded increasing sequence of computable numbers has a computable limit. This may possibly be understood by considering a sequence such as

$$-1, \ -\tfrac{1}{2}, \ -\tfrac{1}{4}, \ -\tfrac{1}{8}, \ -\tfrac{1}{16}, \ \tfrac{1}{2}, \ \ldots .$$

On the other hand, (v) enables us to prove

(vi) If α and β are computable and $\alpha < \beta$ and $\phi(\alpha) < 0 < \phi(\beta)$, where $\phi(\alpha)$ is a computable increasing continuous function, then there is a unique computable number γ, satisfying $\alpha < \gamma < \beta$ and $\phi(\gamma) = 0$.

Computable convergence

We shall say that a sequence β_n of computable numbers *converges computably* if there is a computable integral valued function $N(\varepsilon)$ of the computable variable ε, such that we can show that, if $\varepsilon > 0$ and $n > N(\varepsilon)$ and $m > N(\varepsilon)$, then $|\beta_n - \beta_m| < \varepsilon$.

We can then show that

(vii) A power series whose coefficients form a computable sequence of computable numbers is computably convergent at all computable points in the interior of its interval of convergence.

(viii) The limit of a computably convergent sequence is computable.

And with the obvious definition of "uniformly computably convergent":

(ix) The limit of a uniformly computably convergent computable sequence of computable functions is a computable function. Hence

(x) The sum of a power series whose coefficients form a computable sequence is a computable function in the interior of its interval of convergence.

From (viii) and $\pi = 4(1 - \frac{1}{3} + \frac{1}{5} - \ldots)$ we deduce that π is computable.
From $e = 1 + 1 + \frac{1}{2!} + \frac{1}{3!} + \ldots$ we deduce that e is computable.
From (vi) we deduce that all real algebraic numbers are computable.
From (vi) and (x) we deduce that the real zeros of the Bessel functions are computable.

Proof of (ii)

Let $H(x, y)$ mean "$\eta(x) = y$", and let $K(x, y, z)$ mean "$\phi(x, y) = z$". $\mathfrak{A}_\phi$ is the axiom for $\phi(x, y)$. We take $\mathfrak{A}_\eta$ to be

$$\mathfrak{A}_\phi \ \& \ P \ \& \ \big(F(x, y) \to G(x, y)\big) \ \& \ \big(G(x, y) \ \& \ G(y, z) \to G(x, z)\big)$$
$$\& \ \big(F^{(r)} \to H(u, u^{(r)})\big) \ \& \ \big(F(v, w) \ \& \ H(v, x) \ \& \ K(w, x, z) \to H(w, z)\big)$$
$$\& \ \big[H(w, z) \ \& \ G(z, t) \ \text{v} \ G(t, z) \to (-H(w, t))\big].$$

I shall not give the proof of consistency of $\mathfrak{A}_\eta$. Such a proof may be constructed by the methods used in Hilbert and Bernays, *Grundlagen der Mathematik* (Berlin, 1934), p. 209 *et seq*. The consistency is also clear from the meaning.

Suppose that, for some n, N, we have shown

$$\mathfrak{A}_\eta \ \& \ F^{(N)} \to H\big(u^{(n-1)}, u^{(\eta(n-1))}\big),$$

then, for some M,

$$\mathfrak{A}_\phi \ \& \ F^{(M)} \to K\big(u^{(n)}, u^{(\eta(n-1))}, u^{(\eta(n))}\big),$$
$$\mathfrak{A}_\eta \ \& \ F^{(M)} \to F\big(u^{(n-1)}, u^{(n)}\big) \ \& \ H\big(u^{(n-1)}, u^{(\eta(n-1))}\big)$$
$$\& \ K\big(u^{(n)}, u^{(\eta(n-1))}, u^{(\eta(n))}\big),$$

and

$$\mathfrak{A}_\eta \ \& \ F^{(M)} \to \big[F\big(u^{(n-1)}, u^{(n)}\big) \ \& \ H\big(u^{(n-1)}, u^{(\eta(n-1))}\big)$$
$$\& \ K\big(u^{(n)}, u^{(\eta(n-1))}, u^{(\eta(n))}\big) \to H\big(u^{(n)}, u^{(\eta(n))}\big)\big].$$

Hence $\mathfrak{A}_\eta \ \& \ F^{(M)} \to H\big(u^{(n)}, u^{(\eta(n))}\big).$
Also $\mathfrak{A}_\eta \ \& \ F^{(r)} \to H\big(u, u^{(\eta(0))}\big).$

Hence for each n some formula of the form

$$\mathfrak{A}_\eta \ \& \ F^{(M)} \to H\big(u^{(n)}, u^{(\eta(n))}\big)$$

is provable. Also, if $M' \geqslant M$ and $M' \geqslant m$ and $m \neq \eta(u)$, then

$$\mathfrak{A}_\eta \ \& \ F^{(M')} \to G\big(u^{\eta((n))}, u^{(m)}\big) \ \text{v} \ G\big(u^{(m)}, u^{(\eta(n))}\big)$$

and

$$\mathfrak{A}_\eta \ \& \ F^{(M')} \rightarrow [\{G(u^{(\eta(n))}, u^{(m)}) \ \mathrm{v} \ G(u^{(m)}, u^{(\eta(n))})$$
$$\& \ H(u^{(n)}, u^{(\eta(n))})\} \rightarrow (-H(u^{(n)}, u^{(m)}))].$$

Hence $\qquad \mathfrak{A}_\eta \ \& \ F^{(M')} \rightarrow (-H(u^{(n)}, u^{(m)})).$

The conditions of our second definition of a computable function are therefore satisfied. Consequently η is a computable function.

Proof of a modified form of (iii)

Suppose that we are given a machine $\mathcal{N}$, which, starting with a tape bearing on it ə ə followed by a sequence of any number of letters "F" on F-squares and in the m-configuration b, will compute a sequence γ_n depending on the number n of letters "F". If $\phi_n(m)$ is the m-th figure of γ_n, then the sequence β whose n-th figure is $\phi_n(n)$ is computable.

We suppose that the table for $\mathcal{N}$ has been written out in such a way that in each line only one operation appears in the operations column. We also suppose that Ξ, Θ, $\bar{0}$, and $\bar{1}$ do not occur in the table, and we replace ə throughout by Θ, 0 by $\bar{0}$, and 1 by $\bar{1}$. Further substitutions are then made. Any line of form

$$\mathfrak{A} \qquad \alpha \qquad P\bar{0} \qquad \mathfrak{B}$$

we replace by

$$\mathfrak{A} \qquad \alpha \qquad P\bar{0} \qquad \mathrm{re}(\mathfrak{B}, \mathfrak{u}, h, k)$$

and any line of the form

$$\mathfrak{A} \qquad \alpha \qquad P\bar{1} \qquad \mathfrak{B}$$

by

$$\mathfrak{A} \qquad \alpha \qquad P\bar{1} \qquad \mathrm{re}(\mathfrak{B}, \mathfrak{v}, h, k)$$

and we add to the table the following lines:

$\mathfrak{u}$		$\mathfrak{pe}(\mathfrak{u}_1, 0)$
$\mathfrak{u}_1$	R, Pk, R, PΘ, R, PΘ	$\mathfrak{u}_2$
$\mathfrak{u}_2$		$\mathrm{re}(\mathfrak{u}_3, \mathfrak{u}_3, k, h)$
$\mathfrak{u}_3$		$\mathfrak{pe}(\mathfrak{u}_2, F)$

and similar lines with $\mathfrak{v}$ for $\mathfrak{u}$ and 1 for 0 together with the following line

$$\mathfrak{c} \qquad R, P\Xi, R, Ph \qquad \mathfrak{b}.$$

We then have the table for the machine $\mathcal{N}'$ which computes β. The initial m-configuration is $\mathfrak{c}$, and the initial scanned symbol is the second ə.

11. Application to the Entscheidungsproblem

The results of §8 have some important applications. In particular, they can be used to show that the Hilbert Entscheidungsproblem can have no solution. For the present I shall confine myself to proving this particular theorem. For the formulation of this problem I must refer the reader to Hilbert and Ackermann's *Grundzüge der Theoretischen Logik* (Berlin, 1931), chapter 3.

I propose, therefore, to show that there can be no general process for determining whether a given formula $\mathfrak{A}$ of the functional calculus **K** is provable, *i.e.* that there can be no machine which, supplied with any one $\mathfrak{A}$ of these formulae, will eventually say whether $\mathfrak{A}$ is provable.

It should perhaps be remarked that what I shall prove is quite different from the well-known results of Gödel.[14] Gödel has shown that (in the formalism of Principia Mathematica) there are propositions $\mathfrak{A}$ such that neither $\mathfrak{A}$ nor $-\mathfrak{A}$ is provable. As a consequence of this, it is shown that no proof of consistency of Principia Mathematica (or of **K**) can be given within that formalism. On the other hand, I shall show that there is no general method which tells whether a given formula $\mathfrak{A}$ is provable in **K**, or, what comes to the same, whether the system consisting of **K** with $-\mathfrak{A}$ adjoined as an extra axiom is consistent.

If the negation of what Gödel has shown had been proved, *i.e.* if, for each $\mathfrak{A}$, either $\mathfrak{A}$ or $-\mathfrak{A}$ is provable, then we should have an immediate solution of the Entscheidungsproblem. For we can invent a machine $\mathcal{K}$ which will prove consecutively all provable formulae. Sooner or later $\mathcal{K}$ will reach either $\mathfrak{A}$ or $-\mathfrak{A}$. If it reaches $\mathfrak{A}$, then we know that $\mathfrak{A}$ is provable. If it reaches $-\mathfrak{A}$, then, since **K** is consistent (Hilbert and Ackermann, p. 65), we know that $\mathfrak{A}$ is not provable.

Owing to the absence of integers in **K** the proofs appear somewhat lengthy. The underlying ideas are quite straightforward.

Corresponding to each computing machine $\mathcal{M}$ we construct a formula Un ($\mathcal{M}$) and we show that, if there is a general method for determining whether Un ($\mathcal{M}$) is provable, then there is a general method for determining whether $\mathcal{M}$ ever prints 0.

The interpretations of the propositional functions involved are as follows:

$R_{S_l}(x, y)$ is to be interpreted as "in the complete configuration x (of $\mathcal{M}$) the symbol on the square y is S".

$I(x, y)$ is to be interpreted as "in the complete configuration x the square y is scanned".

---

[14] *Loc. cit.*

$K_{q_m}(x)$ is to be interpreted as "in the complete configuration x the m-configuration is q_m.

$F(x, y)$ is to be interpreted as "y is the immediate successor of x".

Inst $\{q_i S_j S_k L q_l\}$ is to be an abbreviation for

$$(x, y, x', y')\{(R_{S_j}(x, y) \& I(x, y) \& K_{q_i}(x) \& F(x, x') \& F(y', y))$$
$$\rightarrow (I(x', y') \& R_{S_k}(x', y) \& K_{q_l}(x')$$
$$\& (z)[F(y', z) \text{ v } (R_{S_j}(x, z) \rightarrow R_{S_k}(x', z))])\}.$$

Inst $\{q_i S_j S_k R q_l\}$ and Inst $\{q_i S_j S_k N q_l\}$

are to be abbreviations for other similarly constructed expressions.

Let us put the description of $\mathcal{M}$ into the first standard form of §6. This description consists of a number of expressions such as "$q_i S_j S_k L q_l$" (or with R or N substituted for L). Let us form all the corresponding expressions such as Inst $\{q_i S_j S_k L q_l\}$ and take their logical sum. This we call Des ($\mathcal{M}$).

The formula Un ($\mathcal{M}$) is to be

$$(\exists u)[N(u) \& (x)(N(x) \rightarrow (\exists x')F(x, x'))$$
$$\& (y, z)(F(y, z) \rightarrow N(y) \& N(z))$$
$$\& (y)R_{S_0}(u, y) \& I(u, u) \& K_{q_1}(u) \& \text{Des}(\mathcal{M})]$$
$$\rightarrow (\exists s)(\exists t)[N(s) \& N(t) \& R_{S_1}(s, t)].$$

$[N(u) \& \ldots \& \text{Des}(\mathcal{M})]$ may be abbreviated to $A(\mathcal{M})$.

When we substitute the meanings suggested on [pp. 84–85] we find that Un ($\mathcal{M}$) has the interpretation "in some complete configuration of $\mathcal{M}$, S_1 (i.e. 0) appears on the tape". Corresponding to this I prove that

(a) If S_1 appears on the tape in some complete configuration of $\mathcal{M}$, then Un ($\mathcal{M}$) is provable.

(b) If Un ($\mathcal{M}$) is provable, then S_1 appears on the tape in some complete configuration of $\mathcal{M}$.

When this has been done, the remainder of the theorem is trivial.

LEMMA 1. *If S_1 appears on the tape in some complete configuration of $\mathcal{M}$, then* Un ($\mathcal{M}$) *is provable.*

We have to show how to prove Un ($\mathcal{M}$). Let us suppose that in the n-th complete configuration the sequence of symbols on the tape is $S_{r(n, 0)}$, $S_{r(n, 1)}, \ldots, S_{r(n, n)}$, followed by nothing but blanks, and that the scanned symbol is the $i(n)$-th, and that the m-configuration is $q_{k(n)}$. Then we may form the proposition

$$R_{S_{r(n,0)}}(u^{(n)}, u) \& R_{S_{r(n,1)}}(u^{(n)}, u') \& \ldots \& R_{S_{r(n,n)}}(u^{(n)}, u^{(n)})$$
$$\& I(u^{(n)}, u^{(i(n))}) \& K_{q_{k(n)}}(u^{(n)})$$
$$\& (y)F((y, u') \vee F(u, y) \vee F(u', y) \vee \ldots \vee F(u^{(n-1)}, y) \vee R_{S_0}(u^{(n)}, y)),$$

which we may abbreviate to CC_n.

As before, $F(u, u') \& F(u', u'') \& \ldots \& F(u^{(r-1)}, u^{(r)})$ is abbreviated to $F^{(r)}$.

I shall show that all formulae of the form $A(\mathcal{M}) \& F^{(n)} \to CC_n$ (abbreviated to CF_n) are provable. The meaning of CF_n is "The n-th complete configuration of $\mathcal{M}$ is so and so", where "so and so" stands for the actual n-th complete configuration of $\mathcal{M}$. That CF_n should be provable is therefore to be expected.

CF_0 is certainly provable, for in the complete configuration the symbols are all blanks, the m-configuration is q_1, and the scanned square is u, i.e. CC_0 is

$$(y)R_{S_0}(u, y) \& I(u, u) \& K_{q_1}(u).$$

$A(\mathcal{M}) \to CC_0$ is then trivial.

We next show that $CF_n \to CF_{n+1}$ is provable for each n. There are three cases to consider, according as in the move from the n-th to the $(n+1)$-th configuration the machine moves to left or to right or remains stationary. We suppose that the first case applies, i.e. the machine moves to the left. A similar argument applies in the other cases. If $r(n, i(n)) = a$, $r(n+1, i(n+1)) = c$, $k(i(n)) = b$, and $k(i(n+1)) = d$, then Des $(\mathcal{M})$ must include Inst $\{q_a S_b S_d L q_c\}$ as one of its terms, i.e.

$$\text{Des }(\mathcal{M}) \to \text{Inst } \{q_a S_b S_d L q_c\}.$$

Hence $\qquad A(\mathcal{M}) \& F^{(n+1)} \to \text{Inst } \{q_a S_b S_d L q_c\} \& F^{(n+1)}.$

But $\qquad \text{Inst } \{q_a S_b S_d L q_c\} \& F^{(n+1)} \to (CC_n \to CC_{n+1})$

is provable, and so therefore is

$$A(\mathcal{M}) \& F^{(n+1)} \to (CC_n \to CC_{n+1})$$

and

$$\left(A(\mathcal{M}) \& F^{(n)} \to CC_n\right) \to \left(A(\mathcal{M}) \& F^{(n+1)} \to CC_{n+1}\right),$$

i.e. $\qquad CF_n \to CF_{n+1}.$

CF_n is provable for each n. Now it is the assumption of this lemma that S_1 appears somewhere, in some complete configuration, in the sequence of symbols printed by

$\mathcal{M}$; that is, for some integers N, K, CC_N has $R_{S_1}(u^{(N)}, u^{(K)})$ as one of its terms, and therefore $CC_N \to R_{S_1}(u^{(N)}, u^{(K)})$ is provable. We have then

$$CC_N \to R_{S_1}(u^{(N)}, u^{(K)})$$
and
$$A(\mathcal{M}) \,\&\, F^{(N)} \to CC^N.$$

We also have

$$(\exists u)A(\mathcal{M}) \to (\exists u)(\exists u')\ldots(\exists u^{(N')})(A(\mathcal{M}) \,\&\, F^{(N)}),$$

where $N' = \max(N, K)$. And so

$$(\exists u)A(\mathcal{M}) \to (\exists u)(\exists u')\ldots(\exists u^{(N')})R_{S_1}(u^{(N)}, u^{(K)}),$$
$$(\exists u)A(\mathcal{M}) \to (\exists u^{(N)})(\exists u^{(K)})R_{S_1}(u^{(N)}, u^{(K)}),$$
$$(\exists u)A(\mathcal{M}) \to (\exists s)(\exists t)R_{S_1}(s, t),$$

i.e. Un ($\mathcal{M}$) is provable.

This completes the proof of Lemma 1.

LEMMA 2. *If* Un ($\mathcal{M}$) *is provable, then* S_1 *appears on the tape in some complete configuration of* $\mathcal{M}$.

If we substitute any propositional functions for function variables in a provable formula, we obtain a true proposition. In particular, if we substitute the meanings tabulated on pp. [84–85] in Un ($\mathcal{M}$), we obtain a true proposition with the meaning "S_1 appears somewhere on the tape in some complete configuration of $\mathcal{M}$".

We are now in a position to show that the Entscheidungsproblem cannot be solved. Let us suppose the contrary. Then there is a general (mechanical) process for determining whether Un ($\mathcal{M}$) is provable. By Lemmas 1 and 2, this implies that there is a process for determining whether $\mathcal{M}$ ever prints 0, and this is impossible, by § 8. Hence the Entscheidungsproblem cannot be solved.

In view of the large number of particular cases of solutions of the Entscheidungsproblem for formulae with restricted systems of quantors, it is interesting to express Un ($\mathcal{M}$) in a form in which all quantors are at the beginning. Un ($\mathcal{M}$) is, in fact, expressible in the form

$$(u)(\exists x)(w)(\exists u_1)\ldots(\exists u_n)\mathfrak{B}, \tag{I}$$

where $\mathfrak{B}$ contains no quantors, and $n = 6$. By unimportant modifications we can obtain a formula, with all essential properties of Un ($\mathcal{M}$), which is of form (I) with $n = 5$.

Added 28 August, 1936.[15]

Appendix
Computability and effective calculability

The theorem that all effectively calculable (λ-definable) sequences are computable and its converse are proved below in outline. It is assumed that the terms "well-formed formula" (W.F.F.) and "conversion" as used by Church and Kleene are understood. In the second of these proofs the existence of several formulae is assumed without proof; these formulae may be constructed straightforwardly with the help of, e.g., the results of Kleene in "A theory of positive integers in formal logic", *American Journal of Math*, 57 (1935), 153–173, 219–244.

The W.F.F. representing an integer n will be denoted by N_n. We shall say that a sequence γ whose n-th figure is $\phi_\gamma(n)$ is λ-definable or effectively calculable if $1 + \phi_\gamma(u)$ is a λ-definable function of n, i.e. if there is a W.F.F. M_γ such that, for all integers n,

$$\{M_\gamma\}(N_n) \text{ conv } N_{\phi_\gamma(n)+1},$$

i.e. $\{M_\gamma\}(N_n)$ is convertible into $\lambda xy \,.\, x(x(y))$ or into $\lambda xy \,.\, x(y)$ according as the n-th figure of λ is 1 or 0.

To show that every λ-definable sequence γ is computable, we have to show how to construct a machine to compute γ. For use with machines it is convenient to make a trivial modification in the calculus of conversion. This alteration consists in using $x, x', x'', \ldots$ as variables instead of $a, b, c, \ldots$. We now construct a machine $\mathcal{L}$ which, when supplied with the formula M_γ, writes down the sequence γ. The construction of $\mathcal{L}$ is somewhat similar to that of the machine $\mathcal{K}$ which proves all provable formulae of the functional calculus. We first construct a choice machine $\mathcal{L}_1$, which, if supplied with a W.F.F., M say, and suitably manipulated, obtains any formula into which M is convertible. $\mathcal{L}_1$ can then be modified so as to yield an automatic machine $\mathcal{L}_2$ which obtains successively all the formulae into which M is convertible (cf. foot-note p. [77]). The machine $\mathcal{L}$ includes $\mathcal{L}_2$ as a part. The motion of the machine $\mathcal{L}$ when supplied with the formula M_γ is divided into sections of which the n-th is devoted to finding the n-th figure of γ. The first stage in this n-th section is the formation of $\{M_\gamma\}(N_n)$. This formula is then supplied to the machine $\mathcal{L}_2$, which converts it successively into various other formulae. Each formula into which it is convertible eventually appears, and each, as it is found, is compared with

$$\lambda x[\lambda x'[\{x\}(\{x\}(x'))]], \text{ i.e. } N_2,$$

and with $\quad \lambda x[\lambda x'[\{x\}(x')]], \text{ i.e. } N_1.$

If it is identical with the first of these, then the machine prints the figure 1 and the n-th section is finished. If it is identical with the second, then 0 is printed and the section is finished. If it is different from both, then the work of $\mathcal{L}_2$ is resumed.

---

[15] The Graduate College, Princeton University, New Jersey, USA.

By hypothesis, $\{M_\gamma\}(N_n)$ is convertible into one of the formulae N_2 or N_1; consequently the n-th section will eventually be finished, i.e. the n-th figure of γ will eventually be written down.

To prove that every computable sequence γ is λ-definable, we must show how to find a formula M_γ such that, for all integers n,

$$\{M_\gamma\}(N_n) \text{ conv } N_{1+\phi_\gamma(n)}.$$

Let $\mathcal{M}$ be a machine which computes γ and let us take some description of the complete configurations of $\mathcal{M}$ by means of numbers, e.g. we may take the D.N of the complete configuration as described in §6. Let $\xi(n)$ be the D.N of the n-th complete configuration of $\mathcal{M}$. The table for the machine $\mathcal{M}$ gives us a relation between $\xi(n+1)$ and $\xi(n)$ of the form

$$\xi(n+1) = \rho_\gamma(\xi(n)),$$

where ρ_γ is a function of very restricted, although not usually very simple, form: it is determined by the table for $\mathcal{M}$. ρ_γ is λ-definable (I omit the proof of this), i.e. there is a W.F.F. A_γ such that, for all integers n,

$$\{A_\gamma\}(N_{\xi(n)}) \text{ conv } N_{\xi(n+1)}.$$

Let U stand for

$$\lambda u\big[\{\{u\}(A_\gamma)\}(N_r)\big],$$

where $r = \xi(0)$; then, for all integers n,

$$\{U_\gamma\}(N_n) \text{ conv } N_{\xi(n)}.$$

It may be proved that there is a formula V such that

$$\{\{V\}(N_{\xi(n+1)})\}(N_{\xi(n)}) \begin{cases} \text{conv } N_1 & \text{if, in going from the } n\text{-th to} \\ & \text{the } (n+1)\text{-th complete configuration, the} \\ & \text{figure 0 is printed.} \\ \text{conv } N_2 & \text{if the figure 1 is printed.} \\ \text{conv } N_3 & \text{otherwise.} \end{cases}$$

Let W_γ stand for

$$\lambda u\big[\{\{V\}(\{A_\gamma\}(\{U_\gamma\}(u)))\}(\{U_\gamma\}(u))\big],$$

so that, for each integer n,

$$\{\{V\}(N_{\xi(n+1)})\}(N_{\xi(n)}) \text{ conv } \{W_\gamma\}(N_n),$$

and let Q be a formula such that

$$\{\{Q\}(W_\gamma)\}(N_s) \text{ conv } N_{r(z)},$$

where $r(s)$ is the s-th integer q for which $\{W_\gamma\}(N_q)$ is convertible into either N_1 or N_2. Then, if M_γ stands for

$$\lambda w\bigl[\{W_\gamma\}(\{\{Q\}(W_\gamma)\}(w))\bigr],$$

it will have the required property.[16]

---

[16] In a complete proof of the λ-definability of computable sequences it would be best to modify this method by replacing the numerical description of the complete configurations by a description which can be handled more easily with our apparatus. Let us choose certain integers to represent the symbols and the m-configurations of the machine. Suppose that in a certain complete configuration the numbers representing the successive symbols on the tape are $s_1 s_2 \ldots s_n$, that the m-th symbol is scanned, and that the m-configuration has the number t; then we may represent this complete configuration by the formula

$$[[N_{s_1}, N_{s_2}, \ldots, N_{s_{m-1}}], [N_t, N_{s_m}], [N_{s_{m+1}}, \ldots, N_{s_n}]],$$

where $[a, b]$ stands for $\lambda u[\{\{u\}(a)\}(b)]$,

$[a, b, c]$ stands for $\lambda u[\{\{\{u\}(a)\}(b)\}(c)]$,

etc.

CHAPTER 2

On Computable Numbers: Corrections and Critiques

Alan Turing, Emil Post, and Donald W. Davies

Introduction

Jack Copeland

This chapter contains four items:

> 2.1 On Computable Numbers, with an Application to the Entscheidungsproblem. A Correction. *Alan Turing*
> 2.2 On Computable Numbers, with an Application to the Entscheidungsproblem. A Critique. *Emil Post*
> 2.3 Draft of a Letter from Turing to Alonzo Church Concerning the Post Critique
> 2.4 Corrections to Turing's Universal Computing Machine *Donald W. Davies*

As is not uncommon in work of such complexity, there are a number of mistakes in 'On Computable Numbers' (Chapter 1). Turing corrected some of these in his short note 2.1, published in the *Proceedings of the London Mathematical Society* a few months after the original paper had appeared.

The mathematician Emil L. Post's critique of 'On Computable Numbers' was published in 1947 and formed part of Post's paper 'Recursive Unsolvability of a Problem of Thue'.[1] Post is one of the major figures in the development of mathematical logic in the twentieth century, although his work did not gain wide recognition until after his death. (Born in 1897, Post died in the same year as Turing.)

By 1936 Post had arrived independently at an analysis of computability substantially similar to Turing's.[2] Post's 'problem solver' operated in a 'symbol

---

[1] *Journal of Symbolic Logic*, 12 (1947), 1–11.
[2] E. L. Post, 'Finite Combinatory Processes—Formulation 1', *Journal of Symbolic Logic*, 1 (1936), 103–4.

space' consisting of 'a two way infinite sequence of spaces or boxes'. A box admitted 'of but two possible conditions, i.e., being empty or unmarked, and having a single mark in it, say a vertical stroke'. The problem solver worked in accordance with 'a fixed unalterable set of directions' and could perform the following 'primitive acts': determine whether the box at present occupied is marked or not; erase any mark in the box that is at present occupied; mark the box that is at present occupied if it is unmarked; move to the box to the right of the present position; move to the box to the left of the present position.

Later, Post considerably extended certain of the ideas in Turing's 'Systems of Logic Based on Ordinals' (Chapter 3), developing the important field now called *degree theory*.

In his draft letter to Church, Turing responded to Post's remarks concerning 'Turing convention-machines'.[3] It is doubtful whether Turing ever sent the letter. The approximate time of writing can be inferred from Turing's opening remarks: Kleene's review appeared in the issue of the *Journal of Symbolic Logic* dated September 1947 (12: 90–1) and Turing's 'Practical Forms of Type Theory' appeared in the same journal in June 1948.

In his final year at university Donald Davies (1924–2000) heard about Turing's proposed Automatic Computing Engine and the plans to build it at the National Physical Laboratory in London (see Chapter 9). Davies immediately applied to join the National Physical Laboratory and in September 1947 became a member of the small team surrounding Turing. Davies played a leading role in the development and construction of the pilot model of the Automatic Computing Engine, which ran its first programme in May 1950. From 1966 he was head of the computer science division at the National Physical Laboratory. He originated the important concept of 'packet switching' used in the ARPANET, forerunner of the Internet. From 1979 Davies worked on data security and public key cryptosystems.

'On Computable Numbers' contained a number of what would nowadays be called programming errors. Davies described Turing's reaction when he drew Turing's attention to some of these:

I was working more or less under [Turing's] supervision ... I had been reading his famous work on computable numbers ... and I began to question some of the details of his paper. In fact I ... found a number of quite bad programming errors, in effect, in the specification of the machine that he had written down, and I had worked out how to overcome these. I went along to tell him and I was rather cock-a-hoop ... I thought he would say 'Oh fine, I'll send along an addendum' [to the London Mathematical Society]. But in fact he was very annoyed, and pointed out furiously that really it

---

[3] The draft is among the Turing Papers in the Modern Archive Centre, King's College Library, Cambridge; catalogue reference D 2.

didn't matter, the thing was right in principle, and altogether I found him extremely touchy on this subject.[4]

In Section 4 of his 'Corrections to Turing's Universal Computing Machine' Davies mends the errors that he discovered in 1947. He emphasizes that—as Turing said—these programming errors are of no significance for the central arguments of 'On Computable Numbers'.

Davies's lucid commentary forms an excellent introduction to 'On Computable Numbers'.[5]

---

[4] Davies in interview with Christopher Evans ('The Pioneers of Computing: An Oral History of Computing' (London: Science Museum, 1975)).

[5] I am grateful to Diane Davies for her permission to publish this article.

2.1
On Computable Numbers, with an Application to the Entscheidungsproblem. A Correction. (1937)
Alan Turing

In a paper entitled "On computable numbers, with an application to the Entscheidungsproblem" [Chapter 1] the author gave a proof of the insolubility of the Entscheidungsproblem of the "engere Funktionenkalkül". This proof contained some formal errors[1] which will be corrected here: there are also some other statements in the same paper which should be modified, although they are not actually false as they stand.

The expression for Inst $\{q_i S_j S_k L q_l\}$ on p. [85] of the paper quoted should read

$$(x, y, x', y')\{(R_{S_j}(x, y) \& I(x, y) \& K_{q_i}(x) \& F(x, x') \& F(y', y))$$
$$\to (I(x', y') \& R_{S_k}(x', y) \& K_{q_l}(x') \& F(y', z) \vee [(R_{S_0}(x, z) \to R_{S_0}(x', z))$$
$$\& (R_{S_1}(x, z) \to R_{S_1}(x', z)) \& \ldots \& (R_{S_M}(x, z) \to R_{S_M}(x', z))])\},$$

$S_0, S_1, \ldots, S_M$ being the symbols which $\mathcal{M}$ can print. The statement on p. [86], [lines 24–25], viz.

$$\text{"Inst } \{q_a S_b S_d L q_c\} \& F^{(n+1)} \to (CC_n \to CC_{n+1})$$

is provable" is false (even with the new expression for Inst $\{q_a S_b S_d L q_c\}$): we are unable for example to deduce $F^{(n+1)} \to (-F(u, u''))$ and therefore can never use the term

$$F(y', z) \vee [(R_{S_0}(x, z) \to R_{S_0}(x', z)) \& \ldots \& (R_{S_M}(x, z) \to R_{S_M}(x', z))]$$

in Inst $\{q_a S_b S_d L q_c\}$. To correct this we introduce a new functional variable G [$G(x, y)$ to have the interpretation "x precedes y"]. Then, if Q is an abbreviation for

$$(x)(\exists w)(y, z)\{F(x, w) \& (F(x, y) \to G(x, y)) \& (F(x, z) \& G(z, y) \to G(x, y))$$
$$\& [G(z, x) \vee (G(x, y) \& F(y, z)) \vee (F(x, y) \& F(z, y)) \to (-F(x, z))]\}$$

the corrected formula Un ($\mathcal{M}$) is to be

This article first appeared in *Proceedings of the London Mathematical Society*, Series 2, 43 (1937), 544–6. It is reprinted with the permission of the London Mathematical Society and the Estate of Alan Turing.

[1] The author is indebted to P. Bernays for pointing out these errors.

$$(\exists u)A(\mathcal{M}) \rightarrow (\exists s)(\exists t)R_{S_1}(s, t),$$

where $A(\mathcal{M})$ is an abbreviation for

$$Q \,\&\, (y)R_{S_0}(u, y) \,\&\, I(u, u) \,\&\, K_{q_1}(u) \,\&\, \text{Des}\,(\mathcal{M}).$$

The statement on page [86] (line [24]) must then read

$$\text{Inst}\,\{q_a S_b S_d L q_c\} \,\&\, Q \,\&\, F^{(n+1)} \rightarrow (CC_n \rightarrow CC_{n+1}),$$

and [lines 19–20] should read

$$r(n, i(n)) = b, \quad r(n+1, i(n)) = d, \quad k(n) = a, \quad k(n+1) = c.$$

For the words "logical sum" on p. [85], line [13], read "conjunction". With these modifications the proof is correct. Un $(\mathcal{M})$ may be put in the form (I) (p. [87]) with $n = 4$.

Some difficulty arises from the particular manner in which "computable number" was defined (p. [61]). If the computable numbers are to satisfy intuitive requirements we should have:

If we can give a rule which associates with each positive integer n two rationals a_n, b_n satisfying $a_n \leq a_{n+1} < b_{n+1} \leq b_n$, $b_n - a_n < 2^{-n}$, then there is a computable number α for which $a_n \leq \alpha \leq b_n$ each n. (A)

A proof of this may be given, valid by ordinary mathematical standards, but involving an application of the principle of excluded middle. On the other hand the following is false:

There is a rule whereby, given the rule of formation of the sequences a_n, b_n in (A) we can obtain a D.N. for a machine to compute α. (B)

That (B) is false, at least if we adopt the convention that the decimals of numbers of the form $m/2^n$ shall always terminate with zeros, can be seen in this way. Let $\mathcal{N}$ be some machine, and define c_n as follows: $c_n = \frac{1}{2}$ if $\mathcal{N}$ has not printed a figure 0 by the time the n-th complete configuration is reached $c_n = \frac{1}{2} - 2^{-m-3}$ if 0 had first been printed at the m-th complete configuration ($m \leq n$). Put $a_n = c_n - 2^{-n-2}$, $b_n = c_n + 2^{-n-2}$. Then the inequalities of (A) are satisfied, and the first figure of α is 0 if $\mathcal{N}$ ever prints 0 and is 1 otherwise. If (B) were true we should have a means of finding the first figure of α given the D.N. of $\mathcal{N}$: i.e. we should be able to determine whether $\mathcal{N}$ ever prints 0, contrary to the results of §8 of the paper quoted. Thus although (A) shows that there must be machines which compute the Euler constant (for example) we cannot at present describe any such machine, for we do not yet know whether the Euler constant is of the form $m/2^n$.

This disagreeable situation can be avoided by modifying the manner in which computable numbers are associated with computable sequences, the totality of computable numbers being left unaltered. It may be done in many ways of which this is an example.[2] Suppose that the first figure of a computable sequence γ is i and that this is followed by 1 repeated n times, then by 0 and finally by the sequence whose r-th figure is c_r; then the sequence γ is to correspond to the real number

$$(2i-1)n + \sum_{r=1}^{\infty}(2c_r - 1)\left(\tfrac{2}{3}\right)^r.$$

If the machine which computes γ is regarded as computing also this real number then (B) holds. The uniqueness of representation of real numbers by sequences of figures is now lost, but this is of little theoretical importance, since the D.N.'s are not unique in any case.

The Graduate College, Princeton, N.J., U.S.A.

---

[2] This use of overlapping intervals for the definition of real numbers is due originally to Brouwer.

2.2
On Computable Numbers, with an Application to the Entscheidungsproblem. A Critique. (1947)
Emil Post

The following critique of Turing's "computability" paper [Chapter 1] concerns only pp. [58–74] thereof. We have checked the work through the construction of the "universal computing machine" in detail, but the proofs of the two theorems in the section following are there given in outline only, and we have not supplied the formal details. *We have therefore also left in intuitive form the proofs of the statements on recursiveness, and alternative procedures, we make below.*

One major correction is needed. To the instructions for $\mathfrak{con}_1(\mathfrak{C}, \alpha)$ p. [70], add the line: None PD, R, Pα, R, R, R $\mathfrak{C}$. This is needed to introduce the representation D of the blank scanned square when, as at the beginning of the action of the machine, or due to motion right beyond the rightmost previous point, the complete configuration ends with a q, and thus make the $\mathfrak{fmp}$ of p. [71] correct. We may also note the following minor slips and misprints in pp. [58–74]. Page [63], to the instructions for $\mathfrak{f}(\mathfrak{C}, \mathfrak{B}, \alpha)$ add the line: None L $\mathfrak{f}(\mathfrak{C}, \mathfrak{B}, \alpha)$; p. [67] and p. [68], the S.D should begin, but not end, with a semicolon; p. [69], omit the first D in (C_2); p. [70], last paragraph [above skeleton table], add ":" to the first list of symbols; pp. [71–72], replace $\mathfrak{g}$ by $\mathfrak{q}$; p. [71], in the instruction for $\mathfrak{mf}$, $\mathfrak{mf}$ should be $\mathfrak{mf}_1$; p. [71], in the second instruction for $\mathfrak{sim}_2$, replace the first R by L; p. [71], in the first instruction for $\mathfrak{sh}_2$, replace $\mathfrak{sh}_2$ by $\mathfrak{sh}_3$. A reader of the paper will be helped by keeping in mind that the "examples" of pages [63–66] are really parts of the table for the universal computing machine, and accomplish what they are said to accomplish not for all possible printings on the tape, but for certain ones that include printings arising from the action of the universal computing machine. In particular, the tape has ǝ printed on its first two squares, the occurrence of two consecutive blank squares insures all squares to the right thereof being blank, and, usually, symbols referred to are on "F-squares", and obey the convention of p. [63].[1]

Turing's definition of an arbitrary machine is not completely given in his paper, and, at a number of points, has to be inferred from his development. In the first instance his machine is a "computing machine" for obtaining the successive digits of a real number in dyadic notation, and, in that case, starts operating on a blank tape. Where explicitly stated, however, the machine may

---

Post's critique originally formed an untitled appendix occupying pp. 7–11 of 'Recursive Unsolvability of a Problem of Thue', *Journal of Symbolic Logic*, 12 (1947), 1–11. The critique is reprinted here by permission of the Association for Symbolic Logic. All rights reserved. This reproduction is by special permission for this publication only.

[1] Editor's note. This paragraph originally formed a footnote (the first) to Post's appendix.

98 | *Emil Post*

start operating on a tape previously marked. From Turing's frequent references to the beginning of the tape, and the way his universal computing machine treats motion left, we gather that, unlike our tape, this tape is a one-way infinite affair going right from an initial square.

Primarily as a matter of practice, Turing makes his machines satisfy the following convention. Starting with the first square, alternate squares are called *F*-squares, the rest, *E*-squares. In its action the machine then never directs motion left when it is scanning the initial square, never orders the erasure, or change, of a symbol on an *F*-square, never orders the printing of a symbol on a blank *F*-square if the previous *F*-square is blank, and, in the case of a computing machine, never orders the printing of 0 or 1 on an *E*-square. This convention is very useful in practice. However the actual performance, described below, of the universal computing machine, coupled with Turing's proof of the second of the two theorems referred to above, strongly suggests that Turing makes this convention part of the definition of an arbitrary machine. We shall distinguish between a Turing machine and a Turing convention-machine.

By a uniform method of representation, Turing represents the set of instructions, corresponding to our quadruplets,[2] which determine the behavior of a machine by a single string on seven letters called the standard description (S.D) of the machine. With the letters replaced by numerals, the S.D of a machine is considered the arabic representation of a positive integer called the description number (D.N) of the machine. If our critique is correct, a machine is said to be circle-free if it is a Turing computing convention-machine which prints an infinite number of 0's and 1's.[3] And the two theorems of Turing's in question are really the following. There is no Turing convention-machine which, when supplied with an arbitrary positive integer n, will determine whether n is the D.N of a Turing computing convention-machine that is circle-free. There is no Turing convention-machine which, when supplied with an arbitrary positive integer n, will determine whether n is the D.N of a Turing computing convention-machine that ever prints a given symbol (0 say).[4]

---

[2] Our quadruplets are quintuplets in the Turing development. That is, where our standard instruction orders either a printing (overprinting) or motion, left or right, Turing's standard instruction always orders a printing and a motion, right, left, or none. Turing's method has certain technical advantages, but complicates theory by introducing an irrelevant "printing" of a symbol each time that symbol is merely passed over.

[3] "Genuinely prints", that is, a genuine printing being a printing in an empty square. See the previous footnote.

[4] Turing in each case refers to the S.D of a machine being supplied. But the proof of the first theorem, and the second theorem depends on the first, shows that it is really a positive integer n that is supplied. Turing's proof of the second theorem is unusual in that while it uses the unsolvability result of the first theorem, it does not "reduce" [Post (1944)] the problem of the first theorem to that of the second. In fact, the first problem is almost surely of "higher degree of unsolvability" [Post (1944)] than the second, in which case it could not be "reduced" to the second. Despite appearances, that second unsolvability proof, like the first, is a *reductio ad absurdum* proof based on the definition of unsolvability, at the conclusion of which, the first result is used.

Corrections and Critiques | 99

In view of [Turing (1937)], these "no machine" results are no doubt equivalent to the recursive unsolvability of the corresponding problems.[5] But both of these problems are infected by the spurious Turing convention. Actually, the set of n's which are D.N's of Turing computing machines as such is recursive, and hence the condition that n be a D.N offers no difficulty. But, while the set of n's which are not D.N's of convention-machines is recursively enumerable, the complement of that set, that is, the set of n's which are D.N's of convention-machines, is not recursively enumerable. As a result, in both of the above problems, neither the set of n's for which the question posed has the answer yes, nor the set for which the answer is no, is recursively enumerable.

This would remain true for the first problem even apart from the convention condition. But the second would then become that simplest type of unsolvable problem, the decision problem of a non-recursive recursively enumerable set of positive integers [(Post 1944)]. For the set of n's that are D.N's of unrestricted Turing computing machines printing 0, say, is recursively enumerable, though its complement is not. The Turing convention therefore prevents the early appearance of this simplest type of unsolvable problem.

It likewise prevents the use of Turing's second theorem in the ... unsolvability proof of the problem of Thue.[6] For in attempting to reduce the problem of Turing's second theorem to the problem of Thue, when an n leads to a Thue question for which the answer is yes, we would still have to determine whether n is the D.N of a Turing convention-machine before the answer to the question posed by n can be given, and that determination cannot be made recursively for arbitrary n. If, however, we could replace the Turing convention by a convention that is recursive, the application to the problem of Thue could be made. An analysis of what Turing's universal computing machine accomplishes when applied to an arbitrary machine reveals that this can be done.

The universal computing machine was designed so that when applied to the S.D of an arbitrary computing machine it would yield the same sequence of 0's and 1's as the computing machine as well as, and through the intervention of, the successive "complete configurations"—representations of the successive states of tape versus machine—yielded by the computing machine. This it does for a Turing convention-machine.[7] For an arbitrary machine, we have to interpret a direction of motion left at a time when the initial square of the tape is scanned as

---

[5] Our experience with proving that "normal unsolvability" in a sense implicit in [Post (1943)] is equivalent to unsolvability in the sense of Church [(1936)], at least when the set of questions is recursive, suggests that a fair amount of additional labor would here be involved. That is probably our chief reason for making our proof of the recursive unsolvability of the problem of Thue independent of Turing's development.

[6] Editor's note. Thue's problem is that of determining, for arbitrary strings of symbols *A*, *B* from a given finite alphabet, whether or not *A* and *B* are interderivable by means of a succession of certain simple substitutions. (See further Chapter 17.)

[7] Granted the corrections [detailed above].

meaning no motion.[8] The universal computing machine will then yield again the correct complete configurations generated by the given machine. But *the space sequence of 0's and 1's printed by the universal computing machine will now be identical with the time sequence of those printings of 0's and 1's by the given machine that are made in empty squares.* If, now, instead of Turing's convention we introduce the convention that the instructions defining the machine never order the printing of a 0 or 1 except when the scanned square is empty, or 0, 1 respectively, and never order the erasure of a 0 or 1, Turing's arguments again can be carried through. And this "(0, 1) convention", being recursive, allows the application to the problem of Thue to be made.[9] Note that if a machine is in fact a Turing convention-machine, we could strike out any direction thereof which contradicts the (0, 1) convention without altering the behavior of the machine, and thus obtain a (0, 1) convention-machine. But a (0, 1) convention-machine need not satisfy the Turing convention. However, by replacing each internal-configuration q_i of a machine by a pair q_i, q_i' to correspond to the scanned square being an *F*- or an *E*-square respectively, and modifying printing on an *F*-square to include testing the preceding *F*-square for being blank, we can obtain a "(q, q') convention" which is again recursive, and usable both for Turing's arguments and the problem of Thue, and has the property of, in a sense, being equivalent to the Turing convention. That is, every (q, q') convention-machine is a Turing convention-machine, while the directions of every Turing convention-machine can be recursively modified to yield a (q, q') convention-machine whose operation yields the same time sequence and spatial arrangement of printings and erasures as does the given machine, except for reprintings of the same symbol in a given square.

These changes in the Turing convention, while preserving the general outline of Turing's development and at the same [time] admitting of the application to the problem of Thue, would at least require a complete redoing of the formal work of the proof of the second Turing theorem. On the other hand, very little added formal work would be required if the following changes are made in the Turing argument itself, though there would still remain the need of extending the equivalence proof of [Turing (1937)] to the concept of unsolvability. By using the above result on the performance of the universal computing machine when applied to the S.D of an arbitrary machine, we see that Turing's proof of his first theorem, whatever the formal counterpart thereof is, yields the following theorem. There is no Turing convention-machine which, when supplied with an

---

[8] This modification of the concept of motion left is assumed throughout the rest of the discussion, with the exception of the last paragraph.

[9] So far as recursiveness is concerned, the distinction between the Turing convention and the (0, 1) convention is that the former concerns the history of the machine in action, the latter only the instructions defining the machine. Likewise, despite appearances, the later (q, q') convention.

arbitrary positive integer *n*, will determine whether *n* is the D.N of an arbitrary Turing machine that prints 0's and 1's in empty squares infinitely often. Now given an arbitrary positive integer *n*, if that *n* is the D.N of a Turing machine 𝑀, apply the universal computing machine to the S.D of 𝑀 to obtain a machine 𝑀*. Since 𝑀* satisfies the Turing convention, whatever Turing's formal proof of his second theorem is, it will be usable intact in the present proof, and, via the new form of his first theorem, will yield the following usable result. There is no machine which, when supplied with an arbitrary positive integer *n*, will determine whether *n* is the D.N of an arbitrary Turing machine that ever prints a given symbol (0 say).[10]

These alternative procedures assume that Turing's universal computing machine is retained. However, in view of the above discussion, it seems to the writer that Turing's preoccupation with computable numbers has marred his entire development of the Turing machine. We therefore suggest a redevelopment of the Turing machine based on the formulation given in ['Recursive Unsolvability of a Problem of Thue'[11]]. This could easily include computable numbers by defining a computable sequence of 0's and 1's as the *time sequence* of printings of 0's and 1's by an arbitrary Turing machine, provided there are an infinite number of such printings. By adding to Turing's complete configuration a representation of the act last performed, a few changes in Turing's method would yield a universal computing machine which would transform such a time sequence into a space sequence. Turing's convention would be followed as a matter of useful practice in setting up this, and other, particular machines. But it would not infect the theory of arbitrary Turing machines.

---

[10] It is here assumed that the suggested extension of [Turing (1937)] includes a proof to the effect that the existence of an arbitrary Turing machine for solving a given problem is equivalent to the existence of a Turing convention-machine for solving that problem.

[11] Editor's note. See the reference at the foot of p. 97.

References

Church, A. 1936. 'An Unsolvable Problem of Elementary Number Theory', *American Journal of Mathematics*, 58, 345–363.

Post, E. L. 1943. 'Formal Reductions of the General Combinatorial Decision Problem', *American Journal of Mathematics*, 65, 197–215.

Post, E. L. 1944. 'Recursively Enumerable Sets of Positive Integers and their Decision Problems', *Bulletin of the American Mathematical Society*, 50, 284–316.

Turing, A. M. 1937. *Computability and λ-definability*, Journal of Symbolic Logic, 2, 153–163.

2.3
Draft of a Letter from Turing to Alonzo Church Concerning the Post Critique

Dear Professor Church,

I enclose corrected proof of my paper 'Practical forms of type theory' and order for reprints.

Seeing Kleene's review of Post's paper (on problem of Thue) has reminded me that I feel I ought to say a few words somewhere to clear up the points which Post has raised about 'Turing machines' and 'Turing convention machines' [see 2.2]. Post observes that my initial description of a machine differs from the machines which I describe later in that the latter are subjected to a number of conventions (e.g. the use of E and F squares). These conventions are nowhere very clearly enumerated in my paper and cast a fog over the whole concept of a 'Turing machine'. Post has enumerated the conventions and embodied them in a definition of a 'Turing convention machine'.

My intentions in this connection were clear in my mind at the time the paper was written; they were not expressed explicitly in the paper, but I think it is now necessary to do so. It was intended that the 'Turing machine' should always be the machine without attached conventions, and that all general theorems about machines should apply to this definition. To the best of my belief this was adhered to. On the other hand when it was a question of describing particular machines a host of conventions became desirable. Clearly it was best to choose conventions which did not restrict the essential generality of the machine, but one was not called upon to establish any results to this effect. If one could find machines obeying the conventions and able to carry out the desired operations, that was enough. It was also undesirable to keep any fixed list of conventions. At any moment one might wish to introduce a new one.

Published with the permission of the Estate of Alan Turing.

2.4
Corrections to Turing's Universal Computing Machine
Donald W. Davies

1. Introduction

In 1947 I was working in a small team at the National Physical Laboratory in London, helping to build one of the first programmed computers. This had been designed by Turing. (See Chapter 9.)

When I first studied Turing's 'On Computable Numbers, with an Application to the Entscheidungsproblem', it soon became evident to me that there were a number of trivial errors, amounting to little more than typographic errors, in the design of his universal computing machine U. A closer look revealed a—nowadays typical—programming error in which a loop led back to the wrong place. Then I became aware of a more fundamental fault relating to the way U describes the blank tape of the machine it is emulating. Perhaps it is ironic, as well as understandable, that the first emulation program for a computer should have been wrong. I realized that, even though the feasibility of the universal computing machine was not in doubt, the mistakes in Turing's exposition could puzzle future readers and plague anyone who tried to verify Turing's design by implementing his universal machine in practice.

When I told Turing about this he became impatient and made it clear that I was wasting my time and his by my worthless endeavours. Yet I kept in mind the possibility of testing a corrected form of U in the future. It was to be nearly fifty years before I finally did this.

I could not implement exactly Turing's design because this generates a profusion of states when the 'skeleton tables' are substituted by their explicit form (to a depth of 9). Also the way in which U searches for the next relevant instruction involves running from end to end of the tape too many times. Features of Turing's scheme which greatly simplify the description also cause the explicit machine to have many symbols, a considerable number of states and instructions, and to be extremely slow. Turing would have said that this inefficiency was irrelevant to his purpose, which is true, but it does present a practical problem if one is interested in verifying an actual machine. Some fairly simple changes to the design reduce this problem. The final part of this paper outlines a redesign of the universal machine which was tested by simulation and shown to work. There can be reasonable confidence that there are no further significant errors in Turing's design, but a simulation starting directly from Turing's 'skeleton tables' would clinch the matter.

By and large I use Turing's notation and terminology in what follows. Where my notation differs from Turing's the aim has been to make matters clearer. In particular, Turing's Gothic letters are replaced by roman letters. I sometimes introduce words from modern computer technology where this makes things clearer. (There is no special significance to the use of boldface type—this is used simply for increased clarity.)

2. The Turing Machine T

Turing required a memory of unlimited extent and a means of access to that memory. Access by an address would not provide unlimited memory. In this respect the Turing machine goes beyond any existing real machine.

His method, of course, was to store data in the form of symbols written on a tape of unlimited length. Specifically, the tape had a beginning, regarded as its left-hand end, marked with a pair of special symbols 'e e' that can easily be found. To the right of these symbols there are an unlimited number of symbol-positions or 'squares' which can be reached by right and left movements of the machine, shown as 'R' and 'L' respectively in the machine's instructions. By repeated R and L movements any square can be accessed.

Let us consider how the machine's instructions are composed. We are not concerned yet with **U**, the universal machine, but with a specific Turing machine **T**—the 'target machine'—which will later be emulated by **U**.

An instruction for **T** consists of five parts. The purpose of the first two parts is to address the instruction. These give the state of the machine (I shall call this M) and the symbol S that the machine is reading in the scanned square. This state-symbol pair M-S determines the next operation of the machine. Turing called M-S a 'configuration'. For each such pair that can occur (finitely many, since there is a finite number of states and of symbols), the next operation of the machine must be specified in the instructions, by notations in the remaining three parts. The first of these is the new symbol to be written in place of the one that has been read, and I call this S'. Then there is an action A, which takes place after the writing of the new symbol, and this can be a right shift R, a left shift L, or N, meaning no movement. Finally the resultant state M' is given in the last of the five parts of the instruction.

To summarize: an instruction is of the form $MSS'AM'$. The current machine configuration is the pair M-S, which selects an appropriate instruction. The instruction then specifies S', A, and M', meaning that the symbol S' is written in place of S. The machine then moves according to action A (R, L, or N) and finally enters a new state M'. A table of these instructions, each of five parts, specifies the entire behaviour of the target machine **T**. The table should have instructions for all the M-S pairs that can arise during the operation of the machine.

There are some interesting special cases. One of the options is that $S' = S$, meaning that the symbol that was in the scanned square remains in place. In effect, no writing has occurred. Another is that $S' =$ **blank**, meaning that the symbol S which was read has been erased.

As a practical matter, note that the part of the tape which has been used is always finite and should have well-defined ends, so that the machine will not run away down the tape. As already mentioned, Turing specifies that at the left-hand end the pair of special symbols 'e e' is printed. These are never removed or altered. On the extreme right of the used part of the tape there must be a sequence of blank squares. Turing arranges that there will never be two adjacent blanks anywhere in the used part of the tape, so that the right-hand end of the used portion can always be found. This convention is necessary to make U work properly, as is explained in detail later. However, Turing does not necessarily follow this convention in specifying target machines T. This can result in misbehaviour.

The purpose of T is to perform a calculation, therefore T must generate numbers. For this reason the symbols it can print include 0 and 1, which are sufficient to specify a binary result. By convention these two symbols are never erased but remain on the tape as a record of the result of the computation. It happens that they are treated in a special way in the emulation by U, in order to make the result of the calculation more obvious, as I shall explain in due course.

3. The Basic Plan of U

The universal machine U must be provided with the table of instructions of the Turing machine T that it is to emulate. The instructions are given at the start of U's tape, separated by semicolons ';'. At the end of these instructions is the symbol '::' (which is a single symbol). Later I shall settle the question of whether a semicolon should be placed *before* the first instruction or *after* the last one.

Following '::' is the workspace, in which U must place a complete description, in U's own symbols, of machine T. This description consists of all the symbols on T's tape, the position of the machine on that tape, and the state of the machine. I call such a description an *image* or *snapshot* of machine T. As T goes through its computational motions, more and more snapshots are written in U's workspace, so that an entire history of T gradually appears.

In order to make this evolving representation of T's behaviour possible, U must not change any of the images on its tape. U simply adds each new image to the end of the tape as it computes it.

The first action of U is to construct the initial image of T, in the space immediately following the '::' symbol which terminates the set of instructions.

Subsequently, U writes new images of T, separated by colons ':', each image representing a successive step in the evolution of T's computation.

Whenever T is asked to print a 0 or 1, the image in which this happens is followed by the symbols '0 :' or '1 :' respectively. This serves to emphasize the results of the computation. For example, U's tape might look like this, with the instruction set followed by successive images of T. The 'output' character 0 printed in image 3 is highlighted by printing it again after the image, and likewise in the case of the 'output' character 1 that is printed in image 5.

e e ; inst 1 ; inst 2 ; ... inst n :: image 1 : image 2 : image 3 : 0 : image 4 : image 5 : 1 : etc.

Note that each instruction *begins* with a semicolon. This differs from Turing, who has the semicolon *after* each instruction, so that the instructions end with the pair of characters '; ::'. I found this departure from Turing's presentation necessary, as I shall explain later.

The symbols on the tape up to and including the '::'—i.e. T's instructions—are given to U before it starts operating. When U commences its operations, it writes out the first image and then computes successive images from this, using the instructions, and intersperses the images with 'outputs' as required.

To complete this description of the basic plan of U I must specify how the symbols are spaced out in order to allow for *marking* them with other symbols. Looking in more detail at the start of the tape we would find squares associated in pairs. The left square of each pair contains a symbol from the set:

A C D L R N 0 1 ; :: :

These symbols are never erased by U; they form a permanent record of the instructions and the images of T. Thus the first few squares of a tape might contain these symbols representing an instruction, where '−' refers to the blank symbol, meaning an empty square:

$$e\,e\,D - A - D - D - C - R - D - A - A - ; - \ldots \qquad (1)$$

The blank squares leave room for symbols from the set *u v w x y z*, which are reserved for use as markers and serve to mark the symbol to their immediate left. For example, at one stage of the operation of U the parts of this instruction are marked out as follows:

$$e\,e\,D - A - D - D\,u\,C\,u\,R\,u\,D\,y\,A\,y\,A\,y\,; - \ldots$$

Here the '*u*' and '*y*' mark D C R and D A A respectively.

Unlike the symbols A C D L R N 0 1 ; :: : which are never erased, the letters *u v w x y z* are always temporary markings and are erased when they have done their job.

The positions of the A C D L R N 0 1 ; :: : symbols will be called 'non-erasing positions'. Note that the left hand 'e' occupies such a position also.

4. Notation for States, Symbols, and Actions in U's Instructions and in Images of T

The states, symbols, and actions which U represents on its tape are those of T, which U is emulating. We do not know how many states and symbols have to be emulated, yet the set of symbols of U is limited by its design. For economy in U's symbols, the *n*th state of T is represented on U's tape by DAA...A with *n* occurrences of letter 'A'. The blank spaces between these letters are not shown here but are important for the operation of U and will always be assumed. The *n*th symbol of T is shown on U's tape by the string DCC...C with *n* occurrences of letter 'C'. It appears from an example in Turing's text that D by itself (with a marking square to its right, as always) can be one of these 'symbol images'. In fact I shall choose to make this the 'blank' symbol.

Read in accordance with these conventions, the symbols in example (1), above, form the instruction: 'when in state 1—reading a blank—write symbol 0—move right—change to state 2'.

With the semicolons as spacers we now have a complete notation for *instructions* on U's tape. Next we need a notation for an *image* of machine T, which is also made up from these state and symbol images, DAA... and DCC... respectively.

The keys to the next action of T are its present state and the symbol it is reading, which have been stored in the instruction table as the pair M-S. So this same combination is used in the image, by listing in correct sequence all the symbols on T's tape and inserting the state image immediately in front of the symbol which machine T's emulation is currently reading. The placing of the state image indicates which square is currently scanned. So a tape image might look like this:

symbol 1, symbol 2, symbol 3, **state**, symbol 4, symbol 5 :

The current state is given at the position indicated. This emulated tape of T records that T is scanning symbol 4. The combination '**state**, symbol 4' which appears in this string is the M-S pair that U must look up, by searching for it in the instruction table. To get the image of the next state of T, **state** and symbol 4 may have to be changed and the state image, also changed, may have to be moved to the right, or to the left, or not moved. These are the processes which occur in one step of evolution of the machine T. It will be done by building a completely new image to the right of the last ':'. (Consequently the tape space is rapidly used up.)

5. Notation for Machine U

In principle, machine U should be specified in the same formal notation as machine T, but this would be bulky and tedious to read and understand, so Turing used a much more flexible notation. A compiler could be built to take the 'higher-level' notation of machine U's specification as given by Turing and generate the complete set of instructions for U.

The statements which make up Turing's specification of U are similar to procedures with parameters (more correctly they are like macros) and they have four parts. As with the instructions of T, a state and symbol select which statement is to apply, except that the symbol can now be a logical expression such as 'not C', which in the explicit specification would require as many instructions as there are symbols other than 'C'. There are also statements that copy symbols from one part of the (real) tape to another, and this requires one instruction for each allowed symbol variety. The actions, which in the explicit notation can only write a symbol and optionally move one place left or right, are expanded here to allow multiple operations such as 'L, Pu, R, R, R', specifying that symbol 'u' is being printed to the left of the starting point and the machine ends up two places to the right of that point. Turing requires that these specifications be 'compiled' into the standard five-part instructions. I will call these procedures 'routines'. Their parameters are of two kinds, the states that the operations lead to, which are shown as capital letters, and symbol values, which are shown in lower case.

The specification of machine U consists of a collection of 'subroutines' which are then used in a 'program' of nine routines that together perform the evolution of T. We shall first describe the subroutines, then the main program.

6. Subroutines

In principle, the action of a routine depends on the position of the machine when the routine is invoked, i.e. the square on the tape which is being scanned. Also, the position of the scanned square at the end of a routine's operation could be significant for the next operation to come. But Turing's design avoids too much interaction of this kind. It can be assumed that the positions are not significant for the use of the subroutines in U unless the significance is described here. Certain subroutines which are designed to find a particular symbol on the tape (such as f(A, B, a) and q(A, a) and others) leave the machine in a significant position. Con(A, a) has significant starting and finishing positions.

Where a routine uses other routines, these are listed (for convenience in tracing side-effects). Also listed are those routines that use the routine in question. It would be possible to make several of the routines much more efficient, greatly reducing the amount of machine movement, but I have not made such changes here. (For efficiency, routines could be tailored for each of their uses and states and symbols could easily be coded in binary. But this would be a redesign.)

f(A, B, a)
The machine moves left until it finds the start of the tape at an 'e' symbol. Then it moves right, looking for a symbol 'a'. If one is found it rests on that symbol and changes state to A. If there is no symbol 'a' on the whole tape it stops on the first blank non-erasing square to the right of the used portion of tape, going into state B. In general terms, this routine is looking for the leftmost occurrence of the symbol 'a'. The special case **f(A, B, e)** will find the leftmost occurrence of 'e', which is in a non-erasing position. Uses no routines. Used by **b**, **e(A, B, a)**, **f'(A, B, a)**, **cp(A, B, E, a, b)**, **sh**, and **pe(A, b)**.

f'(A, B, a)
As for **f(A, B, a)** except that if a symbol 'a' is found, the machine stops one square to the left, over the square which is marked by the 'a'. Uses **f(A, B, a)** and **1(A)**. Used by **cp(A, B, E, a, b)**, **sim**, and **c(A, b, a)**.

1(A)
Simply shifts one square to the left. Uses none. Used by **f'(A, B, a)**, **mk**, and **inst**.

e(A)
The marks are erased from all marked symbols, leaving the machine in state A. Uses none. Used by **ov**.

e(A, B, a)
The machine finds the leftmost occurrence of symbol 'a', using routine **f(A, B, a)**, then erases it, resting on the blank symbol and changing to state A. If there is no such symbol to erase, it stops on the non-erasing square to the right of the used portion of tape in state B, as for **f(A, B, a)**. Uses **f(A, B, a)**. Used by **kmp**, **cpe(A, B, E, a, b)**, and **e(B, a)**.

e(B, a)
Erases all occurrences of the symbol 'a' on the tape, leaving the machine in state B. Uses **e(A, B, a)**. Used by **sim**.

pe(A, b)
Prints the symbol 'b' in the first blank non-erasing position at the end of the sequence of symbols. Uses **f(A, B, a)**. Used by **pe₂(A, a, b)** and **c(A, B, a)**.

pe₂(A, a, b)

Prints the symbol 'a' and then 'b' in the first blank non-erasing positions. Uses **pe**(A, b). Used by **sh**.

q(A)

Moves to the next non-erasing position after the used portion of tape and goes to state A. Uses none. Used by **q**(A, a).

q(A, a)

Finds the last occurrence of symbol 'a' and stops there in state A. Uses **q**(A). Used by **anf**, **mk**, and **inst**. If the symbol does not exist, the machine will run off the tape to the left. However, in its use in **anf**, **mk**, and **inst**, it will find a colon or '*u*' on which to stop.

c(A, B, a)

Finds the leftmost symbol marked with 'a' and copies it at the end of the tape in the first non-erasing square available, then goes to state A. If no symbol 'a' is found goes to state B. The symbols which this routine (and those that use it) will be required to copy are 'D', 'C', and 'A'. This means that each invocation needs three different states. Uses **f'**(A, B, a) and **pe**(A, b). Used by **ce**(A, B, a).

ce(A, B, a)

Copies at the end of the tape in the first non-erasing square the leftmost symbol marked by 'a', then goes to state A with the single (leftmost) marking 'a' erased. If there is no symbol 'a' on the tape, goes to state B. Uses **c**(A, B, a) and **e**(A, B, a). Used by **ce**(B, a).

ce(B, a)

Copies in the correct sequence, at the end of the tape in non-erasing squares, all the symbols on the tape that are marked with 'a', at the same time erasing each 'a', and then goes to state B. If there are no symbols marked with 'a' it goes straight to B. Uses **ce**(A, B, a). Used by **ce₂**(B, a, b), **ce₃**(B, a, b, c), **ce₄**(B, a, b, c, d), and **ce₅**(B, a, b, c, d, e).

ce₅(B, a, b, c, d, e)

Copies in the sequence given, at the end of the tape in non-erasing squares, all the symbols marked with 'a', then those marked with 'b', then 'c', 'd', and 'e' in turn, ending in state B. Uses **ce₄**(B, a, b, c, d) and **ce**(B, a) and **ce₄** uses **ce₃** which uses **ce₂** and all of these use **ce**(B, a). Only **ce₅**(B, a, b, c, d, e) is used elsewhere, by **inst**.

cp(A, B, E, a, b)

Compares the leftmost symbols marked by 'a' and 'b'. First it finds the symbol marked 'a', by using the routine f'(A, B, a). It enters a different state $cp_2(A, B, x)$ according to the symbol 'x' it finds there. There are three possible symbols: 'D', 'C', and 'A'. Then it finds the symbol marked 'b' in the same way. If they are the same, the resultant state is A, if not the state becomes B. If one of these marked symbols is found, but not the other, the outcome is state B. If neither is found, the outcome is E. Uses f(A, B, a) and f'(A, B, a). Used by cpe(A, B, E, a, b). (I have changed Turing's state-symbol 'U' to 'B' to avoid confusion.)

cpe(A, B, E, a, b)

Action as for cp(A, B, E, a, b) followed by, if the marked symbols are the same, the erasure of both the markings 'a' and 'b'. Uses cp(A, B, E, a, b) and e(A, B, a). Used by cpe(B, E, a, b).

cpe(B, E, a, b)

Comparison of two marked sequences. First it compares the leftmost symbols marked with 'a' and 'b'. If they differ the state B is reached and the process stops. If they are both absent, state E. Otherwise both marked symbols are erased and the process is repeated with the next leftmost marked symbols. So, if the whole sequence of symbols marked with 'a' equals the sequence marked with 'b' (or both are absent) the result is state E and all markings 'a' and 'b' have been removed. If the sequences differ, state B is reached and some of the markings have been removed. Uses cpe(A, B, E, a, b). Used by kmp.

con(A, a)

This routine's action depends on where it starts on the tape. It leaves the machine in a significant position after its action is completed. The purpose is to mark with symbol 'a' the M-S pair next on the right of the start position. The routine must start on a non-erasing square. It seeks a pattern such as:

$$D - A - A - \ldots A - D - C - C - \ldots C -$$

and will replace all the blanks (or any other symbols in these places) by the symbol 'a'. There may be as few as one 'A' after the first 'D', representing the state, and optionally no 'C' following the second 'D', representing the symbol. The starting point can be on the first 'D' of this pattern, or earlier if no other 'A' symbols intervene. For example, in an image there is only one state, so the first 'A' symbol can be sought from anywhere to its left in the image string. The final state is A and the position is two non-erasable squares to the right of the last marked symbol. (Turing refers to this as 'the last square of C' but his own example shows a symbol image with no Cs. Turing's comment 'C is left unmarked' does not seem to make sense.) Uses none. Used by **anf, kom, sim,** and **mk**. The use of **con**(A, a) in **sim** and **mk** employs the final position it reaches.

7. Operation of the Universal Machine U

The routines which comprise the operation of U are entered in succession, except for the process of searching for the relevant instruction, which has its own loop. The initial state of the machine is **b**, remembering that the instruction table for the Turing machine which U is emulating must already be on the tape. The starting position is immaterial.

b

The beginning of U's operation. It writes the symbols ': D A' in the non-erasing squares after the symbol :: that signifies the end of the instructions which are already on the tape. This is the image of the initial state of the emulated machine T and consists of the coding for 'state 1' with no following symbol images, meaning that the initial tape of T is blank.

This operation uses **f**(A, B, a).

anf

Presumably this is from the German *Anfang*, or beginning. It is the start of the process of generating the next image of T. The process will return to **anf** when one new image has been appended on U's tape, so that building of successive images continues. Its action is **q**(**anf**$_1$, :) which finds the last ':' and then **anf**$_1$ uses **con**(**kom**, y) to mark the last M-S pair on the tape with 'y' and go to state **kom**. Thus the machine state and scanned symbol of the last image on the tape have been marked. Initially this results in marking just the 'D' 'A', but there is an error in this design because absence of a symbol image means that **con** will fail, since it looks only for symbols 'A' or 'D' when in its internal state **con**$_1$. This is easily corrected: see Section 8.

This operation uses **con**(A, a) and **q**(A, a).

kom

From its position in the image region, the machine moves left, looking for either ';' or 'z'. It will find ';' at the start of the last instruction, provided that the termination of the instruction area is shown as just '::' and not '; ::' (the latter is implied by Turing). Also, if all instructions are to be available, the first instruction must begin with ';'. The correct designation of a single instruction should be ';' followed by the five parts, and not as shown on p. 68 of Turing's treatment. With these changes, **kom** will ignore any ';' which is marked with 'z'. 'z' signifies that the instruction which follows has already been tried. The rightmost unmarked ';' having been found, this symbol is marked with 'z' and the routine **con**(kmp, x) is used to mark the following M-S pair with 'x'. Each time that **kom** is used, the next instruction to the left will be processed.

If no instruction matches the current state and symbol of T, meaning that the machine is badly defined, the search for a colon will run off the left end of the tape. This bug is fixed in Section 9.

This operation uses **con**(A, a).

kmp

The action of **kmp** is shown by Turing as **cpe**(**e**(**kom**, x, y), **sim**, x, y). This will compare the sequences marked by 'x' and 'y' to discover if the marked instruction actually applies to the M-S pair shown in the current image of T. If it does, the state becomes **sim**, which is the start of building the next image. If not, there is a problem with partial erasure of the markings, so these are erased by the **e**(**kom**, x, y) operation and we try again, this time trying the next instruction to the left of those tried so far, which have the 'z' marking. However, there is a bug, explained and corrected in Section 8.

This operation uses **cpe**(B, E, a, b) and **e**(A, B, a).

sim

This routine marks the parts of the leftmost instruction marked with 'z', which now applies to the next of T's operations to be performed. The leftmost marked colon is located by $\mathbf{f'}(\mathbf{sim}_1, \mathbf{sim}_1, z)$ and then $\mathbf{sim}_1$ is $\mathbf{con}(\mathbf{sim}_2, \ )$, which marks the M-S pair with blanks, since this is not required again in this round. The routine **con** leaves the machine scanning the next non-erasing square to the right of the 'D' in the coding for the new symbol to be written by T. The 'D' is marked with 'u' as are any Cs which follow, so that the new symbol image is marked. The marking with 'u' continues, marking to the left of each square being examined until an 'A' is found, when the marking is changed to 'y'. Consequently, both the new symbol image and the action (L, R, or N) have been marked with 'u' and the new state image has been marked with 'y', which continues until the ';' or '::' terminating the instruction has been reached. Then the 'z' markings are all removed, since the relevant parts of the instruction for the next stage of T's operation have been marked. The line on p. 71 for state $\mathbf{sim}_2$ and symbol 'not A' has an error; see Section 10.

This operation uses **e**(B, a), **con**(A, a), and **f'**(A, B, a).

mk

The last ':' is found by means of an operation which should read $\mathbf{q}(\mathbf{mk}_1, :)$ (see Section 10). Moving right from there, the first 'A' is found, which is at the start of the state image, then two non-erasable squares to the left of this point is the end of the preceding symbol image. The start of the whole image might have been found (':'), if this was the initial state, which finishes this part of the marking. If a 'C' is found, this is marked with 'x' and so are previous Cs moving backwards until a 'D' is found and marked. From this point backward, symbols in non-erasable squares are marked with 'v' until the ':' that begins the image is found. In this way the symbol image preceding the one currently scanned (if any) has been marked with 'x' and all the earlier symbols in the image (if any) are marked with 'v'.

In the second stage of marking, the **con**(A, a) routine is used to find the M-S pair, marking it with a blank and ending two non-erasable squares beyond the last marked 'D' or 'C' of the symbol image. Two left shifts leave the machine at the start of the symbol image which immediately follows the scanned symbol. From there, all symbols are marked with 'w' until the end is reached and finally a ':' is placed at the end of the old image, ready for copying out the new image later.

It may be useful to summarize the marking which is now ready for copying out the new image. The instruction has been marked with 'u' for both the new symbol and the action (R, L, or N) then with 'y' for the new state. The image has been marked in three places. Starting from the left, its symbol images are marked with 'v' until the one before the scanned symbol, which is marked with 'x'. Then the state and symbol scanned, the current M-S pair, are unmarked, because they will be superseded. To the right of the scanned symbol, all symbol images are marked with 'w'.

This operation uses **con**(A, a), 1(A), and **q**(A,).

sh

Before the copying of these marked regions to make a new image, the output of machine T is made visible on the tape. The part marked with 'u' begins with the new symbol (S') to be written on T's tape, which may be 0, coded as D C, or 1, coded as D C C. It is immediately preceded in the instruction by the old symbol that is being overwritten (S). The 0 or 1 is considered as being written if it replaces a space, which is coded as D, so **sh** looks for D–D*u*C*u* or D–D*u*C*u*C*u*, using the '*u*' marking to locate this combination. According to which it finds, it writes either ':−0' or ':−1' at the end of the tape.

The convention that 0 or 1 is *output* if it replaces a space implies that the target machine T conforms to the same convention as U, which is not to overwrite most symbols, only treating special markers in this way. This is in fact an unnecessary restriction on the programming of T. The restriction means that U is not really universal. On the other hand, the use of 0 and 1 as special symbols indicating output goes beyond the fundamental definition of a Turing machine. We must assume that the **sh** operation is designed for use in the special case of a target Turing machine with output, the output action consisting of writing **0** or **1** in a square which previously held a space.

This operation uses f(A, B, a) and **pe**$_2$(A, a, b).

inst

This routine creates the new image of T's machine and tape and writes it at the end of U's tape, after the final colon. The operation $q(1(\textbf{inst}_1), u)$ finds the last 'u' on the tape and the square that it marks, which contains L, R, or N. This single 'u' is deleted because the action symbol L, R, or N must not be copied into the new image. According to the action symbol found, the parts are assembled in one of these sequences, where the machine state 'y' is shown in bold.

$$\begin{array}{cccccc} R & v & x & u & \textbf{y} & w \\ N & v & x & \textbf{y} & u & w \\ L & v & \textbf{y} & x & u & w \end{array}$$

In each case the symbols *preceding* the active part are marked 'v' and are copied first (they may not exist). The symbols *following* the previously scanned symbol are marked 'w' and are copied last. Without a movement of machine T, the new state, marked 'y', is followed by the new symbol, marked 'u'; and the previous symbol, marked 'x', stays in its place. With left movement, the new state is placed before the previous symbol. With right movement, the new state is placed after the new symbol, marked 'u'.

The tape image should always end with a 'blank' symbol, which is simply D. Any rewritten symbols within the used portion of T's tape which are deleted will have been overwritten with 'D', but at the end, if the section marked with ws was empty, the action R may leave the state image at the end of the tape. This will cause a matching failure during the next cycle of the emulation if comparison occurs with the M-S pair of an instruction that has a symbol value of 'blank', represented by D. Repairs are made in Section 8.

This operation uses $1(A)$, $q(A, a)$, $f(A, B, a)$, and $ce_5(B, a, b, c, d, e)$.

ov

This final operation $e(\textbf{anf})$ clears all markings and returns to **anf** to begin once again the process of generating a new tape image. Since the 'z' markings were cleared by **sim** and the ce_5 operation clears all its markings, there seems no need for **ov**, but it does no harm.

This operation uses $e(A)$.

8. The Interesting Errors

The first phase of an evolution of T is to find the relevant instruction. This is done by marking the current state-symbol pair of T with 'y' and the state-symbol pair of an instruction with 'x', then using the **cpe** operation to compare the marked strings. The process of **cpe** deletes some of the 'x' and 'y' markings. When comparison fails on one instruction the machine moves on to the next. This comparing process is shown in the table for U on p. 71 of Turing's paper as

$$\textbf{kmp} \quad \text{cpe}(\text{e}(\textbf{kom}, x, y), \textbf{sim}, x, y)$$

The **e** operation is intended to delete all the remaining 'x' and 'y' markings. In fact this is not quite how erasure works as defined on p. 64 and the correct form would be e(e(**kom**, x), y). But there is a more serious error in returning to **kom**, since the essential 'y' marking will not be restored. Returning to **anf** will repair this error. The correct definition of **kmp** should be

$$\textbf{kmp} \quad \text{cpe}(\text{e}(\text{e}(\textbf{anf}, x), y), \textbf{sim}, x, y)$$

To introduce the second of these interesting errors, it is instructive to look at the penultimate step in the copying out of the evolved new snapshot of T. This has been reduced, by Turing's clever scheme of skeleton tables, to a choice of one of three copy instructions on p. 72, such as, for example

$$\textbf{inst}_1(R) \quad \text{ce}_5(\text{ov}, v, x, u, y, w)$$

This copies five marked areas from the current instruction and the last image of T in the sequence to create the new image, for the case where the machine moves right. The part marked 'y' is the new machine state, 'v' and 'x' form the string of T-symbols to the left, 'u' is the newly printed symbol, and 'w' the string of T-symbols to the right. Because 'u' replaces an existing symbol, the number of symbols (including blanks) on T's conceptual tape has not changed! The same is true for left and null movement. There must be something wrong in an emulation in which the emulated machine can never change the number of symbols on its tape.

The image shows just the occupied part of T's tape, and this is conceptually followed by an unlimited set of blank symbols, which are the tape as yet unused. The number of symbols in the image of T will increase by moving right from the last occupied square and writing on the blank square. After a move right onto blank tape, there will be no string marked 'w', so machine state 'y' will be the last thing in the image.

This will lead to a failure of the emulation at the next evolution because the state-symbol pair of the image, due to be marked with 'y' during the search for the relevant instruction, is incomplete.

The remedy is to print a new blank symbol for T at the end of the image, when the move has been to the right and there is no T-symbol there. The necessary corrections, on p. 72, to the table for U are:

inst$_1$ (R)			ce$_5$(q(inst$_2$, A), v, x, u, y, w)
inst$_2$		R, R	inst$_3$
inst$_3$	{ none	PD	ov
	D		ov

In the case of a move right, after copying the parts of the previous image, the operation **q** finds the last 'A' on the tape, which is the end of the state-symbol copied from markings y. If there is a T-symbol to its right, there is no problem. After two right moves of U, if a 'D' is found there is a T-symbol but, if not, by printing 'D', a new blank tape square is added to the image of T. In this way T's conceptual tape is extended and the state-symbol pair is made complete.

The error perhaps arose because the endless string of blank symbols on U's tape was taken as sufficient for the purpose of T. But for the emulation a blank square is shown as D. Machines U and T represent a blank tape differently.

There is a corresponding error in the way the initial state of T is placed on U's tape. It should contain the U-symbols: DAD with suitable spaces between them, representing T's initial state DA followed by the scanned symbol D, a blank. The correction on p. 70 is:

b_1 R, R, P:, R, R, PD, R, R, PA, R, R, PD   **anf**

9. Diagnostics

With experience of writing programs it is second nature to build in diagnostics. Whether they are needed in U is arguable. Since U is a conceptual tool, its requirements are determined by its use in the argument of Turing's paper. For testing the design of U, diagnostics are certainly needed.

There may be a need for two kinds of failure indication in the program of U.

Suppose that T has a deficient set of instructions, meaning that its latest image has a state-symbol pair which does not appear among the instructions. I believe that Turing would class this as a *circular* machine. The effect on the operation of U is that the search for the relevant instruction fails with U moving left beyond the left-hand end of its tape, and continuing to move left indefinitely. Perhaps this is acceptable for the purpose for which U was intended, but it seems anomalous that a deficiency in T should cause U to misbehave. It can be avoided by adding a line to the definition of **kom** on p. 71:

kom   e   **fail**$_1$ (deficient T instructions)

then changing the next line to respond to symbols *not z nor ; nor e*.

If T moves right without limit, this will be emulated correctly, but moving left beyond the limits of its tape is a problem. The way U works will cause the next image of T to appear as if no shift had occurred. There is no way to represent T as scanning a square to the left of its starting position. This means that the subsequent behaviour of T will differ from what its instructions imply. I think this might affect the use that Turing made of the machine in the main part of the paper.

The changes to deal with this problem are:

inst$_1$(L)   f(**inst**$_4$, **fail**$_2$, x)   (machine T has run off left)
inst$_4$ ce$_5$(ov, v, y, x, u, w)

10. Trivial Errors and Corrections

1. There is potential confusion in the use of the symbol **q** for different states in two places, and it is also confused with state **g**. The best resolution is as follows.

    We can treat the use of **q** in the example on p. 62 as casual, without permanent significance. The same might be said of its use on p. 64, which is unrelated. But from there onwards the examples will form part of the definition of U, so the symbols have global significance.

    On p. 66, the states **q** and q_1 appear but, in their subsequent uses in U, they have been replaced by **g**, for example in the definitions of **anf**, **mk**, and **inst**. I have retained the notation **q**, while remembering that previous uses of this symbol are unrelated.
2. The skeleton tables for **re** and **cr** on p. 65, which comprise five different states, are redundant, serving no illustrative purpose and not being used again.
3. On p. 68, the format of the instruction table of T, as written on the tape of U, is described. Instructions are separated by semicolons. An example DADDCRDDA;DAA...DDRDA; is given. As already explained, this is misleading, because each instruction should be preceded by a semicolon. The example should begin with a semicolon, not end with one.
4. In the explanation of the skeleton table for **con** (p. 70), 'C' is one of the symbols being read and marked. But the words refer to 'the sequence C of symbols describing a configuration'. The final remarks '... to the right of the last square of C. C is left unmarked.' use 'C' in the second sense. It would otherwise seem as if the final symbol 'C' was left unmarked, but this is not so. To clarify, replace by 'the sequence S' and '... last square of S. Configuration S is left unmarked.'
5. On p. 71, a line for sim_2 should read:

    sim_2 not A   L, Pu, R, R, R sim_2
6. On p. 71, the line for **mk** should read:

    mk $q(mk_1, :)$
7. On p. 72, the line for $inst_1(N)$ should read:

    $inst_1(N)$ $ce_5(ov, v, x, y, u, w)$

11. A Redesign of the Universal Machine

To verify, as far as this is possible, that there are no remaining errors in the amended version of Turing's program for U, it would be best to generate the explicit machine instruction table by substitutions and repetitions, then run this machine with one or more examples of a machine T and find if the emulations behaved as they should. But the complexity and slowness of the explicit form of Turing's U makes this difficult.

Therefore I made some changes to the design of U before constructing a simulation of a Turing machine, loading the instructions for U, producing a tape image for a machine T and running the program. After some corrections to my version of U, the simulation behaved correctly. In this section the main features of the redesign are described.

The new version of U follows Turing's methods quite closely. The substitution process introduced with the skeleton tables had been nested to a depth of 9, causing a proliferation of states and instructions in the explicit machine. To avoid this, no skeleton tables were used in the new version and this allowed the procedures to be optimized for each application. The downside is that the 'low-level' description of U which results takes up more space than the original and is harder to understand and check for accuracy. There are 147 states and 295 instructions in the new version, an enormous reduction.

The representation of T's states and symbols in a monadic notation such as DAAAA was replaced by a binary notation. This was an easy change that reduced the length of the workspace used. Because nearly all the time is used moving from end to end of the workspace, this is worthwhile. The small cost of the change is that there are four U-symbols to represent states and symbols instead of three.

The classic Turing machine can move right or left or stay put in each operation. To simplify U a little, the third option was removed, so that a left or right movement became mandatory. For consistency, U was also run on a machine T with this characteristic. In the whole of U's program, a compensating movement became necessary only a few times, so it is not a significant restriction.

Turing's skeleton tables show, for the scanned symbol, such words as *any*, or *not A*. When translated into a list of discrete symbols for the explicit machine these generate many instructions. By introducing a 'wild card' notation and searching instructions in a definite sequence, this proliferation can be avoided. A form of instruction was added which, in its written form, had an asterisk for both the scanned symbol and the written symbol. This acted on any scanned symbol and did not overwrite it. The way that U worked would have made it possible to read a wild card (i.e. any) scanned symbol and write over it or to read a specific symbol and leave it unchanged, but these were never needed in practice. The wild card scanned symbol should only be actioned after all other possibilities (for this particular state) have been tested. Therefore instructions now have a defined sequence and must be tested accordingly. U always did test instructions in sequence but never made use of that fact.

Testing all instructions in sequence to find a match is very time-consuming because it requires marking, then comparing square by square, running from instruction space to work space. It was largely avoided by writing in U's instruction table an *offset* which indicated where the next instruction could be found.

This indication led to a section of instructions dealing with a given state; after this, sequential testing took place. This was a shortcut to speed up U and was not envisaged as a feature of all Turing machines, since it would greatly complicate U. Technically it was a little more complex than I have described, but it has no effect on the design of U, being merely a chore for the programmer and a detail of the computer program which interprets those instructions.

U spends some of its time searching for a region on the tape where it will begin work. To make this easier, additional markers were introduced, for the action symbol (L or R) and for the start of the current snapshot. Also, the end of the workspace was marked, and this marking was placed in one of the squares normally reserved for permanent symbols. Since it had to be overwritten when the workspace extended, this broke one of Turing's conventions.

Finally the two failure-indications described earlier were incorporated, one for a deficient T-instruction set and the other for T running its machine left, beyond the usable tape.

11.1 Testing the redesigned machine

A computer program, which I shall call T*, was written which would simulate the underlying Turing machine, using a set of instructions in its own special code, which had one byte per symbol or state. This code was chosen for convenience of writing U's instructions. It incorporated the wild card feature and the offset associated with each instruction, but the offset did not alter the way it responded to its instructions, only making it faster. When the design of U is complete and its instructions have been loaded, T* will behave as the universal machine U.

A simple editor was written to help the user write and amend the instruction tables for T* and prepare a starting tape for T* which holds the coded instructions for the emulated Turing machine T.

For T, the example given by Turing on p. 62 was used. It prints a sequence of increasing strings of ones, such as 001011011101111011111... This program in its explicit form would have 23 instructions and 18 states. To make it simpler, it was rewritten without the 'alternate squares' principle and it then had 12 instructions and 6 states. It may be interesting to see how the wild card feature operates by studying this example, shown below.

As a first step, the example was loaded into the program space of T* and run, thus testing the mechanism of T* as well as the example in the table below.

Then the example was coded for the initial part of the tape of T*, so that it would cause U to emulate it as the target T. The program of U was loaded in many stages, debugging each by testing its part in the whole operation of U. Two serious program errors were found. One was in the operation **sh** which prints the output of T between the snapshots of T's evolution. The other was in the correction to Turing's scheme which wrote a blank symbol (D) at the end of the tape. It had been inserted at the wrong place. With these and several minor errors corrected the redesigned U performed as expected and the evolution of T agreed with expectation and with its earlier running, directly on T*. Only this one example of T was tried, but it probably does test the universal machine fully. The full results are given below.

Because of the differences between the version of U that was tested and Turing's design with my corrections, the testing must be regarded as incomplete. A compiler could be written to take the design in the form of skeleton tables and generate the explicit machine, which could then be run to emulate examples of the target machine. This would be extraordinarily slow.

12. The Program for T

The instructions for T are given in the standard five-part form: state, scanned symbol, written symbol, movement, and resultant state. The images are shown for the first eleven moves, in the standard form with the state-symbol (a to e, printed bold) preceding the scanned symbol.

The blank space symbol is a hyphen and the other symbols are 0, 1, *x*, and *y*.

The program writes a block of *x*s followed by a *y*, then converts the *x*s successively to 1s and the *y* to a 0, while writing the next block of *x*s and a *y*, increasing the number of *x*s by one.

s	–	0	R	a	print 0	: 0 a –	
a	–	y	R	b	print y at end	: 0 y b –	: 0 0 x a –
a	*	*	R	a			: 0 0 x y b –
b	–	x	L	c	print x at end	: 0 c y x	: 0 0 x c y x
b	*	*	R	b			
c	y	y	L	d	run back to y	: d 0 y x	: 0 0 d x y x
c	*	*	L	c			
d	x	1	R	b	change x to 1		: 0 0 1 b y x
d	0	0	R	e	none left	: 0 e y x	
d	*	*	L	d			
e	y	0	R	a	change y to 0	: 0 0 a x	
e	*	*	R	e			

13. Results of the Test

Here is a copy of the symbols on the tape of T* after 22 evolutions of U. The part up to the symbol % represents the 12 instructions for T. Then follow the 23 images, separated by colons. Whenever U prints a 0 or 1, this is also an output of T. To make this explicit (following Turing's practice) the strings '1 :' or '0 :' are inserted into the tape (bold in our table). So the whole set of evolutions shown has printed '0 0 1 0'. The tape shown is printed on alternate spaces, except for the initial 'e e'. The final F is a device of my own to make it easy to find the end of the written area of tape.

e e ; M C S S C R M D ; M D S S C D R M C C ; M D S E S E R M D ;
M C C S S C C L M C D ; M C C S E S E R M C C ; M C D S C D S C D L M D C ;
M C D S E S E L M C D ; M D C S C C S D R M C C ; M D C S C S C R M D D ;
M D C S E S E L M D C ; M D D S C D S C R M D ; M D D S E S E R M D D % :
M C S : **0** : S C M D S : S C S C D M C C S : S C M C D S C D S C C :
M D C S C S C D S C C : S C M D D S C D S C C : **0** : S C S C M D S C C :
S C S C S C C M D S : S C S C S C C S C D M C C S :
S C S C S C C M C D S C D S C C : S C S C M D C S C C S C D S C C : **1** :
S C S C S D M C C S C D S C C : S C S C S D S C D M C C S C C :
S C S C S D S C D S C C M C C S : S C S C S D S C D M C D S C C S C C :
S C S C S D M C D S C D S C C S C C : S C S C M D C S D S C D S C C S C C :
S C M D C S C S D S C D S C C S C C : S C S C M D D S D S C D S C C S C C :
S C S C S D M D D S C D S C C S C C : **0** : S C S C S D S C M D S C C S C C :
S C S C S D S C S C C M D S C C : S C S C S D S C S C C S C C M D S F

As an aid to understanding this tape, here are the symbols and states of T in U's notation:

–	S	s	MC
0	SC	a	MD
1	SD	b	MCC
x	SCC	c	MCD
y	SCD	d	MDC
*	SE	e	MDD

The first few snapshots therefore read:

s – : 0 : 0 a – : 0 y b – : 0 c y x : d 0 y x : 0 e y x : 0 : 0 0 a x : 0 0 a x – :
0 0 x y b – : 0 0 x c y x : 0 0 d x y x : 1 : 0 0 1 b y x : 0 0 1 y b x : 0 0 1 y x b – :
0 0 1 y c x x : 0 0 1 c y x x :

The final configuration of the above tape is 0 0 1 0 x x a – :

14. The Corrected Tables for U: Summary

The table for f(A, B, a) is unchanged on p. 63.

On p. 64, e(A, B, a) and e(B, a) are unchanged, but note that the state q used in the explanation of e(B, a) is a local notation, unrelated to the states of that name on p. 66.

Corrections and Critiques | 123

On p. 65, pe(A, b), l(A), f′(A, B, a), and c(A, B, a) are unchanged, but r(A) and f″(A, B, a), defined on that page, are not used again.

On pp. 65–66, ce(A, B, a), ce(B, a), cp(A, B, C, a, b), cpe(A, B, C, a, b), and cpe(A, B, a, b) are unchanged but re(A, B, a, b), re(B, a, b), cr(A, B, a), and cr(B, a) are not used again.

On p. 66, q(A), pe$_2$(A, a, b), and e(A) are unchanged. Also, ce$_2$(B, a, b) and ce$_3$(B, a, b, c) are defined, but it is ce$_5$(B, a, b, c, d, e), derived in an analogous way, which is actually used, in the **inst** function.

On p. 70, con(A, a) is unchanged, but the remark that 'C is left unmarked' is confusing and is best ignored.

In the table for U, which begins on p. 70, the state b_1 should have the following action: R, R, P :, R, R, PD, R, R, PA, R, R, PD, in order to print ': D A D' on the F squares, so that a blank symbol D is available for matching with an instruction.

On p. 70 the table for **anf** should lead to q(**anf**$_1$, :).

If the set of instructions for the target machine T is deficient, so that a state-symbol pair is created which has no matching instruction, machine U will attempt to search beyond the left-hand end of its tape. What happens then is undefined. To make it definite, **kom** (p. 71) can be augmented by the line:

kom   e   fail$_1$,

which indicates the failure, and the last line will be:

kom   not z nor ; nor e   **kom**

The table for **kmp** (p. 71) should read:

kmp   cpe(e(e(anf, x), y), sim, x, y),

since e(A, B, a) should return to **anf**, to restore the markings deleted by **cpe**.

On p. 71, **sim**$_2$ with scanned symbol 'not A' should have the action L, Pu, R, R, R.

The first line of **mk** (p. 71) should lead to q(**mk**$_1$, :). On this same page, **sh** is unchanged.

On p. 72, **inst** should lead to q(l(**inst**$_1$), u) and the line for **inst**$_1$(N) should read

inst$_1$(N)   ce$_5$(ov, v, x, y, u, w)

The instruction for **inst**$_1$(L) (p. 72) could try to move the target machine left beyond its end of tape, but there is no way for U to represent this condition, so T will seem not to move. To make this kind of error explicit, these changes can be made:

inst$_1$(L)   f(**inst**$_4$, **fail**$_2$, x)

inst$_4$ ce$_5$(ov, v, y, x, u, w)

To correct the fundamental flaw that a right movement **inst**$_1$(R) (p. 72) could move the state-symbol to the right of all other symbols, making a future match with an instruction impossible, the following change is needed:

inst$_1$ (R)		ce$_5$(q(**inst**$_2$, A), v, x, u, y, w)	finds the last A on the tape
inst$_2$	R, R **inst**$_3$		move to start of scanned symbol
inst$_3$	$\begin{cases} \text{none} & \text{PD} \\ \text{D} & \end{cases}$ ov ov		if blank space, print D but not if a symbol follows

Finally, **ov** (p. 72) is unchanged.

CHAPTER 3

Systems of Logic Based on Ordinals (*1938*)

Alan Turing

Introduction

Jack Copeland

The Princeton Years, 1936–38

On 23 September 1936 Turing left England on a vessel bound for New York.[1] His destination was Princeton University, where the Mathematics Department and the Institute for Advanced Study combined to make Princeton a leading centre for mathematics. Turing had applied unsuccessfully for a Visiting Fellowship to Princeton in the spring of 1935.[2] When a year later he learned of Church's work at Princeton on the *Entscheidungsproblem*, which paralleled his own (see 'Computable Numbers: A Guide'), Turing 'decided quite definitely' to go there.[3] He planned to stay for a year.

In mid-1937 the offer of a Visiting Fellowship for the next academic year persuaded him to prolong his visit, and he embarked on a Ph.D. thesis. Already advanced in his academic career, Turing was an unusual graduate student (in the autumn of 1937, he himself was appointed by Cambridge University to examine a Ph.D. thesis). By October 1937 Turing was looking forward to his thesis being 'done by about Christmas'. It took just a little longer: 'Systems of Logic Based on Ordinals' was accepted on 7 May 1938 and the degree was awarded a few weeks later.[4] The following year the thesis was published in the *Proceedings of the London Mathematical Society*.

'Systems of Logic Based on Ordinals' was written under Church's supervision. His relationship to Turing—whose formalization of the concept of an effective

---

[1] S. Turing, *Alan M. Turing* (Cambridge: Heffer, 1959), 51.

[2] Letter from Turing to Sara Turing, his mother, 24 May 1935 (in the Turing Papers, Modern Archive Centre, King's College Library, Cambridge (catalogue reference K 1)).

[3] Letter from Turing to Sara Turing, 29 May 1936 (Turing Papers, catalogue reference K 1).

[4] Letter from Turing to Sara Turing, 7 May 1938 (see below); Turing, *Alan M. Turing*, 54. The thesis is held in the Seeley G. Mudd Manuscript Library at Princeton University (catalogue reference P685.1938.47).

procedure and work on the *Entscheidungsproblem* was 'possibly more convincing' than Church's own[5]—was hardly the usual one of doctoral supervisor to graduate student. In an interview given in 1984, Church remarked that Turing 'had the reputation of being a loner' and said: 'I forgot about him when I was speaking about my own graduate students—truth is, he was not really mine.'[6] Nevertheless Turing and Church had 'a lot of contact' and Church 'discussed his dissertation with him rather carefully'.[7] Church's influence was not all for the good, however. In May 1938 Turing wrote:

> My Ph.D. thesis has been delayed a good deal more than I had expected. Church made a number of suggestions which resulted in the thesis being expanded to an appalling length. I hope the length of it won't make it difficult to get it published.[8]

Moreover, Turing elected to couch 'Systems of Logic Based on Ordinals' in the notation of Church's lambda calculus, so making his work much less accessible than it might otherwise have been. (By that time even Church's student Kleene, who had contributed importantly to the development of the lambda calculus, had turned away from it. Kleene said: 'I myself, perhaps unduly influenced by rather chilly receptions from audiences around 1933–35 to disquisitions on λ-definability, chose, after general recursiveness had appeared, to put my work in that format.'[9]) In a letter written not long after Turing's death, Turing's friend Robin Gandy said: 'Alan considered that his paper on ordinal logics had never received the attention it deserved (he wouldn't admit that it was a stinker to read).'[10]

Notwithstanding its notational obscurity, 'Systems of Logic Based on Ordinals' is a profound work of first rank importance. Among its achievements are the exploration of a means of circumventing Gödel's incompleteness theorems; the introduction of the concept of an 'oracle machine', thereby opening the field of relative computability; and, in the wake of the demolition of the Hilbert programme (by Gödel, Turing, and Church), an analysis of the place of intuition in mathematics and logic.

Turing's two years at Princeton are the best documented of his life, thanks to a series of letters that he wrote to Sara Turing. (Of the fifty-five letters that he sent her from 1932 until his death, twenty-seven are from the Princeton period.) The following excerpts give a glimpse of his time there. All were written from the Graduate College, Princeton University.[11]

---

[5] See the subsection 'Church's contribution' of 'Computable Numbers: A Guide'.

[6] Church in interview with William Aspray (17 May 1984); transcript no. 5 in the series 'The Princeton Mathematics Community in the 1930s', Princeton University.

[7] Ibid.

[8] See below.

[9] S. C. Kleene, 'Origins of Recursive Function Theory', *Annals of the History of Computing*, 3 (1981), 52–67 (62).

[10] Letter from Gandy to Max Newman, n.d. (Turing Papers, catalogue reference A 8).

[11] All the letters are in the Turing Papers (catalogue reference K 1).

Excerpts from Turing's Letters Home

6 October 1936[12]

I reached here late last Tuesday evening.[13] We were practically in New York at 11:00 a.m. on Tuesday but what with going through quarantine and passing the immigration officers we were not off the boat until 5:30 p.m. Passing the immigration officers involved waiting in a queue for over two hours with screaming children round me. Then, after getting through the customs I had to go through the ceremony of initiation to the U.S.A., consisting of being swindled by a taxi driver. I considered his charge perfectly preposterous, but as I had already been charged more than double English prices for sending my luggage, I thought it was possibly right. However, more knowing people say it was too much. ...

The mathematics department here comes fully up to expectations. There is a great number of the most distinguished mathematicians here. J. v. Neumann, Weyl, Courant, Hardy, Einstein, Lefschetz, as well as hosts of smaller fry. Unfortunately there are not nearly so many logic people here as last year. Church is here of course, but Gödel, Kleene, Rosser and Bernays who were here last year have left. I don't think I mind very much missing any of these except Gödel. Kleene and Rosser are, I imagine, just disciples of Church and have not much to offer that I could not get from Church. Bernays [I] think is getting rather 'vieux jeu': that is the impression I get from his writing, but if I were to meet him I might get a different impression.

The graduate students include a very large number who are working in mathematics, and none of them mind talking shop. It is very different from Cambridge in that way.

I have seen Church two or three times and I get on with him very well. He seems quite pleased with my paper[14] and thinks it will help him to carry out a programme of work he has in mind. I don't know how much I shall have to do with this programme of his, as I am now developping [sic] the thing in a slightly different direction, and shall probably start writing a paper on it in a month or two.[15] After that I may write a book.

The proofs[16] have been sent direct to me here. They arrived last Saturday, and I have just finished them and sent them off. It should not be long now before the paper comes out. I have arranged for the reprints[17] to be sent to you, and will get you, if you would not mind, to send out the ones that are to go to people in Europe, and to send some of the remainder on to me. ...

These Americans have various peculiarities in conversation which catch the ear somehow. Whenever you thank them for any thing they say 'You're welcome'. I rather liked it at first, thinking that I was welcome, but now I find it comes back like a ball thrown against a wall, and become positively apprehensive. Another habit they have is to make the sound

---

[12] Editor's note. Sara Turing's dating of the letters is followed where dates are absent or incomplete.

[13] Editor's note. Turing's previous letter to Sara on 28 Sept. 1936 was written on board the vessel *Berengaria* bound for New York.

[14] Editor's note. Presumably 'On Computable Numbers, with an Application to the Entscheidungsproblem'.

[15] Editor's note. This may refer to the 'development of the theory of functions of a real variable' mentioned by Turing on p. 58 of Chapter 1. No such paper ever appeared, nor a book.

[16] Editor's note. Here the letter is marked '"On Computable Numbers"' in Sara's hand.

[17] Editor's note. The author's copies of 'On Computable Numbers'.

described by authors as 'Aha'. They use it when they have no suitable reply to a remark, but think that silence would be rude.

Maurice Pryce has just got a Fellowship at Trinity.[18]

14 October 1936[19]

I have just discovered a possible application of the kind of thing I am working on at present. It answers the question 'What is the most general kind of code or cipher possible', and at the same time (rather naturally) enables one to construct a lot of particular and interesting codes. One of them is pretty well impossible to decode without the key and very quick to encode. I expect I could sell them to H. M. Government for quite a substantial sum, but am rather doubtful about the morality of such things. ...

Church had me out to dinner the other night. Considering that the guests were all university people I found the conversation rather disappointing. They seem, from what I can remember of it, to have discussed nothing but the different States that they came from. Description of travel and places bores me intensely.

I had a nasty shock when I got into Church's house. I think I had told you that Church was half blind in one eye. Well I saw his father in the house and he was quite blind (and incidentally very deaf). I should have thought very little of it had it not been for Church being rather blind himself. Any hereditary defects of that kind give me the shudders.

Hardy is here for this term. At first he was very standoffish or possibly shy. I met him in Maurice Pryce's rooms the day I arrived, and he didn't say a word to me. But he is getting much more friendly now.

3 November 1936

Church has just suggested to me that I should give a lecture to the Mathematical Club here on my Computable Numbers. I hope I shall be able to get an opportunity to do this, as it will bring the thing to people's attention a bit. ...

I have got one or two things on hand at present not connected with my work in logic, but in theory of groups.[20] One of them is something I did about a year ago and left in cold storage, and which Baer thinks is quite useful; but of course am not taking these things so seriously as the logic.

Tonight is the evening of election day and all results are coming out over the wireless ('radio' they say in the native language). My method of getting the results is to go to bed and read them in the paper next morning.

11 November 1936

One of the Commonwealth Fellows, Francis Price (not to be confused with Maurice Pryce or Bobby Price) arranged a hockey match the other day between the Graduate College and

---

[18] Editor's note. Trinity College, Cambridge.

[19] Editor's note. Sara has written 'probably' against the date.

[20] Editor's note. In 1938 Turing published two papers in group theory: 'Finite Approximations to Lie Groups' (*Annals of Mathematics*, 39: 105–11), which developed a method due to R. Baer, and 'The Extensions of a Group' (*Compositio Mathematica*, 5: 357–67).

Vassar, a women's college (amer.)/university (engl.) some 130 miles away. He got up a team of which only half had ever played before. We had a couple of practice games and went to Vassar in cars on Sunday. It was raining slightly when we arrived, and what was our horror when we were told the ground was not fit for play. However we persuaded them to let us play a pseudo-hockey game in their gymn. at wh. we defeated them 11–3. Francis is trying to arrange a return match, which will certainly take place on a field.

22 November 1936

I am sending you some cuttings about Mrs Simpson as representative sample of what we get over here on this subject. I don't suppose you have even heard of her, but some days it has been 'front page stuff' here.

The hockey here has become a regular fixture three days a week. It's great fun.

1 December 1936

I spent a good deal of my time in New York pottering about Manhattan getting used to their traffic and subways (underground). I went to the Planetarium. ...

I am giving my lecture to the Maths Club tomorrow.

3 December 1936

I am horrified at the way people are trying to interfere with the King's marriage. It may be that the King should not marry Mrs Simpson, but it is his private concern. I should tolerate no interference by bishops myself and I don't see that the King need either.

11 December 1936

I suppose this business of the King's abdication has come as rather a shock to you. I gather practically nothing was known of Mrs Simpson in England till about ten days ago. I am rather divided on my opinion of the whole matter. At first I was wholly in favour of the King retaining the throne and marrying Mrs Simpson, and if this were the only issue it would still be my opinion. However I have heard talk recently which seems to alter it rather. It appears that the King was extremely lax about state documents leaving them about and letting Mrs Simpson and friends see them. There had been distressing leakages. Also one or two other things of same character, but this is the one I mind about most.

December 1936 (no day)

... Talking of Christmas reminds me that as a small child I was quite unable to predict when it would fall, I didn't even realise that it came at regular intervals.

1 January 1937

I have been away with Maurice skiing in New Hampshire. ...

I am sorry that Edward VIII has been bounced into abdicating. I believe the Government wanted to get rid of him and found Mrs Simpson a good opportunity. Whether they were wise to try to get rid of him is another matter. I respect Edward for his courage. As for the Archbishop of Canterbury I consider his behaviour disgraceful. He waited until Edward was safely out of the way and then unloaded a whole lot of quite uncalled-for abuse. He didn't dare do it whilst Edward was King. Further he had no objections to the King having Mrs Simpson as a mistress, but marry her, that wouldn't do at all. I don't see how you can say that Edward was guilty of wasting his ministers time and wits at a critical moment. It was Baldwin who opened the subject.

There was rather bad attendance at the Maths Club for my lecture on Dec 2. One should have a reputation if one hopes to be listened to. The week following my lecture G. D. Birkhoff came down. He has a very good reputation and the room was packed. But his lecture wasn't up to standard at all. In fact everyone was just laughing about it afterwards.

27 January 1937

I have just finished a paper[21] in group theory; not a very exciting one this time. I shall send it off in a day or two to the L. M. S[22] or possibly to the 'Annals of Mathematics', which is the Princeton mathematical journal.

There was a problem in the 'Caliban' volume of the N S & N[23] a few weeks ago set by Eddington. It was phrased in Alice through the Looking Glass language and called 'Looking Glass Zoo'. The solution picked out for publication was also in looking glass language and sent by 'Champ', i.e. Champernowne.[24] It started off 'There couldn't have been more than three girls' reflected Humpty Dumpty 'because a girl is always the square root of minus one, and there are only 12 of those, they taught us that at school'. ...

Maurice and Francis Price arranged a party with a Treasure Hunt last Sunday. There were 13 clues of various kinds, cryptograms, anagrams and others completely obscure to me. It was all very ingenious, but I am not much use at them.

11 February 1937

The printers for the L. M. S have been rather inefficient, sending the reprints straight on to me instead of looking at the address I had filled up on their form. Unfortunately I had not kept a second copy of all the addresses I gave you, so as they are rather tiresome to find I am sending some of the reprints back to you to deal with if you can find time before you go. ... I have dealt with

All King's addresses
Littlewood
Wittgenstein
Newman
Atkins
Eperson

[21] Editor's note. Presumably 'Finite Approximations to Lie Groups' (see above).
[22] Editor's note. London Mathematical Society.
[23] Editor's note. *New Statesman and Nation*.
[24] Editor's note. See the introduction to Chapter 16.

I am told that Bertrand Russell is inclined to be ashamed of his peerage, so the situation calls for tact. I suggest that the correct address for an earl be used on envelope, but that you mark the reprint itself 'Bertrand Russell' on the top right hand corner of the cover.

22 February 1937

I went to the Eisenhart's regular Sunday tea yesterday, and there they took me in relays to try and persuade me to stay another year. Mrs Eisenhart mostly put forward social or semi-moral semi sociological reasons why it would be a good thing to have a second year. The Dean[25] weighed in with hints that the Procter Fellowship[26] was mine for the asking (this is worth $2,000 p.a.). I said I thought King's would probably prefer that I return, but gave some vague promise that I would sound them on the matter. Whether I want to stay is another matter. The people I know here will all be leaving, and I don't much care about the idea of spending a long summer in this country. ... I think it is most likely I shall come back to England.

I have had two letters asking for reprints[27], one from Braithwaite[28] at King's and one from a proffessor [sic] in Germany[29]... They seemed very much interested in the paper. I think possibly it is making a certain amount of impression. I was disappointed by its reception here. I expected Weyl who had done some work connected quite closely with it some years ago at least to have made a few remarks about it.

15 March 1937[30]

I only wrote to the Provost[31] last week so don't expect to hear from him just yet. I was rather diffident and apologetic and told him most probably I should be coming back.

29 March 1937

I have been sent a notice of lecturers in mathematics to be appointed next term, by Philip Hall.[32] Maurice and I are both putting in for it, though I don't suppose either of us will get it: I think it is a good thing to start putting in for these things early, so as to get one's existence recognised.[33] It's a thing I am rather liable to neglect. Maurice is much more conscious of what are the right things to do to help his career. He makes great social efforts with the mathematical bigwigs. ...

---

[25] Editor's note. Luther Eisenhart, Dean of the Graduate College.
[26] Editor's note. The Jane Eliza Procter Visiting Fellowship.
[27] Editor's note. Reprints of 'On Computable Numbers'.
[28] Editor's note. See Chapter 14.
[29] Editor's note. H. Scholz. Scholz's postcard is in the Turing Papers (catalogue reference D 5).
[30] Editor's note. Turing wrote '15 Feb '37'; 'Feb' has been corrected on the letter to 'Mar'.
[31] Editor's note. Provost of King's College, Cambridge.
[32] Editor's note. Hall was a Fellow of King's. In 1938 Hall became Secretary of the London Mathematical Society.
[33] Editor's note. Turing did not get the lectureship.

I am now working out some new ideas in logic.[34] Not so good as the computable numbers, but quite hopeful.

18 April 1937

The temperature here is going right up already. It's almost like June now. Tennis has started. They play on courts of dry clay. Easier on the feet (and probably on the pocket) than our hard courts, but not very quick at recovering from showers.

There was a return hockey match on Sunday against Vassar, who came over to us this time. We defeated them quite easily, but I think only because we could run faster.

I shall certainly be coming back in July.[35]

19 May 1937

I have just made up my mind to spend another year here, but I shall be going back to England for most of summer in accordance with previous programme. Thank you very much for your offer of help with this: I shall not need it, for if I have this Procter as the Dean suggests I shall be a rich man, and otherwise I shall go back to Cambridge. Another year here on the same terms would be rather an extravagance. I don't think there can be any reasonable doubt I shall have the Procter: the Dean would hardly have made any remarks about it unless they meant something. ...

My boat sails June 23. I might possibly do a little travelling here before the boat goes, as there will be very little doing here during the next month and it's not a fearfully good time of year for work. More likely I shall not as I don't usually travel for the sake of travelling.

15 June 1937

Have just been back from Cousin Jack's a couple of days. I went up north with Maurice in his car and Maurice stayed a night with Cousin Jack, and made a good impression there. I enjoyed the time I spent at Cousin Jack's. He is an energetic old bird. He has a little observatory with a telescope that he made for himself. He told me all about the grinding of mirrors. ... I think he comes into competition with Aunt Sybil for the Relations Merit Diploma. Cousin Mary is a little bit of a thing you could pick up and put in your pocket. She is very hospitable and rather timid: she worships Cousin Jack. Cousin Mary's sister Annie also lives in the house. I forgot her surname very soon after I was told it, which put me at rather a disadvantage.

I am just starting in on packing etc.

4 October 1937

Journey[36] completed without any mishap more serious than loss of my fountain pen a few hours after getting on board. ...

A vast parcel of manuscript arrived for me from the L. M. S secretary the other day. It was a paper for me to referee; 135 pages. Also have just heard from Bernays.

---

[34] Editor's note. Presumably the ideas that formed 'Systems of Logic Based on Ordinals'.
[35] Editor's note. To England.
[36] Editor's note. From England to the United States.

19 October 1937

The refereeing business rather petered out. The author's mathematical technique was hopelessly faulty, and his work after about p. 30 was based on so many erroneous notions as to be quite hopeless. So I had to send it back and say so. Rather distressing as the man has apparently been working on it for 18 months or so.

I am working on my Ph.D thesis now. Should have it done by about Christmas.

Scholz of Münster sent me a photolithoprinted reprint the other day, containing the gist of my paper[37] in the L. M. S, apparently as 'vorgetragt'[38] in Münster. It was most delightfully done, with most excellent translations into the German of the expressions I had used.

2 November 1937

Have just been playing in a hockey match, the first we have had this year. The team is not so good as last year's, our two brilliant players from New College no longer being in Princeton. I have found I get involved with making a good deal of arrangements for these games, but it has not yet got to the point of being really tiresome. ...

I am getting rather more competent with the car...

23 November 1937

I had a letter the other day from the Secretary of the Faculty Board of Mathematics at Cambridge asking if I would be a Ph.D examiner: the candidate is the same man whose paper I refereed for L. M. S. After some hesitation I decided to take this thing on. I thought it might be rather unsuitable for me to be connected with it twice, but I talked to Newman (who is here for a term) about it, and he thought such scruples were rather foolish. ...

There is a mysterious woman in Virginia who has invited me to stay for Christmas. She gets the names of Englishmen living in the Graduate College from Mrs Eisenhart.

c. New Year[39]

Did I tell you that a very nice man called Martin (the i is mute in this country) asked me to go and stay with him in South Carolina before Christmas. We drove down from here in two days and then I stayed there for two or three days before I came back to Virginia to stay with Mrs Welbourne. It was quite as far south as I had ever been—about 34°. The people seem to be all very poor down there still, even though it is so long since the civil war.

Mrs Welbourne and her family were all very agreeable, though I didn't make much conversational progress with any of them.

Two short papers of mine have just come out in the Journal of Symbolic Logic.[40]

---

[37] Editor's note. 'On Computable Numbers'.
[38] Editor's note. 'Lectured'.
[39] Editor's note. The letter is undated and marked 'Recd. Jan 14. 38'.
[40] Editor's note. 'Computability and λ-Definability' (*Journal of Symbolic Logic*, 2 (1937), 153–63), and 'The 'ꝭ'-Function in λ-K-Conversion' (*Journal of Symbolic Logic*, 2 (1937), 164).

7 March 1938

I went to 'Murder in the Cathedral' last Saturday with Will James. Was very much impressed with it. It was very much easier to understand when acted than when read. Most particularly this was so with the choruses. ...

I can't say for certain yet when I will be back. I haven't yet booked a passage. Most probably it will be about the same time and I shall be going up for the Long[41] again.

12 April 1938

Have found out now about my Fellowship: it has been renewed. When Daddy wrote me about getting a job here I thought it was time to get King's to say something definite, so I sent them a cable. I can't think why they didn't let me know before. They are usually rather strong on formal notifications: it all seems rather out of character.

I have just been to see the Dean and ask him about possible jobs over here; mostly for Daddy's information, as I think it unlikely I shall take one unless you are actually at war before July. He didn't know of one at present, but said he would bear it all in mind.

Have just been down to Washington and Annapolis with Will James. Will went to visit some people who are running St John's College Annapolis, and we both went to lunch there. They have a scheme in operation for teaching people by making them read a vast syllabus of 100% concentrated classics. Kant's 'Critique of Pure Reason' is a fairly typical example. The trouble about it is that they are so deep that any one of them really needs several years study to be understood. Presumably their undergraduates will only get something very superficial out of them.

We also went and listened to the Senate for a time. They seemed very informal. There were only six or eight of them present and few of them seemed to be attending.

7 May 1938

There was quite a good performance of 'H. M. S. Pinafore' and 'Trial by Jury' here last week-end. The 'Pinafore' didn't seem to be so good as when we saw it in Hertford (and picked up measles). 'Trial by Jury' was very good: I think I like it better than any other Gilbert and Sullivan.

My Ph.D. thesis has been delayed a good deal more than I had expected. Church made a number of suggestions which resulted in the thesis being expanded to an appalling length. I hope the length of it won't make it difficult to get it published. I lost some time too when getting it typed by a professional typist here. I took it to a firm which was very well spoken of, but they put a very incompetent girl onto it. She would copy things down wrong on every page from the original, which was almost entirely in type. I made long lists of corrections to be done and even then it would not be right. ...

The thesis has just been accepted to-day.

I expect to leave here at the beginning of July. Shall probably go direct to Cambridge.

I had an offer of a job here as von Neumann's assistant at $1,500 a year but decided not to take it.

---

[41] Editor's note. The Cambridge long vacation.

The Purpose of Ordinal Logics

Turing explained the purpose of his 'ordinal logics' in Section 11 of 'Systems of Logic Based on Ordinals'. He first distinguished between what he called 'intuition' and 'ingenuity' (a distinction that is discussed again in Chapter 4 in his letters to Newman):

> Mathematical reasoning may be regarded rather schematically as the exercise of a combination of two faculties, which we may call *intuition* and *ingenuity*. The activity of the intuition consists in making spontaneous judgments which are not the result of conscious trains of reasoning. These judgments are often but by no means invariably correct (leaving aside the question what is meant by 'correct'). Often it is possible to find some other way of verifying the correctness of an intuitive judgment. We may, for instance, judge that all positive integers are uniquely factorizable into primes; a detailed mathematical argument leads to the same result. This argument will also involve intuitive judgments, but they will be less open to criticism than the original judgment about factorization. I shall not attempt to explain this idea of 'intuition' any more explicitly.
>
> The exercise of ingenuity in mathematics consists in aiding the intuition through suitable arrangements of propositions, and perhaps geometrical figures or drawings. It is intended that when these are really well arranged the validity of the intuitive steps which are required cannot seriously be doubted.
>
> The parts played by these two faculties differ of course from occasion to occasion, and from mathematician to mathematician. This arbitrariness can be removed by the introduction of a formal logic. The necessity for using the intuition is then greatly reduced by setting down formal rules for carrying out inferences which are always intuitively valid. When working with a formal logic, the idea of ingenuity takes a more definite shape. In general a formal logic will be framed so as to admit a considerable variety of possible steps in any stage in a proof. Ingenuity will then determine which steps are the more profitable for the purpose of proving a particular proposition. (p. 192)

The intuition/ingenuity distinction is illustrated by the Gentzen-style formal logic described in 'Computable Numbers: A Guide' (see the subsection 'A tutorial on first-order predicate calculus'). Most people see intuitively that, for example, the rule

$$\frac{X \vdash Y}{\vdash not(X \,\&\, not\ Y)}$$

is valid. In order to grasp that the rule is valid, it is necessary only to reflect on the rule's meaning, namely:

> If Y can be concluded from X, then it can be concluded that not *both* X and the denial of Y are true.

With only a dozen or so basic rules, this formal logic places very little demand on intuition. Once one has accepted these few rules as valid, proofs can be constructed in the system without the need for any further exercise of intuition. Nevertheless, constructing proofs can place considerable strain on one's ingenuity. To form a proof one must—playing always by the rules—devise a chain of propositions culminating in the proposition that is to be proved. As Turing remarked, at each point in the chain there are always various possibilities for the next move. Typically, most of these possible moves are of no help at all, and it may require significant ingenuity to find a sequence of moves that leads to the desired conclusion.

Important though ingenuity is in practice, it is in principle unnecessary so long as unlimited time and paper are available. This perhaps surprising fact is clear from 'On Computable Numbers'. Once intuition has supplied the materials from which proofs are to be constructed—the basic inference rules, in the case of the logical system under discussion—then a suitably programmed Turing machine is able to grind out all the valid proofs of the system one by one. No ingenuity is required to apply the rules of the system blindly, making legal move after legal move. If a proposition is provable in the system then a machine operating in this 'blind' fashion will sooner or later prove it (so long as the machine is programmed in such a way that no legal moves are missed):

We are always able to obtain from the rules of a formal logic a method of enumerating the propositions proved by its means. We then imagine that all proofs take the form of a search through this enumeration for the theorem for which a proof is desired. In this way ingenuity is replaced by patience. (p. 193)

Intuition, on the other hand, cannot be replaced by patience. This is a lesson of Gödel's incompleteness results.

In pre-Gödel times it was thought by some that it would probably be possible to carry this programme [the setting down of formal rules] to such a point that all the intuitive judgments of mathematics could be replaced by a finite number of these rules. The necessity for intuition would then be entirely eliminated. (pp. 192–3)

Turing is here referring to the Hilbert programme. The Hilbert programme aimed to bring mathematics to order by setting down a finite system of formal rules (a 'concrete basis on which everyone can agree'[42]) by means of which all the infinitely many intuitively true mathematical statements could be proved *without* further appeals to intuition—without 'mysterious arts', as Hilbert put it.[43] (See further 'Computable Numbers: A Guide'.) Following Gödel, it was clear that this cannot be done. No matter which rules are selected, there will always be statements that a mathematician can see intuitively are true but which cannot be proved using the rules.

[42] D. Hilbert, 'Über das Unendliche' [On the Infinite], *Mathematische Annalen*, 95 (1926), 161–90 (180); English translation by E. Putnam and G. Massey in R. L. Epstein and W. A. Carnielli, *Computability: Computable Functions, Logic, and the Foundations of Mathematics* (2nd edn. Belmont, Calif.: Wadsworth, 2000).

[43] D. Hilbert, 'Probleme der Grundlegung der Mathematik' [Problems Concerning the Foundation of Mathematics], *Mathematische Annalen*, 102 (1930), 1–9 (9).

While it had been shown that intuition cannot be replaced by such a system of rules, there remained the question whether it might nevertheless be possible to *circumscribe* the use made of intuition, so that the mathematician is only required to use intuition in judging the truth of (an unlimited number of) propositions of a very specific form. This would not achieve the elimination of intuition desired by Hilbertians, but would achieve something in that direction. The truths of mathematics could be derived, not by using a set of formal rules alone, but by using a set of formal rules together with intuitive judgements of that very specific form. To those wary of intuition, this is certainly preferable to the uncontrolled use of 'mysterious arts'. Although not eliminated, the use of intuition would at least be brought under strict control.

This successor to the defeated Hilbert programme[44] is the subject of investigation of 'Systems of Logic Based on Ordinals':

In our discussions...we have gone to the opposite extreme [to Hilbertians] and eliminated not intuition but ingenuity, and this in spite of the fact that our aim has been in much the same direction. We have been trying to see how far it is possible to eliminate intuition, and leave only ingenuity. (p. 193)

In consequence of the impossibility of finding a formal logic which wholly eliminates the necessity of using intuition, we naturally turn to 'non-constructive' systems of logic with which not all the steps in a proof are mechanical, some being intuitive. . . . What properties do we desire a non-constructive logic to have if we are to make use of it for the expression of mathematical proofs? We want it to show quite clearly when a step makes use of intuition, and when it is purely formal. The strain put on the intuition should be a minimum. (ibid.)

The following extension of the Gentzen-style formal logic just discussed is a simple example of a logical system that incorporates, as well as formal rules, intuitive judgements of a very specific kind. In this extended system, each step in a proof is either a derivation from previous statements in the proof, using a formal rule (as before), or else—and this is the new part—is the assertion of a (true) proposition of form 'Un(**m**)'. 'Un(**m**)' is a construction saying 'At some point, Turing machine **m** prints 0' (see Section 11 of 'On Computable Numbers', and the subsection 'The proof of the undecidability of FOPC' of 'Computable Numbers: A Guide'). Steps of this second sort are non-mechanical, in the sense that no Turing machine—no effective procedure—is able to determine, for all propositions of the form 'Un(**m**)', which are true and which false. Nevertheless, when one is proving theorems in this system, one is allowed to include, at any point in a proof where it is helpful to do so, an intuitively true proposition of the form 'Un(**m**)'. No other type of intuitive step is permitted in the system—the

---

[44] This post-Gödelian programme was also investigated by Church's student Barkley Rosser in his 'Gödel Theorems for Non-Constructive Logics', *Journal of Symbolic Logic*, 2 (1937), 129–37.

strain put on the intuition is 'a minimum'. The result is a system whose theorems go far beyond what is provable in the first-order predicate calculus alone.

The question of *how much* mathematics can be captured by a system in which the use of intuition is strictly controlled in this manner is the crucial one for the success or otherwise of this post-Hilbertian programme.

Ordinal Logics and Gödel's Incompleteness Theorem

To say that a system is complete with respect to some specified set of formulae S, e.g. the set of arithmetical truths, is to say that every formula in S is provable in the system. Gödel showed that not all arithmetical truths can be proved in the formal system of arithmetic set out by Whitehead and Russell in *Principia Mathematica*. Turing's work enabled this result to be extended to *any mechanical procedure* for producing truths of arithmetic (see Section 14 of 'Computable Numbers: A Guide').

Gödel established his incompleteness result by showing how to construct an arithmetical formula—call it G—that is not provable in the system and yet is true. In order to show that G is true, Gödel appealed to the way he constructed G. G is of such a nature that G in effect says that it itself is not provable in the system—and so, since G is *not* provable, what G says is true.

Can the incomplete formal system of arithmetic be made complete by adding G to it as a new axiom (thereby making it the case—trivially—that G is provable in the system)? No. This is because, once G is added, producing a new system, the Gödel construction can be applied once again to produce a true formula G_1 that is unprovable in the new system. And when G_1 is itself added as a new axiom, producing a further system, there is a true but unprovable G_2, and so on *ad infinitum*.

Following Turing's notation (p. 146), let the system of arithmetic that forms the starting point of this infinite progression be called L. The result of adding G to L is called L_1; the result of adding G_1 to L_1 is L_2, and so on. Taken together, the systems in the infinite progression $L, L_1, L_2, L_3, \ldots$ form a non-constructive logic of the sort described by Turing in the above quotation. New axioms are seen to be true by intuition, but otherwise only ingenuity (or patience) is required in proving theorems in any of the systems.

There are *a lot* of systems in the progression $L, L_1, L_2, L_3, \ldots$ Saying merely that there are infinitely many oversimplifies matters. Not only is there a system for each one of the infinitely many finite ordinal numbers 1, 2, 3, … There is a system that contains the theorems of *every one* of the systems L_i, where i is a finite ordinal. This system is called L_ω (ω being the first 'transfinite' ordinal number). The system L_ω is 'bigger' than any one of the systems L_i in the sense that, no matter which L_i is considered, L_ω includes all the theorems of L_i, but not vice versa. If P_1 is the set of provable formulae of L_1, P_2 of L_2, and so on, then P_ω is the union of all the sets P_1, P_2, P_3, ... But even L_ω has a true but unprovable G_ω. Adding G_ω to L_ω produces $L_{\omega+1}$, adding $G_{\omega+1}$ to $L_{\omega+1}$ produces $L_{\omega+2}$, and so on and so on. The progression of systems L, L_1, L_2, L_3, ..., L_ω, $L_{\omega+1}$, $L_{\omega+2}$, ... is an example of an ordinal logic.

As Turing noted in the first paragraph of 'Systems of Logic Based on Ordinals', each L_i is 'more complete' than its predecessor: some of the true formulae unprovable in L are provable in the less incomplete L_1, and so on. This raises the possibility of our being able to construct a progression of systems—an ordinal logic—that is complete, the systems in the progression proving between them all the truths of arithmetic. If so, then not every systematic formulation of arithmetic falls prey to Gödel's theorem. Such an ordinal logic would 'avoid as far as possible the effects of Gödel's theorem', Turing said (p. 178):

> Gödel's theorem shows that such a system cannot be wholly mechanical; but with a complete ordinal logic we should be able to confine the non-mechanical steps... (p. 180)

In his investigations Turing considered sequences of systems in which the non-mechanical steps consist, not of recognizing that Gödel-formulae (the Gs) are true, but of recognizing that certain formulae are what he called *ordinal* formulae. The concept of an ordinal formula is defined in terms of operations of the lambda calculus (p. 162). Roughly, a formula of the lambda calculus is an ordinal formula if it represents a (constructive) ordinal number. The important point is that there is no effective procedure for determining, of any given formula of the calculus, whether or not it is an ordinal formula; Turing proved this on p. 170.

Turing gave examples of ordinal logics of three different types, the logic Λ_P (p. 177), the logic Λ_H (p. 178), and his 'Gentzen type' ordinal logics (Section 12).[45] (The 'P' in 'Λ_P' refers to Gödel's 1931 system P, equivalent to the system of arithmetic given by Whitehead and Russell in *Principia Mathematica*. Seemingly 'H' was for Hilbert.)

Ordinal Logics and Proof-Finding Machines

In one of his letters to Newman (Chapter 4), Turing outlined the relationship between an ordinal logic and a hierarchy of theorem-proving Turing machines (p. 215):

> One imagines different machines allowing different sets of proofs, and by choosing a suitable machine one can approximate 'truth' by 'provability' better than with a less suitable machine, and can in a sense approximate it as well as you please.

---

[45] The 'Gentzen-type' logics of Section 12 are different from the Gentzen-style formulation of first-order predicate calculus discussed above (although, as the names imply, both are suggested by work of Gentzen).

If one wants a particular true statement to be proved by a Turing machine (in the sense described earlier, where 'ingenuity is replaced by patience'), then, since no single Turing machine can prove every true arithmetical statement, one must pick a suitable machine, a machine that actually is able to prove the statement in question. The selection of a suitable machine typically 'involves intuition' (p. 215).

The intuition involved in choosing a suitable proof-finding machine is, Turing went on to say in the letter, 'interchangeable' with the intuition required for selecting a system, from among a progression of systems, in which the statement is provable. Furthermore, if rather than following the rules of a particular logical system, one were to prove the statement free-style, then this too would require intuition, and the necessary intuition would be interchangeable with that required for choosing a suitable proof-finding machine or for choosing a suitable logical system. (See further the introduction to Chapter 12.)

Completeness of Ordinal Logics

Let l_1, l_2, l_3, ... be any progression of logical systems indexed by (expressions for) ordinals. To say that l_1, l_2, l_3, ... is complete with respect to some set of formulae S is to say that for each formula x in S, there is *some* ordinal α such that x is provable in l_α.

Turing proved the following 'completeness theorem': his ordinal logic Λ_P is complete with respect to the set of all true formulae of the form 'for every integer x, $f(x) = 0$', where f is a primitive recursive function (pp. 187–190). Many mathematically interesting theorems are of this form. In modern terminology, formulae of this form are called 'Π^0_1 formulae'. Turing referred to them as being of the form '$f(x)$ vanishes identically' (by '$f(x)$ vanishes' is meant '$f(x) = 0$').

This completeness theorem shows that Λ_P circumvents Gödel's incompleteness result in the way discussed above. Solomon Feferman, who in the 1960s continued Turing's work on ordinal logics, commented on the theorem: '[This] partial completeness result ... could have been regarded as meeting the ... aim of "overcoming" the incompleteness phenomena discovered by Gödel, since these only concerned true but unprovable Π^0_1 statements.'[46]

In his letter to Newman from the Crown (Chapter 4), Turing pointed out that this 'completeness theorem ... is of course completely useless for the purpose of actually producing proofs' (p. 213). Why this is so is explained by means of an example in 'Systems of Logic Based on Ordinals' (p. 191):

---

[46] S. Feferman, 'Turing in the Land of O(z)', in R. Herken (ed.), *The Universal Turing Machine: A Half-Century Survey* (Oxford: Oxford University Press, 1988), 122–3.

Although [the completeness theorem] shows, for instance, that it is possible to prove Fermat's last theorem with Λ_P (if it is true) yet the truth of the theorem would really be assumed by taking a certain formula as an ordinal formula.

Nevertheless, as Turing went on to say in the letter, the completeness theorem does succeed perfectly well in its purpose of providing 'an insurance against certain sorts of "Gödel incompleteness theorems" being proved about the ordinal logic'.

Not all true arithmetical statements are of Π_1^0 form (this is why the completeness theorem is only a partial result). Turing was especially concerned with formulae of the form $(x)(\exists y)f(x, y) = 0$ (where f is a primitive recursive function). In modern terminology, formulae of this form are called 'Π_2^0 formulae'. Turing called (true) formulae of this form 'number-theoretic theorems' (p. 152). (The choice of this term is curious; he defended it in a footnote to p. 152.) In Section 5, Turing explained why he regarded number-theoretic theorems as having 'an importance which makes it worth while to give them special consideration' (mentioning also, on p. 155, that a number of unsolved mathematical problems are number-theoretic). Turing conjectured that Λ_P is complete with respect to the set of all true Π_2^0 formulae, but said 'I cannot at present give a proof of this' (p. 187).

Unfortunately Turing's conjecture that Λ_P is complete with respect to true Π_2^0 formulae was proved incorrect by Feferman, in work published in 1962.[47] Is there nevertheless *some* ordinal logic that is complete with respect to this wider class of truths? A negative answer would represent a spectacular incompleteness result. Commenting on his refutation of Turing's conjecture, Feferman said:

A general incompleteness theorem for recursive progressions ... would have been dramatic proof of the far-reaching extent of incompleteness phenomena. However, the situation has not turned out in this way. ... [A]ll true sentences of elementary number theory are provable in the recursive progression based on [a principle studied by Shoenfield].[48]

At the end of Section 11, Turing gave some reasons for being dissatisfied with the logics Λ_H and Λ_P, and moved on to the Gentzen-type ordinal logics of Section 12. He said of the last of his three examples of Gentzen-type logics, Λ_G^3, that it 'appears to be adequate for most purposes', adding 'How far this is the case can, of course, only be determined by experiment' (p. 202).

Oracle Machines

In Section 4, 'A type of problem which is not number-theoretic', Turing introduced the concept of an *o-machine*. An *o*-machine is like a Turing machine

---

[47] S. Feferman, 'Transfinite Recursive Progressions of Axiomatic Theories', *Journal of Symbolic Logic*, 27 (1962), 259–316.
[48] Ibid. 261; J. R. Shoenfield, 'On a Restricted ω-Rule', *Bulletin de l'Académie Polonaise des Sciences*, 7 (1959), 405–7.

except that the machine is endowed with an additional basic operation of a type that no Turing machine can simulate. For example, the new operation may be that of displaying the answer to any question of the form 'Is Turing machine **m** circle-free?' (A circle-free machine is one that prints an infinite number of binary digits; see Sections 8 and 11 of 'Computable Numbers: A Guide'.) Turing called the new operation the 'oracle'. He did not go into the nature of the oracle: it works by 'some unspecified means' (p. 156).

The question 'Is Turing machine **m** circle-free?' may be presented to the oracle simply by writing out, on successive F-squares of the *o*-machine's tape, the description number of the Turing machine in question (marking the E-squares at the start and finish of the description number with some special symbol, e.g. '@'). As in the case of an ordinary Turing machine, the behaviour of an *o*-machine is governed by a table of instructions. Among the states of the *o*-machine is a state that is used to call in the oracle. When an instruction in the table places the machine in that state, the marked description is 'referred to the oracle' (p. 156). The oracle determines by unspecified means whether or not the Turing machine so numbered is circle-free, and delivers its pronouncement by shifting the machine into one or other of two states, one indicating the affirmative answer and the other the negative.

Turing's aim in Section 4, paralleling his aim in 'On Computable Numbers', was to prove the existence of mathematical problems that cannot be solved by *o*-machine. Just as no Turing machine can decide, of arbitrarily selected Turing-machine description numbers, which are numbers of circle-free machines, no *o*-machine can decide, of arbitrarily selected *o*-machine description numbers, which are numbers of circle-free *o*-machines. Turing showed this by reworking the argument that he gave in 'On Computable Numbers' (p. 72ff of 'On Computable Numbers' and p. 157 of 'Systems of Logic Based on Ordinals').

The connection with number-theoretic problems is via an equivalence pointed out by Turing in Section 3: 'every number-theoretic theorem is equivalent to the statement that a corresponding [Turing] machine is circle free' (p. 154). In the light of this equivalence, an oracle for deciding whether or not Turing machines are circle-free is in effect an oracle for deciding whether or not statements are number-theoretic theorems. (Indeed, Turing introduced *o*-machines in terms of an oracle for 'solving number-theoretic problems' (p. 156).)

Given that an *o*-machine is able to solve all number-theoretic problems, the *o*-machine satisfactoriness problem—the problem of deciding whether arbitrarily selected *o*-machine description numbers are numbers of circle-free machines—is an example of a type of problem that is not number-theoretic.

Turing has shown, then, that there are types of mathematical truth that cannot be proved by means of an effective method augmented by pronouncements of the oracle. If the project is to formalize mathematics by means of Hilbertian inference rules augmented by a strictly circumscribed use of intuition, then the intuitive steps cannot be limited to true propositions of the form 'Turing

Feferman, S., 'Turing in the Land of O(z)', in R. Herken (ed.), *The Universal Turing Machine: A Half-Century Survey* (Oxford: Oxford University Press, 1988).

Kleene, S. C., *Mathematical Logic* (New York: Wiley, 1967).

Rogers, H., *Theory of Recursive Functions and Effective Computability* (New York: McGraw-Hill, 1967).

Shoenfield, J. R., *Degrees of Unsolvability* (Amsterdam: North-Holland, 1971).

Simpson, S. G., 'Degrees of Unsolvability: A Survey of Results', in J. Barwise (ed.), *Handbook of Mathematical Logic* (Amsterdam: North-Holland, 1977).

Soare, R. I., *Recursively Enumerable Sets and Degrees* (Berlin: Springer-Verlag, 1987).

Systems of Logic Based on Ordinals

Introduction *146*
1. The conversion calculus. Gödel representations *147*
2. Effective calculability. Abbreviation of treatment *150*
3. Number-theoretic theorems *152*
4. A type of problem which is not number-theoretic *156*
5. Syntactical theorems as number-theoretic theorems *157*
6. Logic formulae *158*
7. Ordinals *161*
8. Ordinal logics *170*
9. Completeness questions *178*
10. The continuum hypothesis. A digression *191*
11. The purpose of ordinal logics *192*
12. Gentzen type ordinal logics *194*
    Index of definitions *202*
    Bibliography *203*

The well-known theorem of Gödel (Gödel [1], [2]) shows that every system of logic is in a certain sense incomplete, but at the same time it indicates means whereby from a system L of logic a more complete system L' may be obtained.[1] By repeating the process we get a sequence L, $L_1 = L'$, $L_2 = L_1'$, ... each more complete than the preceding. A logic L_ω may then be constructed in which the provable theorems are the totality of theorems provable with the help of the logics L, L_1, L_2, ... We may then form $L_{2\omega}$ related to L_ω in the same way as L_ω was related to L. Proceeding in this way we can associate a system of logic with any constructive ordinal.[2] It may be asked whether a sequence of logics of this kind is complete in the sense that to any problem A there corresponds an ordinal α such that A is solvable by means of the logic L_α. I propose to investigate this question in a rather more general case, and to give some other examples of ways in which systems of logic may be associated with constructive ordinals.

This article first appeared in *Proceedings of the London Mathematical Society*, Series 2, 45 (1939), 161–228. It is reprinted with the permission of the London Mathematical Society and the Estate of Alan Turing.

[1] This paper represents work done while a Jane Eliza Procter Visiting Fellow at Princeton University, where the author received most valuable advice and assistance from Prof. Alonzo Church.

[2] The situation is not quite so simple as is suggested by this crude argument. See pages [162–73], [181–3].

1. The calculus of conversion. Gödel representations

It will be convenient to be able to use the "conversion calculus" of Church for the description of functions and for some other purposes. This will make greater clarity and simplicity of expression possible. I give a short account of this calculus. For detailed descriptions see Church [3], [2], Kleene [1], Church and Rosser [1].

The formulae of the calculus are formed from the symbols {, }, (,) , [,], λ, δ, and an infinite list of others called variables; we shall take for our infinite list $a, b, \ldots, z, x', x'', \ldots$ Certain finite sequences of such symbols are called *well-formed formulae* (abbreviated to W.F.F.); we define this class inductively, and define simultaneously the free and the bound variables of a W.F.F. Any variable is a W.F.F.; it is its only free variable, and it has no bound variables. δ is a W.F.F. and has no free or bound variables. If **M** and **N** are W.F.F. then {**M**}(**N**) is a W.F.F., whose free variables are the free variables of **M** together with the free variables of **N**, and whose bound variables are the bound variables of **M** together with those of **N**. If **M** is a W.F.F. and **V** is one of its free variables, then λ**V**[**M**] is a W.F.F. whose free variables are those of **M** with the exception of **V**, and whose bound variables are those of **M** together with **V**. No sequence of symbols is a W.F.F. except in consequence of these three statements.

In metamathematical statements we use heavy type letters to stand for variable or undetermined formulae, as was done in the last paragraph, and in future such letters will stand for well-formed formulae unless otherwise stated. Small letters in heavy type will stand for formulae representing undetermined positive integers (see below).

A W.F.F. is said to be in normal form if it has no parts of the form {λ**V**[**M**]}(**N**) and none of the form {{δ}(**M**)}(**N**), where **M** and **N** have no free variables.

We say that one W.F.F. is *immediately convertible* into another if it is obtained from it either by:

(i) Replacing one occurrence of a well-formed part λ**V**[**M**] by λ**U**[**N**], where the variable **U** does not occur in **M**, and **N** is obtained from **M** by replacing the variable **V** by **U** throughout.

(ii) Replacing a well-formed part {λ**V**[**M**]}(**N**) by the formula which is obtained from **M** by replacing **V** by **N** throughout, provided that the bound variables of **M** are distinct both from **V** and from the free variables of **N**.

(iii) The process inverse to (ii).

(iv) Replacing a well-formed part {{δ}(**M**)}(**M**) by

$$\lambda f[\lambda x[\{f\}(\{f\}(x))]]$$

if **M** is in normal form and has no free variables.

(v) Replacing a well-formed part $\{\{\delta\}(M)\}(N)$ by

$$\lambda f[\lambda x[\{f\}(x)]]$$

if M and N are in normal form, are not transformable into one another by repeated application of (i), and have no free variables.

(vi) The process inverse to (iv).

(vii) The process inverse to (v).

These rules could have been expressed in such a way that in no case could there be any doubt about the admissibility or the result of the transformation [in particular this can be done in the case of process (v)].

A formula A is said to be *convertible* into another B (abbreviated to "A conv B") if there is a finite chain of immediate conversions leading from one formula to the other. It is easily seen that the relation of convertibility is an equivalence relation, *i.e.* it is symmetric, transitive, and reflexive.

Since the formulae are liable to be very lengthy, we need means for abbreviating them. If we wish to introduce a particular letter as an abbreviation for a particular lengthy formula we write the letter followed by "→" and then by the formula, thus

$$I \to \lambda x[x]$$

indicates that I is an abbreviation for $\lambda x[x]$. We also use the arrow in less sharply defined senses, but never so as to cause any real confusion. In these cases the meaning of the arrow may be rendered by the words "stands for".

If a formula F is, or is represented by, a single symbol we abbreviate $\{F\}(X)$ to $F(X)$. A formula $\{\{F\}(X)\}(Y)$ may be abbreviated to

$$\{F\}(X, Y),$$

or to $F(X, Y)$ if F is, or is represented by, a single symbol. Similarly for $\{\{\{F\}(X)\}(Y)\}(Z)$, etc. A formula $\lambda V_1[\lambda V_2 \ldots [\lambda V_r[M]] \ldots]$ may be abbreviated to $\lambda V_1 V_2 \ldots V_r . M$.

We have not as yet assigned any meanings to our formulae, and we do not intend to do so in general. An exception may be made for the case of the positive integers, which are very conveniently represented by the formulae $\lambda fx.f(x), \lambda fx.f(f(x)), \ldots$ In fact we introduce the abbreviations

$$1 \to \lambda fx.f(x)$$
$$2 \to \lambda fx.f(f(x))$$
$$3 \to \lambda fx.f(f(f(x))), \text{ etc.,}$$

and we also say, for example, that $\lambda fx.f(f(x))$, or in full

$$\lambda f[\lambda x[\{f\}(\{f\}(x))]],$$

represents the positive integer 2. Later we shall allow certain formulae to represent ordinals, but otherwise we leave them without explicit meaning; an implicit meaning may be suggested by the abbreviations used. In any case where any meaning is assigned to formulae it is desirable that the meaning should be invariant under conversion. Our definitions of the positive integers do not violate this requirement, since it may be proved that no two formulae representing different positive integers are convertible the one into the other.

In connection with the positive integers we introduce the abbreviation

$$S \to \lambda ufx . f(u(f, x)).$$

This formula has the property that, if **n** represents a positive integer, $S(\mathbf{n})$ is convertible to a formula representing its successor.[3]

Formulae representing undetermined positive integers will be represented by small letters in heavy type, and we adopt once for all the convention that, if a small letter, n say, stands for a positive integer, then the same letter in heavy type, **n**, stands for the formula representing the positive integer. When no confusion arises from so doing, we shall not trouble to distinguish between an integer and the formula which represents it.

Suppose that $f(n)$ is a function of positive integers taking positive integers as values, and that there is a W.F.F. F not containing δ such that, for each positive integer n, $F(\mathbf{n})$ is convertible to the formula representing $f(n)$. We shall then say that $f(n)$ is λ-*definable* or *formally definable*, and that F *formally defines* $f(n)$. Similar conventions are used for functions of more than one variable. The sum function is, for instance, formally defined by $\lambda abfx . a(f, b(f, x))$; in fact, for any positive integers m, n, p for which $m + n = p$, we have

$$\{\lambda abfx . a(f, b(f, x))\}(\mathbf{m}, \mathbf{n}) \text{ conv } \mathbf{p}.$$

In order to emphasize this relation we introduce the abbreviation

$$X + Y \to \{\lambda abfx . a(f, b(f, x))\}(X, Y)$$

and we shall use similar notations for sums of three or more terms, products, etc.

For any W.F.F. G we shall say that G *enumerates* the sequence $G(\mathbf{1}), G(\mathbf{2}), \ldots$ and any other sequence whose terms are convertible to those of this sequence.

When a formula is convertible to another which is in normal form, the second is described as a *normal form* of the first, which is then said to *have a normal form*. I quote here some of the more important theorems concerning normal forms.

(A) *If a formula has two normal forms they are convertible into one another by the use of* (i) *alone.* (Church and Rosser [1], 479, 481.)

(B) *If a formula has a normal form then every well-formed part of it has a normal form.* (Church and Rosser [1], 480–481.)

---

[3] This follows from (A) below.

(C) *There is (demonstrably) no process whereby it can be said of a formula whether it has a normal form.* (Church [3], 360, Theorem XVIII.)

We often need to be able to describe formulae by means of positive integers. The method used here is due to Gödel (Gödel [1]). To each single symbol s of the calculus we assign an integer $r[s]$ as in the table below.

s	{, (, or [	},), or]	λ	δ	a	...	z	x'	x''	x'''	...
$r[s]$	1	2	3	4	5	...	30	31	32	33	...

If $s_1, s_2, \ldots, s_k$ is a sequence of symbols, then $2^{r[s_1]} 3^{r[s_2]} \ldots p_k^{r[s_k]}$ (where p_k is the k-th prime number) is called the *Gödel representation* (G.R.) of that sequence of symbols. No two W.F.F. have the same G.R.

Two theorems on G.R. of W.F.F. are quoted here.

(D) *There is a W.F.F. "form" such that if a is the G.R. of a W.F.F. A without free variables, then* form (a) conv A. (This follows from a similar theorem to be found in Church [3], 53 66. Metads are used there in place of G.R.)

(E) *There is a W.F.F. Gr such that, if A is a W.F.F. with a normal form without free variables, then* Gr(A) conv a, *where a is the G.R. of a normal form of* A. [Church [3], 53, 66, as (D).]

2. Effective calculability. Abbreviation of treatment

A function is said to be "effectively calculable" if its values can be found by some purely mechanical process. Although it is fairly easy to get an intuitive grasp of this idea, it is nevertheless desirable to have some more definite, mathematically expressible definition. Such a definition was first given by Gödel at Princeton in 1934 (Gödel [2], 26), following in part an unpublished suggestion of Herbrand, and has since been developed by Kleene [2]). These functions were described as "general recursive" by Gödel. We shall not be much concerned here with this particular definition. Another definition of effective calculability has been given by Church (Church [3], 356–358), who identifies it with λ-definability. The author has recently suggested a definition corresponding more closely to the intuitive idea (Turing [1], see also Post [1]). It was stated above that "a function is effectively calculable if its values can be found by some purely mechanical process". We may take this statement literally, understanding by a purely mechanical process one which could be carried out by a machine. It is possible to give a mathematical description, in a certain normal form, of the structures of these machines. The development of these ideas leads to the author's definition of a computable function, and to an identification of computability with effective calculability.[4] It is not difficult, though somewhat laborious, to prove that these three definitions are equivalent (Kleene [3], Turing [2]).

In the present paper we shall make considerable use of Church's identification of effective calculability with λ-definability, or, what comes to the same thing, of the identification with computability and one of the equivalence theorems. In most cases where we have to deal with an effectively calculable function, we shall introduce the corresponding W.F.F. with some such phrase as "the function f is effectively calculable, let F be a formula λ defining it", or "let F be a formula such that $F(\mathbf{n})$ is convertible to ... whenever $\mathbf{n}$ represents a positive integer". In such cases there is no difficulty in seeing how a machine could in principle be designed to calculate the values of the function concerned; and, assuming this done, the equivalence theorem can be applied. A statement of what the formula F actually is may be omitted. We may immediately introduce on this basis a W.F.F. ϖ with the property that

$$\varpi(\mathbf{m}, \mathbf{n}) \text{ conv } \mathbf{r},$$

if r is the greatest positive integer, if any, for which m^r divides n and r is 1 if there is none. We also introduce Dt with the properties

$$\text{Dt}(\mathbf{n}, \mathbf{n}) \text{ conv } 3,$$
$$\text{Dt}(\mathbf{n} + \mathbf{m}, \mathbf{n}) \text{ conv } 2,$$
$$\text{Dt}(\mathbf{n}, \mathbf{n} + \mathbf{m}) \text{ conv } 1.$$

There is another point to be made clear in connection with the point of view that we are adopting. It is intended that all proofs that are given should be regarded no more critically than proofs in classical analysis. The subject matter, roughly speaking, is constructive systems of logic, but since the purpose is directed towards choosing a particular constructive system of logic for practical use, an attempt at this stage to put our theorems into constructive form would be putting the cart before the horse.

Those computable functions which take only the values 0 and 1 are of particular importance, since they determine and are determined by computable properties, as may be seen by replacing "0" and "1" by "true" and "false". But, besides this type of property, we may have to consider a different type, which is, roughly speaking, less constructive than the computable properties, but more so than the general predicates of classical mathematics. Suppose that we have a computable function of the natural numbers taking natural numbers as values, then corresponding to this function there is the property of being a value of the function. Such a property we shall describe as "axiomatic"; the reason for using

---

[4] We shall use the expression "computable function" to mean a function calculable by a machine, and we let "effectively calculable" refer to the intuitive idea without particular identification with any one of these definitions. We do not restrict the values taken by a computable function to be natural numbers; we may for instance have computable propositional functions.

this term is that it is possible to define such a property by giving a set of axioms, the property to hold for a given argument if and only if it is possible to deduce that it holds from the axioms.

Axiomatic properties may also be characterized in this way. A property ψ of positive integers is axiomatic if and only if there is a computable property ϕ of two positive integers, such that $\psi(x)$ is true if and only if there is a positive integer y such that $\phi(x, y)$ is true. Or again ψ is axiomatic if and only if there is a W.F.F. F such that $\psi(n)$ is true if and only if F(n) conv 2.

3. Number-theoretic theorems

By a *number-theoretic theorem*[5] we shall mean a theorem of the form "$\theta(x)$ vanishes for infinitely many natural numbers x", where $\theta(x)$ is a primitive recursive function.[6]

We shall say that a problem is number-theoretic if it has been shown that any solution of the problem may be put in the form of a proof of one or more number-theoretic theorems. More accurately we may say that a class of problems is number-theoretic if the solution of any one of them can be transformed (by a uniform process) into the form of proofs of number-theoretic theorems.

I shall now draw a few consequences from the definition of "number theoretic theorems", and in section 5 I shall try to justify confining our consideration to this type of problem.

An alternative form for number-theoretic theorems is "for each natural number x there exists a natural number y such that $\phi(x, y)$ vanishes", where $\phi(x, y)$ is primitive recursive. In other words, there is a rule whereby, given the

---

[5] I believe that there is no generally accepted meaning for this term, but it should be noticed that we are using it in a rather restricted sense. The most generally accepted meaning is probably this: suppose that we take an arbitrary formula of the functional calculus of the first order and replace the function variables by primitive recursive relations. The resulting formula represents a typical number-theoretic theorem in this (more general) sense.

[6] Primitive recursive functions of natural numbers are defined inductively as follows. Suppose that $f(x_1, \ldots, x_{n-1}), g(x_1, \ldots, x_n), h(x_1, \ldots, x_{n+1})$ are primitive recursive, then $\phi(x_1, \ldots, x_n)$ is primitive recursive if it is defined by one of the sets of equations (a) to (e).

(a) $\phi(x_1, \ldots, x_n) = h(x_1, \ldots, x_{m-1}, g(x_1, \ldots, x_n), x_{m+1}, \ldots, x_{n-1}, x_m)$ $(1 \leq m \leq n)$;
(b) $\phi(x_1, \ldots, x_n) = f(x_2, \ldots, x_n)$;
(c) $\phi(x_1) = a$, where $n = 1$ and a is some particular natural number;
(d) $\phi(x_1) = x_1 + 1$ $(n = 1)$;
(e) $\phi(x_1, \ldots, x_{n-1}, 0) = f(x_1, \ldots, x_{n-1})$;
 $\phi(x_1, \ldots, x_{n-1}, x_n + 1) = h(x_1, \ldots, x_n, \phi(x_1, \ldots, x_n))$.

The class of primitive recursive functions is more restricted than the class of computable functions, but it has the advantage that there is a process whereby it can be said of a set of equations whether it defines a primitive recursive function in the manner described above.

If $\phi(x_1, \ldots, x_n)$ is primitive recursive, then $\phi(x_1, \ldots, x_n) = 0$ is described as a primitive recursive relation between $x_1, \ldots, x_n$.

function $\theta(x)$, we can find a function $\phi(x,y)$, or given $\phi(x,y)$, we can find a function $\theta(x)$, such that "$\theta(x)$ vanishes infinitely often" is a necessary and sufficient condition for "for each x there is a y such that $\phi(x,y) = 0$". In fact, given $\theta(x)$, we define

$$\phi(x,y) = \theta(x) + \alpha(x,y),$$

where $\alpha(x,y)$ is the (primitive recursive) function with the properties

$$\alpha(x,y) = 1 \ (y \leqslant x),$$
$$= 0 \ (y > x).$$

If on the other hand we are given $\phi(x,y)$ we define $\theta(x)$ by the equations

$$\theta_1(0) = 3,$$

$$\theta_1(x+1) = 2^{(1+\varpi_2(\theta_1(x)))\sigma(\phi(\varpi_3(\theta_1(x))-1,\,\varpi_2(\theta_1(x))))} 3^{\varpi_3(\theta_1(x))+1-\sigma(\phi(\varpi_3(\theta_1(x))-1,\,\varpi_2(\theta_1(x))))},$$
$$\theta(x) = \phi(\varpi_3(\theta_1(x)) - 1, \varpi_2(\theta_1(x))),$$

where $\varpi_r(x)$ is defined so as to mean "the largest s for which r^s divides x". The function $\sigma(x)$ is defined by the equations $\sigma(0) = 0$, $\sigma(x+1) = 1$. It is easily verified that the functions so defined have the desired properties.

We shall now show that questions about the truth of the statements of the form "does $f(x)$ vanish identically", where $f(x)$ is a computable function, can be reduced to questions about the truth of number-theoretic theorems. It is understood that in each case the rule for the calculation of $f(x)$ is given and that we are satisfied that this rule is valid, *i.e.* that the machine which should calculate $f(x)$ is circle free ([p. 60]). The function $f(x)$, being computable, is general recursive in the Herbrand–Gödel sense, and therefore, by a general theorem due to Kleene[7], is expressible in the form

$$\psi(\varepsilon y[\phi(x,y) = 0]), \tag{3.2}$$

where $\varepsilon y[\mathfrak{A}(y)]$ means "the least y for which $\mathfrak{A}(y)$ is true" and $\psi(y)$ and $\phi(x,y)$ are primitive recursive functions. Without loss of generality, we may suppose that the functions ϕ, ψ take only the values 0, 1. Then, if we define $\rho(x)$ by the equations (3.1) and

$$\rho(0) = \psi(0)(1 - \theta(0)),$$
$$\rho(x+1) = 1 - (1 - \rho(x)) \, \sigma[1 + \theta(x) - \psi\{\varpi_2(\theta_1(x))\}]$$

it will be seen that $f(x)$ vanishes identically if and only if $\rho(x)$ vanishes for infinitely many values of x.

---

[7] Kleene [3], 727. This result is really superfluous for our purpose, since the proof that every computable function is general recursive proceeds by showing that these functions are of the form (3.2). (Turing [2], 161).

The converse of this result is not quite true. We cannot say that the question about the truth of any number-theoretic theorem is reducible to a question about whether a corresponding computable function vanishes identically; we should have rather to say that it is reducible to the problem of whether a certain machine is circle free and calculates an identically vanishing function. But more is true: every number-theoretic theorem is equivalent to the statement that a corresponding machine is circle free. The behaviour of the machine may be described roughly as follows: the machine is one for the calculation of the primitive recursive function $\theta(x)$ of the number-theoretic problem, except that the results of the calculation are first arranged in a form in which the figures 0 and 1 do not occur, and the machine is then modified so that, whenever it has been found that the function vanishes for some value of the argument, then 0 is printed. The machine is circle free if and only if an infinity of these figures are printed, *i.e.* if and only if $\theta(x)$ vanishes for infinitely many values of the argument. That, on the other hand, questions of circle freedom may be reduced to questions of the truth of number-theoretic theorems follows from the fact that $\theta(x)$ is primitive recursive when it is defined to have the value 0 if a certain machine $\mathcal{M}$ prints 0 or 1 in its $(x+1)$-th complete configuration, and to have the value 1 otherwise.

The conversion calculus provides another normal form for the number-theoretic theorems, and the one which we shall find the most convenient to use. Every number-theoretic theorem is equivalent to a statement of the form "**A**(**n**) is convertible to 2 for every W.F.F. **n** representing a positive integer", **A** being a W.F.F. determined by the theorem; the property of **A** here asserted will be described briefly as "**A** is dual". Conversely such statements are reducible to number-theoretic theorems. The first half of this assertion follows from our results for computable functions, or directly in this way. Since $\theta(x-1)+2$ is primitive recursive, it is formally definable, say, by means of a formula **G**. Now there is (Kleene [1], 232) a W.F.F. $\mathcal{P}$ with the property that, if **T**(**r**) is convertible to a formula representing a positive integer for each positive integer *r*, then $\mathcal{P}$(**T**, **n**) is convertible to *s*, where *s* is the *n*-th positive integer *t* (if there is one) for which **T**(**t**) conv 2; if **T**(**t**) conv 2 for less than *n* values of *t* then $\mathcal{P}$(**T**, **n**) has no normal form. The formula **G**($\mathcal{P}$(**G**, **n**)) is therefore convertible to 2 if and only if $\theta(x)$ vanishes for at least *n* values of *x*, and is convertible to 2 for every positive integer *x* if and only if $\theta(x)$ vanishes infinitely often. To prove the second half of the assertion, we take Gödel representations for the formulae of the conversion calculus. Let $c(x)$ be 0 if x is the G.R. of 2 (*i.e.* if x is $2^3 . 3^{10} . 5 . 7^3 .$ $11^{28} . 13 . 17 . 19^{10} . 23^2 . 29 . 31 . 37^{10} . 41^2 . 43 . 47^{28} . 53^2 . 59^2 . 61^2 . 67^2$) and let $c(x)$ be 1 otherwise. Take an enumeration of the G.R. of the formulae into which **A**(**m**) is convertible: let $a(m, n)$ be the *n*-th number in the enumeration. We can arrange the enumeration so that $a(m, n)$ is primitive recursive. Now the statement that **A**(**m**) is convertible to 2 for every positive integer *m* is equivalent to the statement that, corresponding to each positive integer m, there is a positive integer n such that $c(a(m, n)) = 0$; and this is number-theoretic.

It is easy to show that a number of unsolved problems, such as the problem of the truth of Fermat's last theorem, are number-theoretic. There are, however, also problems of analysis which are number-theoretic. The Riemann hypothesis gives us an example of this. We denote by $\zeta(s)$ the function defined for $\Re s = \sigma > 1$ by the series $\sum_{n=1}^{\infty} n^{-s}$ and over the rest of the complex plane with the exception of the point $s = 1$ by analytic continuation. The Riemann hypothesis asserts that this function does not vanish in the domain $\sigma > \frac{1}{2}$. It is easily shown that this is equivalent to saying that it does not vanish for $2 > \sigma > \frac{1}{2}$, $\Im s = t > 2$, i.e. that it does not vanish inside any rectangle $2 > \sigma > \frac{1}{2} + 1/T$, $T > t > 2$, where T is an integer greater than 2. Now the function satisfies the inequalities

$$\left|\zeta(s) - \sum_{1}^{N} n^{-s} - \frac{N^{1-s}}{s-1}\right| < 2t(N-2)^{-\frac{1}{2}}, \quad 2 < \sigma < \frac{1}{2}, \quad t \geqslant 2,$$
$$|\zeta(s) - \zeta(s')| < 60t|s - s'|, \quad 2 < \sigma' < \frac{1}{2}, \quad t' \geqslant 2,$$

and we can define a primitive recursive function $\xi(l, l', m, m', N, M)$ such that

$$\left|\xi(l, l', m, m', N, M) - M\left|\sum_{1}^{N} n^{-s} + \frac{N^{1-s}}{s-1}\right|\right| < 2, \quad \left(s = \frac{l}{l'} + i\frac{m}{m'}\right),$$

and therefore, if we put

$$\xi(l, M, m, M, M^2 + 2, M) = X(l, m, M),$$

we have

$$\left|\zeta\left(\frac{l+\vartheta}{M} + i\frac{m+\vartheta}{M}\right)\right| \geqslant \frac{X(l, m, M) - 122T}{M},$$

provided that

$$\frac{1}{2} + \frac{1}{T} \leqslant \frac{l-1}{M} < \frac{l+1}{M} < 2 - \frac{1}{M}, \quad 2 < \frac{m-1}{M} < \frac{m+1}{M} < T$$
$$(-1 < \vartheta < 1, \; -1 < \vartheta' < 1).$$

If we define $B(M, T)$ to be the smallest value of $X(l, m, M)$ for which

$$\frac{1}{2} + \frac{1}{T} + \frac{1}{M} \leqslant \frac{l}{M} < 2 - \frac{1}{M}, \quad 2 + \frac{1}{M} < \frac{m}{M} < T - \frac{1}{M},$$

then the Riemann hypothesis is true if for each T there is an M satisfying

$$B(M, T) > 122T.$$

156 | Alan Turing

If on the other hand there is a T such that, for all M, $B(M, T) \leq 122T$, the Riemann hypothesis is false; for let l_M, m_M be such that

$$X(l_M, m_M, M) \leq 122T,$$

then

$$\left| \zeta\left(\frac{l_M + im_M}{M}\right) \right| \leq \frac{244T}{M}.$$

Now if a is a condensation point of the sequence $(l_M + im_M)/M$ then since $\zeta(s)$ is continuous except at $s = 1$ we must have $\zeta(a) = 0$ implying the falsity of the Riemann hypothesis. Thus we have reduced the problem to the question whether for each T there is an M for which

$$B(M, T) > 122T.$$

$B(M, T)$ is primitive recursive, and the problem is therefore number-theoretic.

4. A type of problem which is not number-theoretic[8]

Let us suppose that we are supplied with some unspecified means of solving number-theoretic problems; a kind of oracle as it were. We shall not go any further into the nature of this oracle apart from saying that it cannot be a machine. With the help of the oracle we could form a new kind of machine (call them o-machines), having as one of its fundamental processes that of solving a given number-theoretic problem. More definitely these machines are to behave in this way. The moves of the machine are determined as usual by a table except in the case of moves from a certain internal configuration o. If the machine is in the internal configuration o and if the sequence of symbols marked with l is then the well-formed[9] formula A, then the machine goes into the internal configuration p or t according as it is or is not true that A is dual. The decision as to which is the case is referred to the oracle.

These machines may be described by tables of the same kind as those used for the description of a-machines, there being no entries, however, for the internal configuration o. We obtain description numbers from these tables in the same way as before. If we make the convention that, in assigning numbers to internal configurations, o, p, t are always to be q_2, q_3, q_4, then the description numbers determine the behaviour of the machines uniquely.

Given any one of these machines we may ask ourselves the question whether or not it prints an infinity of figures 0 or 1; I assert that this class of problem is not number-theoretic. In view of the definition of "number-theoretic problem" this means that it is not possible to construct an o-machine which, when supplied

---
[8] Compare Rosser [1].
[9] Without real loss of generality we may suppose that A is always well formed.

with the description of any other o-machine, will determine whether that machine is o-circle free.[10] The argument may be taken over directly from Turing [1], §8. We say that a number is o-satisfactory if it is the description number of an o-circle free machine. Then, if there is an o-machine which will determine of any integer whether it is o-satisfactory, there is also an o-machine to calculate the values of the function $1 - \phi_n(n)$. Let $r(n)$ be the n-th o-satisfactory number and let $\phi_n(m)$ be the m-th figure printed by the o-machine whose description number is $r(n)$. This o-machine is circle free and there is therefore an o-satisfactory number K such that $\phi_K(n) = 1 - \phi_n(n)$ for all n. Putting $n = K$ yields a contradiction. This completes the proof that problems of circle freedom of o-machines are not number-theoretic.

Propositions of the form that an o-machine is o-circle free can always be put in the form of propositions obtained from formulae of the functional calculus of the first order by replacing *some* of the functional variables by primitive recursive relations. Compare foot-note [5] on page [152].

5. Syntactical theorems as number-theoretic theorems

I now mention a property of number-theoretic theorems which suggests that there is reason for regarding them as of particular importance.

Suppose that we have some axiomatic system of a purely formal nature. We do not concern ourselves at all in interpretations for the formulae of this system; they are to be regarded as of interest for themselves. An example of what is in mind is afforded by the conversion calculus (§1). Every sequence of symbols "A conv B", where A and B are well formed formulae, is a formula of the axiomatic system and is provable if the W.F.F. A is convertible to B. The rules of conversion give us the rules of procedure in this axiomatic system.

Now consider a new rule of procedure which is reputed to yield only formulae provable in the original sense. We may ask ourselves whether such a rule is valid. The statement that such a rule is valid would be number-theoretic. To prove this, let us take Gödel representations for the formulae, and an enumeration of the provable formulae; let $\phi(r)$ be the G.R. of the r-th formula in the enumeration. We may suppose $\phi(r)$ to be primitive recursive if we are prepared to allow repetitions in the enumeration. Let $\psi(r)$ be the G.R. of the r-th formula obtained by the new rule, then the statement that this new rule is valid is equivalent to the assertion of

$$(r)(\exists s)[\psi(r) = \phi(s)]$$

(the domain of individuals being the natural numbers). It has been shown in §3 that such statements are number-theoretic.

---

[10] Compare Turing [1], §6, 7.

It might plausibly be argued that all those theorems of mathematics which have any significance when taken alone are in effect syntactical theorems of this kind, stating the validity of certain "derived rules" of procedure. Without going so far as this, I should assert that theorems of this kind have an importance which makes it worth while to give them special consideration.

6. Logic formulae

We shall call a formula L a *logic formula* (or, if it is clear that we are speaking of a W.F.F., simply a *logic*) if it has the property that, if A is a formula such that L(A) conv 2, then A is dual.

A logic formula gives us a means of satisfying ourselves of the truth of number-theoretic theorems. For to each number-theoretic proposition there corresponds a W.F.F. A which is dual if and only if the proposition is true. Now, if L is a logic and L(A) conv 2, then A is dual and we know that the corresponding number-theoretic proposition is true. It does not follow that, if L is a logic, we can use L to satisfy ourselves of the truth of *any* number-theoretic theorem.

If L is a logic, the set of formulae A for which L(A) conv 2 will be called the *extent* of L.

It may be proved by the use of (D), (E), p. [150], that there is a formula X such that, if M has a normal form, has no free variables and is not convertible to 2, then $X(M)$ conv 1, but, if M conv 2, then $X(M)$ conv 2. If L is a logic, then $\lambda x . X(L(x))$ is also a logic whose extent is the same as that of L, and which has the property that, if A has no free variables, then

$$\{\lambda x . X(L(x))\}(A)$$

either is always convertible to 1 or to 2 or else has no normal form. A logic with this property will be said to be *standardized*.

We shall say that a logic L′ is *at least as complete as* a logic L if the extent of L is a subset of the extent of L′. The logic L′ is *more complete than* L if the extent of L is a proper subset of the extent of L′.

Suppose that we have an effective set of rules by which we can prove formulae to be dual; *i.e.* we have a system of symbolic logic in which the propositions proved are of the form that certain formulae are dual. Then we can find a logic formula whose extent consists of just those formulae which can be proved to be dual by the rules; that is to say, there is a rule for obtaining the logic formula from the system of symbolic logic. In fact the system of symbolic logic enables us to obtain a computable function of positive integers whose values run through the Gödel representations of the formulae provable by means of the given rules.[11]

[11] Compare Turing [1], [p. 77, n. 7], [2], 156.

Systems of Logic Based on Ordinals | 159

By the theorem of equivalence of computable and λ-definable functions, there is a formula J such that J(1), J(2), ... are the G.R. of these formulae. Now let

$$W \to \lambda jv \,.\, \mathcal{P}(\lambda u \,.\, \delta(j(u), v), 1, I, 2).$$

Then I assert that W(J) is a logic with the required properties. The properties of $\mathcal{P}$ imply that $\mathcal{P}(C, 1)$ is convertible to the least positive integer **n** for which C(**n**) conv 2, and has no normal form if there is no such integer. Consequently $\mathcal{P}(C, 1, I, 2)$ is convertible to 2 if C(**n**) conv 2 for some positive integer n, and it has no normal form otherwise. That is to say that W(J, A) conv 2 if and only if $\delta(J(n), A)$ conv 2, some n, *i.e.* if J(**n**) conv A some n.

There is conversely a formula W' such that, if L is a logic, then W'(L) enumerates the extent of L. For there is a formula Q such that $Q(L, A, n)$ conv 2 if and only if L(A) is convertible to 2 in less than n steps. We then put

$$W' \to \lambda ln \,.\, \text{form}\, (\varpi(2, \mathcal{P}(\lambda x \,.\, Q(l, \text{form}\, (\varpi(2, x)), \varpi(3, x)), n))).$$

Of course, W'(W(J)) normally entirely different from J and W(W'(L)) from L.

In the case where we have a symbolic logic whose propositions can be interpreted as number-theoretic theorems, but are not expressed in the form of the duality of formulae, we shall again have a corresponding logic formula, but its relation to the symbolic logic is not so simple. As an example let us take the case where the symbolic logic proves that certain primitive recursive functions vanish infinitely often. As was shown in §3, we can associate with each such proposition a W.F.F. which is dual if and only if the proposition is true. When we replace the propositions of the symbolic logic by theorems on the duality of formulae in this way, our previous argument applies and we obtain a certain logic formula L. However, L does not determine uniquely which are the propositions provable in the symbolic logic; for it is possible that "$\theta_1(x)$ vanishes infinitely often" and "$\theta_2(x)$ vanishes infinitely often" are both associated with "A is dual", and that the first of these propositions is provable in the system, but the second not. However, if we suppose that the system of symbolic logic is sufficiently powerful to be able to carry out the argument on pp. [154–5] then this difficulty cannot arise. There is also the possibility that there may be formulae in the extent of L with no propositions of the form "$\theta(x)$ vanishes infinitely often" corresponding to them. But to each such formula we can assign (by a different argument) a proposition p of the symbolic logic which is a necessary and sufficient condition for A to be dual. With p is associated (in the first way) a formula A'. Now L can always be modified so that its extent contains A' whenever it contains A.

We shall be interested principally in questions of completeness. Let us suppose that we have a class of systems of symbolic logic, the propositions of these systems being expressed in a uniform notation and interpretable as number-theoretic theorems; suppose also that there is a rule by which we can assign to each proposition p of the notation a W.F.F. A_p which is dual if and only if p is true, and that to each W.F.F. **A** we can assign a proposition p_A which is a necessary and sufficient condition for **A** to be dual. p_{A_p} is to be expected to differ from p. To each symbolic logic C we can assign two logic formulae L_C and L_C'. A formula **A** belongs to the extent of L_C if p_A is provable in C, while the extent of L_C' consists of all A_p, where p is provable in C. Let us say that the class of symbolic logics is complete if each true proposition is provable in one of them: let us also say that a class of logic formulae is complete if the set-theoretic sum of the extents of these logics includes all dual formulae. I assert that a necessary condition for a class of symbolic logics C to be complete is that the class of logics L_C is complete, while a sufficient condition is that the class of logics L_C' is complete. Let us suppose that the class of symbolic logics is complete; consider p_A, where **A** is arbitrary but dual. It must be provable in one of the systems, C say. **A** therefore belongs to the extent of L_C, *i.e.* the class of logics L_C is complete. Now suppose the class of logics L_C' to be complete. Let p be an arbitrary true proposition of the notation; A_p must belong to the extent of some L_C', and this means that p is provable in C.

We shall say that a single logic formula **L** is complete if its extent includes all dual formulae; that is to say, it is complete if it enables us to prove every true number-theoretic theorem. It is a consequence of the theorem of Gödel (if suitably extended) that no logic formula is complete, and this also follows from (*C*), p. [150], or from the results of Turing [1], §8, when taken in conjunction with §3 of the present paper. The idea of completeness of a logic formula is not therefore very important, although it is useful to have a term for it.

Suppose **Y** to be a W.F.F. such that **Y(n)** is a logic for each positive integer n. The formulae of the extent of **Y(n)** are enumerated by $W(\mathbf{Y(n)})$, and the combined extents of these logics by

$$\lambda r . W(\mathbf{Y}(\varpi(2, r), \varpi(3, r))).$$

If we put

$$\Gamma \to \lambda y . W'(\lambda r . W(y(\varpi(2, r), \varpi(3, r)))),$$

then $\Gamma(\mathbf{Y})$ is a logic whose extent is the combined extent of

$$\mathbf{Y}(1), \mathbf{Y}(2), \mathbf{Y}(3), \ldots$$

To each W.F.F. **L** we can assign a W.F.F. $V(\mathbf{L})$ such that a necessary and sufficient condition for **L** to be a logic formula is that $V(\mathbf{L})$ is dual. Let Nm be a W.F.F. which enumerates all formulae with normal forms and no free variables. Then the condition for **L** to be a logic is that $L(\text{Nm}(r), s)$ conv 2 for all positive integers r, s, *i.e.* that

$$\lambda a . L(\text{Nm}(\varpi(2, a)), \varpi(3, a))$$

Systems of Logic Based on Ordinals | 161

is dual. We may therefore put

$$V \to \lambda la \,.\, l(\mathrm{Nm}(\varpi(2, a)), \varpi(3, a)).$$

7. Ordinals

We begin our treatment of ordinals with some brief definitions from the Cantor theory of ordinals, but for the understanding of some of the proofs a greater amount of the Cantor theory is necessary than is set out here.

Suppose that we have a class determined by the propositional function $D(x)$ and a relation $G(x, y)$ ordering its members, i.e. satisfying

$$\left. \begin{array}{ll} G(x, y) \,\&\, G(y, z) \supset G(x, z), & \text{(i)} \\ D(x) \,\&\, D(y) \supset G(x, y) \text{ v } G(y, x) \text{ v } x = y, & \text{(ii)} \\ G(x, y) \supset D(x) \,\&\, D(y), & \text{(iii)} \\ \sim G(x, x). & \text{(iv)} \end{array} \right\} \quad (7.1)$$

The class defined by $D(x)$ is then called a *series* with the ordering relation $G(x, y)$. The series is said to be *well ordered* and the ordering relation is called an *ordinal* if every sub-series which is not void has a first term, i.e. if

$$(D')\{(\exists x)(D'(x)) \,\&\, (x)(D'(x) \supset D(x)) \\ \supset (\exists z)(y)[D'(z) \,\&\, (D'(y) \supset G(z, y) \text{ v } z = y)]\}. \quad (7.2)$$

The condition (7.2) is equivalent to another, more suitable for our purposes, namely the condition that every descending subsequence must terminate; formally

$$(x)\{D'(x) \supset D(x) \,\&\, (\exists y)(D'(y) \,\&\, G(y, x))\} \supset (x)(\sim D'(x)). \quad (7.3)$$

The ordering relation $G(x, y)$ is said to be similar to $G'(x, y)$ if there is a one-one correspondence between the series transforming the one relation into the other. This is best expressed formally, thus

$$(\exists M)[(x)\{D(x) \supset (\exists x')M(x, x')\} \,\&\, (x')\{D'(x') \supset (\exists x)M(x, x')\} \\ \,\&\, \{(M(x, x') \,\&\, M(x, x'')) \text{ v } (M(x', x) \,\&\, M(x'', x)) \supset x' = x''\} \\ \,\&\, \{M(x, x') \,\&\, M(y, y') \supset (G(x, y) \equiv G(x', y'))\}]. \quad (7.4)$$

Ordering relations are regarded as belonging to the same ordinal if and only if they are similar.

We wish to give names to all the ordinals, but this will not be possible until they have been restricted in some way; the class of ordinals, as at present defined, is more than enumerable. The restrictions that we actually impose are these: $D(x)$ is to imply that x is a positive integer; $D(x)$ and $G(x, y)$ are to be computable properties. Both of the propositional functions $D(x)$, $G(x, y)$ can then be described by means of a single W.F.F. Ω with the properties:

$\Omega(\mathbf{m}, \mathbf{n})$ conv 4 unless both $D(m)$ and $D(n)$ are true,
$\Omega(\mathbf{m}, \mathbf{m})$ conv 3 if $D(m)$ is true,
$\Omega(\mathbf{m}, \mathbf{n})$ conv 2 if $D(m)$, $D(n)$, $G(m, n)$, $\sim (m = n)$ are true,
$\Omega(\mathbf{m}, \mathbf{n})$ conv 1 if $D(m)$, $D(n)$, $\sim G(m, n)$, $\sim (m = n)$ are true.

In consequence of the conditions to which $D(x)$, $G(x, y)$ are subjected, Ω must further satisfy:

(a) if $\Omega(\mathbf{m}, \mathbf{n})$ is convertible to 1 or 2, then $\Omega(\mathbf{m}, \mathbf{m})$ and $\Omega(\mathbf{n}, \mathbf{n})$ are convertible to 3,
(b) if $\Omega(\mathbf{m}, \mathbf{m})$ and $\Omega(\mathbf{n}, \mathbf{n})$ are convertible to 3, then $\Omega(\mathbf{m}, \mathbf{n})$ is convertible to 1, 2, or 3,
(c) if $\Omega(\mathbf{m}, \mathbf{n})$ is convertible to 1, then $\Omega(\mathbf{n}, \mathbf{m})$ is convertible to 2 and conversely,
(d) if $\Omega(\mathbf{m}, \mathbf{n})$ and $\Omega(\mathbf{n}, \mathbf{p})$ are convertible to 1, then $\Omega(\mathbf{m}, \mathbf{p})$ is also,
(e) there is no sequence $m_1, m_2, \ldots$ such that $\Omega(\mathbf{m}_{i+1}, \mathbf{m}_i)$ conv 2 for each positive integer i,
(f) $\Omega(\mathbf{m}, \mathbf{n})$ is always convertible to 1, 2, 3, or 4.

If a formula Ω satisfies these conditions then there are corresponding propositional functions $D(x)$, $G(x, y)$. We shall therefore say that Ω is an *ordinal formula* if it satisfies the conditions (a)–(f). It will be seen that a consequence of this definition is that Dt is an ordinal formula; it represents the ordinal ω. The definition that we have given does not pretend to have virtues such as elegance or convenience. It has been introduced rather to fix our ideas and to show how it is possible in principle to describe ordinals by means of well-formed formulae. The definitions could be modified in a number of ways. Some such modifications are quite trivial; they are typified by modifications such as changing the numbers 1, 2, 3, 4, used in the definition, to others. Two such definitions will be said to be equivalent; in general, we shall say that two definitions are equivalent if there are W.F.F. T, T' such that, if A is an ordinal formula under one definition and represents the ordinal α, then T'(A) is an ordinal formula under the second definition and represents the same ordinal; and, conversely, if A' is an ordinal formula under the second definition representing α, then T(A') represents α under the first definition. Besides definitions equivalent in this sense to our original definition, there are a number of other possibilities open. Suppose for instance that we do not require $D(x)$ and $G(x, y)$ to be computable, but that we require only that $D(x)$ and $G(x, y) \& x < y$ are axiomatic.[12] This leads to a

---

[12] To require $G(x, y)$ to be axiomatic amounts to requiring $G(x, y)$ to be computable on account of (7.1) (ii).

definition of an ordinal formula which is (presumably) not equivalent to the definition that we are using.[13] There are numerous possibilities, and little to guide us in choosing one definition rather than another. No one of them could well be described as "wrong"; some of them may be found more valuable in applications than others, and the particular choice that we have made has been determined partly by the applications that we have in view. In the case of theorems of a negative character, one would wish to prove them for each one of the possible definitions of "ordinal formula". This programme could, I think, be carried through for the negative results of §9, 10.

Before leaving the subject of possible ways of defining ordinal formulae, I must mention another definition due to Church and Kleene (Church and Kleene [1]). We can make use of this definition in constructing ordinal logics, but it is more convenient to use a slightly different definition which is equivalent (in the sense just described) to the Church–Kleene definition as modified in Church [4].

Introduce the abbreviations

$$U \to \lambda ufx \,.\, u(\lambda y \,.\, f(y(I, x))),$$
$$\text{Suc} \to \lambda aufx \,.\, f(a(u, f, x)).$$

We define first a partial ordering relation "<" which holds between certain pairs of W.F.F. [conditions (1)–(5)].

(1) If A conv B, then A < C implies B < C and C < A implies C < B.
(2) A < Suc (A).
(3) For any positive integers m and n, $\lambda ufx \,.\, R(n) < \lambda ufx \,.\, R(m)$ implies $\lambda ufx \,.\, R(n) < \lambda ufx \,.\, u(R)$.
(4) If A < B and B < C, then A < C. (1)–(4) are required for any W.F.F. A, B, C, $\lambda ufx \,.\, R$.
(5) The relation A < B holds only when compelled to do so by (1)–(4).

We define C-K ordinal formulae by the conditions (6)–(10).

(6) If A conv B and A is a C-K ordinal formula, then B is a C-K ordinal formula.
(7) U is a C-K ordinal formula.
(8) If A is a C-K ordinal formula, then Suc (A) is a C-K ordinal formula.
(9) If $\lambda ufx \,.\, R(n)$ is a C-K ordinal formula and

$$\lambda ufx \,.\, R(n) < \lambda ufx \,.\, R(S(n))$$

---

[13] On the other hand, if $D(x)$ is axiomatic and $G(x, y)$ is computable in the modified sense that there is a rule for determining whether $G(x, y)$ is true which leads to a definite result in all cases where $D(x)$ and $D(y)$ are true, the corresponding definition of ordinal formula is equivalent to our definition. To give the proof would be too much of a digression. Probably other equivalences of this kind hold.

for each positive integer n, then $\lambda ufx \,.\, u(R)$ is a C-K ordinal formula.[14]

(10) A formula is a C-K ordinal formula only if compelled to be so by (6)–(9).

The representation of ordinals by formulae is described by (11)–(15).

(11) If **A** conv **B** and **A** represents α, then **B** represents α.
(12) U represents 1.
(13) If **A** represents α, then Suc (**A**) represents $\alpha + 1$.
(14) If $\lambda ufx \,.\, R(n)$ represents α_n for each positive integer n, then $\lambda ufx \,.\, u(R)$ represents the upper bound of the sequence $\alpha_1, \alpha_2, \alpha_3, \ldots$
(15) A formula represents an ordinal only when compelled to do so by (11)–(14).

We denote any ordinal represented by **A** by Ξ_A without prejudice to the possibility that more than one ordinal may be represented by **A**. We shall write $A \leqslant B$ to mean $A < B$ or A conv B.

In proving properties of C-K ordinal formulae we shall often use a kind of analogue of the principle of transfinite induction. If ϕ is some property and we have:

(a) If **A** conv **B** and $\phi(A)$, then $\phi(B)$,
(b) $\phi(U)$,
(c) If $\phi(A)$, then $\phi(\text{Suc }(A))$,
(d) If $\phi(\lambda ufx \,.\, R(n))$ and $\lambda ufx \,.\, R(n) < \lambda ufx \,.\, R(S(n))$ for each positive integer n, then $\phi(\lambda ufx \,.\, u(R))$; \hfill (7.5)

then $\phi(A)$ for each C-K ordinal formula **A**. To prove the validity of this principle we have only to observe that the class of formulae **A** satisfying $\phi(A)$ is one of those of which the class of C-K ordinal formulae was defined to be the smallest. We can use this principle to help us to prove:

(i) Every C-K ordinal formula is convertible to the form $\lambda ufx \,.\, B$, where **B** is in normal form.
(ii) There is a method by which for any C-K ordinal formula, we can determine into which of the forms U, Suc ($\lambda ufx \,.\, B$), $\lambda ufx \,.\, u(R)$ (where

---

[14] If we also allow $\lambda ufx \,.\, u(R)$ to be a C-K ordinal formula when

$$\lambda ufx \,.\, n(R) \text{ conv } \lambda ufx \,.\, S(n, R)$$

for all n, then the formulae for sum, product and exponentiation of C-K ordinal formulae can be much simplified. For instance, if **A** and **B** represent α and β, then

$$\lambda ufx \,.\, B(u, f, A(u, f, x))$$

represents $\alpha + \beta$. Property (6) remains true.

u is free in R) it is convertible, and by which we can determine B, R. In each case B, R are unique apart from conversions.

(iii) If A represents any ordinal, Ξ_A is unique. If Ξ_A, Ξ_B exist and $A < B$, then $\Xi_A < \Xi_B$.

(iv) If A, B, C are C-K ordinal formulae and $B < A$, $C < A$, then either $B < C$, $C < B$, or B conv C.

(v) A formula A is a C-K ordinal formula if:
   (A) $U \leqslant A$,
   (B) If $\lambda ufx \, . \, u(R) \leqslant A$ and n is a positive integer, then
   $$\lambda ufx \, . \, R(n) < \lambda ufx \, . \, R(S(n)),$$
   (C) For any two W.F.F. B, C with $B < A$, $C < A$ we have $B < C$, $C < B$, or B conv C, but never $B < B$,
   (D) There is no infinite sequence $B_1, B_2, \ldots$ for which
   $$B_r < B_{r-1} < A$$
   for each r.

(vi) There is a formula H such that, if A is a C-K ordinal formula, then $H(A)$ is an ordinal formula representing the same ordinal. $H(A)$ is not an ordinal formula unless A is a C-K ordinal formula.

Proof of (i). Take $\phi(A)$ to be "A is convertible to the form $\lambda ufx \, . \, B$, where B is in normal form". The conditions (*a*) and (*b*) are trivial. For (*c*), suppose that A conv $\lambda ufx \, . \, B$, where B is in normal form; then

$$\text{Suc } (A) \text{ conv } \lambda ufx \, . \, f(B)$$

and $f(B)$ is in normal form. For (*d*) we have only to show that $u(R)$ has a normal form, *i.e.* that R has a normal form; and this is true since $R(1)$ has a normal form.

Proof of (ii). Since, by hypothesis, the formula is a C-K ordinal formula we have only to perform conversions on it until it is in one of the forms described. It is not possible to convert it into two of these three forms. For suppose that $\lambda ufx \, . \, f(A(u, f, x))$ conv $\lambda ufx \, . \, u(R)$ and is a C-K ordinal formula; it is then convertible to the form $\lambda ufx \, . \, B$, where B is in normal form. But the normal form of $\lambda ufx \, . \, u(R)$ can be obtained by conversions on R, and that of $\lambda ufx \, . \, f(A(u, f, x))$ by conversions on $A(u, f, x)$ (as follows from Church and Rosser [1], Theorem 2); this, however, would imply that the formula in question had two normal forms, one of form $\lambda ufx \, . \, u(S)$ and one of form $\lambda ufx \, . \, f(C)$, which is impossible. Or let U conv $\lambda ufx \, . \, u(R)$, where R is a well formed formula with u as a free variable. We may suppose R to be in normal form. Now U is $\lambda ufx \, . \, u(\lambda y \, . \, f(y(I, x)))$. By (A), p. [149], R is identical with $\lambda y \, . \, f(y(I, x))$, which does not have u as a free variable. It now remains to show only that if

Suc ($\lambda ufx . B$) conv Suc ($\lambda ufx . B'$) and $\lambda ufx . u(R)$ conv $\lambda ufx . u(R')$, then B conv B' and R conv R'.

If
$$\text{Suc } (\lambda ufx . B) \text{ conv Suc } (\lambda ufx . B'),$$
then
$$\lambda ufx . f(B) \text{ conv } \lambda ufx . f(B');$$

but both of these formulae can be brought to normal form by conversions on B, B' and therefore B conv B'. The same argument applies in the case in which $\lambda ufx . u(R)$ conv $\lambda ufx . u(R')$.

Proof of (iii). To prove the first half, take $\phi(A)$ to be "Ξ_A is unique". Then (7.5) (*a*) is trivial, and (*b*) follows from the fact that U is not convertible either to the form Suc (A) or to $\lambda ufx . u(R)$, where R has *u* as a free variable. For (*c*): Suc (A) is not convertible to the form $\lambda ufx . u(R)$; the possibility that Suc (A) represents an ordinal on account of (12) or (14) is therefore eliminated. By (13), Suc (A) represents $\alpha' + 1$ if A' represents α' and Suc (A) conv Suc (A'). If we suppose that A represents α, then A, A', being C-K ordinal formulae, are convertible to the forms $\lambda ufx . B$, $\lambda ufx . B'$; but then, by (ii), B conv B', *i.e.* A conv A', and therefore $\alpha = \alpha'$ by the hypothesis $\phi(A)$. Then $\Xi_{\text{Suc }(A)} = \alpha' + 1$ is unique. For (*d*): $\lambda ufx . u(R)$ is not convertible to the form Suc (A) or to U if R has *u* as a free variable. If $\lambda ufx . u(R)$ represents an ordinal, it is so therefore in virtue of (14), possibly together with (11). Now, if $\lambda ufx . u(R)$ conv $\lambda ufx . u(R')$, then R conv R', so that the sequence $\lambda ufx . R(1)$, $\lambda ufx . R(2)$, ... in (14) is unique apart from conversions. Then, by the induction hypothesis, the sequence $\alpha_1, \alpha_2, \alpha_3, \ldots$ is unique. The only ordinal that is represented by $\lambda ufx . u(R)$ is the upper bound of this sequence; and this is unique.

For the second half we use a type of argument rather different from our transfinite induction principle. The formulae B for which $A < B$ form the smallest class for which:

Suc (A) belongs to the class.

If C belongs to the class, then Suc (C) belongs to it.

If $\lambda ufx . R(n)$ belongs to the class and
$$\lambda ufx . R(n) < \lambda ufx . R(m),$$
where *m, n* are some positive integers, then $\lambda ufx . u(R)$ belongs to it.

If C belongs to the class and C conv C', then C' belongs to it. (7.6)

It will be sufficient to prove that the class of formulae B for which either Ξ_B does not exist or $\Xi_A < \Xi_B$ satisfies the conditions (7.6). Now

$$\Xi_{\text{Suc }(A)} = \Xi_A + 1 > \Xi_A,$$
$$\Xi_{\text{Suc }(C)} > \Xi_C > \Xi_A \text{ if C is in the class.}$$

If $\Xi_{\lambda ufx \cdot R(n)}$ does not exist, then $\Xi_{\lambda ufx \cdot u(R)}$ does not exist, and therefore $\lambda ufx \cdot u(R)$ is in the class. If $\Xi_{\lambda ufx \cdot R(n)}$ exists and is greater than Ξ_A, and $\lambda ufx \cdot R(n) < \lambda ufx \cdot R(m)$, then

$$\Xi_{\lambda ufx \cdot u(R)} \geq \Xi_{\lambda ufx \cdot R(n)} > \Xi_A,$$

so that $\lambda ufx \cdot u(R)$ belongs to the class.

Proof of (iv). We prove this by induction with respect to A. Take $\phi(A)$ to be "whenever $B < A$ and $C < A$ then $B < C$ or $C < B$ or B conv C". $\phi(U)$ follows from the fact that we never have $B < U$. If we have $\phi(A)$ and $B <$ Suc (A), then either $B < A$ or B conv A; for we can find D such that $B \leq D$, and then $D <$ Suc (A) can be proved without appealing either to (1) or (5); (4) does not apply, so we must have D conv A. Then, if $B <$ Suc (A) and $C <$ Suc (A), we have four possibilities,

B conv A,	C conv A,
B conv A,	$C < A$,
$B < A$,	C conv A,
$B < A$,	$C < A$.

In the first case B conv C, in the second $C < B$, in the third $B < C$, and in the fourth the induction hypothesis applies.

Now suppose that $\lambda ufx \cdot R(n)$ is a C-K ordinal formula, that

$$\lambda ufx \cdot R(n) < \lambda ufx \cdot R(S(n)) \text{ and } \phi(R(n)),$$

for each positive integer n, and that A conv $\lambda ufx \cdot u(R)$. Then, if $B < A$, this means that $B < \lambda ufx \cdot R(n)$ for some n; if we have also $C < A$, then $B < \lambda ufx \cdot R(q)$, $C < \lambda ufx \cdot R(q)$ for some q. Thus, for these B and C, the required result follows from $\phi(\lambda ufx \cdot R(q))$.

Proof of (v). The conditions (C), (D) imply that the classes of interconvertible formulae B, $B < A$ are well-ordered by the relation "<". We prove (v) by (ordinary) transfinite induction with respect to the order type α of the series formed by these classes; (α is, in fact, the solution of the equation $1 + \alpha = \Xi_A$, but we do not need this). We suppose then that (v) is true for all order types less than α. If $E < A$, then E satisfies the conditions of (v) and the corresponding order type is smaller: E is therefore a C-K ordinal formula. This expresses all consequences of the induction hypothesis that we need. There are three cases to consider:

(x) $\alpha = 0$.
(y) $\alpha = \beta + 1$.
(z) α is of neither of the forms (x), (y).

In case (x) we must have A conv U on account of (A). In case (y) there is a formula D such that $D < A$, and $B \leq D$ whenever $B < A$. The relation $D < A$ must hold in virtue of either (1), (2), (3), or (4). It cannot be in virtue of (4); for then there would be B, $B < A$, $D < B$ contrary to (C), taken in conjunction with the definition of D. If it is in virtue of (3), then α is the upper bound of a sequence $\alpha_1, \alpha_2, \ldots$ of ordinals, which are increasing by reason of (iii) and the conditions $\lambda ufx . R(n) < \lambda ufx . R(S(n))$ in (B). This is inconsistent with $\alpha = \beta + 1$. This means that (2) applies [after we have eliminated (1) by suitable conversions on A, D] and we see that A conv Suc (D); but, since $D < A$, D is a C-K ordinal formula, and A must therefore be a C-K ordinal formula by (8). Now take case (z). It is impossible for A to be of the form Suc (D), for then we should have $B < D$ whenever $B < A$, and this would mean that we had case (y). Since $U < A$, there must be an F such that $F < A$ is demonstrable either by (2) or by (3) (after a possible conversion on A); it must of course be demonstrable by (3). Then A is of the form $\lambda ufx . u(R)$. By (3), (B) we see that $\lambda ufx . R(n) < A$ for each positive integer n; each $\lambda ufx . R(n)$ is therefore a C-K ordinal formula. Applying (9), (B) we see that A is a C-K ordinal formula.

Proof of (vi). To prove the first half, it is sufficient to find a method whereby from a C-K ordinal formula A we can find the corresponding ordinal formula Ω. For then there is a formula H_1 such that $H_1(\mathbf{a})$ conv $\mathbf{p}$ if a is the G.R. of A and p is that of Ω. H is then to be defined by

$$H \to \lambda a . \text{form} \, (H_1(\text{Gr}(a))).$$

The method of finding Ω may be replaced by a method of finding $\Omega(\mathbf{m}, \mathbf{n})$, given A and any two positive integers m, n. We shall arrange the method so that, whenever A is not an ordinal formula, either the calculation of the values does not terminate or else the values are not consistent with Ω being an ordinal formula. In this way we can prove the second half of (vi).

Let Ls be a formula such that Ls(A) enumerates the classes of formulae B, $B < A$ [*i.e.* if $B < A$ there is one and only one positive integer n for which Ls(A, n) conv B]. Then the rule for finding the value of $\Omega(\mathbf{m}, \mathbf{n})$ is as follows:

First determine whether $U \leq A$ and whether A is convertible to the form r(Suc, U). This terminates if A is a C-K ordinal formula.

If A conv r(Suc, U) and either $m > r + 1$ or $n > r + 1$, then the value is 4. If $m < n \leq r + 1$, the value is 2. If $n < m \leq r + 1$, the value is 1. If $m = n \leq r + 1$, the value is 3.

If A is not convertible to this form, we determine whether either A or Ls(A, m) is convertible to the form $\lambda ufx . u(R)$; and if either of them is, we verify that $\lambda ufx . R(n) < \lambda ufx . R(S(n))$. We shall eventually come to an affirmative answer if A is a C-K ordinal formula.

Having checked this, we determine concerning m and n whether Ls(A, m) < Ls(A, n), Ls(A, n) < Ls(A, m), or $m = n$, and the value is to be accordingly 1, 2, or 3.

If A is a C-K ordinal formula, this process certainly terminates. To see that the values so calculated correspond to an ordinal formula, and one representing Ξ_A, first observe that this is so when Ξ_A is finite. In the other case (iii) and (iv) show that Ξ_B determines a one–one correspondence between the ordinals β, $1 \leq \beta \leq \Xi_A$, and the classes of interconvertible formulae B, B < A. If we take G(m, n) to be Ls(A, m) < Ls(A, n), we see that G(m, n) is the ordering relation of a series of order type Ξ_A and on the other hand that the values of $\Omega(m, n)$ are related to G(m, n) as on p. [162].[15]

To prove the second half suppose that A is not a C-K ordinal formula. Then one of the conditions (A)–(D) in (v) must not be satisfied. If (A) is not satisfied we shall not obtain a result even in the calculation of $\Omega(1, 1)$. If (B) is not satisfied, we shall have for some positive integers p and q,

$$\text{Ls(A, p) conv } \lambda ufx . u(R)$$

but not $\lambda ufx . R(q) < \lambda ufx . R(S(q))$. Then the process of calculating $\Omega(p, q)$ will not terminate. In case of failure of (C) or (D) the values of $\Omega(m, n)$ may all be calculable, but if so conditions (a)–(f), p. [162], will be violated. Thus, if A is not a C-K ordinal formula, then H(A) is not an ordinal formula.

I propose now to define three formulae Sum, Lim, Inf of importance in connection with ordinal formulae. Since they are comparatively simple, they will for once be given almost in full. The formula Ug is one with the property that Ug(m) is convertible to the formula representing the largest odd integer dividing m: it is not given in full. P is the predecessor function; $P(S(m))$ conv m, $P(1)$ conv 1.

$$\text{Al} \to \lambda pxy . p(\lambda guv . g(v, u), \lambda uv . u(I, v), x, y),$$
$$\text{Hf} \to \lambda m . P(m(\lambda guv . g(v, S(u)), \lambda uv . v(I, u), 1, 2)),$$
$$\text{Bd} \to \lambda ww'aa'x . \text{Al}(\lambda f . w(a, a, w'(a', a', f)), x, 4),$$
$$\text{Sum} \to \lambda ww'pq . \text{Bd}(w, w', \text{Hf}(p), \text{Hf}(q),$$
$$\text{Al}(p, \text{Al}(q, w'(\text{Hf}(p), \text{Hf}(q)), 1), \text{Al}(S(q), w(\text{Hf}(p), \text{Hf}(q)), 2))),$$
$$\text{Lim} \to \lambda zpq . \{\lambda ab . \text{Bd}(z(a), z(b), \text{Ug}(p), \text{Ug}(q), \text{Al}(\text{Dt}(a, b) + \text{Dt}(b, a),$$
$$\text{Dt}(a, b), z(a, \text{Ug}(p), \text{Ug}(q))))\} (\varpi(2, p), \varpi(2, q)),$$
$$\text{Inf} \to \lambda wapq . \text{Al}(\lambda f . w(a, p, w(a, q, f)), w(p, q), 4).$$

The essential properties of these formulae are described by:

---

[15] The order type is β, where $1 + \beta = \Xi_A$; but $\beta = \Xi_A$, since Ξ_A is infinite.

Al(2r − 1, m, n) conv m,   Al(2r, m, n) conv n,
Hf(2m) conv m,   Hf(2m − 1) conv m,
Bd(Ω, Ω', a, a', x) conv 4, unless both

$$\Omega(a, a) \text{ conv } 3 \text{ and } \Omega'(a', a') \text{ conv } 3,$$

it is then convertible to x.

If Ω, Ω' are ordinal formulae representing α, β respectively, then Sum(Ω, Ω') is an ordinal formula representing $\alpha + \beta$. If Z is a W.F.F. enumerating a sequence of ordinal formulae representing $\alpha_1, \alpha_2, \ldots$, then Lim(Z) is an ordinal formula representing the infinite sum $\alpha_1 + \alpha_2 + \alpha_3 \ldots$ If Ω is an ordinal formula representing α, then Inf(Ω) enumerates a sequence of ordinal formulae representing all the ordinals less than α without repetitions other than repetitions of the ordinal 0.

To prove that there is no general method for determining about a formula whether it is an ordinal formula, we use an argument akin to that leading to the Burali-Forti paradox; but the emphasis and the conclusion are different. Let us suppose that such an algorithm is available. This enables us to obtain a recursive enumeration $\Omega_1, \Omega_2, \ldots$ of the ordinal formulae in normal form. There is a formula Z such that Z(n) conv Ω_n. Now Lim (Z) represents an ordinal greater than any represented by an Ω_n, and it has therefore been omitted from the enumeration.

This argument proves more than was originally asserted. In fact, it proves that, if we take any class E of ordinal formulae in normal form, such that, if **A** is any ordinal formula, then there is a formula in E representing the same ordinal as **A**, then there is no method whereby one can determine whether a W.F.F. in normal form belongs to E.

8. Ordinal logics

An ordinal logic is a W.F.F. Λ such that $\Lambda(\Omega)$ is a logic formula whenever Ω is an ordinal formula.

This definition is intended to bring under one heading a number of ways of constructing logics which have recently been proposed or which are suggested by recent advances. In this section I propose to show how to obtain some of these ordinal logics.

Suppose that we have a class W of logical systems. The symbols used in each of these systems are the same, and a class of sequences of symbols called "formulae" is defined, independently of the particular system in W. The rules of procedure of a system C define an axiomatic subset of the formulae, which are to be described as the "provable formulae of C". Suppose further that we have a method whereby, from any system C of W, we can obtain a new system C', also in W, and such that the set of provable formulae of C' includes the provable formulae of C (we shall be most interested in the case in which they are included as a proper subset). It is to be understood that this "method" is an effective procedure for obtaining the rules of procedure of C' from those of C.

Suppose that to certain of the formulae of W we make number-theoretic theorems correspond: by modifying the definition of formula, we may suppose that this is done for all formulae. We shall say that one of the systems C is *valid* if the provability of a formula in C implies the truth of the corresponding number-theoretic theorem. Now let the relation of C' to C be such that the validity of C implies the validity of C', and let there be a valid system C_0 in W. Finally, suppose that, given any computable sequence $C_1, C_2, \ldots$ of systems in W, the "limit system", in which a formula is provable if and only if it is provable in one of the systems C_j, also belongs to W. These limit systems are to be regarded, not as functions of the sequence given in extension, but as functions of the rules of formation of their terms. A sequence given in extension may be described by various rules of formation, and there will be several corresponding limit systems. Each of these may be described as *a* limit system of the sequence.

In these circumstances we may construct an ordinal logic. Let us associate positive integers with the systems in such a way that to each C there corresponds a positive integer m_C, and that m_C completely describes the rules of procedure of C. Then there is a W.F.F. **K**, such that

$$\mathbf{K}(\mathbf{m}_C) \text{ conv } \mathbf{m}_{C'}$$

for each C in W, and there is a W.F.F. Θ such that, if $\mathbf{D}(\mathbf{r})$ conv $\mathbf{m}_{C_r}$ for each positive integer r, then $\Theta(\mathbf{D})$ conv $\mathbf{m}_C$, where C is a limit system of $C_1, C_2, \ldots$ With each system C of W it is possible to associate a logic formula L_C: the relation between them is that, if G is a formula of W and the number-theoretic theorem corresponding to G (assumed expressed in the conversion calculus form) asserts that **B** is dual, then $L_C(\mathbf{B})$ conv **2** if and only if G is provable in C. There is a W.F.F. **G** such that

$$\mathbf{G}(\mathbf{m}_C) \text{ conv } \mathbf{L}_C$$

for each C of W. Put

$$\mathbf{N} \rightarrow \lambda a \,.\, \mathbf{G}(a(\Theta, \mathbf{K}, \mathbf{m}_{C_0})).$$

I assert that $\mathbf{N}(\mathbf{A})$ is a logic formula for each C-K ordinal formula $\mathbf{A}$, and that, if $\mathbf{A} < \mathbf{B}$, then $\mathbf{N}(\mathbf{B})$ is more complete than $\mathbf{N}(\mathbf{A})$, provided that there are formulae provable in C' but not in C for each valid C of W.

To prove this we shall show that to each C-K ordinal formula $\mathbf{A}$ there corresponds a unique system $C[\mathbf{A}]$ such that:

(i) $\mathbf{A}(\Theta, \mathbf{K}, \mathbf{m}_{C_0})$ conv $\mathbf{m}_{C[\mathbf{A}]}$,

and that it further satisfies:

(ii) $C[U]$ is a limit system of C_0', C_0', ...,
(iii) $C\,[\mathrm{Suc}\,(A)]$ is $(C[A])'$,
(iv) $C[\lambda ufx\,.\,u(R)]$ is a limit system of $C[\lambda ufx\,.\,R(1)]$, $C[\lambda ufx\,.\,R(2)]$, ...,

A and $\lambda ufx\,.\,u(R)$ being assumed to be C-K ordinal formulae. The uniqueness of the system follows from the fact that m_C determines C completely. Let us try to prove the existence of $C[A]$ for each C-K ordinal formula A. As we have seen (p. 164) it is sufficient to prove

(a) $C[U]$ exists,
(b) if $C[A]$ exists, then $C[\mathrm{Suc}\,(A)]$ exists,
(c) if $C[\lambda ufx\,.\,R(1)]$, $C[\lambda ufx\,.\,R(2)]$, ... exist, then $C[\lambda ufs\,.\,u(R)]$ exists.

Proof of (a).

$$\{\lambda y\,.\,K(y(I,\,m_{C_0}))\}(n) \text{ conv } K(m_{C_0}) \text{ conv } m_{C_0'}$$

for all positive integers n, and therefore, by the definition of Θ, there is a system, which we call $C[U]$ and which is a limit system of C_0', C_0', ..., satisfying

$$\Theta(\lambda y\,.\,K(y(I,\,m_{C_0}))) \text{ conv } m_{C[U]}.$$

But, on the other hand,

$$U(\Theta,\,K,\,m_{C_0}) \text{ conv } \Theta(\lambda y\,.\,K(y(I,\,m_{C_0}))).$$

This proves (a) and incidentally (ii).

Proof of (b).

$$\mathrm{Suc}\,(A,\,\Theta,\,K,\,m_{C_0}) \text{ conv } K(A(\Theta,\,K,\,m_{C_0}))$$
$$\text{conv } K(m_{C[A]})$$
$$\text{conv } m_{(C[A])'}.$$

Hence $C[\mathrm{Suc}\,(A)]$ exists and is given by (iii).

Proof of (c).

$$\{\{\lambda ufx\,.\,R\}(\Theta,\,K,\,m_{C_0})\}(n) \text{ conv } \{\lambda ufx\,.\,R(n)\}(\Theta,\,K,\,m_{C_0})$$
$$\text{conv } m_{C[\lambda ufx\,.\,R(n)]}$$

by hypothesis. Consequently, by the definition of Θ, there exists a C which is a limit system of

$$C[\lambda ufx\,.\,R(1)], \quad C[\lambda ufx\,.\,R(2)], \ldots,$$

and satisfies

$$\Theta(\{\lambda ufx\,.\,u(R)\}(\Theta,\,K,\,m_{C_0})) \text{ conv } m_C.$$

Systems of Logic Based on Ordinals | 173

We define $C[\lambda u f x . u(R)]$ to be this C. We then have (iv) and

$$\{\lambda u f x . u(R)\}(\Theta, K, m_{C_0}) \text{ conv } \Theta(\{\lambda u f x . R\}(\Theta, K, m_{C_0}))$$
$$\text{conv } m_{C[\lambda u f x . u(R)]}.$$

This completes the proof of the properties (i)–(iv). From (ii), (iii), (iv), the fact that C_0 is valid, and that C' is valid when C is valid, we infer that $C[A]$ is valid for each C-K ordinal formula A: also that there are more formulae provable in $C[B]$ than in $C[A]$ when $A < B$. The truth of our assertions regarding N now follows in view of (i) and the definitions of N and G.

We cannot conclude that N is an ordinal logic, since the formulae A are C-K ordinal formulae; but the formula H enables us to obtain an ordinal logic from N. By the use of the formula Gr we obtain a formula Tn such that, if A has a normal form, then Tn(A) enumerates the G.R.'s of the formulae into which A is convertible. Also there is a formula Ck such that, if h is the G.R. of a formula $H(B)$, then $Ck(h)$ conv B, but otherwise $Ck(h)$ conv U. Since $H(B)$ is an ordinal formula only if B is a C-K ordinal formula, $Ck(Tn(\Omega, n))$ is a C-K ordinal formula for each ordinal formula Ω and each integer n. For many ordinal formulae it will be convertible to U, but, for suitable Ω, it will be convertible to any given C-K ordinal formula. If we put

$$\Lambda \to \lambda w a . \Gamma(\lambda n . N(Ck(Tn(w, n))), a),$$

Λ is the required ordinal logic. In fact, on account of the properties of Γ, $\Lambda(\Omega, A)$ will be convertible to 2 if and only if there is a positive integer n such that

$$N(Ck(Tn(\Omega, n)), A) \text{ conv } 2.$$

If Ω conv $H(B)$, there will be an integer n such that $Ck(Tn(\Omega, n))$ conv B, and then

$$N(Ck(Tn(\Omega, n)), A) \text{ conv } N(B, A).$$

For any n, $Ck(Tn(\Omega, n))$ is convertible to U or to some B, where Ω conv $H(B)$. Thus $\Lambda(\Omega, A)$ conv 2 if Ω conv $H(B)$ and $N(B, A)$ conv 2 or if $N(U, A)$ conv 2, but not in any other case.

We may now specialize and consider particular classes W of systems. First let us try to construct the ordinal logic described roughly in the introduction. For W we take the class of systems arising from the system of *Principia Mathematica*[16] by adjoining to it axiomatic (in the sense described on p. [151]) sets of axioms.[17]

---

[16] Whitehead and Russell [1]. The axioms and rules of procedure of a similar system P will be found in a convenient form in Gödel [1], and I follow Gödel. The symbols for the natural numbers in P are $0, f0, ff0, \ldots, f^{(n)}0 \ldots$ Variables with the suffix "0" stand for natural numbers.

[17] It is sometimes regarded as necessary that the set of axioms used should be computable, the intention being that it should be possible to verify of a formula reputed to be an axiom whether it really is so. We can

Gödel has shown that primitive recursive relations can be expressed by means of formulae in P.[18] In fact, there is a rule whereby, given the recursion equations defining a primitive recursive relation, we can find a formula $\mathfrak{A}[x_0, \ldots, z_0]$ such that

$$\mathfrak{A}[f^{(m_1)}0, \ldots, f^{(m_r)}0]$$

is provable in P if $F(m_1, \ldots, m_r)$ is true, and its negation is provable otherwise.[19] Further, there is a method by which we can determine about a formula $\mathfrak{A}[x_0, \ldots, z_0]$ whether it arises from a primitive recursive relation in this way, and by which we can find the equations which defined the relation. Formulae of this kind will be called *recursion formulae*. We shall make use of a property that they possess, which we cannot prove formally here without giving their definition in full, but which is essentially trivial. $\mathrm{Db}[x_0, y_0]$ is to stand for a certain recursion formula such that $\mathrm{Db}[f^{(m)}0, f^{(n)}0]$ is provable in P if $m = 2n$ and its negation is provable otherwise. Suppose that $\mathfrak{A}[x_0], \mathfrak{B}[x_0]$ are two recursion formulae. Then the theorem which I am assuming is that there is a recursion relation $\mathfrak{C}_{\mathfrak{A}, \mathfrak{B}}[x_0]$ such that we can prove

$$\mathfrak{C}_{\mathfrak{A}, \mathfrak{B}}[x_0] \equiv (\exists y_0)((\mathrm{Db}[x_0, y_0] . \mathfrak{A}[y_0]) \vee (\mathrm{Db}[fx_0, fy_0] . \mathfrak{B}[y_0])) \qquad (8.1)$$

in P.

The significant formulae in any of our extensions of P are those of the form

$$(x_0)(\exists y_0)\mathfrak{A}[x_0, y_0], \qquad (8.2)$$

where $\mathfrak{A}[x_0, y_0]$ is a recursion formula, arising from the relation $R(m, n)$ let us say. The corresponding number-theoretic theorem states that for each natural number m there is a natural number n such that $R(m, n)$ is true.

The systems in W which are not valid are those in which a formula of the form (8.2) is provable, but at the same time there is a natural number, m say, such that, for each natural number n, $R(m, n)$ is false. This means to say that $\sim \mathfrak{A}[f^{(m)}0, f^{(n)}0]$ is provable for each natural number n. Since (8.2) is provable, $(\exists x_0)\mathfrak{A}[f^{(m)}0, y_0]$ is provable, so that

$$(\exists y_0)\mathfrak{A}[f^{(m)}0, y_0], \sim \mathfrak{A}[f^{(m)}0, 0], \sim \mathfrak{A}[f^{(m)}0, f0], \ldots \qquad (8.3)$$

are all provable in the system. We may simplify (8.3). For a given m we may prove a formula of the form $\mathfrak{A}[f^{(m)}0, y_0] \equiv \mathfrak{B}[y_0]$ in P, where $\mathfrak{B}[x_0]$ is a recursion

---

obtain the same effect with axiomatic sets of axioms in this way. In the rules of procedure describing which are the axioms, we incorporate a method of enumerating them, and we also introduce a rule that in the main part of the deduction, whenever we write down an axiom as such, we must also write down its position in the enumeration. It is possible to verify whether this has been done correctly.

[18] A relation $F(m_1, \ldots, m_r)$ is primitive recursive if it is a necessary and sufficient condition for the vanishing of a primitive recursive function $\phi(m_1, \ldots, m_r)$.

[19] Capital German letters will be used to stand for variable or undetermined formulae in P. An expression such as $\mathfrak{A}[\mathfrak{B}, \mathfrak{C}]$ stands for the result of substituting $\mathfrak{B}$ and $\mathfrak{C}$ for x_0 and y_0 in $\mathfrak{A}$.

Systems of Logic Based on Ordinals | 175

formula. Thus we find that a necessary and sufficient condition for a system of W to be valid is that for no recursion formula $\mathfrak{B}[x_0]$ are all of the formulae

$$(\exists x_0)\mathfrak{B}[x_0], \sim \mathfrak{B}[0], \sim \mathfrak{B}[f0], \ldots \qquad (8.4)$$

provable. An important consequence of this is that, if

$$\mathfrak{A}_1[x_0], \mathfrak{A}_2[x_0], \ldots, \mathfrak{A}_n[x_0]$$

are recursion formulae, if

$$(\exists x_0)\mathfrak{A}_1[x_0] \; \mathbf{v} \ldots \mathbf{v} \; (\exists x_0)\mathfrak{A}_n[x_0] \qquad (8.5)$$

is provable in C, and C is valid, then we can prove $\mathfrak{A}_r[f^{(a)}0]$ in C for some natural numbers r, a, where $1 \leq r \leq n$. Let us define $\mathfrak{D}_r$ to be the formula

$$(\exists x_0)\mathfrak{A}_1[x_0] \; \mathbf{v} \ldots \mathbf{v} \; (\exists x_0)\mathfrak{A}_r[x_0]$$

and then define $\mathfrak{E}_r[x_0]$ recursively by the condition that $\mathfrak{E}_1[x_0]$ is $\mathfrak{A}_1[x_0]$ and $\mathfrak{E}_{r+1}[x_0]$ be $\mathfrak{C}_{\mathfrak{E}_r, \mathfrak{A}_{r+1}}[x_0]$. Now I say that

$$\mathfrak{D}_r \supset (\exists x_0)\mathfrak{E}_r[x_0] \qquad (8.6)$$

is provable for $1 \leq r \leq n$. It is clearly provable for $r = 1$: suppose it to be provable for a given r. We can prove

$$(y_0)(\exists x_0)\mathrm{Db}[x_0, y_0]$$

and $$(y_0)(\exists x_0)\mathrm{Db}[fx_0, fy_0],$$

from which we obtain

$$\mathfrak{E}_r[y_0] \supset (\exists x_0)((\mathrm{Db}[x_0, y_0] \cdot \mathfrak{E}_r[y_0]) \; \mathbf{v} \; (\mathrm{Db}[fx_0, fy_0] \cdot \mathfrak{A}_{r+1}[y_0]))$$

and

$$\mathfrak{A}_{r+1}[y_0] \supset (\exists x_0)((\mathrm{Db}[x_0, y_0] \cdot \mathfrak{E}_r[y_0]) \; \mathbf{v} \; (\mathrm{Db}[fx_0, fy_0] \cdot \mathfrak{A}_{r+1}[y_0])).$$

These together with (8.1) yield

$$(\exists y_0)\mathfrak{E}_r[y_0] \; \mathbf{v} \; (\exists y_0)\mathfrak{A}_{r+1}[y_0] \supset (\exists x_0)\mathfrak{C}_{\mathfrak{E}_r, \mathfrak{A}_{r+1}}[x_0],$$

which is sufficient to prove (8.6) for $r + 1$. Now, since (8.5) is provable in C, $(\exists x_0)\mathfrak{E}_n[x_0]$ must also be provable, and, since C is valid, this means that $\mathfrak{E}_n[f^{(m)}0]$ must be provable for some natural number m. From (8.1) and the definition of $\mathfrak{E}_n[x_0]$ we see that this implies that $\mathfrak{A}_r[f^{(a)}0]$ is provable for some natural numbers a and r, $1 \leq r \leq n$.

To any system C of W we can assign a primitive recursive relation $P_C(m, n)$ with the intuitive meaning "*m* is the G.R. of a proof of the formula whose G.R. is *n*". We call the corresponding recursion formula Proof$_C[x_0, y_0]$ (*i.e.* Proof$_C[f^{(m)}0, f^{(n)}0]$ is provable when $P_C(m, n)$ is true, and its negation is provable otherwise). We can now explain what is the relation of a system C' to its predecessor C. The set of axioms which we adjoin to P to obtain C' consists of those adjoined in obtaining C, together with all formulae of the form

$$(\exists x_0) \, \text{Proof}_C[x_0, f^{(m)}0] \supset \mathfrak{F}, \tag{8.7}$$

where m is the G.R. of $\mathfrak{F}$.

We want to show that a contradiction can be obtained by assuming C' to be invalid but C to be valid. Let us suppose that a set of formulae of the form (8.4) is provable in C'. Let $\mathfrak{A}_1, \mathfrak{A}_2, \ldots, \mathfrak{A}_k$ be those axioms of C' of the form (8.7) which are used in the proof of $(\exists x_0) \mathfrak{B}[x_0]$. We may suppose that none of them is provable in C. Then by the deduction theorem we see that

$$(\mathfrak{A}_1 . \mathfrak{A}_2 \ldots \mathfrak{A}_k) \supset (\exists x_0) \mathfrak{B}[x_0] \tag{8.8}$$

is provable in C. Let $\mathfrak{A}_l$ be $(\exists x_0) \, \text{Proof}_C[x_0, f^{(m_l)}0] \supset \mathfrak{F}_l$. Then from (8.8) we find that

$$(\exists x_0) \, \text{Proof}_C[x_0, f^{(m_1)}0] \, \vee \ldots \vee \, (\exists x_0) \, \text{Proof}_C[x_0, f^{(m_k)}0] \, \vee \, (\exists x_0) \, \mathfrak{B}[x_0]$$

is provable in C. It follows from a result which we have just proved that either $\mathfrak{B}[f^{(c)}0]$ is provable for some natural number c, or else $\text{Proof}_C[f^{(n)}0, f^{(m_l)}0]$ is provable in C for some natural number u and some $l, 1 \leq l \leq k$: but this would mean that $\mathfrak{F}_l$ is provable in C (this is one of the points where we assume the validity of C) and therefore also in C', contrary to hypothesis. Thus $\mathfrak{B}[f^{(c)}0]$ must be provable in C'; but we are also assuming $\sim \mathfrak{B}[f^{(c)}0]$ to be provable in C'. There is therefore a contradiction in C'. Let us suppose that the axioms $\mathfrak{A}_1', \ldots, \mathfrak{A}_{k'}'$, of the form (8.7), when adjoined to C are sufficient to obtain the contradiction and that none of these axioms is that provable in C. Then

$$\sim \mathfrak{A}_1' \vee \sim \mathfrak{A}_2' \vee \ldots \vee \sim \mathfrak{A}_{k'}'$$

is provable in C, and if $\mathfrak{A}_l'$ is $(\exists x_0) \, \text{Proof}_C[x_0, f^{(m_l')}0] \supset \mathfrak{F}_l'$ then

$$(\exists x_0) \, \text{Proof}_C[x_0, f^{(m_1')}0] \, \vee \ldots \vee \, (\exists x_0) \, \text{Proof}[x_0, f^{(m_{k'}')}0]$$

is provable in C. But, by repetition of a previous argument, this means that $\mathfrak{A}_l'$ is provable for some $l, 1 \leq l \leq k'$, contrary to hypothesis. This is the required contradiction.

We may now construct an ordinal logic in the manner described on pp. [171–3]. We shall, however, carry out the construction in rather more detail, and with some modifications appropriate to the particular case. Each system C of our set W may be described by means of a W.F.F. M_C which enumerates the G.R.'s of the axioms of C. There is a W.F.F. E such that, if a is the G.R. of some proposition $\mathfrak{F}$, then $E(M_C, a)$ is convertible to the G.R. of

$(\exists x_0) \operatorname{Proof}_C[x_0, f^{(a)}0] \supset \mathfrak{F}.$

If **a** is not the G.R. of any proposition in P, then $E(M_C, \mathbf{a})$ is to be convertible to the G.R. of $0 = 0$. From E we obtain a W.F.F. K such that $K(M_C, 2\mathbf{n} + 1)$ conv $M_C(\mathbf{n})$, $K(M_C, 2\mathbf{n})$ conv $E(M_C, \mathbf{n})$. The successor system C' is defined by $K(M_C)$ conv $M_{C'}$. Let us choose a formula G such that $G(M_C, \mathbf{A})$ conv 2 if and only if the number-theoretic theorem equivalent to "**A** is dual" is provable in C. Then we define Λ_P by

$$\Lambda_P \to \lambda wa \,.\, \Gamma(\lambda y \,.\, G(\operatorname{Ck}(\operatorname{Tn}(w,y), \lambda mn \,.\, m(\varpi(2,n), \varpi(3,n)), K, M_P)), a).$$

This is an ordinal logic provided that P is valid.

Another ordinal logic of this type has in effect been introduced by Church.[20] Superficially this ordinal logic seems to have no more in common with Λ_P than that they both arise by the method which we have described, which uses C-K ordinal formulae. The initial systems are entirely different. However, in the relation between C and C' there is an interesting analogy. In Church's method the step from C to C' is performed by means of subsidiary axioms of which the most important (Church [2], p. 88, 1_m) is almost a direct translation into his symbolism of the rule that we may take any formula of the form (8.4) as an axiom. There are other extra axioms, however, in Church's system, and it is therefore not unlikely that it is in some respects more complete than Λ_P.

There are other types of ordinal logic, apparently quite unrelated to the type that we have so far considered. I have in mind two types of ordinal logic, both of which can be best described directly in terms of ordinal formulae without any reference to C-K ordinal formulae. I shall describe here a specimen Λ_H of one of these types of ordinal logic. Ordinal logics of this kind were first considered by Hilbert (Hilbert [1], 183 ff.), and have also been used by Tarski (Tarski [1], 395 ff.); see also Gödel [1], foot-note 48[a].

Suppose that we have selected a particular ordinal formula Ω. We shall construct a modification P_Ω of the system P of Gödel (see foot-note [16] on p. [173]). We shall say that a natural number n is a *type* if it is either even or $2p - 1$, where $\Omega(\mathbf{p}, \mathbf{p})$ conv 3. The definition of a variable in P is to be modified by the condition that the only admissible subscripts are to be the types in our sense. Elementary expressions are then defined as in P: in particular the definition of an elementary expression of type 0 is unchanged. An elementary formula is defined to be a sequence of symbols of the form $\mathfrak{A}_m \mathfrak{A}_n$, where $\mathfrak{A}_m, \mathfrak{A}_n$ are elementary expressions of types m, n satisfying one of the conditions (a), (b), (c).

(a) m and n are both even and m exceeds n,
(b) m is odd and n is even,
(c) $m = 2p - 1, n = 2q - 1$, and $\Omega(\mathbf{p}, \mathbf{q})$ conv 2.

---

[20] In outline Church [1], 279–280. In greater detail Church [2], Chap. X.

With these modifications the formal development of P_Ω is the same as that of P. We want, however, to have a method of associating number-theoretic theorems with certain of the formulae of P_Ω. We cannot take over directly the association which we used in P. Suppose that G is a formula in P interpretable as a number-theoretic theorem in the way described in the course of constructing Λ_P (p. [174]). Then, if every type suffix in G is doubled, we shall obtain a formula in P_Ω which is to be interpreted as the same number-theoretic theorem. By the method of §6 we can now obtain from P_Ω a formula L_Ω which is a logic formula if P_Ω is valid; in fact, given Ω there is a method of obtaining L_Ω, so that there is a formula Λ_H such that $\Lambda_H(\Omega)$ conv L_Ω for each ordinal formula Ω.

Having now familiarized ourselves with ordinal logics by means of these examples we may begin to consider general questions concerning them.

9. Completeness questions

The purpose of introducing ordinal logics was to avoid as far as possible the effects of Gödel's theorem. It is a consequence of this theorem, suitably modified, that it is impossible to obtain a complete logic formula, or (roughly speaking now) a complete system of logic. We were able, however, from a given system to obtain a more complete one by the adjunction as axioms of formulae, seen intuitively to be correct, but which the Gödel theorem shows are unprovable in the original system; from this we obtained a yet more complete system by a repetition of the process, and so on.[21] We found that the repetition of the process gave us a new system for each C-K ordinal formula. We should like to know whether this process suffices, or whether the system should be extended in other ways as well. If it were possible to determine about a W.F.F. in normal form whether it was an ordinal formula, we should know for certain that it was necessary to make extensions in other ways. In fact for any ordinal formula Λ it would then be possible to find a single logic formula L such that, if $\Lambda(\Omega, A)$ conv 2 for some ordinal formula Ω, then L(A) conv 2. Since L must be incomplete, there must be formulae A for which $\Lambda(\Omega, A)$ is not convertible to 2 for any ordinal formula Ω. However, in view of the fact, proved in §7, that there is no method of determining about a formula in normal form whether it is an ordinal formula, the case does not arise, and there is still a possibility that some ordinal logics may be complete in some sense. There is a quite natural way of defining completeness.

Definition of completeness of an ordinal logic. We say that an ordinal logic Λ is complete if corresponding to each dual formula A there is an ordinal formula Ω_A such that $\Lambda(\Omega_A, A)$ conv 2.

---

[21] In the case of P we adjoined all of the axioms $(\exists x_0)$ Proof $[x_0, f^{(m)}0] \supset \mathfrak{F}$, where m is the G.R. of $\mathfrak{F}$; the Gödel theorem shows that *some* of them are unprovable in P.

Systems of Logic Based on Ordinals | 179

As has been explained in §2, the reference in the definition to the existence of Ω_A for each A is to be understood in the same naïve way as any reference to existence in mathematics.

There is room for modification in this definition: we might require that there is a formula X such that X(A) conv Ω_A, X(A) being an ordinal formula whenever A is dual. There is no need, however, to discuss the relative merits of these two definitions, because in all cases in which we prove an ordinal logic to be complete we shall prove it to be complete even in the modified sense; but in cases in which we prove an ordinal logic to be incomplete, we use the definition as it stands.

In the terminology of §6, Λ is complete if the class of logics $\Lambda(\Omega)$ is complete when Ω runs through all ordinal formulae.

There is another completeness property which is related to this one. Let us for the moment describe an ordinal logic Λ as *all inclusive* if to each logic formula L there corresponds an ordinal formula $\Omega_{(L)}$ such that $\Lambda(\Omega_{(L)})$ is as complete as L. Clearly every all inclusive ordinal logic is complete; for, if A is dual, then $\delta(A)$ is a logic with A in its extent. But, if Λ is complete and

$$\text{Ai} \to \lambda kw \,.\, \Gamma(\lambda ra \,.\, \delta(4, \delta(2, k(w, V(\text{Nm}(r)))) + \delta(2, \text{Nm}(r, a)))),$$

then Ai(Λ) is an all inclusive ordinal logic. For, if A is in the extent of $\Lambda(\Omega_A)$ for each A, and we put $\Omega_{(L)} \to \Omega_{V(L)}$, then I say that, if B is in the extent of L, it must be in the extent of Ai($\Lambda, \Omega_{(L)}$). In fact, we see that Ai($\Lambda, \Omega_{V(L)}, B$) is convertible to

$$\Gamma(\lambda ra \,.\, \delta(4, \delta(2, \Lambda(\Omega_{V(L)}, V(\text{Nm}(r)))) + \delta(2, \text{Nm}(r, a))), B).$$

For suitable n, Nm(n) conv L and then

$$\Lambda(\Omega_{V(L)}, V(\text{Nm}(n))) \text{ conv } 2,$$

$$\text{Nm}(n, B) \text{ conv } 2,$$

and therefore, by the properties of Γ and δ

$$\text{Ai}(\Lambda, \Omega_{V(L)}, B) \text{ conv } 2.$$

Conversely Ai($\Lambda, \Omega_{V(L)}, B$) can be convertible to 2 only if both Nm(n, B) and $\Lambda(\Omega_{V(L)}, V(\text{Nm}(n)))$ are convertible to 2 for some positive integer n; but, if $\Lambda(\Omega_{V(L)}, V(\text{Nm}(n)))$ conv 2, then Nm(n) must be a logic, and, since Nm(n, B) conv 2, B must be dual.

It should be noticed that our definitions of completeness refer only to number-theoretic theorems. Although it would be possible to introduce formulae analogous to ordinal logics which would prove more general theorems than number-theoretic ones, and have a corresponding definition of completeness, yet, if our theorems are too general, we shall find that our (modified) ordinal logics are never complete. This follows from the argument of §4. If our "oracle" tells us, not whether any given number-theoretic statement is true, but whether a given formula is an ordinal formula, the argument still applies, and we find that there are classes of problem which cannot be solved by a uniform process even with the help of this oracle. This is equivalent to saying that there is no ordinal logic of the proposed modified type which is complete with respect to these problems. This situation becomes more definite if we take formulae satisfying conditions (a)–(e), (f') (as described at the end of §12) instead of ordinal formulae; it is then not possible for the ordinal logic to be complete with respect to any class of problems more extensive than the number-theoretic problems.

We might hope to obtain some intellectually satisfying system of logical inference (for the proof of number-theoretic theorems) with some ordinal logic. Gödel's theorem shows that such a system cannot be wholly mechanical; but with a complete ordinal logic we should be able to confine the non-mechanical steps entirely to verifications that particular formulae are ordinal formulae.

We might also expect to obtain an interesting classification of number-theoretic theorems according to "depth". A theorem which required an ordinal α to prove it would be deeper than one which could be proved by the use of an ordinal β less than α. However, this presupposes more than is justified. We now define

Invariance of ordinal logics. An ordinal logic Λ is said to be *invariant up to* an ordinal α if, whenever Ω, Ω' are ordinal formulae representing the same ordinal less than α, the extent of $\Lambda(\Omega)$ is identical with the extent of $\Lambda(\Omega')$. An ordinal logic is *invariant* if it is invariant up to each ordinal represented by an ordinal formula.

Clearly the classification into depths presupposes that the ordinal logic used is invariant.

Among the questions that we should now like to ask are

(a) Are there any complete ordinal logics?
(b) Are there any complete invariant ordinal logics?

To these we might have added "are all ordinal logics complete?"; but this is trivial; in fact, there are ordinal logics which do not suffice to prove any number-theoretic theorems whatever.

We shall now show that (a) must be answered affirmatively. In fact, we can write down a complete ordinal logic at once. Put

$$\text{Od} \to \lambda a \,.\, \{\lambda fmn \,.\, \text{Dt}(f(m), f(n))\}(\lambda s \,.\, \mathscr{P}(\lambda r \,.\, r(I, a(s)), 1, s)))$$

and $$\text{Comp} \to \lambda wa \,.\, \delta(w, \text{Od}(a)).$$

I shall show that Comp is a complete ordinal logic.

For if, Comp(Ω, **A**) conv 2, then

Systems of Logic Based on Ordinals | 181

$$\Omega \text{ conv Od (A)}$$

$$\text{conv } \lambda mn \, . \, \mathrm{Dt}(\mathscr{P}(\lambda r \, . \, r(I, \mathrm{A}(m)), 1, m), \mathscr{P}(\lambda r \, . \, r(I, \mathrm{A}(n)), 1, n))).$$

$\Omega(m, n)$ has a normal form if Ω is an ordinal formula, so that then

$$\mathscr{P}(\lambda r \, . \, r(I, \mathrm{A}(m)), 1)$$

has a normal form; this means that $r(I, \mathrm{A}(m))$ conv 2 some r, i.e. $\mathrm{A}(m)$ conv 2. Thus, if Comp(Ω, A) conv 2 and Ω is an ordinal formula, then A is dual. Comp is therefore an ordinal logic. Now suppose conversely that A is dual. I shall show that Od(A) is an ordinal formula representing the ordinal ω. For

$$\mathscr{P}(\lambda r \, . \, r(I, \mathrm{A}(m)), 1, m) \text{ conv } \mathscr{P}(\lambda r \, . \, r(I, 2), 1, m)$$

$$\text{conv } 1(m) \text{ conv } m,$$

$$\mathrm{Od}(\mathrm{A}, m, n) \text{ conv } \mathrm{Dt}(m, n),$$

i.e. Od(A) is an ordinal formula representing the same ordinal as Dt. But

$$\mathrm{Comp}(\mathrm{Od}(\mathrm{A}), \mathrm{A}) \text{ conv } \delta(\mathrm{Od}(\mathrm{A}), \mathrm{Od}(\mathrm{A})) \text{ conv } 2.$$

This proves the completeness of Comp.

Of course Comp is not the kind of complete ordinal logic that we should really wish to use. The use of Comp does not make it any easier to see that A is dual. In fact, if we really want to use an ordinal logic, a proof of completeness for that particular ordinal logic will be of little value; the ordinals given by the completeness proof will not be ones which can easily be seen intuitively to be ordinals. The only value in a completeness proof of this kind would be to show that, if any objection is to be raised against an ordinal logic, it must be on account of something more subtle than incompleteness.

The theorem of completeness is also unexpected in that the ordinal formulae used are all formulae representing ω. This is contrary to our intentions in constructing Λ_P for instance; implicitly we had in mind large ordinals expressed in a simple manner. Here we have small ordinals expressed in a very complex and artificial way.

Before trying to solve the problem (b), let us see how far Λ_P and Λ_H are invariant. We should certainly not expect Λ_P to be invariant, since the extent of $\Lambda_P(\Omega)$ will depend on whether Ω is convertible to a formula of the form $H(A)$: but suppose that we call an ordinal logic Λ "C-K invariant up to α" if the extent of $\Lambda(H(A))$ is the same as the extent of $\Lambda(H(B))$ whenever A and B are C-K ordinal formulae representing the same ordinal less than α. How far is Λ_P C-K invariant? It is not difficult to see that it is C-K invariant up to any finite ordinal, that is to say up to ω. It is also C-K invariant up to $\omega + 1$, as follows from the fact that the extent of

$$\Lambda_P(H(\lambda ufx \, . \, u(R)))$$

is the set-theoretic sum of the extents of

$$\Lambda_P(H(\lambda ufx . R(1))), \qquad \Lambda_P(H(\lambda ufx . R(2))), \ldots$$

However, there is no obvious reason for believing that it is C-K invariant up to $\omega + 2$, and in fact it is demonstrable that this is not the case (see the end of this section). Let us find out what happens if we try to prove that the extent of

$$\Lambda_P(H(\text{Suc } (\lambda ufx . u(\mathbf{R}_1))))$$

is the same as the extent of

$$\Lambda_P(H(\text{Suc } (\lambda ufx . u(\mathbf{R}_2)))),$$

where $\lambda ufx . u(\mathbf{R}_1)$ and $\lambda ufx . u(\mathbf{R}_2)$ are two C-K ordinal formulae representing ω. We should have to prove that a formula interpretable as a number-theoretic theorem is provable in $C[\text{Suc }(\lambda ufx . u(\mathbf{R}_1))]$ if, and only if, it is provable in $C[\text{Suc }(\lambda ufx . u(\mathbf{R}_2))]$. Now $C[\text{Suc }(\lambda ufx . u(\mathbf{R}_1))]$ is obtained from $C[\lambda ufx . u(\mathbf{R}_1)]$ by adjoining all axioms of the form

$$(\exists x_0) \text{ Proof}_{C[\lambda ufx . u(\mathbf{R}_1)]}[x_0, f^{(m)}0] \supset \mathfrak{F}, \tag{9.1}$$

where m is the G.R. of $\mathfrak{F}$, and $C[\text{Suc }(\lambda ufx . u(\mathbf{R}_2))]$ is obtained from $C[\lambda ufx . u(\mathbf{R}_2)]$ by adjoining all axioms of the form

$$(\exists x_0) \text{ Proof}_{C[\lambda ufx . u(\mathbf{R}_2)]}[x_0, f^{(m)}0] \supset \mathfrak{F}. \tag{9.2}$$

The axioms which must be adjoined to P to obtain $C[\lambda ufx . u(\mathbf{R}_1)]$ are essentially the same as those which must be adjoined to obtain the system $C[\lambda ufx . u(\mathbf{R}_2)]$: however the *rules of procedure which have to be applied before these axioms can be written down are in general quite different in the two cases*. Consequently (9.1) and (9.2) are quite different axioms, and there is no reason to expect their consequences to be the same. A proper understanding of this will make our treatment of question (*b*) much more intelligible. See also footnote [17] on page [173].

Now let us turn to Λ_H. This ordinal logic is invariant. Suppose that Ω, Ω' represent the same ordinal, and suppose that we have a proof of a number-theoretic theorem G in P_Ω. The formula expressing the number-theoretic theorem does not involve any odd types. Now there is a one–one correspondence between the odd types such that if $2m - 1$ corresponds to $2m' - 1$ and $2n - 1$ to $2n' - 1$ then $\Omega(m, n)$ conv 2 implies $\Omega'(m', n')$ conv 2. Let us modify the odd type-subscripts occurring in the proof of G, replacing each by its mate in the one–one correspondence. There results a proof in $P_{\Omega'}$ with the same end formula G. That is to say that if G is provable in P_Ω it is provable in $P_{\Omega'}$. Λ_H is invariant.

The question (*b*) must be answered negatively. Much more can be proved, but we shall first prove an even weaker result which can be established very quickly, in order to illustrate the method.

I shall prove that an ordinal logic Λ cannot be invariant and have the property that the extent of $\Lambda(\Omega)$ is a strictly increasing function of the ordinal represented by Ω. Suppose that Λ has these properties; then we shall obtain a contradiction. Let A be a W.F.F. in normal form and without free variables, and consider the process of carrying out conversions on A(1) until we have shown it convertible to 2, then converting A(2) to 2, then A(3) and so on: suppose that after r steps we are still performing the conversion on $A(m_r)$. There is a formula Jh such that Jh(A, r) conv m_r for each positive integer r. Now let Z be a formula such that, for each positive integer n, Z(n) is an ordinal formula representing ω^n, and suppose B to be a member of the extent of $\Lambda(\mathrm{Suc}\,(\mathrm{Lim}(Z)))$ but not of the extent of $\Lambda(\mathrm{Lim}(Z))$. Put

$$K^* \to \lambda a \,.\, \Lambda(\mathrm{Suc}\,(\mathrm{Lim}(\lambda r \,.\, Z(\mathrm{Jh}(a,\, r))))),\, B);$$

then K^* is a complete logic. For, if A is dual, then

$$\mathrm{Suc}\,(\mathrm{Lim}(\lambda r \,.\, Z(\mathrm{Jh}(A,\, r)))))$$

represents the ordinal $\omega^\omega + 1$, and therefore $K^*(A)$ conv 2; but, if $A(c)$ is not convertible to 2, then

$$\mathrm{Suc}\,(\mathrm{Lim}(\lambda r \,.\, Z(\mathrm{Jh}(A,\, r)))))$$

represents an ordinal not exceeding $\omega^c + 1$, and $K^*(A)$ is therefore not convertible to 2. Since there are no complete logic formulae, this proves our assertion.

We may now prove more powerful results.

Incompleteness theorems. (A) If an ordinal logic Λ is invariant up to an ordinal α, then for any ordinal formula Ω representing an ordinal β, $\beta < \alpha$, the extent of $\Lambda(\Omega)$ is contained in the (set-theoretic) sum of the extents of the logics $\Lambda(P)$, where P is finite.

(B) If an ordinal logic Λ is C-K invariant up to an ordinal α, then for any C-K ordinal formula A representing an ordinal β, $\beta < \alpha$, the extent of $\Lambda(H(A))$ is contained in the (set-theoretic) sum of the extents of the logics $\Lambda(H(F))$, where F is a C-K ordinal formula representing an ordinal less than ω^2.

Proof of (A). It is sufficient to prove that, if Ω represents an ordinal γ, $\omega \leqslant \gamma < \alpha$, then the extent of $\Lambda(\Omega)$ is contained in the set-theoretic sum of the extents of the logics $\Lambda(\Omega')$, where Ω' represents an ordinal less than γ. The ordinal γ must be of the form $\gamma_0 + \rho$, where ρ is finite and represented by P say, and γ_0 is not the successor of any ordinal and is not less than ω. There are two cases to consider; $\gamma_0 = \omega$ and $\gamma_0 \geqslant 2\omega$. In each of them we shall obtain a contradiction from the assumption that there is a W.F.F. B such that $\Lambda(\Omega, B)$ conv 2 whenever Ω represents γ, but is not convertible to 2 if Ω represents a smaller ordinal. Let us take first the case $\gamma_0 \geqslant 2\omega$. Suppose that $\gamma_0 = \omega + \gamma_1$, and that Ω_1 is an ordinal formula representing γ_1. Let A be any W.F.F. with a normal form and no free variables, and let Z be the class of those positive integers which are exceeded by all integers n for which A(n) is not convertible to 2. Let E be the class of integers $2p$ such that $\Omega(p, n)$ conv 2 for some n belonging to Z. The class E, together with the class Q of all odd integers, is constructively enumerable. It is evident that the class can be enumerated with repetitions, and since it is infinite the required enumeration can be obtained by striking out the repetitions. There is, therefore, a formula En such that En(Ω, A, r) runs through the formulae of the class $E + Q$ without repetitions as r runs through the positive integers. We define

$$\text{Rt} \to \lambda wamn \,.\, \text{Sum}(\text{Dt}, w, \text{En}(w, a, m), \text{En}(w, a, n)).$$

Then Rt(Ω_1, A) is an ordinal formula which represents γ_0 if A is dual, but a smaller ordinal otherwise. In fact

$$\text{Rt}(\Omega_1, \text{A}, \text{m}, \text{n}) \text{ conv } \{\text{Sum}(\text{Dt}, \Omega_1)\}(\text{En}(\Omega_1, \text{A}, \text{m}), \text{En}(\Omega_1, \text{A}, \text{n})).$$

Now, if A is dual, $E + Q$ includes all integers m for which

$$\{\text{Sum}(\text{Dt}, \Omega_1)\}(\text{m}, \text{m}) \text{ conv } 3.$$

(This depends on the particular form that we have chosen for the formula Sum.) Putting "En(Ω_1, A, p) conv q" for $M(p, q)$, we see that condition (7.4) is satisfied, so that Rt(Ω_1, A) is an ordinal formula representing γ_0. But, if A is not dual, the set $E + Q$ consists of all integers m for which

$$\{\text{Sum}(\text{Dt}, \Omega_1)\}(\text{m}, \text{r}) \text{ conv } 2,$$

where r depends only on A. In this case Rt(Ω_1, A) is an ordinal formula representing the same ordinal as Inf(Sum(Dt, Ω_1), r), and this is smaller than γ_0. Now consider K:

$$\text{K} \to \lambda a \,.\, \Lambda(\text{Sum}(\text{Rt}(\Omega_1, \text{A}), \text{P}), \text{B}).$$

If A is dual, K(A) is convertible to 2 since Sum(Rt(Ω_1, A), P) represents γ. But, if A is not dual, it is not convertible to 2, since Sum(Rt(Ω_1, A), P) then represents an ordinal smaller than γ. In K we therefore have a complete logic formula, which is impossible.

Now we take the case $\gamma_0 = \omega$. We introduce a W.F.F. Mg such that if n is the D.N. of a computing machine $\mathcal{M}$, and if by the m-th complete configuration of $\mathcal{M}$ the figure 0 has been printed, then Mg(n, m) is convertible to $\lambda pq \,.\, \text{Al}(4(P, 2p + 2q), 3, 4)$ (which is an ordinal formula representing the ordinal 1), but if 0 has not been printed it is convertible to $\lambda pq \,.\, p(q, I, 4)$ (which represents 0). Now consider

$$\text{M} \to \lambda n \,.\, \Lambda(\text{Sum}(\text{Lim}(\text{Mg}(n)), \text{P}), \text{B}).$$

If the machine never prints 0, then $\text{Lim}(\lambda r\,.\,\text{Mg}(\mathbf{n},r))$ represents ω and $\text{Sum}(\text{Lim}(\text{Mg}(\mathbf{n})),\mathbf{P})$ represents γ. This means that $\mathbf{M}(\mathbf{n})$ is convertible to 2. If, however, $\mathcal{M}$ never prints 0, $\text{Sum}(\text{Lim}(\text{Mg}(\mathbf{n})),\mathbf{P})$ represents a finite ordinal and $\mathbf{M}(\mathbf{n})$ is not convertible to 2. In M we therefore have means of determining about a machine whether it ever prints 0, which is impossible (Turing [1], §8).[22] This completes the proof of (A).

Proof of (B). It is sufficient to prove that, if C represents an ordinal γ, $\omega^2 \leqslant \gamma < \alpha$, then the extent of $\Lambda(H(\mathbf{C}))$ is included in the set-theoretic sum of the extents of $\Lambda(H(\mathbf{G}))$, where G represents an ordinal less than γ. We obtain a contradiction from the assumption that there is a formula B which is in the extent of $\Lambda(H(\mathbf{G}))$ if G represents γ, but not if it represents any smaller ordinal. The ordinal γ is of the form $\delta + \omega^2 + \xi$, where $\xi < \omega^2$. Let D be a C-K ordinal formula representing δ and $\lambda ufx\,.\,\mathbf{Q}(u,f,\mathbf{A}(u,f,x))$ one representing $\alpha + \xi$ whenever A represents α.

We now define a formula Hg. Suppose that A is a W.F.F. in normal form and without free variables; consider the process of carrying out conversions on $\mathbf{A}(1)$ until it is brought into the form 2, then converting $\mathbf{A}(2)$ to 2, then $\mathbf{A}(3)$, and so on. Suppose that at the r-th step of this process we are doing the n_r-th step in the conversion of $\mathbf{A}(\mathbf{m}_r)$. Thus, for instance, if A is not convertible to 2, m_r can never exceed 3. Then $\text{Hg}(\mathbf{A},\mathbf{r})$ is to be convertible to $\lambda f\,.\,f(\mathbf{m}_r,\mathbf{n}_r)$ for each positive integer r. Put

$$\text{Sq} \rightarrow \lambda dmn\,.\,n(\text{Suc}, m(\lambda aufx\,.\,u(\lambda y\,.\,y(\text{Suc}, a(u,f,x))), d(u,f,x))),$$
$$\mathbf{M} \rightarrow \lambda aufx\,.\,\mathbf{Q}(u,f,u(\lambda y\,.\,\text{Hg}(a,y,\text{Sq}(\mathbf{D}))))$$
$$\mathbf{K}_1 \rightarrow \lambda a\,.\,\Lambda(\mathbf{M}(a),\mathbf{B}),$$

then I say that $\mathbf{K}_1$ is a complete logic formula. $\text{Sq}(\mathbf{D}, \mathbf{m}, \mathbf{n})$ is a C-K ordinal formula representing $\delta + m\omega + n$, and therefore $\text{Hg}(\mathbf{A},\mathbf{r},\text{Sq}(\mathbf{D}))$ represents an ordinal ζ_r which increases steadily with increasing r, and tends to the limit $\delta + \omega^2$ if A is dual. Further

$$\text{Hg}(\mathbf{A},\mathbf{r},\text{Sq}(\mathbf{D})) < \text{Hg}(\mathbf{A}, S(\mathbf{r}),\text{Sq}(\mathbf{D}))$$

for each positive integer r. Therefore $\lambda ufx\,.\,u(\lambda y\,.\,\text{Hg}(\mathbf{A},y,\text{Sq}(\mathbf{D})))$ is a C-K ordinal formula and represents the limit of the sequence $\zeta_1,\zeta_2,\zeta_3,\ldots$ This is $\delta + \omega^2$ if A is dual, but a smaller ordinal otherwise. Likewise $\mathbf{M}(\mathbf{A})$ represents γ if A is dual, but is a smaller ordinal otherwise. The formula B therefore belongs to the extent of $\Lambda(H(\mathbf{M}(\mathbf{A})))$ if and only if A is dual, and this implies that $\mathbf{K}_1$ is a complete logic formula, as was asserted. But this is impossible and we have the required contradiction.

---

[22] This part of the argument can equally well be based on the impossibility of determining about two W.F.F. whether they are interconvertible. (Church [3], 363.)

As a corollary to (A) we see that Λ_H is incomplete and in fact that the extent of $\Lambda_H(\text{Dt})$ contains the extent of $\Lambda_H(\Omega)$ for any ordinal formula Ω. This result, suggested to me first by the solution of question (*b*), may also be obtained more directly. In fact, if a number-theoretic theorem can be proved in any particular P_Ω, it can also be proved in $P_{\lambda mn.m(n,I,4)}$. The formulae describing number-theoretic theorems in P do not involve more than a finite number of types, type 3 being the highest necessary. The formulae describing the number-theoretic theorems in any P_Ω will be obtained by doubling the type subscripts. Now suppose that we have a proof of a number-theoretic theorem G in P_Ω and that the types occurring in the proof are among $0, 2, 4, 6, t_1, t_2, t_3, \ldots$ We may suppose that they have been arranged with all the even types preceding all the odd types, the even types in order of magnitude, and the type $2m - 1$ preceding $2n - 1$ if $\Omega(\mathbf{m}, \mathbf{n})$ conv 2. Now let each t_r be replaced by $10 + 2r$ throughout the proof of G. We thus obtain a proof of G in $P_{\lambda mn.(n,I,4)}$.

As with problem (*a*), the solution of problem (*b*) does not require the use of high ordinals [*e.g.* if we make the assumption that the extent of $\Lambda(\Omega)$ is a steadily increasing function of the ordinal represented by Ω we do not have to consider ordinals higher than $\omega + 2$]. However, if we restrict what we are to call ordinal formulae in some way, we shall have corresponding modified problems (*a*) and (*b*); the solutions will presumably be essentially the same, but will involve higher ordinals. Suppose, for example, that Prod is a W.F.F. with the property that $\text{Prod}(\Omega_1, \Omega_2)$ is an ordinal formula representing $\alpha_1 \alpha_2$ when Ω_1, Ω_2 are ordinal formulae representing α_1, α_2 respectively, and suppose that we call a W.F.F. a 1-ordinal formula when it is convertible to the form Sum $(\text{Prod}(\Omega, \text{Dt}), P)$, where Ω, P are ordinal formulae of which P represents a finite ordinal. We may define 1-ordinal logics, 1-completeness and 1-invariance in an obvious way, and obtain a solution of problem (*b*) which differs from the solution in the ordinary case in that the ordinals less than ω^2 take the place of the finite ordinals. More generally the cases that I have in mind are covered by the following theorem.

Suppose that we have a class V of formulae representing ordinals in some manner which we do not propose to specify definitely, and a subset U of the class V such that:

(i) There is a formula Φ such that if T enumerates a sequence of members of U representing an increasing sequence of ordinals, then $\Phi(\mathbf{T})$ is a member of U representing the limit of the sequence.[23]
(ii) There is a formula E such that $E(\mathbf{m}, \mathbf{n})$ is a member of U for each pair of positive integers m, n and, if it represents $\varepsilon_{m,n}$, then $\varepsilon_{m,n} < \varepsilon_{m',n'}$ if either $m < m'$ or $m = m', n < n'$.

---

[23] The subset U wholly supersedes V in what follows. The introduction of V serves to emphasise the fact that the set of ordinals represented by members of U may have gaps.

Systems of Logic Based on Ordinals | 187

(iii) There is a formula **G** such that, if **A** is a member of *U*, then **G**(**A**) is a member of *U* representing a larger ordinal than does **A**, and such that **G**(**E**(**m**, **n**)) always represents an ordinal not larger than $\varepsilon_{m,\,n+1}$.

We define a *V*-ordinal logic to be a W.F.F. Λ such that $\Lambda(A)$ is a logic whenever **A** belongs to *V*. Λ is *V*-invariant if the extent of $\Lambda(A)$ depends only on the ordinal represented by **A**. Then it is not possible for a *V*-ordinal logic Λ to be *V*-invariant and have the property that, if C_1 represents a greater ordinal than C_2 (C_1 and C_2 both being members of *U*), then the extent of $\Lambda(C_1)$ is greater than the extent of $\Lambda(C_2)$.

We suppose the contrary. Let **B** be a formula belonging to the extent of $\Lambda((\Phi(\lambda r\,.\,E(r,1))))$ but not to the extent of $\Lambda(\Phi(\lambda r\,.\,E(r,1)))$, and let

$$\mathbf{K}' \rightarrow \lambda a\,.\,\Lambda(G(\Phi(\lambda r\,.\,Hg(a,r,E)),B).$$

Then **K**′ is a complete logic. For

$$\mathrm{Hg}(\mathbf{A},\mathbf{r},\mathbf{E})\ \mathrm{conv}\ \mathbf{E}(\mathbf{m}_r,\mathbf{n}_r).$$

$\mathbf{E}(\mathbf{m}_r,\mathbf{n}_r)$ is a sequence of *V*-ordinal formulae representing an increasing sequence of ordinals. Their limit is represented by $\Phi(\lambda r\,.\,Hg(A,r,E))$; let us see what this limit is. First suppose that **A** is dual: then m_r tends to infinity as *r* tends to infinity, and $\Phi(\lambda r\,.\,Hg(A,r,E))$ therefore represents the same ordinal as $\Phi(\lambda r\,.\,E(r,1))$. In this case we must have

$$\mathbf{K}'(\mathbf{A})\ \mathrm{conv}\ 2.$$

Now suppose that **A** is not dual: m_r is eventually equal to some constant number, *a* say, and $\Phi(\lambda r\,.\,Hg(A,r,E))$ represents the same ordinal as $\Phi(\lambda r\,.\,E(a,r))$, which is smaller than that represented by $\Phi(\lambda r\,.\,E(r,1))$. **B** cannot therefore belong to the extent of $\Lambda(G(\Phi(\lambda r\,.\,Hg(A,r,E))))$, and **K**′(**A**) is not convertible to 2. We have proved that **K**′ is a complete logic, which is impossible.

This theorem can no doubt be improved in many ways. However, it is sufficiently general to show that, with almost any reasonable notation for ordinals, completeness is incompatible with invariance.

We can still give a certain meaning to the classification into depths with highly restricted kinds of ordinals. Suppose that we take a particular ordinal logic Λ and a particular ordinal formula Ψ representing the ordinal α say (preferably a large one), and that we restrict ourselves to ordinal formulae of the form Inf(Ψ, a). We then have a classification into depths, but the extents of all the logics which we so obtain are contained in the extent of a single logic.

We now attempt a problem of a rather different character, that of the completeness of Λ_P. It is to be expected that this ordinal logic is complete. I cannot at present give a proof of this, but I can give a proof that it is complete as regards a simpler type of theorem than the number-theoretic theorems, viz. those of form "$\theta(x)$ vanishes identically", where $\theta(x)$ is primitive recursive. The proof will have to be much abbreviated since we do not wish to go into the formal details of the system P. Also there is a certain lack of definiteness in the problem as at present stated, owing to the fact that the formulae G, E, M_P were not completely defined. Our attitude here is that it is open to the sceptical reader to give detailed definitions for these formulae and then verify that the remaining details of the proof can be filled in, using his definition. It is not asserted that these details can be filled in whatever be the definitions of G, E, M_P consistent with the properties already required of them, only that they can be filled in with the more natural definitions.

I shall prove the completeness theorem in the following form. If $\mathfrak{B}[x_0]$ is a recursion formula and if $\mathfrak{B}[0], \mathfrak{B}[f0], \ldots$ are all provable in P, then there is a C-K ordinal formula A such that $(x_0)\mathfrak{B}[x_0]$ is provable in the system P^A of logic obtained from P by adjoining as axioms all formulae whose G.R.'s are of the form

$$A(\lambda mn . m(\varpi(2, n), \varpi(3, n)), K, M_P, r)$$

(provided they represent propositions).

First let us define the formula A. Suppose that **D** is a W.F.F. with the property that **D**(n) conv 2 if $\mathfrak{B}\left[f^{(n-1)}0\right]$ is provable in P, but **D**(n) conv 1 if $\sim\mathfrak{B}\left[f^{(n-1)}0\right]$ is provable in P (P is being assumed consistent). Let Θ be defined by

$$\Theta \rightarrow \{\lambda vu . u(v(v, u))\}(\lambda vu . u(v(v, u))),$$

and let Vi be a formula with the properties

$$\text{Vi}(2) \text{ conv } \lambda u . u(\text{Suc}, U),$$

$$\text{Vi}(1) \text{ conv } \lambda u . u(I, \Theta(\text{Suc})).$$

The existence of such a formula is established in Kleene [1], corollary on p. 220. Now put

$$A^* \rightarrow \lambda ufx . u(\lambda y . \text{Vi}(D(y), y, u, f, x)),$$
$$A \rightarrow \text{Suc } (A^*).$$

I assert that A^*, A are C-K ordinal formulae whenever it is true that $\mathfrak{B}[0], \mathfrak{B}[f0]$, ... are all provable in P. For in this case A^* is $\lambda ufx . u(R)$, where

$$R \rightarrow \lambda y . \text{Vi}(D(y), y, u, f, x),$$

and then

$\lambda ufx . R(n)$ conv $\lambda ufx . \text{Vi}(D(n), n, u, f, x)$
    conv $\lambda ufx . \text{Vi}(2, n, u, f, x)$
    conv $\lambda ufx . \{\lambda n . n(\text{Suc}, U)\}(n, u, f, x)$
    conv $\lambda ufx . n(\text{Suc}, U, u, f, x)$, which is a C-K ordinal formula,

and
$$\lambda ufx \,.\, S(\mathbf{n}, \mathrm{Suc}, U, u, f, x) \text{ conv } \mathrm{Suc}\,(\lambda ufx \,.\, \mathbf{n}(\mathrm{Suc}, U, u, f, x)).$$

These relations hold for an arbitrary positive integer n and therefore $\mathbf{A}^*$ is a C-K ordinal formula [condition (9) p. [163]]: it follows immediately that $\mathbf{A}$ is also a C-K ordinal formula. It remains to prove that $(x_0)\mathfrak{B}[x_0]$ is provable in P^A. To do this it is necessary to examine the structure of $\mathbf{A}^*$ in the case in which $(x_0)\mathfrak{B}[x_0]$ is false. Let us suppose that $\sim \mathfrak{B}[f^{(a-1)}0]$ is true, so that $\mathbf{D}(\mathbf{a})$ conv 1, and let us consider $\mathbf{B}$ where

$$\mathbf{B} \to \lambda ufx \,.\, \mathrm{Vi}(\mathbf{D}(\mathbf{a}), \mathbf{a}, u, f, x).$$

If $\mathbf{A}^*$ was a C-K ordinal formula, then $\mathbf{B}$ would be a member of its fundamental sequence; but

$$\begin{aligned}
\mathbf{B} &\text{ conv } \lambda ufx \,.\, \mathrm{Vi}(1, \mathbf{a}, u, f, x) \\
&\text{ conv } \lambda ufx \,.\, \{\lambda u \,.\, u(I, \Theta(\mathrm{Suc}))\}(\mathbf{a}, u, f, x) \\
&\text{ conv } \lambda ufx \,.\, \Theta(\mathrm{Suc}, u, f, x) \\
&\text{ conv } \lambda ufx \,.\, \{\lambda u \,.\, u(\Theta(u))\}(\mathrm{Suc}, u, f, x) \\
&\text{ conv } \lambda ufx \,.\, \mathrm{Suc}\,(\Theta(\mathrm{Suc}), u, f, x) \\
&\text{ conv } \mathrm{Suc}\,(\lambda ufx \,.\, \Theta(\mathrm{Suc}, u, f, x)) \\
&\text{ conv } \mathrm{Suc}\,(\mathbf{B}). \tag{9.3}
\end{aligned}$$

This, of course, implies that $\mathbf{B} < \mathbf{B}$ and therefore that $\mathbf{B}$ is no C-K ordinal formula. This, although fundamental in the possibility of proving our completeness theorem, does not form an actual step in the argument. Roughly speaking, our argument amounts to this. The relation (9.3) implies that the system P^B is inconsistent and therefore that P^{A^*} is inconsistent and indeed we can prove in P (and *a fortiori* in P^A) that $\sim (x_0)\mathfrak{B}[x_0]$ implies the inconsistency of P^{A^*}. On the other hand in P^A we can prove the consistency of P^{A^*}. The inconsistency of P^B is proved by the Gödel argument. Let us return to the details.

The axioms in P^B are those whose G.R.'s are of the form

$$B(\lambda mn \,.\, m(\varpi(2, n), \varpi(3, n)), K, M_P, \mathbf{r}).$$

When we replace $\mathbf{B}$, by $\mathrm{Suc}\,(\mathbf{B})$, this becomes

$$\mathrm{Suc}\,(\mathbf{B}, \lambda mn \,.\, m(\varpi(2, n), \varpi(3, n)), K, M_P, \mathbf{r})$$
$$\text{conv } K(B(\lambda mn \,.\, m(\varpi(2, n), \varpi(3, n)), K, M_P, \mathbf{r}))$$
$$\text{conv } B(\lambda mn \,.\, m(\varpi(2, n), \varpi(3, n)), K, M_P, \mathbf{p})$$

if $\mathbf{r}$ conv $2\mathbf{p}+1$,

$$\text{conv } E(B(\lambda mn \,.\, m(\varpi(2, n), \varpi(3, n)), K, M_P), \mathbf{p})$$

if $\mathbf{r}$ conv $2\mathbf{p}$.

When we remember the essential property of the formula E, we see that the axioms of P^B include all formulae of the form

$$(\exists x_0) \, \text{Proof}_{P^B}[x_0, f^{(q)}0] \supset \mathfrak{F},$$

where q is the G.R. of the formula $\mathfrak{F}$.

Let b be the G.R. of the formula $\mathfrak{A}$.

$$\sim (\exists x_0)(\exists y_0)\{\text{Proof}_{P^B}[x_0, y_0] \cdot \text{Sb}[z_0, z_0, y_0]\}. \tag{$\mathfrak{A}$}$$

$\text{Sb}[x_0, y_0, z_0]$ is a particular recursion formula such that $\text{Sb}[f^{(l)}0, f^{(m)}0, f^{(n)}0]$ holds if and only if n is the G.R. of the result of substituting $f^{(m)}0$ for z_0 in the formula whose G.R. is l at all points where z_0 is free. Let p be the G.R. of the formula $\mathfrak{C}$.

$$\sim (\exists x_0)(\exists y_0)\{\text{Proof}_{P^B}[x_0, y_0] \cdot \text{Sb}[f^{(b)}0, f^{(b)}0, y_0]\}. \tag{$\mathfrak{C}$}$$

Then we have as an axiom in P

$$(\exists x_0)\text{Proof}_{P^B}[x_0, f^{(p)}0] \supset \mathfrak{C},$$

and we can prove in P^A

$$(x_0)\{\text{Sb}[f^{(b)}0, f^{(b)}0, x_0] \equiv x_0 = f^{(p)}0\}, \tag{9.4}$$

since $\mathfrak{C}$ is the result of substituting $f^{(b)}0$ for z_0 in $\mathfrak{A}$; hence

$$\sim (\exists y_0)\text{Proof}_{P^B}[y_0, f^{(p)}0] \tag{9.5}$$

is provable in P. Using (9.4) again, we see that $\mathfrak{C}$ can be proved in P^B. But, if we can prove $\mathfrak{C}$ in P^B, then we can prove its provability in P^B, the proof being in P; i.e. we can prove

$$(\exists x_0)\text{Proof}_{P^B}[x_0, f^{(p)}0]$$

in P (since p is the G.R. of $\mathfrak{C}$). But this contradicts (9.5), so that, if

$$\sim \mathfrak{B}[f^{(a-1)}0]$$

is true, we can prove a contradiction in P^B or in P^{A^*}. Now I assert that the whole argument up to this point can be carried through formally in the system P, in fact, that, if c is the G.R. of $\sim (0 = 0)$, then

$$\sim (x_0)\mathfrak{B}[x_0] \supset (\exists v_0)\text{Proof}_{P^{A^*}}[v_0, f^{(c)}0] \tag{9.6}$$

is provable in P. I shall not attempt to give any more detailed proof of this assertion.

The formula

$$(\exists x_0)\text{Proof}_{P^{A^*}}[x_0, f^{(c)}0] \supset \sim (0 = 0) \tag{9.7}$$

is an axiom in P^A. Combining (9.6), (9.7) we obtain $(x_0)\mathfrak{B}[x_0]$ in P^A.

This completeness theorem as usual is of no value. Although it shows, for instance, that it is possible to prove Fermat's last theorem with Λ_P (if it is true) yet the truth of the theorem would really be assumed by taking a certain formula as an ordinal formula.

That Λ_P is not invariant may be proved easily by our general theorem; alternatively it follows from the fact that, in proving our partial completeness theorem, we never used ordinals higher than $\omega + 1$. This fact can also be used to prove that Λ_P is not C-K invariant up to $\omega + 2$.

10. The continuum hypothesis. A digression

The methods of §9 may be applied to problems which are constructive analogues of the continuum hypothesis problem. The continuum hypothesis asserts that $2^{\aleph_0} = \aleph_1$, in other words that, if ω_1 is the smallest ordinal α greater than ω such that a series with order type α cannot be put into one–one correspondence with the positive integers, then the ordinals less than ω_1 can be put into one–one correspondence with the subsets of the positive integers. To obtain a constructive analogue of this proposition we may replace the ordinals less than ω_1 either by the ordinal formulae, or by the ordinals represented by them; we may replace the subsets of the positive integers either by the computable sequences of figures 0, 1, or by the description numbers of the machines which compute these sequences. In the manner in which the correspondence is to be set up there is also more than one possibility. Thus, even when we use only one kind of ordinal formula, there is still great ambiguity concerning what the constructive analogue of the continuum hypothesis should be. I shall prove a single result in this connection.[24] A number of others may be proved in the same way.

We ask "Is it possible to find a computable function of ordinal formulae determining a one-one correspondence between the ordinals represented by ordinal formulae and the computable sequences of figures 0, 1?" More accurately, "Is there a formula **F** such that if Ω is an ordinal formula and n a positive integer then $F(\Omega, n)$ is convertible to 1 or to 2, and such that $F(\Omega, n)$ conv $F(\Omega', n)$ for each positive integer n, if and only if Ω and Ω' represent the same ordinal?" The answer is "No", as will be seen to be a consequence of the following argument: there is no formula **F** such that $F(\Omega)$ enumerates one sequence of integers (each being 1 or 2) when Ω represents ω and enumerates another sequence when Ω represents 0. If there is such an **F**, then there is an a such that $F(\Omega, a)$ conv (Dt, a) if Ω represents ω but $F(\Omega, a)$ and $F(Dt, a)$ are convertible to different integers (1 or 2) if Ω represents 0. To obtain a contradiction from this we introduce

---

[24] A suggestion to consider this problem came to me indirectly from F. Bernstein. A related problem was suggested by P. Bernays.

a W.F.F. Gm not unlike Mg. If the machine $\mathcal{M}$ whose D.N. is n has printed 0 by the time the m-th complete configuration is reached then

$$\text{Gm}(\mathbf{n}, \mathbf{m}) \text{ conv } \lambda mn \,.\, m(n, I, 4);$$

otherwise $\text{Gm}(\mathbf{n}, \mathbf{m})$ conv $\lambda pq \,.\, \text{Al}(4(P, 2p + 2q), 3, 4)$. Now consider $\text{F}(\text{Dt}, \mathbf{a})$ and $\text{F}(\text{Lim}(\text{Gm}(\mathbf{n})), \mathbf{a})$. If $\mathcal{M}$ never prints 0, $\text{Lim}(\text{Gm}(\mathbf{n}))$ represents the ordinal ω. Otherwise it represents 0. Consequently these two formulae are convertible to one another if and only if $\mathcal{M}$ never prints 0. This gives us a means of determining about any machine whether it ever prints 0, which is impossible.

Results of this kind have of course no real relevance for the classical continuum hypothesis.

11. The purpose of ordinal logics

Mathematical reasoning may be regarded rather schematically as the exercise of a combination of two faculties, which we may call *intuition* and *ingenuity*.[25] The activity of the intuition consists in making spontaneous judgments which are not the result of conscious trains of reasoning. These judgments are often but by no means invariably correct (leaving aside the question what is meant by "correct"). Often it is possible to find some other way of verifying the correctness of an intuitive judgment. We may, for instance, judge that all positive integers are uniquely factorizable into primes; a detailed mathematical argument leads to the same result. This argument will also involve intuitive judgments, but they will be less open to criticism than the original judgment about factorization. I shall not attempt to explain this idea of "intuition" any more explicitly.

The exercise of ingenuity in mathematics consists in aiding the intuition through suitable arrangements of propositions, and perhaps geometrical figures or drawings. It is intended that when these are really well arranged the validity of the intuitive steps which are required cannot seriously be doubted.

The parts played by these two faculties differ of course from occasion to occasion, and from mathematician to mathematician. This arbitrariness can be removed by the introduction of a formal logic. The necessity for using the intuition is then greatly reduced by setting down formal rules for carrying out inferences which are always intuitively valid. When working with a formal logic, the idea of ingenuity takes a more definite shape. In general a formal logic, will be framed so as to admit a considerable variety of possible steps in any stage in a proof. Ingenuity will then determine which steps are the more profitable for the purpose of proving a particular proposition. In pre-Gödel times it was thought

---

[25] We are leaving out of account that most important faculty which distinguishes topics of interest from others; in fact, we are regarding the function of the mathematician as simply to determine the truth or falsity of propositions.

by some that it would probably be possible to carry this programme to such a point that all the intuitive judgments of mathematics could be replaced by a finite number of these rules. The necessity for intuition would then be entirely eliminated.

In our discussions, however, we have gone to the opposite extreme and eliminated not intuition but ingenuity, and this in spite of the fact that our aim has been in much the same direction. We have been trying to see how far it is possible to eliminate intuition, and leave only ingenuity. We do not mind how much ingenuity is required, and therefore assume it to be available in unlimited supply. In our metamathematical discussions we actually express this assumption rather differently. We are always able to obtain from the rules of a formal logic a method of enumerating the propositions proved by its means. We then imagine that all proofs take the form of a search through this enumeration for the theorem for which a proof is desired. In this way ingenuity is replaced by patience. In these heuristic discussions, however, it is better not to make this reduction.

In consequence of the impossibility of finding a formal logic which wholly eliminates the necessity of using intuition, we naturally turn to "non-constructive" systems of logic with which not all the steps in a proof are mechanical, some being intuitive. An example of a non-constructive logic is afforded by any ordinal logic. When we have an ordinal logic, we are in a position to prove number-theoretic theorems by the intuitive steps of recognizing formulae as ordinal formulae, and the mechanical steps of carrying out conversions. What properties do we desire a non-constructive logic to have if we are to make use of it for the expression of mathematical proofs? We want it to show quite clearly when a step makes use of intuition, and when it is purely formal. The strain put on the intuition should be a minimum. Most important of all, it must be beyond all reasonable doubt that the logic leads to correct results whenever the intuitive steps are correct.[26] It is also desirable that the logic shall be adequate for the expression of number-theoretic theorems, in order that it may be used in metamathematical discussions (cf. §5).

Of the particular ordinal logics that we have discussed, Λ_H and Λ_P certainly will not satisfy us. In the case of Λ_H we are in no better position than with a constructive logic. In the case of Λ_P (and for that matter also Λ_H) we are by no means certain that we shall never obtain any but true results, because we do not know whether all the number-theoretic theorems provable in the system P are true. To take Λ_P as a fundamental non-constructive logic for metamathematical

---

[26] This requirement is very vague. It is not of course intended that the criterion of the correctness of the intuitive steps be the correctness of the final result. The meaning becomes clearer if each intuitive step is regarded as a judgment that a particular proposition is true. In the case of an ordinal logic it is always a judgment that a formula is an ordinal formula, and this is equivalent to judging that a number-theoretic proposition is true. In this case then the requirement is that the reputed ordinal logic *is* an ordinal logic.

arguments would be most unsound. There remains the system of Church which is free from these objections. It is probably complete (although this would not necessarily mean much) and it is beyond reasonable doubt that it always leads to correct results.[27] In the next section I propose to describe another ordinal logic, of a very different type, which is suggested by the work of Gentzen and which should also be adequate for the formalization of number-theoretic theorems. In particular it should be suitable for proofs of metamathematical theorems (cf. §5).

12. Gentzen type ordinal logics

In proving the consistency of a certain system of formal logic Gentzen (Gentzen [1]) has made use of the principle of transfinite induction for ordinals less than ε_0, and has suggested that it is to be expected that transfinite induction carried sufficiently far would suffice to solve all problems of consistency. Another suggestion of basing systems of logic on transfinite induction has been made by Zermelo (Zermelo [1]). In this section I propose to show how this method of proof may be put into the form of a formal (non-constructive) logic, and afterwards to obtain from it an ordinal logic.

We can express the Gentzen method of proof formally in this way. Let us take the system P and adjoin to it an axiom $\mathfrak{A}_\Omega$ with the intuitive meaning that the W.F.F. Ω is an ordinal formula, whenever we feel certain that Ω *is* an ordinal formula. This is a non-constructive system of logic which may easily be put into the form of an ordinal logic. By the method of §6 we make correspond to the system of logic consisting of P with the axiom $\mathfrak{A}_\Omega$ adjoined a logic formula L_Ω: L_Ω is an effectively calculable function of Ω, and there is therefore a formula Λ_G^1 such that $\Lambda_G^1(\Omega)$ conv L_Ω for each formula Ω. Λ_G^1 is certainly not an ordinal logic unless P is valid, and therefore consistent. This formalization of Gentzen's idea would therefore not be applicable for the problem with which Gentzen himself was concerned, for he was proving the consistency of a system weaker than P. However, there are other ways in which the Gentzen method of proof can be formalized. I shall explain one, beginning by describing a certain logical calculus.

The symbols of the calculus are f, x, $'$, $_1$, 0, S, R, Γ, Δ, E, $|$, $\odot$, $!$, $($, $)$, $=$, and the comma ",". For clarity we shall use various sizes of brackets $(,)$ in the following. We use capital German letters to stand for variable or undetermined sequences of these symbols.

---

[27] This ordinal logic arises from a certain system C_0 in essentially the same way as Λ_P arose from P. By an argument similar to one occurring in §8 we can show that the ordinal logic leads to correct results if and only if C_0 is valid; the validity of C_0 is proved in Church [1], making use of the results of Church and Rosser [1].

Systems of Logic Based on Ordinals | 195

It is to be understood that the relations that we are about to define hold only when compelled to do so by the conditions that we lay down. The conditions should be taken together as a simultaneous inductive definition of all the relations involved.

Suffixes

$_1$ is a suffix. If $\mathfrak{S}$ is a suffix then $\mathfrak{S}_1$ is a suffix.

Indices

^{1} is an index. If $\mathfrak{I}$ is an index then $\mathfrak{I}^1$ is an index.

Numerical variables

If $\mathfrak{S}$ is a suffix then $x\mathfrak{S}$ is a numerical variable.

Functional variables

If $\mathfrak{S}$ is a suffix and $\mathfrak{I}$ is an index, then $f\mathfrak{S}\mathfrak{I}$ is a functional variable of index $\mathfrak{I}$.

Arguments

(,) is an argument of index ^{1}. If ($\mathfrak{A}$) is an argument of index $\mathfrak{I}$ and $\mathfrak{T}$ is a term, then ($\mathfrak{A}\mathfrak{T}$,) is an argument of index $\mathfrak{I}^1$.

Numerals

0 is a numeral.
If $\mathfrak{N}$ is a numeral, then $S(, \mathfrak{N},)$ is a numeral.
In metamathematical statements we shall denote the numeral in which S occurs r times by $S^{(r)}(, 0,)$.

Expressions of a given index

A functional variable of index $\mathfrak{I}$ is an expression of index $\mathfrak{I}$.
R, S are expressions of index ^{111}, ^{11} respectively.
If $\mathfrak{N}$ is a numeral, then it is also an expression of index ^{1}.
Suppose that $\mathfrak{G}$ is an expression of index $\mathfrak{I}$, $\mathfrak{H}$ one of index $\mathfrak{I}^1$ and $\mathfrak{K}$ one of index $\mathfrak{I}^{111}$; then $(\Gamma\mathfrak{G})$ and $(\Delta\mathfrak{G})$ are expressions of index $\mathfrak{I}$, while $(E\mathfrak{G})$ and $(\mathfrak{G}|\mathfrak{H})$ and $(\mathfrak{G}\odot\mathfrak{K})$ and $(\mathfrak{G}!\mathfrak{H}!\mathfrak{K})$ are expressions of index $\mathfrak{I}^1$.

Function constants

An expression of index $\mathfrak{I}$ in which no functional variable occurs is a function constant of index $\mathfrak{I}$. If in addition R does not occur, the expression is called a *primitive function constant*.

Terms

0 is a term.
Every numerical variable is a term.

If $\mathfrak{G}$ is an expression of index $\mathfrak{I}$ and $(\mathfrak{A})$ is an argument of index $\mathfrak{I}$, then $\mathfrak{G}(\mathfrak{A})$ is a term.

Equations

If $\mathfrak{T}$ and $\mathfrak{T}'$ are terms, then $\mathfrak{T} = \mathfrak{T}'$ is an equation.

Provable equations

We define what is meant by the provable equations relative to a given set of equations as axioms.

(a) The provable equations include all the axioms. The axioms are of the form of equations in which the symbols $\Gamma, \Delta, E, |, \odot, !$ do not appear.

(b) If $\mathfrak{G}$ is an expression of index $\mathfrak{I}^{11}$ and $(\mathfrak{A})$ is an argument of index $\mathfrak{I}$, then

$$(\Gamma\mathfrak{G})(\mathfrak{A}x_1, x_{11},) = \mathfrak{G}(\mathfrak{A}x_{11}, x_1,)$$

is a provable equation.

(c) If $\mathfrak{G}$ is an expression of index $\mathfrak{I}^1$, and $(\mathfrak{A})$ is an argument of index $\mathfrak{I}$, then

$$(\Delta\mathfrak{G})(\mathfrak{A}x_1,) = \mathfrak{G}(, x_1\ \mathfrak{A})$$

is a provable equation.

(d) If $\mathfrak{G}$ is an expression of index $\mathfrak{I}$, and $(\mathfrak{A})$ is an argument of index $\mathfrak{I}$, then

$$(E\mathfrak{G})(\mathfrak{A}x_1,) = \mathfrak{G}(\mathfrak{A})$$

is a provable equation.

(e) If $\mathfrak{G}$ is an expression of index $\mathfrak{I}$ and $\mathfrak{H}$ is one of index $\mathfrak{I}^1$, and $(\mathfrak{A})$ is an argument of index $\mathfrak{I}$, then

$$(\mathfrak{G}|\mathfrak{H})(\mathfrak{A}) = \mathfrak{H}\left(\mathfrak{A}\mathfrak{G}(\mathfrak{A}),\right)$$

is a provable equation.

(f) If $\mathfrak{N}$ is an expression of index ^{1}, then $\mathfrak{N}(,) = \mathfrak{N}$ is a provable equation.

(g) If $\mathfrak{G}$ is an expression of index $\mathfrak{I}$ and $\mathfrak{K}$ one of index $\mathfrak{I}^{111}$, and $(\mathfrak{A})$ an argument of index $\mathfrak{I}^1$, then

$$(\mathfrak{G} \odot \mathfrak{K})(\mathfrak{A}0,) = \mathfrak{G}(\mathfrak{A})$$

and $\quad (\mathfrak{G} \odot \mathfrak{K})\left(\mathfrak{A}S(, x_1,),\right) = \mathfrak{K}\left(\mathfrak{A}x_1, S(, x_1,), (\mathfrak{G} \odot \mathfrak{K})(\mathfrak{A}x_1,),\right)$

are provable equations. If in addition $\mathfrak{H}$ is an expression of index $\mathfrak{I}^1$ and

$$R\left(,\mathfrak{G}\left(\mathfrak{A}S(, x_1,),\right), x_1,\right) = 0$$

is provable, then

$$(\mathfrak{G}!\mathfrak{K}!\mathfrak{H})(\mathfrak{A}0,) = \mathfrak{G}(\mathfrak{A})$$

and

$$(\mathfrak{G}!\mathfrak{K}!\mathfrak{H})(\mathfrak{A}S(,x_1,),)$$
$$=\mathfrak{K}\bigg((\mathfrak{A}\mathfrak{H}(\mathfrak{A}S(,x_1,),),S(,x_1,),(\mathfrak{G}!\mathfrak{K}!\mathfrak{H})(\mathfrak{A}\mathfrak{H}(\mathfrak{A}S(,x_1,),),),\bigg)$$

are provable.

(h) If $\mathfrak{T} = \mathfrak{T}'$ and $\mathfrak{U} = \mathfrak{U}'$ are provable, where $\mathfrak{T}$, $\mathfrak{T}'$, $\mathfrak{U}$ and $\mathfrak{U}'$ are terms, then $\mathfrak{U}' = \mathfrak{U}$ and the result of substituting $\mathfrak{U}'$ for $\mathfrak{U}$ at any particular occurrence in $\mathfrak{T} = \mathfrak{T}'$ are provable equations.

(i) The result of substituting any term for a particular numerical variable throughout a provable equation is provable.

(j) Suppose that $\mathfrak{G}$, $\mathfrak{G}'$ are expressions of index $\mathfrak{I}^1$, that $(\mathfrak{A})$ is an argument of index $\mathfrak{I}$ not containing the numerical variable $\mathfrak{X}$ and that $\mathfrak{G}(\mathfrak{A}0,) = \mathfrak{G}'(\mathfrak{A}0,)$ is provable. Also suppose that, if we add

$$\mathfrak{G}(\mathfrak{A}\mathfrak{X},) = \mathfrak{G}'(\mathfrak{A}\mathfrak{X},)$$

to the axioms and restrict (i) so that it can never be applied to the numerical variable $\mathfrak{X}$, then

$$\mathfrak{G}\bigg(\mathfrak{A}S(,\mathfrak{X},),\bigg) = \mathfrak{G}'\bigg(\mathfrak{A}S(,\mathfrak{X}),\bigg)$$

becomes a provable equation; in the hypothetical proof of this equation this rule (j) itself may be used provided that a different variable is chosen to take the part of $\mathfrak{X}$.

Under these conditions $\mathfrak{G}(\mathfrak{A}\mathfrak{X},) = \mathfrak{G}'(\mathfrak{A}\mathfrak{X},)$ is a provable equation.

(k) Suppose that $\mathfrak{G}$, $\mathfrak{G}'$, $\mathfrak{H}$ are expressions of index $\mathfrak{I}^1$, that $(\mathfrak{A})$ is an argument of index $\mathfrak{I}$ not containing the numerical variable $\mathfrak{X}$ and that

$$\mathfrak{G}(\mathfrak{A}0,) = \mathfrak{G}'(\mathfrak{A}0,) \quad \text{and} \quad R\bigg(,\mathfrak{H}(\mathfrak{A}S(,\mathfrak{X},),),S(,\mathfrak{X},),\bigg) = 0$$

are provable equations. Suppose also that, if we add

$$\mathfrak{G}\bigg(\mathfrak{A}\mathfrak{H}\big(\mathfrak{A}S(,\mathfrak{X},),\big),\bigg) = \mathfrak{G}'\bigg(\mathfrak{A}\mathfrak{H}\big(\mathfrak{A}S(,\mathfrak{X},),\big),\bigg)$$

to the axioms, and again restrict (i) so that it does not apply to $\mathfrak{X}$, then

$$\mathfrak{G}(\mathfrak{A}\mathfrak{X},) = \mathfrak{G}'(\mathfrak{A}\mathfrak{X},) \tag{12.1}$$

becomes a provable equation; in the hypothetical proof of (12.1) the rule (k) may be used if a different variable takes the part of $\mathfrak{X}$.

Under these conditions (12.1) is a provable equation.

We have now completed the definition of a provable equation relative to a given set of axioms. Next we shall show how to obtain an ordinal logic from this calculus. The first step is to set up a correspondence between some of the equations and number-theoretic theorems, in other words to show how they can be interpreted as number-theoretic theorems. Let $\mathfrak{G}$ be a primitive function constant of index [111]. $\mathfrak{G}$ describes a certain primitive recursive function $\phi(m, n)$, determined by the condition that, for all natural numbers m, n, the equation

$$\mathfrak{G}\Big(, S^{(m)}(, 0,), S^{(n)}(, 0,), \Big) = S^{(\phi(m, n))}(, 0,)$$

is provable without using the axioms (a). Suppose also that $\mathfrak{H}$ is an expression of index $\mathfrak{I}$. Then to the equation

$$\mathfrak{G}\Big(, x_1, \mathfrak{H}(, x_1,), \Big) = 0$$

we make correspond the number-theoretic theorem which asserts that for each natural number m there is a natural number n such that $\phi(m, n) = 0$. (The circumstance that there is more than one equation to represent each number-theoretic theorem could be avoided by a trivial but inconvenient modification of the calculus.)

Now let us suppose that some definite method is chosen for describing the sets of axioms by means of positive integers, the null set of axioms being described by the integer 1. By an argument used in §6 there is a W.F.F. Σ such that, if r is the integer describing a set A of axioms, then $\Sigma(r)$ is a logic formula enabling us to prove just those number-theoretic theorems which are associated with equations provable with the above described calculus, the axioms being those described by the number r.

I explain two ways in which the construction of the ordinal logic may be completed.

In the first method we make use of the theory of general recursive functions (Kleene [2]). Let us consider all equations of the form

$$R\Big(, S^{(m)}(, 0,), S^{(n)}(, 0,), \Big) = S^{(p)}(, 0,) \qquad (12.2)$$

which are obtainable from the axioms by the use of rules $(h), (i)$. It is a consequence of the theorem of equivalence of λ-definable and general recursive functions (Kleene [3]) that, if $r(m, n)$ is any λ-definable function of two variables, then we can choose the axioms so that (12.2) with $p = r(m, n)$ is obtainable in this way for each pair of natural numbers m, n, and no equation of the form

$$S^{(m)}(, 0,) = S^{(n)}(, 0,) \quad (m \neq n) \qquad (12.3)$$

is obtainable. In particular, this is the case if $r(m, n)$ is defined by the condition that

$$\Omega(m, n) \text{ conv } S(p) \quad \text{implies } p = r(m, n),$$
$$r(0, n) = 1, \quad \text{all } n > 0, r(0, 0) = 2,$$

where Ω is an ordinal formula. There is a method for obtaining the axioms given the ordinal formula, and consequently a formula Rec such that, for any ordinal formula Ω, Rec (Ω) conv m, where m is the integer describing the set of axioms corresponding to Ω. Then the formula

$$\Lambda_G{}^2 \to \lambda w . \Sigma\Big((\text{Rec }(w))\Big)$$

is an ordinal logic. Let us leave the proof of this aside for the present.

Our second ordinal logic is to be constructed by a method not unlike the one which we used in constructing Λ_P. We begin by assigning ordinal formulae to all sets of axioms satisfying certain conditions. For this purpose we again consider that part of the calculus which is obtained by restricting "expressions" to be functional variables or R or S and restricting the meaning of "term" accordingly; the new provable equations are given by conditions (a), (h), (i), together with an extra condition (l).

(l) The equation

$$R\Big(, 0, S(, x_1,), \Big) = 0$$

is provable.

We could design a machine which would obtain all equations of the form (12.2), with $m \neq n$, provable in this sense, and all of the form (12.3), except that it would cease to obtain any more equations when it had once obtained one of the latter "contradictory" equations. From the description of the machine we obtain a formula Ω such that

$$\Omega(\mathbf{m}, \mathbf{n}) \text{ conv 2 if } R\Big(, S^{(m-1)}(, 0,), S^{(n-1)}(, 0,), \Big) = 0$$

is obtained by the machine,

$$\Omega(\mathbf{m}, \mathbf{n}) \text{ conv 1 if } R\Big(, S^{(n-1)}(, 0,), S^{(m-1)}(, 0,), \Big) = 0$$

is obtained by the machine, and

$$\Omega(\mathbf{m}, \mathbf{m}) \text{ conv 3 always.}$$

The formula Ω is an effectively calculable function of the set of axioms, and therefore also of m: consequently there is a formula M such that M(m) conv Ω when m describes the set of axioms. Now let Cm be a formula such that, if b is the G.R. of a formula M(m), then Cm(b) conv m, but otherwise Cm(b) conv 1. Let

$$\Lambda_G{}^3 \to \lambda wa . \Gamma\Big(\lambda n . \Sigma\big(\text{Cm}(\text{Tn}(w, n)), a\big)\Big).$$

Then $\Lambda_G{}^3$ (Ω, A) conv 2 if and only if Ω conv M(m), where m describes a set of axioms which, taken with our calculus, suffices to prove the equation which is, roughly speaking, equivalent to "A is dual". To prove that $\Lambda_G{}^3$ is an ordinal logic, it is sufficient to prove that the calculus with the axioms described by m proves only true number-theoretic theorems when Ω is an ordinal formula. This condition on

m may also be expressed in this way. Let us put $m \ll n$ if we can prove $R\big(, S^{(m)}(, 0,), S^{(n)}(, 0,), \big) = 0$ with (a), (h), (i), (l): the condition is that $m \ll n$ is a well ordering of the natural numbers and that no contradictory equation (12.3) is provable with the same rules (a), (h), (i), (l). Let us say that such a set of axioms is *admissible*. $\Lambda_G{}^3$ is an ordinal logic if the calculus leads to none but true number-theoretic theorems when an admissible set of axioms is used.

In the case of $\Lambda_G{}^2$, Rec (Ω) describes an admissible set of axioms whenever Ω is an ordinal formula. $\Lambda_G{}^2$ therefore is an ordinal logic if the calculus leads to correct results when admissible axioms are used.

To prove that admissible axioms have the required property, I do not attempt to do more than show how interpretations can be given to the equations of the calculus so that the rules of inference (a)–(k) become intuitively valid methods of deduction, and so that the interpretation agrees with our convention regarding number-theoretic theorems.

Each expression is the name of a function, which may be only partially defined. The expression S corresponds simply to the successor function. If $\mathfrak{G}$ is either R or a functional variable and has $p+1$ symbols in its index, then it corresponds to a function g of p natural numbers defined as follows. If

$$\mathfrak{G}\big(, S^{(r_1)}(, 0,), S^{(r_2)}(, 0,), \ldots, S^{(r_p)}(, 0,), \big) = S^{(l)}(, 0,)$$

is provable by the use of (a), (h), (i), (l) only, then $g(r_1, r_2, \ldots, r_p)$ has the value p. It may not be defined for all arguments, but its value is always unique, for otherwise we could prove a "contradictory" equation and $M(\mathbf{m})$ would then not be an ordinal formula. The functions corresponding to the other expressions are essentially defined by (b)–(f). For example, if g is the function corresponding to $\mathfrak{G}$ and g' that corresponding to $(\Gamma\mathfrak{G})$, then

$$g'(r_1, r_2, \ldots, r_p, l, m) = g(r_1, r_2, \ldots, r_p, m, l).$$

The values of the functions are clearly unique (when defined at all) if given by one of (b)–(e). The case (f) is less obvious since the function defined appears also in the definiens. I do not treat the case of $(\mathfrak{G} \odot \mathfrak{K})$, since this is the well-known definition by primitive recursion, but I shall show that the values of the function corresponding to $(\mathfrak{G}\,!\,\mathfrak{K}\,!\,\mathfrak{H})$ are unique. Without loss of generality we may suppose that $(\mathfrak{A})$ in (f) is of index [1]. We have then to show that, if $h(m)$ is the function corresponding to $\mathfrak{H}$ and $r(m, n)$ that corresponding to R, and $k(u, v, w)$ is a given function and a a given natural number, then the equations

$$l(0) = a, \tag{α}$$

$$l(m+1) = k\bigg(h(m+1), m+1, l\big(h(m+1)\big)\bigg) \tag{β}$$

do not ever assign two different values for the function $l(m)$. Consider those values of r for which we obtain more than one value of $l(r)$, and suppose that there is at least one such. Clearly 0 is not one, for $l(0)$ can be defined only by (α). Since the relation $\ll$ is a well ordering, there is an integer r_0 such that $r_0 > 0$, $l(r_0)$ is not unique, and if $s \neq r_0$ and $l(s)$ is not unique then $r_0 \ll s$. We may put $s = h(r_0)$, for, if $l\big(h(r_0)\big)$ were unique, then $l(r_0)$, defined by (β), would be unique. But $r\big(h(r_0), r_0\big) = 0$ i.e. $s \ll r_0$. There is, therefore, no integer r for which we obtain more than one value for the function $l(r)$.

Our interpretation of expressions as functions gives us an immediate interpretation for equations with no numerical variables. In general we interpret an equation with numerical variables as the (infinite) conjunction of all equations obtainable by replacing the variables by numerals. With this interpretation (h), (i) are seen to be valid methods of proof. In (j) the provability of

$$\mathfrak{G}\Big(\mathfrak{A}S(,x_1,),\Big) = \mathfrak{G}'\Big(\mathfrak{A}S(,x_1,),\Big)$$

when $\mathfrak{G}(\mathfrak{A}x_1,) = \mathfrak{G}'(\mathfrak{A}x_1,)$ is assumed to be interpreted as meaning that the implication between these equations holds for all substitutions of numerals for x_1. To justify this, one should satisfy oneself that these implications always hold when the hypothetical proof can be carried out. The rule of procedure (j) is now seen to be simply mathematical induction. The rule (k) is a form of transfinite induction. In proving the validity of (k) we may again suppose $(\mathfrak{A})$ is of index¹. Let $r(m, n), g(m), g_1(m), h(n)$ be the functions corresponding respectively to R, $\mathfrak{G}$, $\mathfrak{G}'$, $\mathfrak{H}$. We shall prove that, if $g(0) = g'(0)$ and $r\big(h(n), n\big) = 0$ for each positive integer n and if $g(n+1) = g'(n+1)$ whenever $g\big(h(n+1)\big) = g'\big(h(n+1)\big)$, then $g(n) = g'(n)$ for each natural number n. We consider the class of natural numbers for which $g(n) = g'(n)$ is not true. If the class is not void it has a positive member n_0 which precedes all other members in the well ordering $\ll$. But $h(n_0)$ is another member of the class, for otherwise we should have

$$g\Big(h(n_0)\Big) = g'\Big(h(n_0)\Big)$$

and therefore $g(n_0) = g'(n_0)$, i.e. n_0 would not be in the class. This implies $n_0 \ll h(n_0)$ contrary to $r\big(h(n_0), n_0\big) = 0$. The class is therefore void.

It should be noticed that we do not really need to make use of the fact that Ω is an ordinal formula. It suffices that Ω should satisfy conditions (a)–(e) (p. [162]) for ordinal formulae, and in place of (f) satisfy (f').

(f') There is no formula T such that $T(n)$ is convertible to a formula representing a positive integer for each positive integer n, and such that $\Omega\Big(T(n), n\Big)$ conv 2, for each positive integer n for which $\Omega(n, n)$ conv 3.

The problem whether a formula satisfies conditions (a)–(e), (f') is number-theoretic. If we use formulae satisfying these conditions instead of ordinal formulae with Λ_G^2 or Λ_G^3, we have a non-constructive logic with certain advantages over ordinal logics. The intuitive judgments that must be made are all judgments of the truth of number-theoretic theorems. We have seen in §9 that the connection of ordinal logics with the classical theory of ordinals is quite superficial. There seem to be good reasons, therefore, for giving attention to ordinal formulae in this modified sense.

The ordinal logic Λ_G^3 appears to be adequate for most purposes. It should, for instance, be possible to carry out Gentzen's proof of consistency of number theory, or the proof of the uniqueness of the normal form of a well-formed formula (Church and Rosser [1]) with our calculus and a fairly simple set of axioms. How far this is the case can, of course, only be determined by experiment.

One would prefer a non-constructive system of logic based on transfinite induction rather simpler than the system which we have described. In particular, it would seem that it should be possible to eliminate the necessity of stating explicitly the validity of definitions by primitive recursions, since this principle itself can be shown to be valid by transfinite induction. It is possible to make such modifications in the system, even in such a way that the resulting system is still complete, but no real advantage is gained by doing so. The effect is always, so far as I know, to restrict the class of formulae provable with a given set of axioms, so that we obtain no theorems but trivial restatements of the axioms. We have therefore to compromise between simplicity and comprehensiveness.

Index of definitions

No attempt is being made to list heavy type formulae since their meanings are not always constant throughout the paper. Abbreviations for definite well-formed formulae are listed alphabetically.

	Page		Page		Page
Ai	179	Gr	150	M	199
Al	169	H	165, 168	M_P	177
Bd	169	H_1	168	Mg	184
Ck	173	Hf	169	Nm	160
Cm	199	Hg	185	Od	180
Comp	180	I	148	P	169
Dt	151	Inf	169	Prod	186
E	176	Jh	183	Q	159
form	150	K	177	Rec	198
G	177	Lim	169	Rt	184
Gm	192	Ls	168	S	149

Sum ... 169	X ... 158	$\Lambda_G{}^3$ ... 199
Sq. ... 185	Z ... 183	Λ_H ... 178
Tn ... 173		Λ_P ... 177
Ug ... 169	Γ ... 160	ϖ ... 151
V ... 160	δ ... 147	Σ ... 198
Vi ... 188	Θ ... 188	1, 2, 3, ... 148
W ... 159	$\Lambda_G{}^1$ ... 194	𝒫 ... 154[28]
W' ... 159	$\Lambda_G{}^2$ ... 199	

Bibliography

Alonzo Church, [1]. "A proof of freedom from contradiction", *Proc. Nat. Acad. Sci.* 21 (1935), 275–281.

—— [2]. *Mathematical logic*, Lectures at Princeton University (1935–6), mimeographed, 113 pp.

—— [3]. "An unsolvable problem of elementary number theory", *American J. of Math.* 58 (1936), 345–363.

—— [4]. "The constructive second number class", *Bull. American Math. Soc.* 44 (1938), 224–238.

G. Gentzen, [1]. "Die Widerspruchsfreiheit der reinen Zahlentheorie", *Math. Annalen*, 112 (1936), 493–565.

K. Gödel, [1]. "Über formal unentscheidbare Sätze der Principia Mathematica und verwandter Systeme, I", *Monatshefte für Math. und Phys.* 38 (1931), 173–189.

—— [2]. *On undecidable propositions of formal mathematical systems*, Lectures at the Institute for Advanced Study, Princeton, N.J., 1934, mimeographed, 30 pp.

D. Hilbert, [1]. "Über das Unendliche", *Math. Annalen*, 95 (1926), 161–190.

S. C. Kleene, [1]. "A theory of positive integers in formal logic", *American J. of Math.* 57 (1935), 153–173 and 219–244.

—— [2]. "General recursive functions of natural numbers", *Math. Annalen*, 112 (1935–6), 727–742.

—— [3]. "λ-definability and recursiveness", *Duke Math. Jour.* 2 (1936), 340–353.

E. L. Post, [1]. "Finite combinatory processes—formulation 1", *Journal Symbolic Logic*, 1 (1936), 103–105.

J. B. Rosser, [1]. "Gödel theorems for non-constructive logics", *Journal Symbolic Logic*, 2 (1937), 129–137.

A. Tarski, [1]. "Der Wahrheitsbegriff in den formalisierten Sprachen", *Studia Philosophica*, 1 (1936), 261–405 (translation from the original paper in Polish dated 1933).

A. M. Turing, [1]. "On computable numbers, with an application to the Entscheidungsproblem". [Chapter 1].

—— [2]. "Computability and λ-definability", *Journal Symbolic Logic*, 2 (1937), 153–163.

---

[28] Editor's note. The remainder of Turing's index has been incorporated into the general index at the rear of the book.

E. Zermelo, [1]. "Grundlagen einer allgemeiner Theorie der mathematischen Satzsysteme, I", *Fund. Math.* 25 (1935), 136–146.

Alonzo Church and S. C. Kleene, [1]. "Formal definitions in the theory of ordinal numbers", *Fund. Math.* 28 (1936), 11–21.

Alonzo Church and J. B. Rosser, [1]. "Some properties of conversion", *Trans. American Math. Soc.* 39 (1936), 472–482.

D. Hilbert and W. Ackermann, [1]. *Grundzüge der theoretischen Logik* (2nd edition revised, Berlin, 1938), 130 pp.

A. N. Whitehead and Bertrand Russell, [1]. *Principia Mathematica* (2nd edition, Cambridge, 1925–1927), 3 vols.

[Received 31 May, 1938.—Read 16 June, 1938.]

CHAPTER 4

Letters on Logic to Max Newman (c.1940)
Alan Turing

Introduction
Jack Copeland

At the outbreak of war with Germany in September 1939, Turing left Cambridge to take up work as a codebreaker at Bletchley Park, the wartime headquarters of the Government Code and Cypher School (see 'Enigma', below). In the early months of 1940, Turing received a letter from the Cambridge mathematician M. H. A. Newman, his teacher, colleague, and friend. Turing replied on 23 March, writing from his lodgings at the Crown Inn (situated in the small village of Shenley Brook End): 'Dear Newman, Very glad to get your letter, as I needed some stimulus to make me start thinking about logic.' This was to be the first of five letters that Turing wrote to Newman during the seventeen months before Newman too left Cambridge for Bletchley Park.

In his first letter Turing agreed (presumably at Newman's request—Newman's letters seem not to have been preserved) to 'let [Newman] in on... the tricks of the conversion calculus'. The conversion calculus, or 'λ-calculus', is due to Alonzo Church, with whom Turing studied in Princeton from 1936 to 1938 (see the introduction to Chapter 3).[1] Turing's letters consist for the most part of detailed remarks on the conversion calculus, often elucidating material from what Turing calls 'Church's notes'—a substantial typescript entitled 'Mathematical Logic' which was in circulation at Princeton and elsewhere and which Newman was evidently reading.[2]

Their correspondence on Church's work issued in their joint paper 'A Formal Theorem in Church's Theory of Types', which was submitted to Church's *Journal*

---

[1] Turing and Church also corresponded at this time. A letter from Church addressed to Turing at the Crown is dated 15 May 1940 and replies to Turing's of 15 April. I am grateful to Alonzo Church's son, Alonzo Church Jnr, for sending me a copy of Church's letter.

[2] The title page reads: 'MATHEMATICAL LOGIC   Lectures by Alonzo Church   Princeton University, October 1935–January 1936. (Notes by F. A. Ficken, H. G. Landau, H. Ruja, R. R. Singleton, N. E. Steenrod, J. H. Sweer, F. J. Weyl).' I am grateful to Alonzo Church Jnr for information concerning this typescript.

of Symbolic Logic in May 1941 and published in March 1942.[3] The paper was written while Turing played a leading role in the battle to break Naval Enigma (see 'Enigma' and Chapters 5–8). Turing would spend his occasional nights off duty 'coming in as usual..., doing his own mathematical research at night, in the warmth and light of the office, without interrupting the routine of daytime sleep'.[4]

The two most interesting items of the correspondence, which are printed here, contain substantial passages in which Turing departs from his commentary on Church's work and expounds his own views. These elegant passages provide information about Turing's thoughts on the logical foundations of mathematics which is not to be found elsewhere in his writings.

Of particular importance are the sections headed 'Intuition. Inspiration. Ingenuity', in which he discusses the unsolvability and incompleteness results in logic and explains the basic idea underlying his ordinal logics (Chapter 3); 'Ingenuity and Intuition', discussing the extent to which provability by Turing machine approximates mathematical truth; 'The Completeness Theorem', concerning the completeness theorem established in Chapter 3; and 'Consequences', in which two notions of logical consequence are compared. These sections contain occasional formulae of the conversion calculus, but the formulae are not necessary to Turing's points, and readers unfamiliar with the notation of the calculus should not be deterred.

M. H. A. Newman: Mathematician, Codebreaker, and Computer Pioneer

Max Newman played an important part in Turing's intellectual life over many years. It was Newman who, in a lecture in Cambridge in 1935, launched Turing on the research that led to the universal Turing machine:

I believe it all started because he attended a lecture of mine on foundations of mathematics and logic... I think I said in the course of this lecture that what is meant by saying that [a] process is constructive is that it's a purely mechanical machine—and I may even have said, a machine can do it.

And this of course led [Turing] to the next challenge, what sort of machine, and this inspired him to try and say what one would mean by a perfectly general computing machine.[5]

---

[3] M. H. A. Newman and A. M. Turing, 'A Formal Theorem in Church's Theory of Types', *Journal of Symbolic Logic*, 7 (1942), 28–33.

[4] J. Murray, 'Hut 8 and Naval Enigma, Part I', in H. Hinsley and A. Stripp (eds.), *Codebreakers: The Inside Story of Bletchley Park* (Oxford: Oxford University Press, 1993), 117.

[5] Newman in interview with Christopher Evans ('The Pioneers of Computing: An Oral History of Computing' (London: Science Museum)).

In April 1936, Turing presented Newman with the draft typescript of 'On Computable Numbers'.[6] Not long after, an offprint of Church's paper proving the undecidability of first-order predicate calculus arrived in Cambridge.[7] Newman proved a staunch ally at what must have been a painful time for Turing. On 29 May 1936 Turing wrote in a letter to his mother:

> Meanwhile a paper has appeared in America, written by Alonzo Church, doing the same things in a different way. Mr Newman and I have decided however that the method is sufficiently different to warrant the publication of my paper too.[8]

It was clear to Newman that 'Turing's "machine" had a significance going far beyond this particular application [the *Entscheidungsproblem*]'.[9] Turing's paper contained (in Newman's words) 'this extraordinary definition of a perfectly general...computable function, thus giving the first idea...of a perfectly general computing machine.'[10] Newman advised Turing during the final stages of preparation of 'On Computable Numbers', and he wrote to the Secretary of the London Mathematical Society saying that Church's prior publication should not stand in the way of Turing's paper appearing in the *Proceedings*.[11]

In 1942 Newman received a letter from Frank Adcock, another Cambridge man and a veteran of Room 40 (the forerunner of the Government Code and Cypher School): 'Dear Newman, There is some work going at a government institution which would I think interest you and which is certainly important for the War ...'.[12] Newman wrote to the Master of St John's to request leave of absence and at the end of August 1942 he joined the Research Section at Bletchley Park.

The Research Section was attempting to break the German cipher machine they nicknamed 'Tunny'. Used mainly by the German Army, Tunny was one of three types of German machine—collectively referred to as 'Fish' by the British—for enciphering the binary teleprinter alphabet (the other two were 'Sturgeon', used mainly by the German Air Force, and 'Thrasher'). From the autumn of 1942 Tunny was used in preference to Enigma for the encryption of messages between the German High Command and the various Army Group commanders in the field—intelligence of the highest grade.

---

[6] A. Hodges, *Alan Turing: The Enigma* (London: Vintage, 1992), 109.

[7] M. H. A. Newman, 'Alan Mathison Turing, 1912–1954', *Biographical Memoirs of Fellows of the Royal Society*, 1 (1955), 253–63 (258). On the paper by Church, 'A Note on the Entscheidungsproblem', see 'Computable Numbers: A Guide'.

[8] The letter is among the Turing Papers in the Modern Archive Centre, King's College Library, Cambridge (catalogue reference K 1).

[9] Newman, 'Alan Mathison Turing, 1912–1954', 258.

[10] Newman in interview with Christopher Evans (see n. 5).

[11] See further 'Max Newman: Mathematician, Codebreaker and Computer Pioneer', by William Newman (Max's son), to appear in B. J. Copeland (ed.), *Colossus: The First Electronic Computer* (Oxford University Press).

[12] Quoted in W. Newman, 'Max Newman: Mathematician, Codebreaker and Computer Pioneer'.

In November 1942 William Tutte found a way of breaking Tunny messages known as the 'Statistical Method'.[13] The rub was that the method seemed impractical, involving a very large amount of time-consuming work—basically, the comparing of two streams of 0s and 1s, counting the number of times that each had 0 in the same position. If the comparing and counting were done by hand, the intelligence in the message would be stale before the work was completed. Tutte explained his method to Newman and Newman suggested using electronic counters. It was a brilliant idea. In December 1942 Newman was given the job of developing the necessary machinery.[14] The electronic counters were designed by C. E. Wynn-Williams at the Telecommunications Research Establishment (TRE) in Malvern. Construction of the new machine was carried out at the Post Office Research Station at Dollis Hill in London and at TRE. In June 1943 the completed machine began work in the 'Newmanry', a newly created section at Bletchley Park headed by Newman.

This first machine—known as 'Heath Robinson', after a popular cartoonist who drew bizarre contraptions—was relay-based, with some electronic circuits for counting and for performing simple logical (i.e. boolean) operations. Heath Robinson was unreliable and slow, and its high-speed paper tapes tended to stretch and tear, but it proved the worth of Newman's approach. Newman ordered a dozen more Robinsons from the Post Office.

During the design phase of Heath Robinson there had been difficulties with the logic unit—the 'combining unit' in the terminology of 1942. At Turing's suggestion Newman had approached the Post Office engineer Thomas H. Flowers for help (Flowers had previously assisted Turing with the design of a machine for use against Enigma).[15] Flowers and his switching group at Dollis Hill successfully redesigned the combining unit; but Flowers did not think much of the overall design of the Robinson, and in February 1943 presented Newman with the alternative of a fully electronic machine. This idea received little encouragement from Bletchley Park, however, where opinion was that a machine containing as many electronic valves (vacuum tubes) as Flowers was proposing—about 2,000—would not work reliably. Flowers, with over ten years' experience of electronic valves, knew better, and on his own initiative began building the machine he could see was necessary, working independently at the Post Office Research Station. Flowers has said that he was probably the only person in

---

[13] W. T. Tutte, 'At Bletchley Park', to appear in Copeland (ed.), *Colossus: The First Electronic Computer*.

[14] Part 1 of 'General Report on Tunny'. 'General Report on Tunny' was written in 1945 by Jack Good, Donald Michie, and Geoffrey Timms, all members of Newman's section at GC & CS. This document was released by the British government in 2000 to the Public Record Office at Kew (document reference HW 25/ 4, HW 25/5). A digital facsimile of the document is available in The Turing Archive for the History of Computing <www.AlanTuring.net/tunny_report>.

[15] Flowers in interview with Copeland (July 1996, July 1998).

Britain who understood at this time that electronic valves could be used in large numbers for high-speed digital computing.[16]

Flowers' 'Colossus', the first large-scale electronic digital computing machine, was installed in the Newmanry on 8 December 1943 (see the introduction to Chapter 9). By the end of the war, there were nine more Colossi working in the Newmanry. The Colossi gave the Allies access to the most secret German radio communications, including messages from Hitler to his front-line generals. Intelligence obtained via Colossus was vital to the planning of the D-day landings and played a major role in the subsequent defeat of Hitler.[17]

In September 1945 Newman took up the Fielden Chair of Mathematics at the University of Manchester. Five months later he wrote the following to the Princeton mathematician and computer pioneer John von Neumann:

I am...hoping to embark on a computing machine section here, having got very interested in electronic devices of this kind during the last two or three years. By about eighteen months ago I had decided to try my hand at starting up a machine unit when I got out. ... I am of course in close touch with Turing.[18]

Newman lost no time in establishing the Royal Society Computing Machine Laboratory at the University. He introduced the engineers Frederick Williams and Thomas Kilburn—newly recruited to Manchester University from the Telecommunications Research Establishment, where they had worked on radar (they knew nothing of the top-secret Colossus)—to Turing's idea of a stored-programme computer and explained to them what facilities were necessary in a computer (see the introduction to Chapter 9).[19] It was in Newman's Computing Machine Laboratory that Kilburn and Williams built the world's first electronic stored-programme digital computer. Their prototype ran its first programme on 21 June 1948 (see further the introduction to Chapter 9).

That same year Newman recruited Turing to Manchester from the National Physical Laboratory, appointing him Deputy Director of the Computing Machine Laboratory (see the introduction to Chapter 10). Turing remained at Manchester until his death in 1954.

---

[16] Flowers in interview with Copeland (July 1996).

[17] F. H. Hinsley et al., *British Intelligence in the Second World War*, vol. iii, part 2 (London: Her Majesty's Stationery Office, 1988), 53, 799.

[18] Letter from Newman to von Neumann, 8 Feb. 1946 (in the von Neumann Archive at the Library of Congress, Washington, DC; a digital facsimile is in The Turing Archive for the History of Computing <www.AlanTuring.net/newman_vonneumann_8feb46>).

[19] Williams in interview with Christopher Evans in 1976 ('The Pioneers of Computing: An Oral History of Computing' (London: Science Museum)).

Further reading

Barendregt, H. P., *The Lambda-Calculus, its Syntax and Semantics* (Amsterdam: North-Holland, 1984).

Church, A., 'A Set of Postulates for the Foundation of Logic', *Annals of Mathematics*, 33 (1932), 346–66.

—— 'An Unsolvable Problem of Elementary Number Theory', *American Journal of Mathematics*, 58 (1936), 345–63.

—— *The Calculi of Lambda-Conversion* (Princeton: Princeton University Press, 1941).

Copeland, B. J., 'Colossus and the Dawning of the Computer Age', in R. Erskine and M. Smith (eds.), *Action This Day* (London: Bantam, 2001).

Lalement, R., *Computation as Logic* (Hemel Hempstead: Prentice Hall, 1993).

Provenance

What follows are transcriptions of Turing's letters. The original letters are among the Turing Papers in the Modern Archive Centre, King's College Library, Cambridge.[20]

---

[20] Catalogue reference D 2. The letters are published with the permission of the Estate of Alan Turing.

Letter from The Crown, Shenley Brook End

April 21

The Crown
Shenley Brook End
Bletchley

Dear Newman,

<u>The δ-function</u>. One certainly can manage without δ for defining computable functions. The purpose for which it is really brought in in Church's notes is to enable one to 'describe the syntax of the system within itself' i.e. at any rate to define the formula Gr (or something like it) such that Gr($\underline{A}$) conv G.R of a certain normal form of $\underline{A}$ (if there is one & if $\underline{A}$ has no free variables, otherwise Gr($\underline{A}$) has no normal form)

$$\text{Gr} \to \lambda a \cdot \mathcal{P}(1, \lambda u \cdot \delta(\text{form}(u), a) \ \& \ \text{Norm}(u))^1$$

where Norm($\underline{u}$) conv 2 if u is G.R of a formula in normal form
conv 1 otherwise

(form can be defined without δ)

2 & 1 conv 1
1 & 2 conv 1
1 & 1 conv 1
2 & 2 conv 2

I do not know that it has been proved that a Gr cannot be obtained without δ, but at any rate defining Gr without δ would be equivalent to defining a formula without δ which would have the properties of δ. I haven't got Church's notes with me, but I think most of his bracket technique was in connection with his 'metads' (sort of G.Rs).[2]

There can be no very general picking out function, even using δ. The formulae to be picked out must certainly have no free variables (if Δ_1 picks out first term of two then $\Delta_1(x, y)$ conv x and l.h.s[3] has different free variables from right) but also they must have normal forms, for a formula without normal form will poison any formula in which it enters. If all the formulae involved have normal forms one can pick out with δ e.g. in this way –

$$\text{Gr}([\underline{A}, \underline{B}]) \text{ is G.R of } [\underline{A}, \underline{B}]$$

---

[1] Editor's note. See S. C. Kleene, 'A Theory of Positive Integers in Formal Logic, Part II', *American Journal of Mathematics*, 57 (1935), 219–44 (231–2).

[2] Editor's note. 'Metad' is defined in Kleene, 'A Theory of Positive Integers in Formal Logic', 233.

[3] Editor's note. Presumably 'left-hand side'.

there is a λ-definable function (defined by D say) which gives G.R of A as function of G.R of [A, B]

Then A conv form(D(Gr([A, B])))

Intuition. Inspiration. Ingenuity

I am not sure whether my use of the word 'intuition' is right or whether your 'inspiration' would be better. I rather think that you are using 'inspiration' to cover what I had called 'ingenuity'. To give a concrete example of ingenuity, suppose I want a formula Θ with the property

$$\Theta(x) \text{ conv } x(\Theta(x))^4$$

I can of course search through an enumeration of all formulae Θ and perform conversions on $\Theta(x)$ (saving time over the possibly infinite conversion processes, by the 'diagonal process'), but if while I am doing this some bystander writes down

$$\Theta \rightarrow \{\lambda w \cdot w(w)\} \, (\lambda v u \cdot u(v(v, u)))$$

and says 'try that', I should say he had found a formula by 'ingenuity'.[5] In such cases there is no need to worry about how the formula is arrived at. That it is right is verified by a simple conversion, or something equally uncontroversial. Isn't this what you would call inspiration?

The straightforward unsolvability or incompleteness results about systems of logic amount to this

α) One cannot expect to be able to solve the Entscheidungsproblem for a system

β) One cannot expect that a system will cover all possible methods of proof (does not apply to 'restricted function calculus')

It seemed to me that in your account of what we want a system of logic to do you had α) in mind but not β). I should agree with your point of view, in so far as we can shut our eyes to β) i.e. we do not really want to make proofs by hunting through enumerations for them, but by hitting on one and then checking up to see that it is right. However this method is always theoretically, though not practically, replacable by the longer method if one has got a method of checking up. The enumeration of proofs is for instance obtained from an enumeration of all possible sequences of symbols by striking out those which do not pass the test. When one takes β) into account one has to admit that not one but many methods of checking up are needed. In writing about ordinal logics

---

[4] Editor's note. Presumably this should read: $\Theta(x)$ conv $\lambda x \cdot x(\Theta(x))$.

[5] Editor's note. Turing's example concerns a formula that he himself found and published in his 'The p-Function in λ-K-Conversion' (*Journal of Symbolic Logic*, 2 (1937), 164). The published formula is $\Theta \rightarrow \{\lambda v u \cdot u(v(v, u))\} \, (\lambda v u \cdot u(v(v, u)))$.

I had this kind of idea in mind.[6] In proofs there is actually an enormous amount of sheer slogging, a certain amount of ingenuity, while in most cases the actual 'methods of proof' are quite well known. Cannot we make it clearer where the slogging comes in, where there is ingenuity involved, and what are the methods of proof? In fact can we not express quite shortly what is the status of each proof? The ordinals were meant to give concise notations for the status of proofs.

The Completeness Theorem

The proof of my completeness theorem (P_A etc) is of course completely useless for the purpose of actually producing proofs. P_A will only be a convincing logic if $\underline{A}$ is rather simple, and easily recognized as an ordinal formula. The completeness theorem was written from a rather different point of view from most of the rest, and therefore tends to lead to confusion. I think that all this proof does is to provide an insurance against certain sorts of 'Gödel incompleteness theorems' being proved about the ordinal logic.

As soon as any question arises of having to prove that the formulae one is using are ordinal formulae one is returning to the single logic point of view, unless the kind of proof to be used is something different, being a kind of propaganda rather than formal proof.

The exercise

I have no complaints at all about this. You have evidently got the tools necessary for barging through anywhere where one can get through. I don't remember, even if I ever knew, what the standard way of doing this job is. I have toyed for half an hour or so with trying to do it with things of form

$$[\underline{A}, [\underline{B}, [\underline{C}, \ldots]] \ldots] \text{ instead of } [\underline{A}, \underline{B}, \ldots]$$

One might in that way avoid the trouble of looking after the number of variables. One will need a 'picking out function' L which will satisfy

$$L([\underline{m}, [\underline{n}, \ldots]]) \text{ conv } \underline{m}$$

but I cannot find one independent of the number of variables.

Church tells me he is going to publish his form of Principia involving the use of λ, and simple theory of types. I am very glad of it, as the system makes things much clearer than any other I know and is not too cumbrous to be used.

Gödel's paper has reached me at last. I am very suspicious of it now but will have to swot up the Zermelo–v. Neumann system a bit before I can put objections down in black & white.

Yours sincerely
A. M. Turing

[6] Editor's note. Turing is referring to 'Systems of Logic Based on Ordinals' (Chapter 3).

Letter from King's College, Cambridge

Sunday
                                                        King's College
                                                               Cambridge

Dear Newman,

   Church's notes certainly are rather a mouthful. I have never worked steadily through them myself, but have taken them in much the same spirit as you are doing. Fortunately I was able to go to the fountainhead for information.

   i) Metads certainly are a form of 'Gödel representation' which Ch. finds it convenient to use in his system.

   ii) I think the point of using the peculiar form of negation is that one wants Th 1, 2, 3 p 48[7] to hold in the form in which they stand. If one has $\sim \to \lambda x \cdot 3 \dot{-} x$ (i.e. $\sim \to \lambda x \cdot 3$ (p, x)), $\sim \underline{A}$ will have a normal form sometimes when $\underline{A}$ is not convertible to 1 or 2 e.g. if $\underline{A}$ conv 3 (also incidentally, as there is no 0, $3 \dot{-} x$ has the value 1 in this case and this would be bad apart from the normal form difficulty, but this is more easily corrected).

   iii) <u>Consequences</u>. I think one wants here to distinguish two ideas a) 'consequences of an assumption' (p. 82, 14)[8], b) consequences of an assumption relative to a set of rules of procedure. The first of these is an 'intuitive' idea which one tries to approximate by the second with suitable sets of rules of procedure. To get the idea of 'consequences of an assumption' imagine that the underlined letters are admitted as parts of formulae in a new system. I will use only letters for variable underlined letters, and variable formulae involving underlined letters. Then if $\mathfrak{A}, \mathfrak{B}$ are such formulae involving underlined letters $\mathfrak{X}, \mathfrak{Y}, \ldots, \mathfrak{Z}$, we say that $\mathfrak{B}$ is a consequence of $\mathfrak{A}$ if, for all substitutions of formulae (in original sense, with δ) for $\mathfrak{X}, \mathfrak{Y}, \ldots, \mathfrak{Z}$ it happens that $\mathfrak{B}$ conv 2 whenever $\mathfrak{A}$ conv 2. This of course implies two different uses of underlined letters from Chap X[9] onwards, but I think Church is really doing this. The idea of consequences of an assumption relative to given rules of procedure I think explains itself. One tries of course to make the rules of procedure such that the consequences will be consequences in the sense a), but also to get as many consequences as one can consistent with this. Of course one cannot get all such with one set of rules.

   iv) The Π_m's. These certainly are much the same as my ordinal logics, that is to say that the rule by which the Π_m's are formed can easily be used to help one construct an ordinal logic. They are better and better approximations by

---

    [7] Editor's note. A reference to Church's 'Mathematical Logic' (see p. 205 n. 2).
    [8] Editor's note. A reference to Church's 'Mathematical Logic'.
    [9] Editor's note. Chapter X of Church's 'Mathematical Logic' is entitled 'The Universal Quantifier'.

consequences of type b) above to the consequences of type a). The meaning of the Π_m's is this. One has defined the rules $1_m \ldots 7_m$ and $1_r \ldots 7_r$ with $r < m$. Taking all these and $1 \ldots 63$ we have a set of rules of procedure $Proc_m$ say. From them we get Π_m which is such that $\Pi_m(\underline{F}, \underline{G})$ conv 2 if and only if $\underline{F}$ and $\underline{G}$ are metads of formulae $\mathfrak{A}, \mathfrak{B}$ such that $\mathfrak{B}$ is a consequence of $\mathfrak{A}$ relative to $Proc_m$. (There is some confusion between underlined letters and ordinary variables in this definition of the Π_m's, as metads are names of formulae without underlined letters. Probably you have to regard all the free variables in the formulae described by the metads as replaced by underlined letters if we are to follow my description under iii).)

v) <u>Ingenuity and Intuition.</u> I think you take a much more radically Hilbertian attitude about mathematics than I do. You say 'If all this whole formal outfit is not about finding proofs which can be checked on a machine it's difficult to know what it is about.' When you say 'on a machine' do you have in mind that there is (or should be or could be, but has not been actually described anywhere) some fixed machine on which proofs are to be checked, and that the formal outfit is, as it were, about this machine. If you take this attitude (and it is this one that seems to me so extreme Hilbertian) there is little more to be said: we simply have to get used to the technique of this machine and resign ourselves to the fact that there are some problems to which we can never get the answer. On these lines my ordinal logics would make no sense. However I don't think you really hold quite this attitude because you admit that in the case of the Gödel example one can decide that the formula is true i.e. you admit that there is a fairly definite idea of a true formula which is quite different from the idea of a provable one. Throughout my paper on ordinal logics I have been assuming this too.[10] It mostly takes the form of talking about such things as a formula $\underline{A}$ such that $\underline{A}(\underline{n})$ conv 2 for all pos. integers $\underline{n}$.

If you think of various machines I don't see your difficulty. One imagines different machines allowing different sets of proofs, and by choosing a suitable machine one can approximate 'truth' by 'provability' better than with a less suitable machine, and can in a sense approximate it as well as you please. The choice of a proof checking machine involves intuition, which is interchangeable with the intuition required for finding an $\underline{\Omega}$ if one has an ordinal logic Λ, or as a third alternative one may go straight for the proof and this again requires intuition: or one may go for a proof finding machine. I am rather puzzled why you draw this distinction between proof finders and proof checkers. It seems to me rather unimportant as one can always get a proof finder from a proof checker, and the converse is almost true: the converse fails if for instance one allows the proof finder to go through a proof in the ordinary way, and then, rejecting the steps, to write down the final formula as a 'proof' of itself. One can easily think

[10] Editor's note. Turing is referring to 'Systems of Logic Based on Ordinals' (Chapter 3).

up suitable restrictions on the idea of proof which will make this converse true and which agree well with our ideas of what a proof should be like.

I am afraid this may be more confusing to you than enlightening. If so I will try again.

Yours sincerely
A. M. Turing

Enigma

Jack Copeland

1. Turing Joins the Government Code and Cypher School *217*
2. The Enigma Machine *220*
3. The Polish Contribution, 1932–1940 *231*
4. The Polish Bomba *235*
5. The Bombe and the Spider *246*
6. Naval Enigma *257*
7. Turing Leaves Enigma *262*

1. Turing Joins the Government Code and Cypher School

Turing's personal battle with the Enigma machine began some months before the outbreak of the Second World War.[1] At this time there was no more than a handful of people in Britain tackling the problem of Enigma. Turing worked largely in isolation, paying occasional visits to the London office of the Government Code and Cypher School (GC & CS) for discussions with Dillwyn Knox.[2] In 1937, during the Spanish Civil War, Knox had broken the type of Enigma machine used by the Italian Navy.[3] However, the more complicated form of Enigma used by the German military, containing the *Steckerbrett* or plug-board, was not so easily defeated.

On 4 September 1939, the day following Chamberlain's announcement of war with Germany, Turing took up residence at the new headquarters of the Government Code and Cypher School, Bletchley Park.[4] GC & CS was a tiny organization

---

[1] Letters from Peter Twinn to Copeland (28 Jan. 2001, 21 Feb. 2001). Twinn himself joined the attack on Enigma in February 1939. Turing was placed on Denniston's 'emergency list' (see below) in March 1939, according to 'Staff and Establishment of G.C.C.S.' (undated), held in the Public Record Office: National Archives (PRO), Kew, Richmond, Surrey (document reference HW 3/82). (I am grateful to Ralph Erskine for drawing my attention to this document.)

[2] Letters from Twinn to Copeland (see n. 1).

[3] M. Batey, 'Breaking Italian Naval Enigma', in R. Erskine and M. Smith (eds.), *Action This Day* (London: Bantam, 2001), 98.

[4] Letter from A. G. Denniston to T. J. Wilson of the Foreign Office (7 Sept. 1939). PRO document reference FO 366/1059.

218 | Jack Copeland

ill prepared for war. By 1942, however, Bletchley Park had become a veritable factory, and with the help of the codebreaking machines called 'bombes'— designed by Turing, Gordon Welchman, and, on the engineering side, Harold Keen—GC & CS was deciphering about 39,000 Enigma messages each month.[5] By 1945 almost 9,000 people were employed at Bletchley Park.[6] It is estimated that the breaking of Enigma—and in particular the breaking of Home Waters Naval Enigma, in which Turing played the crucial role—may have shortened the war in Europe by some two years.[7]

Figure 1. The Mansion, Bletchley Park.
Source: Bletchley Park Trust.

The Government Code and Cypher School had developed from the old 'Room 40', established by the Admiralty during the First World War for the purpose of reading enemy ciphers.[8] A branch of the Foreign Office, GC & CS was located in

---

[5] F. H. Hinsley et al., *British Intelligence in the Second World War*, vol. ii (London: Her Majesty's Stationery Office, 1981), 29.

[6] F. H. Hinsley et al., *British Intelligence in the Second World War*, vol. iii, part 1 (London: Her Majesty's Stationery Office, 1984), 461.

[7] This estimate was given by Hinsley, official historian of the British Secret Service, on p. 12 of his and Alan Stripp's edited volume *Codebreakers: The Inside Story of Bletchley Park* (Oxford: Oxford University Press, 1993). If, wrote Hinsley, the achievements of GC & CS 'had not prevented the U-boats from dominating the Atlantic...it is not unreasonable to believe that...Overlord [the invasion of Normandy, 1944] would have had to be deferred till 1946'.

[8] The older spelling 'cypher' and the newer 'cipher' were both in use at GC & CS during 1939–45. Mahon used 'cypher' in a 1945 document, part of which forms Chapter 5, and Turing used 'cipher' in a 1940 document, parts of which appear in Chapters 5 and 6.

Whitehall until the summer of 1939.[9] By the beginning of 1938 the Director of Naval Intelligence, Admiral Hugh Sinclair, was looking for premises outside London to which GC & CS could move in the event of war. Bletchley Park—a large Victorian mansion with ample grounds situated in the town of Bletchley, a major railway junction linking London, Oxford, and Cambridge—was purchased in the spring of 1938 (out of Sinclair's own pocket, it is said).

In the course of 1937 and 1938 Commander Alastair Denniston, Head of GC & CS and a veteran of Room 40, supervised a clandestine programme of recruitment, centred largely on Oxford and Cambridge. Denniston's aim was to build up what he described as an 'emergency list [of] men of the Professor type'[10].

At certain universities ... there were men now in senior positions who had worked in our ranks during 1914–18. These men knew the type required. Thus it fell out that our most successful recruiting occurred from these universities. During 1937 and 1938 we were able to arrange a series of courses to which we invited our recruits to give them even a dim idea of what would be required of them ... These men joined up in September 1939.[11]

(Frank Adcock and Frank Birch, the two veterans of Room 40 who were most active in recruitment as the new war approached, were both from the same college as Turing, King's.[12]) In the days following the outbreak of war in September 1939 a group of about thirty people assembled at Bletchley Park, many of them—including Turing—drawn from Denniston's 'emergency list'.[13]

An organizational structure rapidly began to emerge at Bletchley, newly formed sections being known simply as 'Hut 4', 'Hut 6', and so on. The 'huts' were single-storey wooden structures hastily constructed in the grounds of the mansion. Here dons worked among uniformed Naval and Army personnel. Military discipline never took root among the 'men of the Professor type' and parts of Bletchley Park had something of the atmosphere of an Oxbridge college. There were some notable eccentrics among the codebreakers. Dilly Knox, another fellow of King's and veteran of Room 40, liked to work in a hot bath. Once, at his lodgings, Knox

stayed so long in the bathroom that his fellow-lodgers at last forced the door. They found him standing by the bath, a faint smile on his face, his gaze fixed on abstractions, both taps full on and the plug out. What then was passing in his mind could possibly have solved a problem that was to win a battle.[14]

---

[9] Probably in August (R. Erskine, 'GC and CS Mobilizes "Men of the Professor Type"', *Cryptologia*, 10 (1986), 50–9 (50)).

[10] Letter from Denniston to Wilson (3 Sept. 1939). PRO document reference FO 366/1059.

[11] A. G. Denniston, 'The Government Code and Cypher School between the Wars', in C. W. Andrew (ed.), *Codebreaking and Signals Intelligence* (London: Cass, 1986), 52.

[12] Andrew, *Codebreaking and Signals Intelligence*, 4.

[13] S. Milner-Barry, 'Hut 6: Early Days', in Hinsley and Stripp (eds.), *Codebreakers*, 90; 'Staff and Establishment of G.C.C.S.'; Erskine, 'GC and CS Mobilizes "Men of the Professor Type"', 50.

[14] E. R. Vincent, Unpublished Memoirs, Corpus Christi College Archives, Cambridge; quoted in C. W. Andrew, *Secret Service: The Making of the British Intelligence Community* (London: Guild, 1985), 94.

2. The Enigma Machine

The Enigma machine had something of the appearance of an old-fashioned typewriter. Designed by the Berlin engineer Arthur Scherbius, Enigma was marketed commercially from 1923.[15] In 1926 the German Navy adopted Enigma, followed by the German Army in 1928 and the German Air Force in 1935.[16] At the outbreak of war with Britain, Enigma was the Germans' principal method for protecting their military communications. In 1930, the German military had considerably enhanced the security of the machine by adding the *Steckerbrett* or *plug-board* (see Figure 4).[17] It is this form of Enigma—German military, or *Wehrmacht*, Enigma—that is dealt with here. Successive modifications were made to the operating procedures of the military machine, resulting in substantial variation both over time and from one branch of the armed services to another.

Battery powered and highly portable, the *Wehrmacht* Enigma machine could be used from a general's office in Berlin, an armoured vehicle, a submarine, or a trench. The machine's keyboard had twenty-six keys, each marked with a letter (Figure 4). Instead of an arrangement for typing letters onto paper, the machine had a lampboard consisting of twenty-six bulbs, each of which shone through a stencil on which a letter of the alphabet was marked. The operator of the Enigma machine would be handed a message in plain text. His job was to type the message at the keyboard of the machine. Each time he pressed a key, a letter on the lampboard would light up. The operator's assistant kept a note of which letters lit up on the lampboard. This enciphered form of the message was then sent to its recipient, if by radio then in Morse code. The sending radio operator would preface the message with his radio call-sign, followed by that of the intended receiver. The Germans also sent Enigma messages by land-lines; for these messages, Morse was not used. (Land-lines are not mentioned further in this introduction, since German message traffic sent in this way was not intercepted in Britain.)

Each time the operator pressed a key, one or more wheels turned inside the machine, and each time a wheel moved it altered the wiring between the keyboard and the lampboard. So if, for example, the operator repeatedly depressed the O-key, the connections between the key and the lampboard would change with each key press, resulting in a succession of different letters lighting up, for example Q M P W A J Y R.

[15] F. L. Bauer, *Decrypted Secrets: Methods and Maxims of Cryptology* (Berlin: Springer-Verlag, 2nd edn. 2000), 107.

[16] F. H. Hinsley et al. *British Intelligence in the Second World War*, vol. iii, part 2 (London: Her Majesty's Stationery Office, 1988), 946.

[17] M. Rejewski, 'Remarks on Appendix 1 to British Intelligence in the Second World War by F. H. Hinsley', *Cryptologia*, 6 (1982), 75–83 (76).

Figure 2. A three-wheel Enigma with the plug-board (at the front of the machine) exposed. The lampboard is behind the keyboard. The three wheel-slots are visible behind the lampboard. Beside each wheel-slot is a window through which letters marked on the wheels are visible to the operator.

Source: Science and Society Picture Library, National Museum of Science and Industry.

222 | Jack Copeland

Figure 3. Enigma machine with the three wheels exposed.
Source: Science and Society Picture Library, National Museum of Science and Industry.

The letter O itself would never appear in this succession of letters, however. Because of the action of the reflector, a letter was never enciphered as itself (see Figure 4). This rule was very useful to the codebreakers at Bletchley Park.

At the receiving end of the radio link, the message would be converted from Morse into ordinary letters. This cipher text was then typed at the keyboard of the recipient's Enigma machine. The letters that lit up on the lampboard would be the very same letters that the sender had keyed in—the plain text with which the process had begun. The design of the Enigma machines was such that if a key was pressed on one machine, say O, and the letter that lit up on the machine's

Enigma | 223

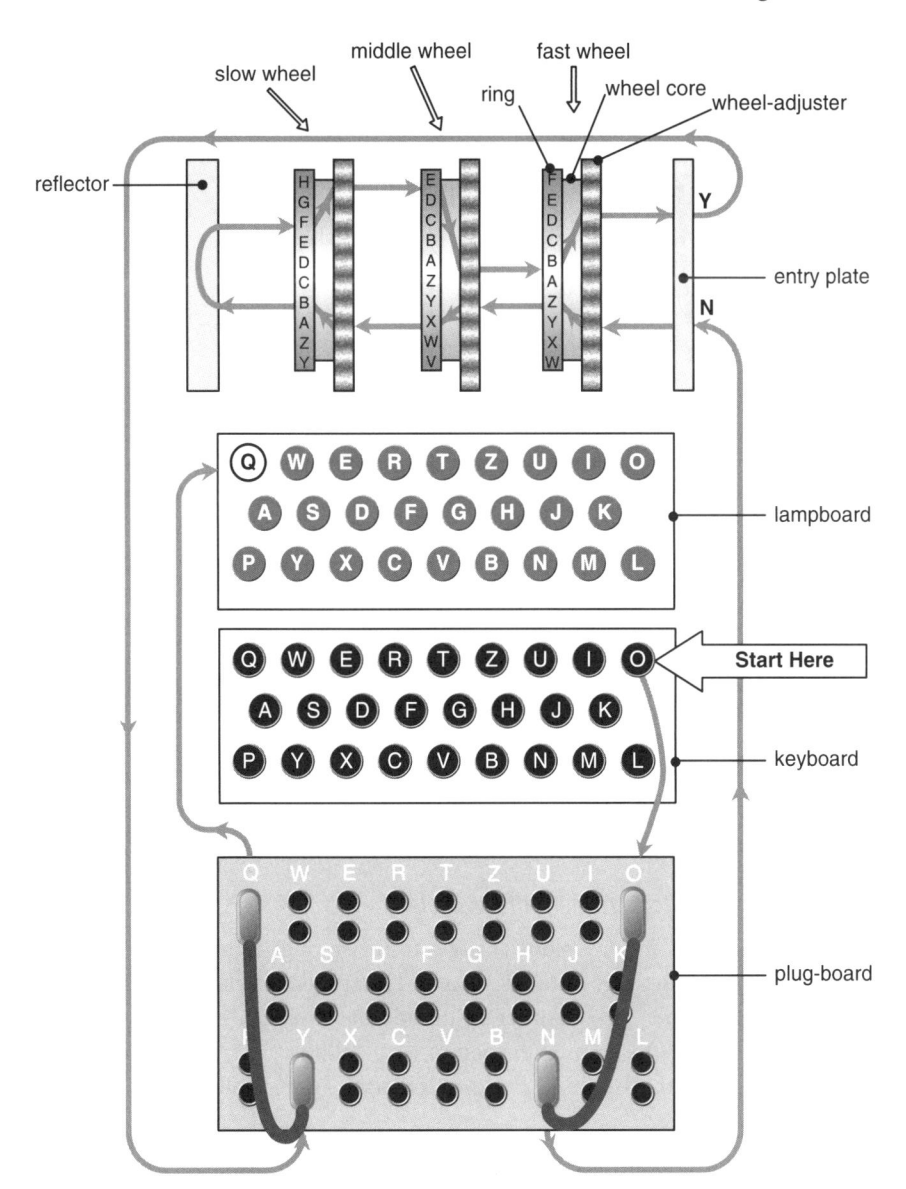

Figure 4. Path of electric current through the Enigma. Pressing a key at the keyboard causes a letter to light up at the lampboard. The core of each wheel contains a maze of 26 insulated wires, with each wire joining one of 26 contacts on the right hand side of the wheel to one of 26 contacts on the left-hand side. The wiring is different in each wheel.

Diagram by Dustin A. Barrett.

lampboard was keyed into a second machine, then—provided the two machines had been set up in exactly the same way by their respective operators—the second machine would light up O on its lampboard.

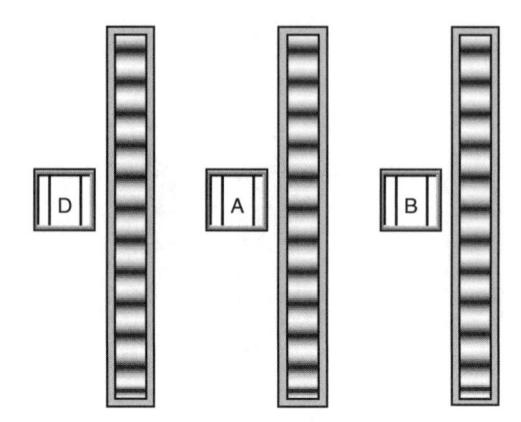

Figure 5. View of the wheels with the case closed. The three wheel-adjusters protrude through slots in the case. The windows allow the operator to see one letter from the ring of each wheel. The 'message setting' is the triple of letters visible at the start of typing a message.

Diagram by Dustin A. Barrett.

In a word, the letter-substitutions were *reversible*: if O produced Q (for example) then, at the same machine-settings, Q produced O. This was the basic principle of the Enigma system, hard-wired into the machine. Figure 4 indicates how this was achieved. If Q were pressed at the keyboard, current would flow along a wire leading to Q at the plug-board, then across the plug-board to Y and through the wheels in the reverse direction to that shown, exiting the wheels at N, crossing the plug-board to O, and lighting O at the lampboard.

The Plug-Board (Steckerbrett) and Wheels

The operator could make various changes to the settings of his machine before he began typing a message at the keyboard. The recipient would set up his own machine in the same way in order to decode the message. How the recipient knew which settings to use is explained in what follows.

The settings of the machine could be changed in the following ways. (See Figure 4.)

1. The operator could make alterations to the plug-board (*Steckerbrett*) on the front of the machine, pulling electrical leads out of sockets and plugging them back into different sockets. This altered some of the connections between the keyboard and the lampboard. (The plug-board was absent from the commercial version of the machine.[18])

---

[18] The commercial model remained on sale after the German military adopted Enigma. The Germans knew how to break the commercial model and from 1938 several hundred were sold to neutral Switzerland by the German manufacturers. The commercial model was also sold by Germany to Hungary during the war. Commercial model Enigmas sold to Spain were used during the Spanish Civil War. (I am grateful to Frode Weierud for this information (personal communication).)

2. The operator could alter the positions of the rotating wheels inside the machine (sometimes also called 'rotors') by turning them manually. Part of the circumference of each wheel protruded through the case of the machine enabling the operator to click the wheels round with his thumb or finger (Figure 5). In the early years of the war there were three rotatable wheels inside the machine; in 1941, the first Naval machines with a fourth rotatable wheel came into use (see the introduction to Chapter 8).[19] (Another two components of the Enigma are sometimes referred to as wheels or rotors, the *Umkehrwalze* (described by Mahon on p. 269 of Chapter 5) and the *Eintrittwalze*. In the forms of German military Enigma discussed here, both these components were stationary, and they will be referred to as the *reflector* and the *entry plate* respectively (Figure 4).)
3. The operator could open the case of the machine, lift out two or more of the wheels, and replace them in a different order. For example, he might switch the left- and right-hand wheels, leaving the centre wheel untouched. Each wheel was wired differently inside. Since the electrical pathways from the keyboard to the lampboard passed through the wheels, changing the order of the wheels altered the pathways. Alternatively, rather than simply switching the order of the wheels in the machine, the operator might replace one or more of them with different wheels from a box that accompanied the machine. From December 1938 until about the beginning of the war, there were a total of five wheels, numbered I–V, and any three of the five might be inside the machine at any one time. For example, the wheels in use might be I, II, and IV, in the order IV/I/II. From 1940 (or possibly as early as 1939) Enigma machines used by the German Navy were equipped with additional wheels and the operator would select three from a total of eight (numbered I–VIII).

The wheels were somewhat analogous to the wheels of a combination lock, turning through a number of discrete positions. Each wheel had a total of twenty-six possible rotational positions, A–Z. The wheel on the right, the first on the path from keyboard to lampboard, would always turn on one 'click' each time a key was pressed. Hence the term 'fast wheel' (Figure 4). After a certain number of clicks, this wheel would cause the centre wheel to turn one click. Likewise, the centre wheel would at some point cause the wheel on the left—the 'slow wheel'—to move one click. (An extra complication: when this happened, the centre wheel would itself turn forward one click also.[20])

---

[19] The fourth wheel differed from the other three in that once the operator had set it to one of its twenty-six positions, it remained stationary during the encipherment of the message. (That the fourth wheel came into Naval use in 1941 is documented in R. Erskine, 'Breaking German Naval Enigma on Both Sides of the Atlantic', in Erskine and Smith (eds.), *Action this Day*, 181.

[20] H. Alexander, 'Cryptographic History of Work on the German Naval Enigma' (no date (c.1945), PRO document reference HW 25/1), 3; a digital facsimile of Alexander's typescript is available in The Turing Archive for the History of Computing <www.AlanTuring.net/alexander_naval_enigma>.

Figure 6. A dismantled wheel.
Source: Science and Society Picture Library, National Museum of Science and Industry.

Precisely when a wheel would cause its neighbour to turn was determined by the position of a notch cut into the ring of the wheel. Since wheels I–V all had their notches in different places, changing or rearranging wheels could affect the 'turn-overs' (Bletchley's term for the points at which wheels would cause their neighbours to turn). The Naval wheels VI–VIII were slightly different. These had their notches in the same places as one another, and moreover each had *two* notches (see pp. 268, 285 below). The extra notch meant that in the course of one revolution, the doubly notched wheel would cause its neighbour to move twice.

Which letter lit up on the lampboard depended, therefore, not only on which key was depressed, but also on how the plug-board was connected up, which of the possible wheels were inside the machine, what order these wheels were arranged in, and which of its twenty-six rotational positions each wheel occupied at the time the key was pressed. In fact, by altering these variables, the operator was able to set up a machine with a total of three wheels in excess of a thousand million million different ways. The message remained protected even if the enemy captured an Enigma machine of the type that the sender was using. In order for a recipient to decipher the message, he or she needed to know which of the astronomically many possible settings the sender had used to encipher the text.

Enigma Keys

The sender and the (authorized) recipient were issued with printed tables of settings so that they could set up their machines in the same way. A group of Enigma-users operating with the same tables is called a *network*. A set of tables covered a period of one month and specified how, on any given day, the members of the network should set up their machines. Different networks used different tables.

GC & CS referred to a network of Enigma-users as a 'key'. Each key was given a name—Yellow, Red, Green, Light Blue, Shark, Dolphin, Porpoise, Kestrel, Phoenix, Locust, Snowdrop, etc. At the beginning of the war, the number of known keys was small enough for GC & CS to be able to represent them on a chart by means of coloured pencils, the colour used becoming the name of the key. As the war progressed, the number of keys became much larger.

The term 'network' is perhaps clearer than 'key', especially since at Bletchley, 'key' was used ambiguously for a network of Enigma-users and in the term 'daily key' (whose meaning is explained below). Some writers prefer 'crypto-net' to 'network', since the former term makes it clear that it is an Enigma network and not a radio network that is being described.[21] One and the same radio network could carry the message traffic of several crypto-nets.

Wheel Order, Stecker, and Ringstellung

The wheel order for a particular day for a certain network or key might be III/I/II, for example.

Stecker is short for *Steckerverbindungen*, meaning 'plug connections'. The *Stecker*, or plug-board configuration, for a particular day might be A/C, D/V, F/M, H/W, L/X, R/I. Corresponding to each letter on the plug-board is a pair of sockets, one for a cable leading *to* another letter, and one for a cable leading *from* another letter (Figure 4). The operator would set up the plug-board by connecting together the pair of sockets labelled 'A' and the pair of sockets labelled 'C' by means of a short cable with a double plug at each end. Likewise for the 'D' sockets and the 'V' sockets, and so on. The Germans' use of double plugs meant that if A is steckered to C, then C is steckered to A—a fatal simplification, as we shall see.

Ringstellung means 'ring position'. The ring is like a tyre mounted round the core of each wheel. It is marked with the letters of the alphabet, one for each of the twenty-six rotational positions of the wheel (Figure 4). (Sometimes the numerals '01' to '26' were used instead of letters.) The ring could be moved around the wheel core to a selected position and then fixed in position with a clip. The day's ring position for a given wheel was specified by a single letter, say X. The operator would turn the ring until the letter X was aligned against a fixed

---

[21] See, for example, G. Welchman, *The Hut Six Story: Breaking the Enigma Codes* (Kidderminster: M. & M. Baldwin, 2nd edn. 1997), 205.

index mark embossed on the wheel and then would fix the ring in this position. The complete *Ringstellung* for the day would consist of a trigram, say XYZ, one letter for each wheel in the machine.

The Daily Key

The daily wheel order, *Stecker*, and *Ringstellung* for the machine were specified in the tables issued to each Enigma network. *Stecker*, wheel order, and *Ringstellung* were elements of the *daily key*, or basic settings for the day for a given network of Enigma users.

The reason for changing the basic settings daily was to minimize the number of messages encoded at the same settings. The Germans knew that security could be compromised if too many messages were encoded at the same basic settings. During the later years of the war, some networks changed the *Stecker*, wheel order, and *Ringstellung* not daily but every eight hours.[22]

The Message Setting

Setting up the sender's and recipient's machines in accordance with the specified *Stecker*, wheel order, and *Ringstellung* did not suffice to place the two machines completely in register. There was also the question of the rotational positions of the three wheels at the start of the message.

Once the ring position was set, the rotational position of a wheel could be described by saying which of the letters on the ring was uppermost when the wheel was in place inside the machine. The machine's case was fitted with three small windows, one above each wheel, so that the operator could see the uppermost letter (Figure 5).

The positions occupied by the wheels at the start of typing a message were specified by a trigram, for example QVZ, meaning that Q is visible in the window over the left-hand wheel, V in the window over the middle wheel, and Z in the window over the right-hand wheel. QVZ was known as the *message setting*.[23]

Notice that knowing the message setting does not reveal the rotational positions of the wheels at the start of the message unless the *Ringstellung* is also known—QVZ may specify any one of the 26 × 26 × 26 possible positions, depending on which ring positions have been selected.

---

[22] M. Rejewski, 'Summary of our Methods for Reconstructing Enigma and Reconstructing Daily Keys, and of German Efforts to Frustrate Those Methods', in W. Kozaczuk, *Enigma: How the German Machine Cipher Was Broken, and How It Was Read by the Allies in World War Two*, trans. C. Kasparek (London: Arms and Armour Press, 1984), 243.

[23] Rejewski's accounts of the work of the Polish cryptanalysts use 'message key' instead of the Bletchley term 'message setting'. See, for example, M. Rejewski, 'Jak Matematycy polscy rozszyfrowali Enigme' [How the Polish Mathematicians Broke Enigma], *Annals of the Polish Mathematical Society, Series II: Mathematical News*, 23 (1980), 1–28. (This article appears in an English translation by C. Kasparek as appendix D of Kozaczuk, *Enigma*; another translation, by J. Stepenske, appears in *Annals of the History of Computing*, 3 (1981), 213–34, under the title 'How Polish Mathematicians Deciphered the Enigma'.)

Operating Procedures

In order to decode the message, a recipient needs the wheel order, the *Stecker*, the *Ringstellung*, and the message setting. The most direct way to make the message setting available to the authorized recipient would be to make it an element of the daily key printed in the monthly tables. The operator would then simply look up the specified trigram for the day in question, and ensure that it was visible in the windows at the start of each message. This was the procedure used with the commercial form of Enigma.[24] But this method provided very weak security, reducing the problem of breaking a day's messages to that of solving a number of *substitution* ciphers.

The substitution cipher is an ancient and simple form of cipher in which the alphabet is paired with a 'scrambled' alphabet. For example:

A	B	C	D	E	F	G	H	I	J	K	L	M	N	O	P	Q	R	S	T	U	V	W	X	Y	Z
Z	Y	X	W	V	U	T	S	R	Q	P	O	N	M	L	K	J	I	H	G	F	E	D	C	B	A

THE ESSENTIAL TURING = GSV VHHVMGRZO GFIRMT

The great Polish cryptanalyst Marian Rejewski explained the weakness of enciphering a day's Enigma traffic at the same message setting:

> the first letters of all the messages... constituted an ordinary substitution cipher, a very primitive cipher easily soluable given sufficient material, and all the second letters of the messages... constituted another substitution cipher, and so on. These are not merely theoretical deliberations. It was in that very way that in France in 1940 we solved the Swiss Enigma cipher machine.[25]

The German armed forces employed more secure methods for making the message setting known to the intended recipient. The method adopted varied from service to service and from time to time, generally speaking with increasingly secure methods being used as time went on. From 1937 the German Navy used a particularly complicated method—although Turing did manage to break it. This method is described by Patrick Mahon in Chapter 5, which is an extract from Mahon's previously unpublished 'The History of Hut 8'. (Written in 1945, Mahon's 'History' was kept secret by the British and American governments until 1996.[26])

From the autumn of 1938 until May 1940 the German Army and Air Force used the following—as it turned out, highly insecure—method for sending the

---

[24] Rejewski, 'Remarks on Appendix 1 to British Intelligence in the Second World War by F. H. Hinsley', 79.

[25] Rejewski, 'How the Polish Mathematicians Broke Enigma', trans. Kasparek, 251.

[26] Mahon's 'The History of Hut 8' is in the US National Archives and Records Administration (NARA) in Washington, DC (document reference: RG 457, Historic Cryptographic Collection, Box 1424, NR 4685) and in the UK Public Record Office (document reference HW 25/2). A digital facsimile of the original typescript is available in The Turing Archive for the History of Computing <www.AlanTuring.net/mahon_hut_8>.

message setting to the recipient.[27] The sender would select two trigrams at random, say RBG and VAK. RBG is the message setting. VAK specifies the starting positions of the wheels that will be used not when encoding the message itself but when encoding the message setting prior to broadcasting it to the recipient. VAK would be broadcast to the recipient as part of an unencoded *preamble* to the encoded message. (The preamble could also include, for example, the time of origin of the message, the number of letters in the encoded message, and a group of letters called a *discriminant*, identifying the Enigma network to which the message belonged (e.g. Red).[28] The preamble might also contain an indication that the message was the second (or later) part of a two-part or multi-part message; see Mahon's discussion of 'forts' on pp. 278–9 below.)

The Indicator and Indicator Setting

Having selected the two trigrams, the sender would first set up VAK in the windows of his machine. He would then type RBGRBG. The group of six letters that lit up, say PRUKAC, is called the *indicator*. VAK is called the *indicator setting* (or '*Grundstellung*').[29] The indicator would be broadcast immediately before the enciphered message. The reason for sending the encipherment of RBGRBG, rather than simply of RBG, was to provide the recipient with a check that the message setting had been correctly received, radio reception sometimes being poor.

Once the sender had enciphered the message setting to form the indicator, he would set up RBG in the windows of his machine and type the plain text. Then the whole thing would be sent off to the recipient—preamble, indicator, and enciphered text.

The authorized recipient of the message would first rotate the wheels of his machine (already set up in accordance with the daily key) until VAK appeared in the windows. He would then type the indicator PRUKAC and the letters RBGRBG would light up at the lampboard. Now equipped with the message setting, he would set his wheels to RBG and retrieve the plain text by typing the encoded message.

---

[27] Rejewski, 'How the Polish Mathematicians Broke Enigma', trans. Kasparek, 265–6; Hinsley, *British Intelligence in the Second World War*, vol. iii, part 2, 949, 953.

[28] G. Bloch and R. Erskine, 'Enigma: The Dropping of the Double Encipherment', *Cryptologia*, 10 (1986), 134–41.

[29] The term 'indicator' is used by Mahon and Turing in the next chapter and is listed in 'A Cryptographic Dictionary', GC & CS (1944). ('A Cryptographic Dictionary' was declassified in 1996 (NARA document reference: RG 457, Historic Cryptographic Collection, Box 1413, NR 4559); a digital facsimile is available in The Turing Archive for the History of Computing <www.AlanTuring.net/crypt_dic_1944>.) However, the term 'indicator setting', which is from Welchman (*The Hut Six Story*, 36, 46) may not have been in use at Bletchley Park, where the German term *Grundstellung* (or 'Grund') was used (see e.g. pp. 272–3, below), as it was by the Poles (letter from Rejewski to Woytak, quoted on p. 237 of Kozaczuk, *Enigma*).

The method just described of selecting and making known the message setting is an example of what is called an *indicator system*.

3. The Polish Contribution, 1932–1940[30]

Unknown to GC & CS, the Biuro Szyfrów—the Polish Cipher Bureau—had already broken *Wehrmacht* Enigma, with assistance from the French secret service. The Biuro read the message traffic of the German Army regularly from 1933 to the end of 1938, and at other times during this period read the message traffic of other branches of the military, including the Air Force. Statistics gathered by the Biuro early in 1938 showed that, at that time, about 75 per cent of all intercepted Enigma material was being successfully decoded by the Biuro Szyfrów.

Towards the end of 1932 Rejewski had devised a method for reconstructing a day's message settings from the indicators, given about sixty messages sent on the day. He was helped by the fact that, in this early period, the indicator system was simpler than the later system just described. The daily key included an indicator setting *for the day*, e.g. VAK. The sender would choose his own message setting for each message, e.g. RBG. With the wheels in the positions specified in the daily key (VAK), he would type RBGRBG to produce the indicator. Then he would set the wheels to RBG and type the plain text of the message. The encoded message was sent prefaced by the preamble and the indicator—but, of course, there was no need to send the indicator setting.

Using information obtained from his attack on the indicators, Rejewski devised a method that enabled him to determine the internal wiring of wheels I–III (in those early days there were no additional wheels). This was one of the most far-reaching achievements in the history of cryptanalysis. Rejewski was assisted by the French secret service, whose agent Hans-Thilo Schmidt, a German employed in the cipher branch of the German Army, supplied photographs of two tables setting out the daily keys—*Stecker*, wheel order, *Ringstellung*, and the daily indicator setting—for September and October 1932. Rejewski describes this material as the 'decisive factor in breaking the machine's secrets'.[31]

---

[30] The sources for this section are: 'A Conversation with Marian Rejewski' (in Kozaczuk, *Enigma*), Rejewski's articles 'How the Polish Mathematicians Broke Enigma', 'Summary of our Methods for Reconstructing Enigma and Reconstructing Daily Keys, and of German Efforts to Frustrate Those Methods', 'The Mathematical Solution of the Enigma Cipher' (in Kozaczuk, *Enigma*), and 'Remarks on Appendix 1 to British Intelligence in the Second World War by F. H. Hinsley', together with Hinsley, vol. iii, part 2, appendix 30 'The Polish, French and British Contributions to the Breaking of the Enigma: A Revised Account'. (Appendix 30 replaces the sometimes very inaccurate appendix 1, 'The Polish, French and British Contributions to the Breaking of the Enigma', of Hinsley et al., *British Intelligence in the Second World War*, vol. i (London: Her Majesty's Stationery Office, 1979).

[31] Rejewski, 'How Polish Mathematicians Deciphered the Enigma', trans. Stepenske, 221.

In 1931 the French had attempted to interest the British in documents obtained by Schmidt, including operating manuals for German military Enigma. It is said that the British showed little interest, however, and declined to help the French meet the costs of obtaining them. It was not until 1936 that GC & CS began to study Enigma seriously. By the middle of 1939, Knox had discovered something like the Polish method for obtaining the message settings from the indicators (for German Army traffic).[32] However, he was unable to determine the internal wiring of the wheels. Without the wiring, it was impossible to use the method to decode the messages. GC & CS probably discovered a version of the same method that Rejewski had used to determine the wiring of the wheels, calling the method a 'Saga' (Mahon mentions it briefly on p. 278 of the next chapter). Knox is said to have outlined a 'more complicated version' of the Rejewski method at a meeting in Paris in January 1939.[33] However, he was never able to use this method to find the wiring of the wheels. This was because he was never able to discover the pattern of fixed wiring leading from the plugboard to the right-hand wheel via the entry plate (see Figure 4)—the 'QWERTZU', as he liked to call this unknown pattern, after the letters along the top row of the Enigma keyboard. This entirely humdrum feature of the military machine was what defeated Knox. Rejewski himself discovered the pattern by a lucky guess.

Once Rejewski had worked out the internal wiring of the wheels, he attacked the problem of how to determine the daily keys. This he solved early in 1933. At this stage, Rejewski was joined by Henryk Zygalski and Jerzy Różycki. Zygalski, Różycki, and Rejewski had graduated together from a course in cryptology that the Biuro Szyfrów had given in 1928–9. (Rejewski said later that it could have been the Biuro's fruitless efforts to break Enigma during 1928—the year in which the first messages were intercepted—that prompted the organization of the course at which the three were recruited.[34])

Now that the Polish cryptanalysts were able to find the daily keys on a regular basis, they needed access to Enigma machines in order to decipher the daily traffic. Using what Rejewski had found out concerning the wiring of the wheels, copies of the Wehrmacht Enigma were built by a Warsaw factory. Initially about half a dozen clerical staff were employed by the Biuro Szyfrów to operate the replica Enigmas. The clerical staff were 'put into a separate room, with the sole assignment of deciphering the stream of messages, the daily keys to which we soon began supplying'.[35] The number of replica Enigmas in use at the Biuro increased to about a dozen by mid-1934.

---

[32] Hinsley, *British Intelligence in the Second World War*, vol. iii, part 2, appendix 30, 951.
[33] Ibid.
[34] Letter from Rejewski to Richard Woytak, 15 Apr. 1979; the letter is printed in Kozaczuk, *Enigma*, 237–8.
[35] Rejewski, 'How the Polish Mathematicians Broke Enigma', trans. Kasparek, 261.

This state of affairs persisted until September 1938, when the German Army and Air Force abandoned the indicator system that Rejewski had broken in 1932. They switched to the indicator system described above: the indicator setting was no longer supplied in the tables giving the daily key, but was made up by the sender himself. Overnight the Poles' methods for determining the daily keys and message settings became useless. (In German Naval Enigma, the system broken by Rejewski had been abandoned in May 1937, when the complicated indicator system described by Mahon in the next chapter was adopted. Mahon outlines the Polish work on Naval Enigma to 1937.)

Within a few weeks of the September change, however, the Poles had devised two new methods of attack. One involved the use of perforated sheets of paper to determine the daily key, starting from a sufficient number of messages whose indicators displayed certain patterns of repeated letters. (Knox devised a similar method and was planning to use marks on photographic film rather than perforations, but was unable to put the method into practice without knowing the internal wiring of the wheels.[36]) The Poles' other method involved an electromechanical apparatus, designed by Rejewski and (on the engineering side) Antoni Palluth.[37] This was the *bomba* (plural 'bomby'), forerunner of the Bletchley Park bombe. How the bomba worked is explained in the next section. Six bomby were in operation by mid-November 1938.

The bomby and the perforated sheets depended on the fact that the indicator was formed by enciphering the message setting *twice* (e.g. enciphering RGBRGB rather than simply RGB). If the indicator system were changed so that the message setting was enciphered only once, the bomby and the perforated sheets would become unusable. This is precisely what was to happen in May 1940. Well before this, however, the bomby became overwhelmed by other changes designed to make Enigma more secure.

In December 1938 the Germans introduced the two extra wheels, IV and V. The Poles were able to determine the internal wiring of the new wheels by the method used in 1932 (thanks to the fact that one Enigma network—the intelligence service of the Nazi party—had not adopted the indicator system that came into force on other networks in September 1938 and was still using the system that the Poles could break by their earlier methods). But the material resources of the Biuro Szyfrów were insufficient to enable the Poles to cope with the increase in the number of wheel orders that the two new wheels produced. Where previously there had been only six possible wheel orders, there were now sixty. In order to investigate the new wheel orders, at least thirty-six replicas of each new wheel were required. The factory could not produce replicas fast enough.

---

[36] Hinsley, *British Intelligence in the Second World War*, vol. iii, part 2, appendix 30, 951.
[37] Rejewski, 'How the Polish Mathematicians Broke Enigma', 267.

Work with the perforated sheets was affected in the same way. The drawback of the sheet method had always been that the manufacture of a single sheet required the cutting of about 1,000 tiny perforations in exactly the right positions, with twenty-six sheets being required for each possible wheel order. Suddenly a huge number of additional sheets was required.

The result of the addition of the new wheels was that the Poles were able to read German Army and Air Force messages on only those days when it happened that wheels I, II, and III were in the machine—on average one day in ten.

Pyry and After

In July 1939 the Poles invited members of the British and French intelligence services to a meeting at Pyry near Warsaw. Denniston and Knox represented GC & CS. At this meeting, Rejewski relates, 'we told everything that we knew and showed everything that we had'—a replica Enigma, the bomba, the perforated sheets, and of course the all-important internal wiring of the wheels, which Knox still had not been able to work out.[38] Without the Poles, Knox and Turing might not have found out the wiring of the wheels until May 1940, when the British captured several intact Enigma machines from the German Army in Norway.

Knox's first question to the Poles was 'What is the QWERTZU?'[39] The answer was almost a joke—the connections were in *alphabetical* order, with the A-socket of the plug-board connected to the first terminal inside the entry plate, the B-socket to the second, and so on. Knox was ecstatic to know the answer at last, chanting in a shared taxi 'Nous avons le QWERTZU, nous marchons ensemble' ('We have the QWERTZU, we march along together').[40]

At Pyry the Poles also undertook to supply their British and French allies with two replica Enigma machines. The replica destined for GC & CS was couriered from Paris to London on 16 August 1939 by two men, Gustave Bertrand, head of the codebreaking section of the French Intelligence Service, and 'Uncle Tom', a diplomatic courier for the British Embassy in Paris. On the platform of Victoria Station they handed the machine over to Admiral Sinclair's deputy, Colonel Stewart Menzies. Menzies, on his way to an evening engagement, was dressed in a dinner jacket and he sported the rosette of the Légion d'Honneur in his buttonhole. *Accueil triomphal*—a triumphant welcome, Bertrand declared.[41]

Following the invasion of Poland, Rejewski and his colleagues moved to France. By January 1940 GC & CS, with its superior resources, had produced two complete sets of perforated sheets. The Poles received one of the sets in instalments. Turing delivered some of the sheets himself.

---

[38] Ibid. 269.
[39] Ibid. 257; P. Twinn, 'The *Abwehr* Enigma', in Hinsley and Stripp (eds.), *Codebreakers*, 126.
[40] Twinn, 'The *Abwehr* Enigma', 126–7.
[41] G. Bertrand, *Enigma, ou la plus grande énigme de la guerre 1939–1945* (Paris: Plon, 1973), 60–1.

Rejewski recollected: 'We treated [Turing] as a younger colleague who had specialized in mathematical logic and was just starting out in cryptology. Our discussions, if I remember correctly, pertained to the commutator [plugboard] and plug connections (Steckerverbindungen) that were Enigma's strong point.'[42] Little did Rejewski know that Turing had already devised the brilliant method of dealing with the *Steckerverbindungen* on which the British bombe was based.

For several months the British and the Poles worked in cooperation. The first break of wartime traffic since September 1939 was achieved by the Poles in mid-January 1940, followed a few days later by further breaks at GC & CS. During the period of fruitful collaboration that ensued, the Poles with their lesser resources were responsible for about 17 per cent of the daily keys broken.

Then, in May 1940, everything changed. The new indicator system introduced by the German Army and Air Force on 1 May made the perforated sheets useless for all networks except one, Yellow, which continued to employ the old system. Even Yellow, an inter-services key in use during the Norway campaign, went out of service on 14 May.[43] The change of indicator system and the German occupation of France effectively ended the attack on Enigma by the exiled Biuro Szyfrów.

The British were able to continue reading German Air Force messages (from 20 May) by means of methods developed at GC & CS which exploited the bad habits of some German Enigma operators. One was the habit of enciphering the message setting at the position that the wheels happened to be in at the end of the previous message, or at a closely neighbouring position (obtained e.g. by lazily turning only one wheel some small number of clicks).

From the summer of 1940 the codebreakers at GC & CS began to receive assistance from Turing's radically redesigned version of the Polish bomba.

4. The Polish Bomba

Origin of the Name 'Bomba'

In Chapter 5, Mahon says that the British bombe 'was so called because of the ticking noise it made, supposedly similar to that made by an infernal machine regulated by a clock' (p. 291). This story was well entrenched among Bletchleyites. The need-to-know principle meant that few were aware of the Polish bomba. Similarly, the explanation that circulated at Bletchley Park of why certain patterns, involving repetitions of letters at the same places, were known as 'females' took no account of the fact that the terminology had been borrowed

---

[42] Quoted in Kozaczuk, *Enigma*, 97. On Turing's visit to the Poles, see ibid 96–7; Welchman, *The Hut Six Story*, 220; and R. Erskine, 'Breaking Air Force and Army Enigma', in Erskine and Smith, *Action This Day*, 54.

[43] Erskine, 'Breaking Air Force and Army Enigma', 55.

from the Poles. The equivalent Polish term 'samiczki', meaning 'females', was quite likely the result of a play on words, 'samiczki' being used as short for a Polish phrase meaning 'the same places.'[44]

Why the Poles chose the name 'bomba' seems not to have been recorded. Rejewski's only comment was that the name was used 'for lack of a better idea'.[45] As well as meaning 'bomb', 'bomba' is the Polish word for a type of ice-cream dessert—*bombe* in French. Tadeusz Lisicki, who corresponded with Rejewski during the years before the latter's death in 1980, is quoted as saying: 'The name "bomba" was given by Różycki... [T]here was in Warsaw [an] ice-cream called [a] bomba... [T]he idea [for] the machine came while they were eating it.'[46]

A different story is told in recently declassified American documents. As explained later in this section, the bomba is required to stop immediately it detects a certain feature. How this was achieved by the Polish engineers is not known for sure. The American documents suggest that the stopping mechanism involved the dropping of weights, and the claim is made that this is how the name arose.

> [A] bank of Enigma Machines now has the name 'bombe'. This term was used by the Poles and has its origin in the fact that on their device when the correct position was reached a weight was dropped to give the indication.[47]

> When a possible solution was reached a part would fall off the machine onto the floor with a loud noise. Hence the name 'bombe'.[48]

It is not implausible that falling weights were used to disengage the bomba's drive mechanism (a printer designed by Babbage as part of his Difference Engine used a similar idea). However, the two American documents in question were written some years after Rejewski and his colleagues destroyed all six bomby in 1939[49] and neither cites a source for the claim quoted (the documents are dated 1943 and 1944). Moreover, both documents contain inaccurate claims concerning the Polish attack on Enigma (for example, that the bomba was 'hand operated', and that the military Enigma machine had no plug-board until 'about 1938').[50] The sketch of the bomba that accompanies Rejewski's 'The Mathematical Solution

---

[44] Kozaczuk, *Enigma*, 63.
[45] Rejewski, 'How the Polish Mathematicians Broke Enigma', 267.
[46] Tadeusz Lisicki quoted in Kozaczuk, *Enigma*, 63.
[47] Untitled typescript dated 11 Oct. 1943 (NARA, document reference RG 457, Historic Cryptographic Collection, Box 705, NR 4584), 1.
[48] 'Operations of the 6312th Signal Security Detachment, ETOUSA', 1 Oct. 1944 (NARA, document reference: RG 457, Historic Cryptographic Collection, Box 970, NR 2943), 5. (Thanks to Ralph Erskine for drawing my attention to this quotation and to Frode Weierud for sending me a copy of the document.)
[49] Rejewski, 'Remarks on Appendix 1 to British Intelligence in the Second World War by F. H. Hinsley', 81.
[50] Untitled typescript dated 11 Oct. 1943, 2; 'Operations of the 6312th Signal Security Detachment, ETOUSA', 5.

of the Enigma Cipher' shows no system of falling weights—although nor is an alternative system for stopping the bomba depicted.[51]

Simple Enigma and a Mini Bomba

Let us suppose, for purposes of illustration, that we are dealing with an imaginary, highly simplified, version of the Enigma machine called Simple Enigma. Simple Enigma has one wheel rather than three and no plug-board; in other respects it is the same as a full-scale Enigma.

Suppose that we have a message to decode beginning NYPN... Suppose further that we have a *crib*. A crib is a series of letters or words that are thought likely to occur in the plain language message that the cipher text encrypts. Say we have good reason to believe that the first and fourth letters of the plain text are both E (perhaps a prisoner gasped out the first four letters of the plain text before he died, but his second and third gasps were inaudible). We will use a machine to help us find the message setting—i.e. the rotational position of the wheel at which the sender began typing the message.

Our code-breaking machine consists of two replicas of the Simple Enigma machine plus some additional devices. There is a mechanism for holding down any selected key at the keyboards of the replicas, thereby keeping the current flowing from key to wheel. The wheel of each replica can be locked in step with the other, and there is an electric motor that will click the wheels round in unison through their twenty-six rotational positions, one position at a time. Additional circuitry bridging the two lampboards detects whether a selected letter—E, for example—lights up simultaneously at each lampboard. A switch or relay is wired in such a way that if the selected letter does light simultaneously, the electric motor is turned off, with the result that the wheels stop turning at exactly the position that caused the simultaneous lighting of the letter. This is called a 'stop'.

Assuming that the crib is correct, we know that if the intended recipient of the message sets the wheel of Simple Enigma to the message setting and types the first letter of the cipher text, N, the letter E will light up at the lampboard. The recipient will then type the next two letters of the cipher text, YP, causing unknown letters to light, followed by the fourth letter of the cipher text, N, which will cause E to light up again. Each time the recipient presses a key at the keyboard, the wheel advances one click. So the position of the wheel at which the fourth letter of the cipher text decodes as E is three clicks on from the position at which the first letter of the cipher text decodes as E. This is expressed by saying that these two positions are at a *distance of three* from each other. What we want our codebreaking machine to do is to search through the twenty-six possible positions of the wheel, looking for a position p that satisfies these two conditions:

[51] Kozaczuk, *Enigma*, figure E-8, 289.

1. At position p, keying N causes E to light;
2. At position $p + 3$ (i.e. the position three on from p), keying N again causes E to light.

We set up the codebreaking machine to perform this search by turning one of the two identical wheels so that it is three positions ahead of the other. For example, we might turn the wheel on the right so that, of the twenty-six letters marked around its ring, Z is uppermost, and then position the wheel on the left three clicks further on, i.e. with C uppermost. The two wheels are then locked together so that they will maintain their position relative to one another while the motor rotates them. The locked wheels are described as being at an *offset* of three clicks.

Next we set up the additional circuitry at the lampboards so that the simultaneous lighting of the letter E at each board will produce a stop. Finally, we clamp down the N-key at each of the two keyboards and start the electric motor.

The motor turns the wheels from position to position. If all goes well, a point is reached where E lights at both boards and the machine stops. If at that stage the wheels have not yet completed a full revolution, we note the position at which the stop occurred and then start the motor again, since there might be more than one position at which conditions 1 and 2 are jointly satisfied. (If, after a complete revolution, there are no stops, our crib was incorrect.)

If a complete revolution brings only one stop, then the position of the right-hand wheel of the pair must be the position at which the sender began encoding the message. We pass this setting to a clerk sitting at another replica of the Simple Enigma, who turns the wheel to that position and keys in the cipher text, producing the plain text at the lampboard. If there were several stops, then the clerk has to try each of the possible settings in turn until one is found that yields German at the lampboard.

Notice that we have not discovered the actual message setting—the letter visible in the window of the sender's machine at the start of typing the message (and enciphered to form the indicator). Which letter is visible in the window depends on how the sender has positioned the ring around the 'core' of the wheel. Leaving the core in one position, the operator could make any one of the twenty-six letters appear in the window by twisting the ring around the core. What we have found is the position *of the wheel core* at the start of the message. At GC & CS this was called the 'rod-position' of the wheel. The rod-position is all we need to be able to decipher the message.

Of course, with only twenty-six positions to search through, there is hardly any need for the electric motor, the detector circuitry at the lampboards, and so forth, because one could quite quickly conduct the search simply by turning the wheels of two replica machines manually. However, the additional equipment is certainly necessary when it is the full-scale Enigma machine that is being attacked, since the existence of three wheels and six possible wheel orders means that one must search through not 26 but $6 \times 26 \times 26 \times 26 = 105{,}456$ possible positions. (This figure ignores the small complications introduced by double-notching and by the extra movement of the middle wheel described above.)

The Actual Bomba

The Polish bomba was a more complicated version of the mini bomba just described. It consisted in effect of six replica Enigma machines, with six sets of duplicates of wheels I, II, and III—eighteen wheels in all. Each of the six replica Enigmas in a single bomba was usually set up with the same wheel order, for example III/I/II. The wheels used in a bomba had no rings (and so no notches for producing a 'turnover' of the adjacent wheel).

The six replica Enigmas were linked in pairs to form three double-Enigmas—just as in the example of the mini bomba, where two Simple Enigmas are linked to form a double Simple Enigma. Each of these double-Enigmas included three pairs of wheels and equipment equivalent to two keyboards and two lampboards. The complete bomba consisted of the three double-Enigmas plus the electric motor, a mechanism for detecting simultaneities and producing stops, and arrangements for holding constant the letter going into each double-Enigma.

At this point it may be helpful to repeat that the first, or outermost, of the three wheels in an Enigma machine—the wheel linked directly to the keyboard and plug-board and which moved once with every key-stroke—was always the right-hand member of the trio. For example, if the wheel order is I/II/III, it is wheel III that is the outermost of the three wheels.

As in the mini bomba, the identical wheels of a double-Enigma were locked in step, sometimes with one member of a pair a number of positions ahead of the other member. For example, the two IIIs might be locked in step at an offset of three clicks (as above), while the two IIs are locked in step with no offset, and likewise the two Is.

The corresponding wheels of different double-Enigmas in the same bomba were also locked in step with one another. For example, the locked pair of III wheels of one double-Enigma might be locked in step (at an offset of twelve clicks, say) with the locked pair of III wheels of another double-Enigma.

Once all the wheels were appropriately linked, the electric motor would be started and the bomba's six replica Enigmas would move in synchronization, each passing through $26 \times 26 \times 26$ positions. This took about two hours, each outer wheel moving through 676 revolutions, each middle wheel through 26 revolutions, and each left-hand wheel through one revolution. In the space of roughly two hours, the bomba could do the same work that would occupy a human computer for about 200 hours.[52]

[52] See p. 40.

The Indicator Method

In the previous example, we imagined using a mini bomba to discover wheel positions consistent with a crib concerning the first and fourth letters of the cipher text. The method employed by the Poles was different and did not involve text-cribbing (although the method that Turing would later devise for the British bombe did). The Poles focused on the *indicator* (to recapitulate: the six-letter group preceding the cipher text and produced by enciphering the message setting twice, at an indicator setting that the sender broadcast 'in clear' as part of the preamble to the message).

In a proportion of the intercepted messages, the first and fourth letters of the indicator would be the same, as for example in the indicator WAHWIK.[53] Since an indicator is produced by typing a three-letter message setting twice, the first and fourth letters of any indicator both encode the same letter as each other. This is true also of the second and fifth letters of any indicator, and the third and sixth. So both the occurrences of W in WAHWIK encode the same letter; and moreover three clicks of the right-hand wheel separate the two positions at which W encodes this unknown letter.

Let me use 'p_R' when referring to a position of the Enigma's right-hand wheel, and similarly 'p_M' in the case of the middle wheel and 'p_L' in the case of the left-hand wheel. We could attempt to use the bomba to search for rod-positions p_L, p_M, and p_R such that at position p_R and position $p_R + 3$, W encodes the same letter. As I will explain, this is not in fact an effective way to proceed, but in order to get the feel of the bomba, let's briefly consider how to carry out this search.

We select one of the double-Enigmas, pick a wheel order, say I/II/III, and put the three pairs of wheels into this order. We then lock the right-hand pair, the IIIs, in step at an offset of three (just as in the example of the mini bomba). The wheels in the middle pair (the IIs) are locked in step at the same position as one another, and likewise the wheels in the left-hand pair (the Is). Finally, we set the detector circuits to produce a stop whenever the same letter—any letter—lights simultaneously in both Enigmas. (The remaining two double-Enigmas are not needed for this search.) The motor is switched on and each replica Enigma moves through its $26 \times 26 \times 26$ positions. Any stops give pairs of positions, three clicks of the right-hand wheel apart, at which typing W produces the same letter at the lampboard. Another five runs of the bomba are required to explore all six wheel orders. (Alternatively we might use all three double-Enigmas, each

---

[53] The indicators and indicator settings used in this example are adapted from p. 266 of Kasparek's translation of Rejewski's 'Jak Matematycy polscy rozszyfrowali Enigme' in Kozaczuk, *Enigma*. The present description of the bomby has been reconstructed from Rejewski's rather compressed account appearing on that page. Unfortunately, Stepenske's translation of these same paragraphs in the *Annals of the History of Computing* is marred by an error that seriously affects the sense. The phrase that Stepenske translates 'by striking key W three times in a row, the same lamp would light' (p. 226) should be translated 'if key W is struck the same lamp will light again after three more strokes'.

with a different wheel order, so enabling the bomba to explore three wheel orders simultaneously. In this case only two runs of the bomba are necessary to cover all the possible wheel orders.)

Notice that an assumption is being made here concerning 'turnovers'. As previously explained, the movement of the right-hand wheel of the Enigma machine at some point causes the centre wheel to turn forward one click; and the movement of the centre wheel at some point causes the left-hand wheel to advance one click. The positions at which these turnovers occur are determined by the *Ringstellung*. In locking the pair of II wheels (the middle wheels) of the double-Enigma together in the *same* position as one another, we are assuming that, as the sender's machine lights up the letters WAHWIK, no movement of the middle wheel occurs during the three clicks forward of the right-hand wheel that lie between the production of the first and second occurrences of W. And in locking the left-hand wheels of the double-Enigma together in the same position, we are making the same assumption about the left-hand wheel of the sender's machine.

Of course, these assumptions might be wrong, in which case the search will fail. This is no less true in the case of the full-blooded search described below involving three indicators. However, the assumption that only the right-hand wheel moves in the course of typing a group of six letters is true much more often than not, and so searches based on this assumption will, other things being equal, succeed much more often than not.

The problem with the method of searching just described is that it would typically produce excessively many stops—many triples of positions p_L, p_M, p_R are liable to satisfy the rather mild constraint that W encodes the same letter at both p_R and $p_R + 3$. It would take the clerk who tries out each stop by hand on a further replica Enigma far, far too long to winnow out the correct wheel positions. It is necessary to find additional indicators from the same day's traffic that can be used to narrow the focus of the bomba's search. Here is what the Poles actually did.

In order to put a bomba to work effectively, it is necessary to find in a single day's traffic (i.e. traffic encoded with the same wheel order and *Stecker*) three messages whose indicators exhibit the following patterns of repetitions. One indicator must display the pattern just discussed—the same letter repeated at the first and fourth positions, as in the example

WAHWIK.

A second indicator must have the selfsame letter that is at positions 1 and 4 in the first indicator at its second and fifth positions, as in

DWJMWR.

A third indicator must have that same letter at its third and sixth positions, as in

RAWKTW.

The Poles called these patterns 'females' (see above). At Bletchley Park the three patterns were referred to as a 1–4 female, a 2–5 female, and a 3–6 female respectively. It is because this indicator system admits three types of female that the bomba contains three double-Enigmas, each one utilizing the information contained in one of the three females.

Let the position of the right-hand wheel when the first letter of the first indicator was produced be p_R and the position of the right-hand wheel when the first letter of the second indicator was produced be q_R, and likewise r_R in the case of the third indicator. We know from the patterns of repeated letters in the indicators that:

Keying W produces a simultaneity at p_R and $p_R + 3$ (i.e. at p_R and $p_R + 3$ the same letter lights). Keying W produces another simultaneity at $q_R + 1$ and $q_R + 4$ (possibly involving a different letter at the lampboard). Keying W produces a third simultaneity at $r_R + 2$ and $r_R + 5$.

In fact we know more than this. A rich source of information has not yet been used—the indicator settings which appear in clear in the preambles to the messages. Suppose these are as follows.

indicator setting	indicator
RTJ	WAHWIK
DQY	DWJMWR
HPB	RAWKTW

Without the wheel order and the *Ringstellung* for the day in question, which of course we do not yet possess, the indicator setting cannot be used straightforwardly to decode the indicator. Nevertheless, the indicator settings are far from useless to us, because they contain information about the *relative* positions of the wheels when the indicators were produced; and using this information, we can deduce the relationship between p_R, q_R, and r_R.

The right-hand letter of each indicator setting specifies the position of the right-hand wheel when the encryption—or equivalently the decryption—of each message setting begins. Similarly, the middle letter specifies the position of the middle wheel when the encryption of the message setting begins, and the left-hand letter the position of the left-hand wheel. Picture the letters of the alphabet arranged evenly around the circumference of a circle, as on the ring of a wheel. The right-hand letter of the second indicator setting, Y, is fifteen letters further on than the right-hand letter of the first indicator setting, J. Therefore the position of the right-hand wheel at which the first letter of the second indicator was produced, q_R, is fifteen clicks on from the position at which the first letter of the first indicator was produced, p_R:

$$q_R = p_R + 15$$

The right-hand letter of the third indicator setting, B, is eighteen letters on from J (JKLMNOPQRSTUVWXYZAB). Therefore the position of the right-hand wheel at which the first letter of the third indicator was produced, r_R, is eighteen clicks on from p_R:

$$r_R = p_R + 18$$

Inserting this additional information into the above statement about simultaneities gives:

Keying W produces a simultaneity at p_R and $p_R + 3$; another simultaneity at $(p_R + 15)+1$ and $(p_R + 15)+ 4$; and a third simultaneity at $(p_R+18)+2$ and $(p_R+18) + 5$.

Or more simply:

Keying W produces a simultaneity at p_R and $p_R + 3$; another simultaneity at $p_R + 16$ and $p_R + 19$; and a third simultaneity at $p_R + 20$ and $p_R + 23$.

Now we have a much stronger constraint on p_R and can use the bomba to search for p_R and the accompanying positions of the other wheels in the expectation that the number of stops will be small enough to be manageable.

Using the Bomba

The bomba is set up for the search as follows. The stopping mechanism is arranged to produce a stop whenever the eighteen wheels move into a configuration that causes a simultaneity at each of the three double-Enigmas at once. The three simultaneities need not involve the same lampboard letter as each other. W is input continuously into the Enigmas.

One double-Enigma is set up as above: the wheel order is I/II/III, the III wheels are locked together at an offset of three, and the other pairs of wheels are locked with no offset (the assumption being, as before, that neither the middle nor the left-hand wheel of the sender's machine moved during the production of WAH-WIK). Call this double-Enigma's III wheels l_1 and r_1 (for the left and right members of the pair); r_1 is three clicks ahead of l_1.

The second double-Enigma is set up with the same wheel order. Call its III wheels l_2 and r_2. l_2 is locked in step with l_1 at an offset of 16, and r_2 is locked in step with l_2 at an offset of 3 (so r_2 is nineteen clicks ahead of l_1). As with the first double-Enigma, the II wheels are locked in step with no offset, and likewise the Is. The third double-Enigma is also set up with wheel order I/II/III. Its III wheels are l_3 and r_3. l_3 is locked in step with l_1 at an offset of 20, and r_3 is locked in step with l_3 at an offset of 3 (so r_3 is twenty-three clicks ahead of l_1). Again, the II wheels are locked in step with no offset, and the same for the Is.

Next, each double-Enigma must have its pair of II wheels suitably synchronized with those of its neighbours, and similarly its I wheels. This is achieved as in the case of the III wheels by making use of the information contained in the indicator settings about the relative positions of the wheels of the sender's machine when the indicators were produced.

The middle letter of the second indicator setting, Q, is twenty-three places ahead of the middle letter of the first indicator setting, T. So the middle wheels of the second double-Enigma—the IIs—are locked in step with the middle wheels of the first at an offset of 23. The middle letter of the third indicator setting, P, is twenty-five places ahead of the middle letter of the second indicator setting, Q, so the middle wheels of the third double-Enigma are locked in step with the middle wheels of the second at an offset of 25. The left-hand letter of the second indicator setting, D, is twelve places ahead of the left hand letter of the first indicator setting, R, so the left-hand wheels of the second double-Enigma—the Is—are locked in step with the left-hand wheels of the first double-Enigma at an offset of 12. Finally, the left-hand letter of the third indicator setting, H, is four places ahead of the left-hand letter of the second indicator setting, D, so the left-hand wheels of the third double-Enigma are locked in step with the left-hand wheels of the second at an offset of 4.

The motor is switched on. As before, the stops that are produced during a run through all $26 \times 26 \times 26$ positions are noted and then tested by a clerk. If none works, it is necessary to set up the bomba again with a different wheel order. Six runs are required to search through all the wheel orders—approximately twelve hours of bomba time in total. By running six bomby simultaneously, one for each wheel order, the Poles reduced the search time to no more than two hours.

The clerk at the replica Enigma tests the various positions at which the stops occurred. He or she eventually finds one that deciphers each indicator into something of the form XYZXYZ. The cryptanalysts now know the message settings and the rod-positions of the wheels at which the message settings were enciphered.

To use the message settings to decode the messages it is necessary to know the *Ringstellung* (since a message setting XYZ could specify any one of the $26 \times 26 \times 26$ positions, depending on the position of the ring). However, the *Ringstellung* lies only a step away. It can be deduced by comparing the rod-positions of the wheels at which the first letter of any of the indicators was produced with the corresponding indicator setting.

For example, if the *Ringstellung* is set correctly, then what should appear in the windows when the wheel cores lie in the positions at which the first W of WAHWIK was produced is RTJ. Since these rod-positions are know, it is a simple matter to take replicas of the wheels and to twist the rings until the letters R, T and J are uppermost at these rod-positions. Once the rings are correctly positioned, a wheel's ring setting is given by the position of the ring against the embossed index mark on the wheel core: whatever letter lies against the index mark is the ring setting for that wheel. The complete *Ringstellung* is the trigram consisting of the letter for each wheel arranged in the wheel order for the day.

Now the messages can be decoded on a replica Enigma, as can other intercepted messages with the same wheel order and *Ringstellung*.

The Plug-Board Problem

It remains to explain how the permutations introduced by the plug-board were dealt with. In the military Enigma machine, the plug-board or stecker-board lay in the path both of current flowing from the keyboard to the wheels and of current flowing from the wheels to the lampboard (see Figure 4). Not every keyboard key was affected by the plug-board. When the bomba first came into operation, the Germans were using the plug-board to scramble between ten and sixteen of the twenty-six keys (in effect by swapping the output wires of pairs of keys). The remaining keys were unaffected, being 'self-steckered'.

It was specified in the daily key which (keyboard) keys were to be affected on any given day and how the affected (keyboard) keys were to be paired up. For example, suppose the daily key says that T and K are to be 'steckered'. The operator connects together the plug-board sockets labelled T and K (by means of a cord with a plug at each end). The result of this extra twist is that pressing the T-key at the keyboard produces the effect at the wheels which pressing the K-key would have produced had there been no scrambling of the letters at the plug-board. Likewise pressing the K-key produces the effect which pressing the T-key would have produced in the unsteckered case.

The plug-board comes into play a second time, in between the wheels and the lampboard. If K lights up in the steckered case, then the selfsame output from the wheels would have caused T to light up had T been one of the letters unaffected by the plug-board. Likewise if T lights up, the output would have caused K to light up had K been unaffected by the plug-board.

The bomba took no account at all of *Stecker*. If the females in the chosen indicators had been produced without interference from the plug-board (i.e. if all the letters in the indicators were self-steckered), then the bomba could produce the correct message setting. But if stecker-substitutions were involved, the bomba would be looking for the wrong thing. Returning to the above example, it would not be W that produces simultaneities at p_R and $p_R + 3$, and so on, but the letter to which W happened to be steckered; and so the bomba's search would fail.

The success of the bomby depended on the fact that, with between ten and sixteen letters unaffected by the plug-board, there was a reasonable chance of the day's traffic containing three indicators unpolluted by *Stecker* and displaying the requisite females.

Once the wheel order and *Ringstellung* had been discovered, messages could be deciphered using a replica Enigma on which all letters were self-steckered. The result would be German words peppered with incorrect letters produced by plug-board substitutions. These incorrect letters gave away the plug-board connections of the sender's machine.

On 1 January 1939 the Germans increased the number of letters affected by *Stecker* (from between five and eight pairs of letters to between seven and ten pairs). The effectiveness of the bomba—already severely compromised by the introduction of wheels IV and V in December 1938—diminished still further.

5. The Bombe and the Spider

At Pyry, Knox observed that the indicator system exploited by the bomba might 'at any moment be cancelled'—as did indeed happen in May 1940 (see above).[54] It was clear to Knox that even if the problems engendered by the increases in the number of wheels and the number of steckered letters could be solved, the modified bomba might become unusable overnight. After the Warsaw meeting Knox and Turing considered the possibility of using a bomba-like machine to attack not the indicators but the message text itself, via cribs.[55] The decision was taken to build a flexible machine that could be used both in the Polish manner against the indicators and also with cribs.

Turing was responsible for the logical design of the machine—the 'bombe'. He passed his design to Harold 'Doc' Keen at the factory of the British Tabulating Machine Company in Letchworth. Keen handled the engineering side of the design. Notes dated 1 November 1939 signed by Knox, Turing, Twinn, and Welchman refer to 'the machine now being made at Letchworth, resembling but far larger than the Bombe of the Poles (superbombe machine)' and state: 'A large 30 enigma bomb [*sic*] machine, adapted to use for cribs, is on order and parts are being made at the British Tabulating Company.'[56]

Knox himself appears to have made little or no contribution to the design and development of the bombe. His greatest achievements during the war were breaking the versions of Enigma used by the Italian Navy and by the *Abwehr*, the secret intelligence service of the German High Command.[57] He died in February 1943.

In its mature form the bombe contained thirty-six replica Enigmas. (The replicas were made at Letchworth and in Chapter 6 Turing refers to them as 'Letchworth Enigmas'.) The intricate bombe contained some ten miles of wire and one million soldered connections. Enclosed in a cabinet, the bombe stood 6 feet 6½ inches tall (5 feet 10 inches without its 8½ inch castors), 7 feet 3¾ inches

---

[54] Hinsley, *British Intelligence in the Second World War*, vol. iii, part 2, appendix 30, p. 954.
[55] Ibid.
[56] 'Enigma—Position' and 'Naval Enigma Situation', notes dated 1 Nov. 1939 and signed by Knox, Twinn, Welchman, and Turing. Both notes are in the Public Record Office (document reference HW 14/2).
[57] Batey, 'Breaking Italian Naval Enigma'; Twinn, 'The *Abwehr* Enigma'.

Figure 7. A Bletchley bombe.

Source: Science and Society Picture Library, National Museum of Science and Industry.

long, and 2 feet 7 inches deep.[58] From the front, nine rows of rotating drums were visible. Each drum mimicked a single Enigma wheel.[59] The drums (which were almost 5 inches in diameter and $1\frac{3}{4}$ inches deep) were removable and could be arranged to correspond to different wheel orders. Colour-coding was used to indicate which wheel, e.g. IV, a particular drum mimicked. The drums were interconnected by means of a large panel at the rear of the bombe (a panel that 'almost defies description—a mass of dangling plugs on rows of letters and numbers', according to one WRN operator; Mahon says that when viewed from the rear, the bombe appeared to consist 'of coils of coloured wire, reminiscent of a Fair Isle sweater' (p. 291, below)).[60] The replica Enigmas in the bombe could be connected together arbitrarily, according to the demands of whatever crib was being run.

[58] 'Operations of the 6312th Signal Security Detachment, ETOUSA', 60. (Thanks to John Harper for additional information.)
[59] 'Operations of the 6312th Signal Security Detachment, ETOUSA', 67.
[60] D. Payne, 'The Bombes', in Hinsley and Stripp (eds.), *Codebreakers*, 134. The coils of wire described by Mahon were probably red in colour. Red wire and very rarely black wire were used by the Letchworth bombe factory (letter from John Harper to Copeland (25 Feb. 2003), reporting interviews with engineers who worked on the bombes at the Letchworth factory).

Figure 8. Rear panel of a bombe.

Source: Science and Society Picture Library, National Museum of Science and Industry.

Cribs

Cribs resulted both from the stereotyped nature of the messages sent by the Germans and from the thoughtlessly insecure habits of some operators. For example, weather stations regularly sent messages beginning in stereotyped ways, such as 'WETTER FUER DIE NACHT' ('Weather for the night') and 'ZUSTAND OST WAERTIGER KANAL' ('Situation Eastern Channel'). In Chapter 5 Mahon relates how a certain station transmitted the confirmation 'FEUER BRANNTEN WIE BEFOHLEN' each evening ('Beacons lit as ordered').

The position of the cribbed phrase within the cipher text could often be found by making use of the fact that the Enigma never encoded a letter as itself. The cryptanalyst would slide a suspected fragment of plain text (e.g. ZUSTAND) along the cipher text, looking for positions at which there were no matches.

In order to uncover cribs, a 'cribster' often had to read through large quantities of decrypts, keeping meticulous records. As the war progressed, 'cribbing' developed to a fine art. The discovering of cribs presupposes that the message traffic is already being read: the period of work from January 1940 with the perforated sheets and other hand methods was an essential preliminary to the success of the bombe.

In the earlier fictitious example, a mini bomba was used in conjunction with a two-letter crib. One replica of the Simple Enigma was dedicated to the first letter of the crib and another to the second (with the two replicas being set in step at a distance of three, as dictated by the crib). Setting all complications to one side—and in particular *Stecker*—the bombe functions in its bare essentials like the mini bomba in that example.

Suppose we have a message whose first seven letters are

    1   2   3   4   5   6   7
    B   I   M   Q   E   R   P

and the one-word crib

    Z   U   S   T   A   N   D

In a world without *Stecker*, we can exploit the crib by connecting seven replica Enigmas together in such a way that the right-hand wheel (or drum) of the second is one position further on than the right-hand wheel of the first, the right-hand wheel of the third is one position further on than the right-hand wheel of the second, and so on. The seven middle wheels are locked in step in the same position as one another, and likewise the left-hand wheels. As with the set-up procedure for the bomba, this assumes that the middle and left-hand wheels of the sender's machine do not turn over during the first seven letters of the message.

During each run, B is input continuously into the first replica Enigma, I into the second, and so on. The electric motor moves the wheels of each replica Enigma through all their possible positions, one by one. The bombe is set up to stop whenever the letters Z U S T A N D light simultaneously at the seven replica Enigmas. When this happens, the positions of the wheels of the first of the seven are noted. These are candidates for the rod-positions of the wheels at the start of the message.

Each stop is tested by hand, using either a replica Enigma or a British Typex cipher machine set up to emulate an Enigma. (The Typex—also written 'Type X'—was in effect an improved form of the Enigma.[61]) If the rest of the message decodes—or at any rate that part of it up to the point where a turnover of the middle or left wheel occurred—then the correct rod-positions have been found.

The seven replica Enigmas all have the same wheel order. By using more replicas, set up in the same way but with different wheel orders, several wheel orders can be tested simultaneously. Several runs of the bombe are required to test all the possible wheel orders.

If there is no success under the assumption that there were no turn-overs of the middle and left-hand wheels during the enciphering of ZUSTAND, then it is necessary to carry out more runs of the bombe, testing the various possibilities

---

[61] Bauer, *Decrypted Secrets*, 112, 135.

for when a turnover occurred. Turing describes this procedure in Chapter 6, p. 316.

Once the correct rod-positions are discovered, the wheel order is known and with some trial and error the *Ringstellung* can be worked out. In a steckerless world, the codebreakers now have the daily key and all the intercepted messages encoded on that key can be deciphered. This was done by Typex operators. The messages were decoded by following exactly the same steps that the intended recipient would: the indicator setting, transmitted in clear in the message preamble, was used to decrypt the three-letter indicator, producing the message setting.

Turing's Method for Finding the Plug-Board Settings

Turing employed a simple but brilliant idea in order to deal with the substitutions brought about by the plug-board. He describes this in Chapter 6, which is an extract from his 'Treatise on the Enigma'.[62] (Released in 1996, this material has not previously been published.) 'Treatise on the Enigma' was written in the summer or autumn of 1940 and seems to have been intended for use as a form of training manual.[63] It was known affectionately at Bletchley Park as 'Prof's Book' ('Prof' being Turing's nickname among his colleagues).

Turing's method for finding the plug-board settings dates from 1939. In the example just given, the replica Enigmas are connected 'in parallel'. Turing's idea was to make provision for replica Enigmas (without plug-boards) to be connected nose to tail, with the letter that exits from the wheels of the first being fed into the next in the chain as if it were unsteckered keyboard input. These chains of replica Enigmas could be of varying length, as demanded by the crib.

Each chain exploited a feature of the cribbed message that Turing called a 'closure', but which might equally well be called a 'loop'. There are no closures in the ZUSTAND example. The following, longer, crib (discussed by Turing in Chapter 6, pp. 315ff) contains several examples of closures. (The meaning of the crib is 'No additions to preliminary report'.)

---

[62] The title 'Treatise on the Enigma' was probably added to Turing's document by a third party outside GC & CS and quite probably in the United States. The copy of the otherwise untitled document held in the US National Archives and Records Administration (document reference RG 457, Historic Cryptographic Collection, Box 201, NR 964) is prefaced by a page typed some years later than the document itself. It is this page that bears the title 'Turing's Treatise on the Enigma'. Another copy of the document held in the British Public Record Office (document reference HW 25/3) carries the title 'Mathematical theory of ENIGMA machine by A M Turing'; this, too, was possibly added at a later date. Mahon refers to the document simply as 'Prof's Book'. The PRO copy is complete, and much more legible than the incomplete NARA copy, which lacks many figures. A digital facsimile of the PRO typescript is available in The Turing Archive for the History of Computing <www.AlanTuring.net/profs_book>. A retyped version of the complete work, prepared by Ralph Erskine, Philip Marks, and Frode Weierud, is available at <http://home.cern.ch/frode/crypto>.

[63] See J. Murray, 'Hut 8 and Naval Enigma, Part I', in Hinsley and Stripp (eds.), *Codebreakers*, 116. The date of composition of the document, summer 1940, is given by Hinsley, *British Intelligence in the Second World War*, vol. iii, part 2, appendix 30, 955.

```
 1  2  3  4  5  6  7  8  9 10 11 12 13 14 15 16 17 18 19 20 21 22 23 24 25
 D  A  E  D  A  Q  O  Z  S  I  Q  M  M  K  B  I  L  G  M  P  W  H  A  I  V
 K  E  I  N  E  Z  U  S  A  E  T  Z  E  Z  U  M  V  O  R  B  E  R  I  Q  T
```

One closure or loop occurs at positions 2 and 5 and is shown in Figure 9. At position 2, E *encodes* as A and at position 5, A *decodes* as E. Using an upward-pointing arrow to mean 'encodes' and a downward-pointing arrow to mean 'decodes', the loop is as shown in Figure 9.

(Notice that it is equally true that at position 2, A *decodes* as E, and at position 5, E *encodes* as A. It is also true—because the letter substitutions produced by the Enigma are *reversible* (see p. 224)—that at position 2, A encodes as E, and at position 5, E decodes as A. Any of these equivalent ways of describing the loop will do.)

Another closure, this time involving three letters, occurs at positions 5, 10, and 23 (Figure 10). At position 5, E encodes as A, at position 23, A decodes as I, and at position 10, I decodes as E.

E is called the *central* letter of these two closures. The crib contains a number of other closures with central letter E (see Turing's Figure 59 on p. 317).

The point about closures is that they are, as Turing says, 'characteristics of the crib which are independent of the Stecker' (p. 316). Figure 9 tells us that there is

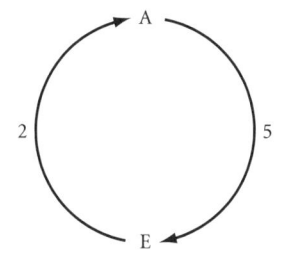

Figure 9. A loop or 'closure'.

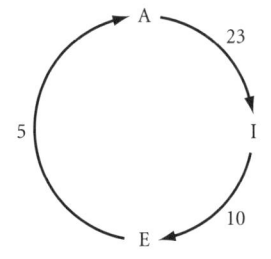

Figure 10. A closure involving three letters.

some letter which, when fed into the wheels at position 2, produces a letter which, if fed into the wheels at position 5, gives the original letter again. At the present stage, we have no idea which letter this is, since unless the central letter E happens to be self-steckered, the letter that goes into the wheels at position 2 is not E itself but whichever letter it is that E is connected to at the plug-board. Turing calls E's mate at the plug-board the 'stecker value' of E. Equally, we have no idea which letter it is that comes out of the wheels at position 2—unless A is self-steckered, the letter that emerges will not be A but A's stecker value.

Figure 10 also represents an assertion about the wheels that is true independently of how the plug-board is set up. There is some letter, x, which when fed into the wheels at position 5—that is to say, with the right-hand wheel four clicks further on than at the start of the message—produces some letter, y, which when fed into the wheels at position 23, produces some letter, z, which when fed into the wheels at position 10, produces x again.

As explained below, these closures are used in determining the stecker value of E. Once E's stecker-mate has been found, then the stecker values of the other letters in the loops are easily found out. For example, A's stecker value is whatever letter emerges from the wheels at position 2 when E's stecker-mate is fed in.

Using the Turing Bombe

In Turing's bombe, replica Enigmas without plug-boards are connected into chains that mimic the loops in the crib. In general, a crib containing three or more loops was necessary for Turing's bombe to work successfully.

In the case of the loop in Figure 9, two replica Enigmas are connected nose to tail. The right-hand wheel of the second machine is three clicks further on than the right-hand wheel of the first (because three clicks separate positions 2 and 5). As usual, the wheels are locked in step. To deal with the loop in Figure 10, three replica Enigmas are connected nose to tail. The right-hand wheel of the first machine in the chain is set three clicks ahead of the right-hand wheel of the first machine in the chain that corresponds to Figure 9 (three clicks separating positions 2 and 5). The right-hand wheel of the second machine in the chain of three is eighteen clicks ahead of the right-hand wheel of the first machine in that chain (since eighteen clicks separate positions 5 and 23). The right-hand wheel of the third machine in the chain is five clicks ahead of the right-hand wheel of the first (since five clicks separate positions 5 and 10). Other chains are set up for other closures in the crib also having E as central letter (see Turing's Figure 59 on p. 317).

The bombe works like this. We are going to input the same letter into each of the chains. What we are looking for is the stecker value of the central letter, E. We are going to set about finding it by trying out each of the twenty-six possibilities in turn. First we try the hypothesis that E's stecker-mate is A. So we input A into each of the chains.

The bombe is set up to stop whenever the wheels move into a configuration that produces the input letter—A, during the first run—as the output letter of each of the chains. At any stops during the run, we note not only the positions of the wheels, but also the output letter of each of the replica Enigmas in each chain. If the input letter is indeed E's stecker-mate, and the wheel positions are correct, then these 'interior' letters are the stecker-mates of the intermediate letters of the various closures.

If, in searching through all the possible wheel positions, we find no case in which the last machine in every chain produces A as output, then the hypothesis that E is steckered to A must be incorrect. If, however, we do manage to get A lighting up at the end of every chain, the hypothesis that A is E's stecker-mate remains in the running, and is passed on to someone else to investigate further by hand.

Once the first run is complete, we proceed to the hypothesis that E's stecker-mate is B, and again the wheels are moved through all their positions. And so on, taking each of the twenty-six stecker hypotheses in turn.

Additional runs may be required to test various hypotheses concerning the turnover of the middle and left wheels (as mentioned above). There is also the question of the wheel orders. Typically several different wheel orders will be tested simultaneously. (A thirty-six-Enigma bombe could usually test three wheel orders simultaneously, assuming that no more than twelve Enigmas were required for the loops in the crib.) In the case of an 'all wheel order crib', where no information is available to rule out some of the wheel orders, a number of successive runs, or simultaneous runs on several bombes, will be required in order to examine each possible wheel order.

Unless the data provided by a crib is especially scanty, in which case there might be many stops, this procedure would usually produce a manageably small number of stops. These were tried out manually in another building on a replica Enigma or Typex. Usually the stops were tested more or less as they occurred. As soon as one was found that turned part of the remaining ciphertext into German—albeit German peppered with incorrect letters—the instruction would be telephoned back to the bombe operators to strip the bombe and ready it for the next cribbed message in the queue.

The prototype Turing bombe, named 'Victory', was installed at Bletchley Park on 18 March 1940.[64] It seems to have been used exclusively by Turing and other members of Hut 8 in their attempt to break Naval Enigma.[65]

---

[64] Hinsley, *British Intelligence in the Second World War*, vol. iii, part 2, appendix 30, 954. When Mahon says in the next chapter that the 'first bombe arrived in April 1940' (p. 292), he is probably referring to the time at which the bombe became available to the codebreakers.

[65] 'Squadron Leader Jones, Section' (Public Record Office, document reference HW 3/164). (Thanks to Ralph Erskine for sending me a copy of this document.)

Simultaneous Scanning

The efficiency of the bombe could be increased greatly by—instead of, as just described, trying out one stecker hypothesis at all positions of the wheels before moving on to the next hypothesis—allowing all the twenty-six possible stecker hypotheses for the central letter (E to A, E to B, etc.) to be tried out together in the short interval before the wheels (drums) shifted from one position to the next. This is 'simultaneous scanning'. Turing's original intention was to include additional electrical apparatus in the prototype bombe to implement simultaneous scanning and he outlines a way of doing this in Chapter 6 (see the section 'Pye simultaneous scanning'—Pye was an electronics company located in Cambridge). However, the problem proved difficult for the engineers and the additional apparatus was not ready in time to be incorporated in Victory.[66] Turing explains in Chapter 6 that the method the engineers were proposing would 'probably have worked if they had had a few more months experimenting', but that their work was in the end overtaken by the discovery of a solution 'which was more along mathematical than along electrical engineering lines' (p. 319).

Turing presents this mathematical solution in two stages in Chapter 6. First he explains (what will in this introduction be called) his *feedback method* (see his section 'The Spider'). This Turing describes as 'a way of getting simultaneous scanning on the Bombe' (p. 323). Then he goes on to explain the role of Welchman's dazzlingly ingenious invention, the *diagonal board* (see Turing's section 'The Spider. A Second Description. Actual Form'). Welchman's diagonal board brought about a dramatic increase in the effectiveness of the bombe.

Turing's Feedback Method

Let us reconsider the previous search for the stecker value of the central letter E. We first tried the hypothesis that E's stecker-mate is A. Inputting A, we rotated the wheels looking for a position at which the letter to emerge is again A. The feedback method is this.[67] Before the wheels are shifted from the current position to the next, whichever letter emerges from the suitably interconnected Enigmas—which will in all probability not be A—is fed back in as the new input letter. (This is done automatically via a braid of twenty-six wires.) This step is then repeated: whichever letter emerges is fed back in, and so on. Unless the first attempt produced A, the effect of these cycles of feedback is that different stecker hypotheses are tested at the current position of the wheels.

If the wheel position is not the starting position for the message then, given a crib with sufficient loops, all twenty-six letters will usually be produced as output during the cycles of feedback. So if the emerging letters are imagined as appearing at a lampboard, all twenty-six lamps will light. At some positions, however—the

---

[66] Hinsley, *British Intelligence in the Second World War*, vol. iii, part 2, appendix 30, p. 954.
[67] Welchman gives an account of the method, *The Hut Six Story*, 237–41.

interesting positions—not all the lamps light. At these positions it is usually true (again given a crib with sufficient loops) either that only one lamp lights, or that only one remains unlit (a reflection of the fact that the Enigma's letter-substitutions are reversible). Either way, the letter on the odd lamp out is a candidate for the stecker value of the central letter, and the position of the wheels is a candidate for the starting position. Letters produced by other Enigmas within the chains are candidates for the stecker values of other letters of the loops.

The Diagonal Board

Welchman conceived the diagonal board as a way of increasing the effectiveness of the bombe by further exploiting the reciprocal character of the stecker-substitutions. (The substitutions are reciprocal in the sense that if letter L_1 is steckered to L_2 then—owing to the design of the plug-board—L_2 is inevitably steckered to L_1.) With the diagonal board in operation, the bombe could work cribs containing fewer than three closures and even cribs containing no closures at all (as in the ZUSTAND example) provided the length of the crib was sufficient. (If Welchman's diagonal board had never been conceived, bombes of the earlier type could have been used successfully against Enigma networks producing enough cribs with at least three closures—although at the expense of greater amounts of bombe time.[68])

Once Welchman had thought of the diagonal board, Turing quickly saw that it could be used to implement simultaneous scanning. Joan Clarke, who worked alongside Turing in Hut 8, said: 'Turing soon jumped up, saying that Welchman's diagonal board would provide simultaneous scanning.'[69] (Clarke was one of Welchman's mathematics students at Cambridge. For a short period in 1941, she and Turing were engaged to be married.)

The new form of bombe with the diagonal board was initially called the 'Spider' to distinguish it from Turing's earlier form, but soon simply 'bombe' prevailed. (Possibly the name 'Spider' arose in virtue of the practice of using 'web' as a term to refer to the connected parts of a diagram depicting the loops in a crib; see Chapter 6, pp. 325, 329.[70]) The first Spider was installed on 8 August 1940.[71] It was known as 'Agnus', short for 'Agnus Dei' (the name later became corrupted to 'Agnes' and 'Aggie').[72] Agnus contained thirty replica Enigmas, six fewer than in later models. Both Hut 8 (Naval Enigma) and Hut 6 (Army and Air Force Enigma) were given access to the new machine.[73]

---

[68] C. A. Deavours and L. Kruh, 'The Turing Bombe: Was It Enough?', *Cryptologia*, 14 (1990), 331–49 (346–8).

[69] Murray (née Clarke), 'Hut 8 and Naval Enigma, Part I', 115.

[70] I am indebted to Frank Carter for this suggestion.

[71] Hinsley, *British Intelligence in the Second World War*, vol. iii, part 2, appendix 30, 955.

[72] 'Squadron Leader Jones, Section' (see n. 65); R. Erskine, 'Breaking Air Force and Army Enigma', in Erskine and Smith, *Action this Day*, 56.

[73] 'Squadron Leader Jones, Section'.

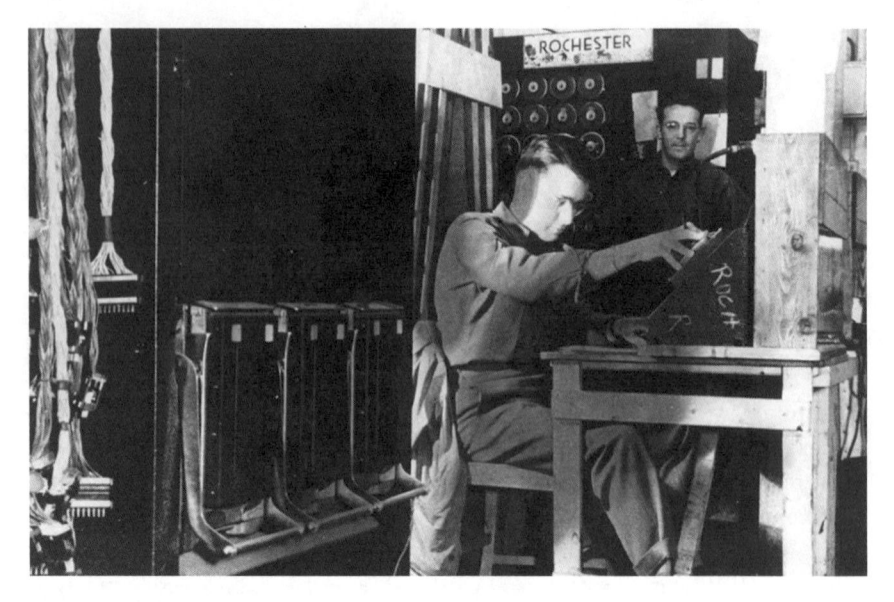

Figure 11. Working in a bombe room at Out Station Eastcote. 'Menus' for the outstation bombes were received from Bletchley Park over a teleprinter line.

Source: Photograph from 'Operations of the 6312th Signal Security Detachment, ETOUSA', 1 October 1944 (NARA, document reference: RG 457, Historic Cryptographic Collection, Box 970, NR 2943).

Subsequent Developments

At first the number of bombes increased relatively slowly, and much of the codebreakers' energy went into the use of hand methods—such as Turing's method of *Banburismus*—designed to reduce the amount of bombe time required to break a crib. By June 1941 there were only five bombes in operation, rising to fifteen by November.[74] The picture changed markedly when a new factory dedicated to the production of bombes came into operation at Letchworth. The output of Enigma decrypts produced by GC & CS more than doubled during 1942 and 1943, rising to some 84,000 per month by the autumn of 1943.[75] Groups of bombes were housed in 'outstations' in the district surrounding Bletchley Park, and then subsequently at three large satellite sites in the suburbs of London, with dedicated teletype and telephone links to Bletchley Park.[76] By the end of the war there were around 200 bombes in continuous operation at these various sites.[77] From August 1943, US Navy bombes began to go into operation in Washington, DC. About 125 were in operation by the time Germany fell.[78] Good cable communications

---

[74] Alexander 'Cryptographic History of Work on the German Naval Enigma', 31, 35.
[75] Hinsley, *British Intelligence in the Second World War*, ii. 29.
[76] Welchman, *The Hut Six Story*, 139–41, 147.
[77] Ibid. 147.
[78] Erskine, 'Breaking German Naval Enigma on Both Sides of the Atlantic', 192–3.

enabled Bletchley to use the Washington bombes 'almost as conveniently as if they had been at one of our outstations 20 or 30 miles away'.[79]

6. Naval Enigma

Turing's Break

Between 1934 and 1937 the Poles had enjoyed some success against German Naval Enigma. However, on 1 May 1937 a major change of indicator procedure rendered Naval Enigma impenetrable.

During much of 1940 German Air Force traffic was being read in large quantities by GC & CS, but Naval traffic—including the all-important messages to and from the wolf-packs of U-boats in the North Atlantic—remained cloaked. The German strategy was to push Britain toward defeat by sinking the convoys of merchant ships that were Britain's lifeline, bringing food, raw materials, and other supplies across the Atlantic from North America. From the outbreak of war to December 1940 a devastating total of 585 merchant ships were sunk by U-boats, compared to 202 merchant vessels sunk by aircraft during the same period.[80] If Home Waters Naval Enigma (*Heimische Gewässe*)—called 'Dolphin' at Bletchley Park—could be broken, the positions of the wolf-packs in the North Atlantic would be known and convoys could be routed around them.

When Turing took up residence at Bletchley Park in September 1939 no work was being done on Naval Enigma, which some thought unbreakable. As late as the summer of 1940 Denniston declared to Birch, the head of the Naval Section at GC & CS (Hut 4): 'You know, the Germans don't mean you to read their stuff, and I don't suppose you ever will.'[81] This was never the opinion of Birch and Turing. Alexander's history of the attack on Naval Enigma (written at the end of the war and kept secret by the British government until very recently) recounted:

Birch thought it could be broken because it had to be broken and Turing thought it could be broken because it would be so interesting to break it... Turing first got interested in the problem for the typical reason that 'no one else was doing anything about it and I could have it to myself'.[82]

The chief reason why Dolphin was so difficult to break was that the indicator system required the sender to encipher the message setting by two different methods before broadcasting it—once by means of the Enigma machine, as was usual, and once by hand. Mahon describes the procedure in detail in Chapter 5. The hand encipherment was performed by means of a set of *bigram tables*. These tables specified substitutions for pairs of letters, such as 'DS' for

---

[79] Alexander, 'Cryptographic History of Work on the German Naval Enigma', 90.
[80] S. W. Roskill, *The War at Sea 1939–1945* (London: HMSO, 1954), 615–16.
[81] C. Morris, 'Navy Ultra's Poor Relations', in Hinsley and Stripp (eds.), *Codebreakers*, 237.
[82] Alexander, 'Cryptographic History of Work on the German Naval Enigma', 19–20.

'HG' and 'YO' for 'NB'. Enigma operators were issued with a set of nine complete tables, each table giving substitutions for all the 676 possible bigrams.[83] Which table was to be used on any given day was set out in a calendar issued with the tables. New sets of tables came into force periodically. Crews were under strict instructions to destroy the tables before abandoning ship or if the enemy was about to board.

Turing started his attack exactly where the Poles had left off over two years before, studying 100 or so messages from the period 1–8 May 1937 whose message settings were known. Before the end of 1939 he had fathomed out exactly how the complicated indicator system worked. Chapter 5 contains an extract from Turing's 'Treatise on the Enigma' (published here for the first time) in which Turing explains how he performed this remarkable piece of cryptanalysis.

Hut 8

In 1940 Turing established Hut 8, the section devoted to breaking Naval Enigma. Initially the Naval Enigma group consisted of Turing, Twinn, and 'two girls'.[84] Early in 1940 they were joined by Tony Kendrick, followed by Joan Clarke in June of that year, and then in 1941 by Shaun Wylie, Hugh Alexander, Jack Good, Rolf Noskwith, Patrick Mahon, and others.[85] Turing was 'rightly recognized by all of us as the authority on any theoretical matter connected with the machine', said Alexander (himself later head of Hut 8).[86] In Chapter 5 Mahon recounts how, under Turing's leadership, Hut 8 slowly gained control of Dolphin during 1940 and 1941.

Unlike *Heimische Gewässe* (Dolphin), *Ausserheimische Gewässe*—meaning 'Distant Waters'—would never be broken by Hut 8, and several other Naval Enigma networks also resisted attack.[87] *Süd*, on the other hand, used in the Mediterranean from mid-1941, was a much easier proposition than *Heimische Gewässe*. As Mahon mentions in the next chapter (p. 273), *Süd* employed a version of the indicator system broken by the Poles. *Süd*'s procedure of enciphering the message setting twice (on which the bomby had depended) meant that Hut 8 was able to read *Süd* traffic without any need for cribs.[88]

Pinches

Turing's discovery of how the indicator system worked could not be used to read the German traffic until the bigram tables were known. Materials obtained by the

[83] Alexander, 'Cryptographic History of Work on the German Naval Enigma', 7.
[84] Chapter 5, p. 285.
[85] Alexander, 'Cryptographic History of Work on the German Naval Enigma', 26, 28, 30; Murray, 'Hut 8 and Naval Enigma, Part I', 112.
[86] Alexander, 'Cryptographic History of Work on the German Naval Enigma', 33.
[87] Erskine, 'Breaking German Naval Enigma on Both Sides of the Atlantic', and 'Naval Enigma: The Breaking of Heimisch and Triton', *Intelligence and National Security*, 3 (1988), 162–83.
[88] *Süd* is discussed in Erskine, 'Naval Enigma: An Astonishing Blunder', *Intelligence and National Security*, 11 (1996), 468–73, and 'Breaking German Naval Enigma on Both Sides of the Atlantic', 186–9.

Royal Navy from enemy vessels enabled the codebreakers to reconstruct the all-important tables. (Many of the captures are described in Hugh Sebag-Montefiore's fast-paced book *Enigma: The Battle for the Code*; see the section of further reading.)

The first capture, or 'pinch', of Home Waters daily keys—which Alexander described as 'long awaited'—was on 26 April 1940.[89] A party from the British destroyer HMS *Griffin* boarded an armed German trawler disguised as a Dutch civilian vessel (bearing the false name 'Polares').[90] The trawler was bound for the Norwegian port of Narvik to deliver munitions. The 'Narvik Pinch', as it became known, yielded various documents, including notes containing letter-for-letter cribs for 25 and 26 April (see Mahon's account in the next chapter).[91] Among the documents was a loose scrap of paper (overlooked at first) on which were scribbled the *Stecker* and the indicator setting for 23 and 24 April.[92] Also captured were exact details of the indicator system, confirming Turing's deductions.[93]

The crib for 26 April was tried on the recently arrived Victory, and according to Alexander 'after a series of misadventures and a fortnight's work the machine triumphantly produced the answer' (see also p. 286, below).[94] Alexander reports that 27 April could then also be broken, the 26th and 27th being 'paired days'—days with the same wheel order and *Ringstellung*.[95] Thanks to the Narvik Pinch, the days 22–5 April were also broken (not on the bombe but by hand methods).[96]

Another pinch was needed if Dolphin was to be broken for any substantial period. Various plans were discussed. One, code-named 'Operation Ruthless', was masterminded by Lieutenant Commander Ian Fleming of Naval Intelligence, who later created the character James Bond. Mahon describes the plan, which he credits to Birch, in the next chapter. In the event, Operation Ruthless was not carried out. Turing's reaction is described in a letter by Birch dated 20 October 1940:

[89] Alexander, 'Cryptographic History of Work on the German Naval Enigma', 24.

[90] The report of the engagement, 'Second and Last War Cruise', is in PRO (document reference ADM 186/805). See also R. Erskine, 'The First Naval Enigma Decrypts of World War II', *Cryptologia*, 21 (1997), 42–6.

[91] Alexander, 'Cryptographic History of Work on the German Naval Enigma', 24; Ralph Erskine (personal communication).

[92] Alexander, 'Cryptographic History of Work on the German Naval Enigma', 24; Murray, 'Hut 8 and Naval Enigma, Part I', 113.

[93] Alexander, 'Cryptographic History of Work on the German Naval Enigma', 24.

[94] Ibid. 25.

[95] Ibid. 5, 25.

[96] Alexander's statements on pp. 24–5 (or possibly Mahon's on p. 286, below) have been interpreted, probably incorrectly, by the authors of *British Intelligence in the Second World War* as implying that materials obtained from the Narvik Pinch enabled Hut 8 to read Naval Enigma traffic for the six days 22–7 April *during May* (Hinsley et al., *British Intelligence in the Second World War*, i. 163, 336). In 1993 Joan Clarke stated that some of these days were not broken until June (Murray, 'Hut 8 and Naval Enigma, Part I', 113; see also Erskine, 'The First Naval Enigma Decrypts of World War II', 43).

Turing and Twinn came to me like undertakers cheated of a nice corpse two days ago, all in a stew about the cancellation of operation Ruthless. The burden of their song was the importance of a pinch. Did the authorities realise that...there was very little hope, if any, of their deciphering current, or even approximately current, enigma for months and months and months—if ever? Contrariwise, if they got a pinch...they could be pretty sure, after an initial delay, of keeping going from day to day from then on...because the level of traffic now is so much higher and because the machinery has been so much improved.[97]

Turing did not get what he wanted until the 'Lofoten Pinch' of March 1941, which Mahon describes as 'one of the landmarks in the history of the Section' (p. 290). On 4 March, during a commando raid on the Norwegian coast—planned with a pinch in mind—the Royal Navy destroyer HMS *Somali* opened fire on the German armed trawler *Krebs* near the Lofoten Islands.[98] *Krebs* was boarded and tables giving the daily keys for the complete month of February 1941 were captured.[99] Short of obtaining the bigram tables as well, this was exactly what was needed. A month's daily keys were sufficient to enable Hut 8 to reconstruct the tables.[100] Suddenly Hut 8 was properly open for business and by the beginning of April was looking forward to breaking the Naval traffic 'as nearly currently as possible'.[101]

Eager to follow up on the *Krebs* success, Harry Hinsley in Hut 4 put forward a plan to capture a German weather ship, *München*, operating north-east of Iceland.[102] On 7 May 1941 *München* was duly boarded by a party from the *Somali*.[103] The booty included the daily keys for the month of June. The July keys soon followed, captured from the weather ship *Lauenburg* in another raid planned by Hinsley.[104] The capture of the June and July keys helped Hut 8 reconstruct the new bigram tables issued on 15 June (see the next chapter).[105] The new tables were current until November 1941.[106]

During June and July Hut 8 was producing decrypts of Enigma messages within one hour of their being received. Mahon says on pp. 290–291, 'There can be no doubt that at this stage the battle was won and the problem was simply

[97] Birch's letter is included in a contemporary report entitled 'Operation Ruthless' by C. Morgan (PRO document reference ADM 223/463).
[98] Alexander, 'Cryptographic History of Work on the German Naval Enigma', 27. An official report of the operation is in PRO (document reference DEFE 2/142). The operation is described in Erskine 'Breaking German Naval Enigma on Both Sides of the Atlantic', 178.
[99] Chapter 5, p. 290; Alexander, 'Cryptographic History of Work on the German Naval Enigma', 27.
[100] Chapter 5, p. 290.
[101] Alexander, 'Cryptographic History of Work on the German Naval Enigma', 28.
[102] Hinsley, *British Intelligence in the Second World War*, i. 337.
[103] A report of the capture is in PRO (document reference ADM 199/447).
[104] Hinsley, *British Intelligence in the Second World War*, i. 337.
[105] Alexander, 'Cryptographic History of Work on the German Naval Enigma', 31.
[106] Ibid. 7.

one of perfecting methods, of gaining experience, and of obtaining and above all of training staff.'

Probably the most dramatic pinch of all occurred to the south of Iceland on 9 May 1941, during the pursuit of the submarine *U-110* by several Royal Naval vessels acting as convoy escorts.[107] Sub-Lieutenant David Balme, of the destroyer HMS *Bulldog*, led the party that boarded the stricken submarine. In an interview Balme described the depth-charging of the *U-110*:

Suddenly two ships were torpedoed one after the other. It was obvious where the attack had come from and the corvette *Aubretia* made a very accurate attack on the U-boat. Must have got the depth-charges just at the right depth. It was a classic attack: depth-charges underneath the U-boat blew it to the surface. It was the dream of every escort vessel to see a U-boat blown to the surface. Usually they just sink when you have a successful attack.

The German crew abandoned ship shortly before Balme boarded the U-boat. He continued:

I couldn't imagine that the Germans would have abandoned this U-boat floating in the Atlantic without someone down below trying to sink her. But at any rate I got on and got my revolver out. Secondary lighting, dim blue lighting, was on and I couldn't see anybody, just a nasty hissing noise that I didn't like the sound of.[108]

But the U-boat was deserted and, inexplicably, the Germans had made no attempt to destroy the Enigma materials on board. Balme and his men carried off the Enigma machine and the bigram tables. However, the tables had already been reconstructed laboriously by Turing and co. (see p. 290). Balme's pinch was not of major significance to Hut 8 and does not even rate a mention by Mahon or Alexander.

Banburismus

Another of Turing's pivotal contributions to the breaking of Naval Enigma was his invention of the hand method called *Banburismus*. The name arose because the method involved the use of specially made sheets bearing the alphabet which, being printed in the nearby town of Banbury, came to be called 'Banburies'. Mahon records that Turing invented the method the same night in 1939 that he worked out the indicator system (see Chapter 5).

The aim of Banburismus was to identify the day's right-hand and middle wheels. This meant that fewer wheel orders had to be tried on the bombe, thereby saving large amounts of bombe time. During the years when so few bombes were available, it was Banburismus which made it possible to read Dolphin. As Mahon

---

[107] The official account of the pursuit is in PRO (document reference ADM 1/11133). See also R. Erskine, 'Naval Enigma: A Missing Link', *International Journal of Intelligence and Counterintelligence*, 3 (1989), 493–508.
[108] Balme interviewed on British Channel 4 TV, 1998.

says, for two or three years Banburismus was 'the fundamental process which Hut 8 performed' (p. 281). Banburismus was discontinued in September 1943, bombes being plentiful enough by that stage.

The Battle of the Atlantic

Hut 8's ability to decode the U-boat messages had an immediate effect on the course of the war.

At the beginning of June 1941 Churchill had been informed by his planners that, as a result of the attacks on convoys, Britain's predicted imports amounted to substantially less than the minimum quantity of food necessary to keep the population fed during the remainder of 1941.[109] Oil and other imports would also arrive in insufficient quantities for war production to be maintained. The U-boats were crippling Britain. However, during June 1941—when Dolphin was read currently for the first time—reroutings based on Hut 8 decrypts were so successful that for the first twenty-three days of the month, the North Atlantic U-boats made not a single sighting of a convoy.[110]

The pattern continued in subsequent months. The Admiralty's Operational Intelligence Centre (OIC) became increasingly skilled at evasive routing based on Bletchley's Ultra intelligence, and the wolf-packs spent more and more time searching fruitlessly.[111] Although Hut 8's battle with the U-boats was to see-saw—for eleven long months of 1942, Hut 8 was blacked out of the North Atlantic U-boat traffic by the new fourth wheel inside the Enigma—the intelligence from Naval Enigma decrypts played a crucial role in the struggle for supremacy in the North Atlantic.

7. Turing Leaves Enigma

Mahon records that towards the end of 1941 Turing was running out of theoretical problems to solve concerning Naval Enigma (p. 312). Soon Turing was taking little part in Hut 8's activities. His talent for groundbreaking work was needed elsewhere.

For a period during 1942 Turing rejoined the Research Section to work on the new problem of 'Tunny'.[112] From June 1941 GC & CS had begun to receive enciphered messages that were very different from the Enigma traffic. These were carried by an experimental radio link between Berlin and Greece. Numerous other links soon came into existence, connecting Berlin to German Army Group commands throughout Europe. Unlike Enigma radio transmissions, which were

---

[109] Hinsley, *British Intelligence in the Second World War*, ii. 168–71.
[110] Ibid. 171.
[111] Ibid. 169–70, 172–5.
[112] W. Tutte, 'Bletchley Park Days', in B. J. Copeland (ed.), *Colossus: The First Electronic Computer* (Oxford: Oxford University Press, 2005).

Enigma | 263

in Morse code, the messages on these links were broadcast in binary teleprinter code. The British code-named the machine encrypting the new traffic 'Tunny'. Tunny was one of three different types of non-Morse 'Fish' traffic known to Bletchley (the others were codenamed 'Sturgeon' and 'Thrasher').

It was not until July 1942 that up-to-date Tunny traffic was read for the first time, by means of a paper-and-pencil method invented by Turing and known simply as 'Turingery'.[113] The Germans used Tunny for high-level Army communications and sometimes messages signed by Hitler himself would be deciphered.[114] With the arrival of the 'Heath Robinson' in June 1943, followed a few months later by the first of the electronic Colossus computers, the Tunny traffic, like Enigma before it, succumbed to the Bletchley machines (see further the introductions to Chapters 4 and 9).

Alexander gradually took over the running of Hut 8. In November 1942, Turing departed for the United States, where he liased with the US Navy's codebreakers and bombe-builders.[115] He was never to do any more work in Hut 8.[116] Following his return to Bletchley, in March 1943, he held a wider brief, acting as scientific policy adviser.[117] Turing eventually left Bletchley Park at the end of 1943, moving to Hanslope Park to work on the problem of automatically enciphering speech. He remained at Hanslope until the end of the war.[118]

In his history of Bletchley's attack on Naval Enigma, Alexander included the following appreciation of Turing's 'great contribution':

There should be no question in anyone's mind that Turing's work was the biggest factor in Hut 8's success. In the early days he was the only cryptographer who thought the problem worth tackling and not only was he primarily responsible for the main theoretical work within the Hut (particularly the developing of a satisfactory scoring technique for dealing with Banburismus) but he also shared with Welchman and Keen the chief credit for the invention of the Bombe. It is always difficult to say that anyone is absolutely indispensable but if anyone was indispensable to Hut 8 it was Turing. The pioneer work always tends to be forgotten when experience and routine later make everything seem easy and many of us in Hut 8 felt that the magnitude of Turing's contribution was never fully realized by the outside world.[119]

[113] I. J. Good, D. Michie, and G. Timms, 'General Report on Tunny' (1945), 458. 'General Report on Tunny' was released by the British government in 2000 to the Public Record Office (document reference HW 25/4, HW 25/5). A digital facsimile is in The Turing Archive for the History of Computing <www.AlanTuring.net/tunny_report>.

[114] Peter Hilton in interview with Copeland (July 2001).

[115] S. Turing, *Alan M. Turing* (Cambridge: Heffer, 1959), 71. Turing's report 'Visit to National Cash Register Corporation of Dayton, Ohio' (n.d.; *c.* Dec. 1942) is now declassified (document reference: NARA, RG 38, CNSG Library, 5750/441). A digital facsimile of the report is in The Turing Archive for the History of Computing <www.AlanTuring.net/turing_ncr>.

[116] Alexander, 'Cryptographic History of Work on the German Naval Enigma', 42.

[117] S. Turing, *Alan M. Turing*, 72; Don Horwood in interview with Copeland (Oct. 2001).

[118] There is an account of Turing's Hanslope period on pp. 269–90 of Hodges's biography (see the section of further reading in 'Alan Turing 1912–1954', above).

[119] Alexander, 'Cryptographic History of Work on the German Naval Enigma', 42–3.

In July 1941 Turing, Alexander, and Welchman were summoned to the Foreign Office in London to be thanked for what they had done.[120] Each was given £200 (a sizeable sum in those days—Turing's Fellowship at King's paid him less than twice this amount per annum). At the end of the war, Turing received the Order of the British Empire for the role he had played in defeating Hitler—a role that, after more than half a century of secrecy, has only now come fully into the light of day.[121]

Further reading

Bauer, F. L., *Decrypted Secrets: Methods and Maxims of Cryptology* (Berlin: Springer-Verlag, 2nd edn. 2000).

Budiansky, S., *Battle of Wits: The Complete Story of Codebreaking in World War II* (New York: Free Press, 2000).

Erskine, R., and Smith, M. (eds.), *Action This Day* (London: Bantam, 2001).

Hinsley, F. H., and Stripp, A. (eds), *Codebreakers: The Inside Story of Bletchley Park* (Oxford: Oxford University Press, 1993).

Kahn, D., *Seizing the Enigma: The Race to Break the German U-Boat Codes, 1939–1943* (Boston: Houghton Mifflin, 1991).

Sebag-Montefiore, H., *Enigma: The Battle for the Code* (London: Weidenfeld and Nicolson, 2000).

Smith, M., *Station X: The Codebreakers of Bletchley Park* (London: Channel 4 Books, 1998).

Welchman, G., *The Hut Six Story: Breaking the Enigma Codes* (Cleobury Mortimer: M&M Baldwin, 2nd edn. 2000).

---

[120] Diary of Sir Alexander Cadogan, Permanent Under-Secretary at the Foreign Office, 15 July 1941: Andrew, *Codebreaking and Signals Intelligence*, 3.

[121] I am grateful to Friedrich Bauer, Frank Carter, Ralph Erskine, John Harper, Diane Proudfoot, and Frode Weierud for their comments on a draft of this chapter.

CHAPTER 5

History of Hut 8 to December 1941 (*1945*)

Patrick Mahon

Introduction

Jack Copeland

Patrick Mahon (A. P. Mahon) was born on 18 April 1921, the son of C. P. Mahon, Chief Cashier of the Bank of England from 1925 to 1930 and Comptroller from 1929 to 1932. From 1934 to 1939 he attended Marlborough College before going up to Clare College, Cambridge, in October 1939 to read Modern Languages. In July 1941, having achieved a First in both German and French in the Modern Languages Part II,[1] he joined the Army, serving as a private (acting lance-corporal) in the Essex Regiment for several months before being sent to Bletchley. He joined Hut 8 in October 1941, and was its head from the autumn of 1944 until the end of the war. On his release from Bletchley in early 1946 he decided not to return to Cambridge to obtain his degree but instead joined the John Lewis Partnership group of department stores. John Spedan Lewis, founder of the company, was a friend of Hut 8 veteran Hugh Alexander, who effected the introduction. At John Lewis, where he spent his entire subsequent career, Mahon rapidly achieved promotion to director level, but his health deteriorated over a long period. He died on 13 April 1972.[2]

This chapter consists of approximately the first half of Mahon's 'The History of Hut Eight, 1939–1945'. Mahon's typescript is dated June 1945 and was written at Hut 8. It remained secret until 1996, when a copy was released by the US government into the National Archives and Records Administration (NARA) in Washington, DC.[3] Subsequently another copy was released by the British

---

[1] With thanks to Elizabeth Stratton (Edgar Bowring Archivist at Clare College) for information.
[2] This paragraph by Elizabeth Mahon.
[3] Document reference RG 457, Historic Cryptographic Collection, Box 1424, NR 4685.

government into the Public Record Office at Kew.[4] Mahon's 'History' is published here for the first time.[5]

Mahon's account is first-hand from October 1941. Mahon says, 'for the early history I am indebted primarily to Turing, the first Head of Hut 8, and most of the early information is based on conversations I have had with him'.

---

[4] Document reference HW 25/2.

[5] I have made as few editorial changes as possible. Mahon's chapter headings have been replaced by section headings. Errors in typing have been corrected and some punctuation marks have been added. Sometimes sentences which Mahon linked by 'and', or a punctuation mark, have been separated by a full stop. Mahon preferred the lower-case 'enigma', which has been altered to 'Enigma' for the sake of consistency with other chapters. In Mahon's introductory remarks 'account' has been substituted for 'book' (and the second word of that section has been changed from 'reading' to 'writing'). Occasionally a word or sentence has been omitted (indicated '...') and sometimes a word or phrase has been added (indicated '[]'). In every case, the omitted material consists of a reference to later parts of the 'History' not reproduced here.

History of Hut 8 to December 1941

> The king hath note of all that they intend,
> By interception which they dream not of.
>
> King Henry V

Introduction

Before writing this history I have not had the unpleasant task of reading voluminous records and scanning innumerable documents.[1] We have never been enthusiastic keepers of diaries and log books and have habitually destroyed records when their period of utility was over, and it is the merest chance that has preserved a few documents of interest; hardly any of these are dated. Since March 1943, the Weekly Report to the Director provides a valuable record of our activities, but it is naturally this more recent period which human memory most easily recalls, and it is the lack of documentary evidence about early days which is the most serious. A very large portion of this history is simply an effort of memory confirmed by referring to other members of the Section. I joined the Section myself in October 1941 and have fairly clear personal recollections from that time; for the early history I am indebted primarily to Turing, the first Head of Hut 8, and most of the early information is based on conversations I have had with him. I also owe a considerable debt of gratitude to Mr. Birch who lent me the surviving 1939–1940 Naval Section documents which yielded several valuable pieces of information and afforded an interesting opportunity of seeing Hut 8 as others saw us. Many past members of the Section and many people from elsewhere in B. P.[2] have been kind enough to answer questions. ...

With the exception of Turing, whose position as founder of the Section is a very special one, I have adopted the policy of not mentioning individuals by name. Attributing this or that accomplishment to an individual would be an

---

All footnotes are editorial; Mahon's typescript contains none. Some footnotes are indebted to a document entitled 'A Cryptographic Dictionary', produced by GC & CS in 1944 (declassified in 1996; NARA document reference: RG 457, Historic Cryptographic Collection, Box 1413, NR 4559). A digital facsimile of 'A Cryptographic Dictionary' is available in The Turing Archive for the History of Computing <www.AlanTuring.net/crypt_dic_1944>.

I am grateful to Rolf Noskwith (who worked with Mahon in Hut 8) for providing translations of signals, cribs, and German technical terms.

[1] Mahon's 'The History of Hut Eight, 1939–1945' is Crown copyright. This extract is published with the permission of the Public Record Office and Elizabeth Mahon. The extract from Turing's 'Treatise on the Enigma' is Crown copyright and is published with the permission of the Public Record Office and the Estate of Alan Turing. A digital facsimile of the complete original typescript of Mahon's 'The History of Hut Eight, 1939–1945' is available in The Turing Archive for the History of Computing <www.AlanTuring.net/mahon_hut_8>.

[2] Editor's note. Bletchley Park. Also B/P.

invidious process contrary to the traditional attitude of the Section towards its work and it would inevitably give a misleading impression of the relative contributions of the different members of the Section.

It is impossible to write a truly objective history of a Section which has been one's principal interest in life for the last $3\frac{1}{2}$ years and so this account is written on a comparatively personal note—mostly in the first person plural with occasionally a purely personal recollection or opinion included: to the best of my ability I have only introduced by 'we' opinions with which the Section as a whole would have agreed. We have always prided ourselves on not trying to conceal our failures, and on admitting where we might have done better, and I have attempted to avoid any tendency to 'whitewash' our efforts for the benefit of posterity.

This history is intended for the layman. Our work has been traditionally incomprehensible—the last distinguished visitor I remember had barely sat down before he announced that he was not a mathematician and did not expect to understand anything. (Anyone wishing to probe the more abstruse mathematical aspect of it should turn to the technical volume which is being compiled in collaboration with Hut 6.) In fact, there is nothing very difficult to understand in the work we did, although it was confusing at first sight. I have attempted to explain only the basic principles involved in the methods we used. As a result of this I hope that anyone interested in Hut 8 and willing to read the semi-technical passages with some care will get a fairly clear idea of our work and I make no excuse for having deliberately avoided mentioning many of the complications which arose—thus when describing the machine I say that after pressing the keys $26 \times 26 \times 26$ times the machine has returned to its starting place, but the mathematician will realize that the introduction of wheels 6, 7, and 8 with 2 turnovers each is liable to split this cycle into several smaller cycles.

The account starts with a description of the machine and the methods of sending messages. [This] is followed by some information as to where the machine was used and the volume of traffic carried. After these tedious but rather necessary pages of background [the account] follows the course of events more or less chronologically, starting some time before the war. Certain subjects—like Banburismus...—required whole sections to themselves outside the historical narrative and these are the subjects of a series of digressions. ...

The Machine and the Traffic

If the history of Hut 8 is to be understood, it is essential to understand roughly how the [Enigma] machine works and thus obtain some idea of the problem which had to be tackled. Developments which have taken place during the war have complicated the problem but have left the machine fundamentally the same.

History of Hut 8 to December 1941 | 269

The process of cyphering is simple and quick. The message is 'typed' on a normal keyboard and as each letter is pressed, another letter is illuminated on a lampboard containing the 26 letters of the alphabet. The series of letters illuminated on the lampboard form the cypher text and the recipient of the cypher message, in possession of an identical machine, types out the cypher text and the decoded message appears on the lampboard.

Wheels

The main scrambler unit consists of 3 (later 4) wheels and an Umkehrwalze which I shall refer to henceforth as the Reflector—an admirable American translation. These wheels have on each side 26 contacts which we will for convenience label A to Z. The contacts on the one side are wired in an arbitrary and haphazard fashion to the contacts on the other. Each wheel is, of course, wired differently. The reflector has 26 contacts which are wired together arbitrarily in pairs. What happens when one of the letters of the keyboard is pressed may be seen from the following diagram.

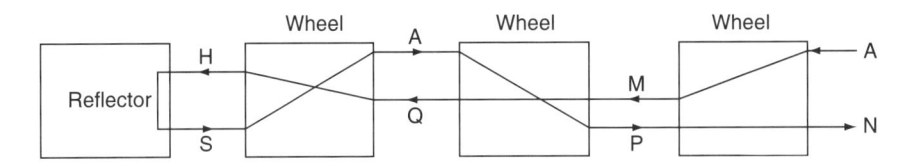

The current in this example enters the right hand wheel at A and leaves it at M, A being wired to M in this wheel: it enters the middle wheel at M and leaves it at Q, and so on until it reaches the reflector, where it turns around and returns through the wheels in a similar fashion, eventually leaving the right hand wheel at position N and lighting the appropriate lamp in the lampboard. Pressing a key may light up any bulb except that which is the same as the key pressed—for a letter to light up itself it would be necessary for the current to return through the wheels by the same route as it entered and, from the nature of the reflector, this is clearly impossible. This inability of the machine to encypher a letter as itself is a vital factor in the breaking of Enigma. It should also be noted at this point that the machine is reciprocal, that is to say that, if at a given position of the machine N lights up A, then A will light up N.

Each time a key is pressed the right-hand wheel moves on one so that if, in the position immediately following our example above, the same key is pressed, the current will enter the right-hand wheel at B and not A, and will pursue an entirely different course. Once in every 26 positions, the right hand wheel moves the middle wheel over one so that when the right hand wheel returns to position A, the middle wheel is in a new position. Similarly, the middle wheel turns over the left hand wheel once for every complete revolution it makes. Thus it will be seen that 26 × 26 × 26 (about 17,000) letters have to be encyphered before the machine returns to the position at which it started.

For most of the period with which we are concerned, there have been 8 wheels. Wheels 1 to 5 turn over the wheel next to them once per revolution, wheels 6, 7, 8, twice per revolution, this somewhat complicating the cycle of the machine as described in the previous paragraph. The turnovers (by which I mean the position of the wheel at which it turns over its next door neighbour) on wheels 1 to 5 are all in different places; in 6, 7, and 8 they are always at M and Z. As we shall see later, this was an important development.

Ringstellung

On each wheel is a tyre, marked with the letters of the alphabet. One of these letters can be seen through a window on top of the machine and the position of the wheel is referred to by the letter shown in the window. The tyre is completely independent of the core of the wheel, which contains the wiring, the relative position being fixed by the Ringstellung or clip which connects the tyre with the core. Thus even if the starting position of the message is known, it still cannot be decoded unless the clips, which fix the relative position of tyre and core, are known also.

Stecker

The Enigma machine would be a comparatively simple affair if it were not for the Stecker. This is a substitution process affecting 20 of the 26 letters before and after the current travels through the wheels. Let us return to our original example and assume for the moment that A is steckered to F and N to T. In our example we pressed key A and entered the right-hand wheel at a position we called A, but if we now press A the current will be sidetracked before entering the wheel, and will in fact enter at F and pursue a quite different course. If, on the other hand, we press F, the current will enter at A and proceed as before, coming out at N. N will not, however, light up on the lampboard, but rather T, because N has been steckered to T. This steckering process affects 20 letters; the remaining 6 are referred to as self-steckered and, when they are involved, the current proceeds directly to or from the wheels.

Set-Up

In order to decode a message one has, then, to know wheel order, clips, starting position of message, and Stecker. Any 3 of the 8 wheels may be chosen—336 possibilities. There are 17,000 possible clip combinations and 17,000 possible starting positions—in the 4-wheeled machine half a million. The number of possible Stecker combinations is in the region of half a billion. In fact, the number of ways the machine may be set up is astronomical, and it is out of the question to attempt to get messages out by a process of trial and error. I mention this as it is a hypothetical solution to the problem often put forward by the uninitiated, when in fact all the coolies in China could experiment for months without reading a single message. On most keys the wheel order and clips were changed every two days—this is the so-called 'innere Einstellung'[3] which was supposed to be changed only by officers and which was printed on a separate sheet of paper. The Stecker and Grundstellung (of which more later) normally changed every 24 hours. In a 31-day month, the odd day was coped with by having 3 days on 1 wheel order—a triplet. This sometimes came at the end of the month and sometimes in the middle. Triplets happened also in 30-day months, with the result that the last day was a 'singleton' with a wheel order of its own. These rules were not obeyed by all keys, but we shall meet the exceptions as we proceed with the historical survey. The later 4-wheeled keys had a choice of 2 reflectors and 2 reflector wheels; these were only changed once a month.

Indicating System

The next essential is to understand the indicating system. Various indicating systems were used simultaneously, but if we examine the most complicated fairly carefully the others will be simple to explain.

[The keys] Dolphin, Plaice, Shark, Narwhal, and Sucker all built up their indicators with the help of bigram tables and K book.[4]

K Book

One half of the K book consists of a Spaltenliste[5] containing all 17,576 existing trigrams,[6] divided into 733 numbered columns of 24 trigrams chosen at random. The second half consists of a Gruppenliste[7] where the trigrams are sorted into alphabetical order; after each trigram are 2 numbers, the first giving the number of the column in the Spaltenliste in which the trigram occurs, the second giving the position of the trigram in the column.

By means of a Zuteilungsliste[8] the columns of the K book are divided amongst the various keys, the large keys being given several blocks of columns, small keys as few as 10. The K book is a large document which has probably changed only once—the current edition having come into force in 1941—but the Zuteilungsliste was changed fairly frequently.

---

[3] Editor's note. *Innere Einstellung* = inner settings.
[4] Editor's note. K book = *Kenngruppenbuch* or *Kennbuch* = Identification Group Book. 'Group' refers to groups of letters.
[5] Editor's note. *Spaltenliste* = column list.
[6] Editor's note. A trigram is any trio of letters, e.g. ABC, XYZ.
[7] Editor's note. *Gruppenliste* = group list.
[8] Editor's note. *Zuteilungsliste* = assignation list.

Bigram Tables

A set of bigram tables consisted of 9 tables, each giving a series of equivalents for the 676 existing bigrams. These tables were reciprocal, i.e. if AN=OD then OD=AN, a useful property as we shall see later. Which bigram table was in force on any given day was determined by means of a calendar which was issued with the tables. New sets of bigram tables were introduced in June 1941, November 1941, March 1943, July 1944.

Now that we are familiar with all the necessary documents—key sheets, K book, bigram tables, etc.—it will be profitable to follow in detail the steps taken by a German operator wishing to send a message.

He is on board a U-boat and has Shark keys and, after consulting his Zuteilungsliste, goes to columns 272–81, which have been allotted to Shark. Here he selects the trigram HNH to serve as his Schluesselkenngruppe.[9] He then selects from anywhere in the book another trigram (PGB) and writes them down like this:

. H N H
P G B .

He then fills in the 2 blanks with dummy letters of his own choice:

Q H N H
P G B L

Taking the bigram table which is in force at the time, [he] substitutes for each vertical pair of letters QP, HG, etc.:

I D Y B
N S O I

The indicator groups of his message will then be IDYB NSOI.

The trigram which he chose at random (Verfahrenkenngruppe[10]) now provides him with the starting position of his message. To obtain this he sets up the Grundstellung in the window of his machine and taps out PGB. The three encyphered letters which result are the set-up for his message.

The first step in decyphering the message is, of course, to decypher the indicator groups by means of the bigram tables. At this stage the Schluesselkenngruppe can be looked up in the K book, and it can then be established by each station whether the message is on a key which it possesses. The Grundstellung is then set up, the trigram tapped out, and the message decoded.

Some form of Grundstellung procedure was common to all keys, but there were a number of keys not using bigram tables and K book. In the Mediterranean

---

[9] Editor's note. *Schluesselkenngruppe* = key identification group.
[10] Editor's note. *Verfahrenkenngruppe* = procedure identification group.

area Kenngruppenverfahren Sued[11] was used and discrimination between keys was dependent on the first letters of the first and second groups—the resulting bigram indicating the key. In this case, the operator chose any trigram he wished and encyphered it twice at the Grundstellung and the resulting 6 letters formed the last three letters of the first 2 groups. ... Of the other keys, Bonito and Bounce relied for recognition on the fact that in external appearance their traffic was unlike any other and so used no discriminating procedure, simply sending as indicators a trigram encyphered at the Grundstellung and filling the groups up with dummy letters as required.

This machine was in general use by the German Navy in all parts of the world; it was used alike for communication between ship and shore and shore and shore and by all vessels from mine sweepers and MTBs up to U-boats and major units.

Traffic

The history of Hut 8 is conditioned very largely by the rapid growth of the German Naval communications system and the resulting increases of traffic and increasing number of keys. ... It is not necessary to go into details here, but a few figures will illustrate very clearly that the gradual contraction of German occupied territory in no way signified a decrease of traffic and a simplification of the problem. Our traffic figures do not go back to 1940 but the following are the daily averages of messages for March 1941–5:

1941	465
1942	458
1943	981
1944	1,560
1945	1,790

The largest number of messages ever registered in Hut 8 on one day was on March 13, 1945, when 2,133 were registered.

As a general rule German Naval traffic was sent out in 4 letter groups with the indicator groups at the beginning and repeated at the end. Both from the German and our own point of view this made Naval traffic easily recognizable, and gave a check on the correct interception of the indicator groups. Messages were normally broadcast on fixed frequencies which changed comparatively rarely, so that it was possible for the cryptographer without W/T[12] knowledge to keep the different frequencies and the areas to which they belonged in his head. The principal exception to this was the U-boats, which used a complicated W/T programme, but from our point of view identification of services was made easy by the use of an independent set of serial numbers for each service. We

---

[11] Editor's note. *Kenngruppenverfahren Sued* = identification group procedure South.
[12] Editor's note. W/T = wireless telegraphy. R/T or radio-telephony involves the transmission of speech, W/T the transmission of e.g. Morse code or Baudot-Murray (teleprinter) code.

should have had some difficulty also with the Mediterranean area if the intercept stations had not given a group letter to each W/T service and appended it to the frequency when teleprinting the traffic.

Except for a short period in the early days of Bonito, fixed call signs were always used, though it was unfortunately by no means always possible to tell from whom a message originated. 'Addressee' call signs—a very great help to the cryptographer—were little used except in the Mediterranean where, if we were fortunate, it might be possible to tell both the originator and the destination of a message. Like the Mediterranean keys, Bonito gave a lot away by its call signs, but other keys stuck to the old procedure. The only exception of any note to this rule was the emergency W/T links which replaced teleprinter communications if the latter broke down.

From early days Scarborough was our chief intercept station and was responsible for picking up most of the traffic. Other stations were brought in if interception at Scarborough was unsatisfactory, as was often the case in the Mediterranean and North Norwegian areas. For a considerable time North Norwegian traffic was being sent back from Murmansk, while the W/T station at Alexandria played an important part in covering many Mediterranean frequencies. The principal disadvantage of traffic from distant intercept stations was the length of time it took to reach us. Traffic came from Alexandria by cable and an average delay of 6 hours between time of interception and time of receipt at B/P was considered good.

Unlike Hut 6 we never controlled the disposition of the various receiving sets at our disposal but made our requests, which were considerable, through Naval Section. As I do not ever recollect an urgent request having been refused, this system worked very satisfactorily from our point of view.

The reason for our numerous demands for double, treble, and even quadruple banking[13] of certain frequencies was that, for cryptographic reasons which I will explain later, it was absolutely essential to have a 100% accurate text of any message that might be used for crib purposes. In our experience it was most unwise to believe in the accuracy of single text, even if it was transmitted with Q.S.A. 5,[14] and so we asked automatically for double banking on frequencies likely to be used for cribbing. Especially on the Mediterranean keys and Bonito, interception was extremely unreliable and quadruple banking on crib frequencies was often necessary. On Bonito the assistance of R.S.S.,[15] who did not normally work on Naval traffic, was enlisted, and for some Bounce traffic originating from weak transmitters in Northern Italy, we relied on R.A.F. stations in Italy.

---

[13] Editor's note. Where two independent operators are assigned to monitor the same radio frequency, the frequency is described as 'double-banked'.

[14] Editor's note. In International Q-Code (radio operators' code) the strength of the received signal is represented on a scale of 1 to 5. 'Q.S.A. 5' means 'signal strength excellent'. (I am grateful to Stephen Blunt for this information.)

[15] Editor's note. Radio Security Service: a branch of the SIS (Secret Intelligence Service) which intercepted Enigma and other enciphered traffic.

Requests for double banking for cryptographic reasons became extremely numerous by the end of the war. In 1941, with only one key, special cover on a small group of frequencies was sufficient, but by late 1944 we were normally breaking some 9 or 10 keys, for each of which special cover on some frequencies was required. On the whole the policy was to ask for double banking if it was at all likely that it might be useful and accept the fact that a certain amount of unnecessary work was being done by intercept stations. There were at any rate sufficient hazards involved in breaking keys and it was felt that it would be foolhardy policy to take risks in the matter of interception when these could be obviated by double banking.

Most traffic arrived from the intercept stations by teleprinter, being duplicated by carbon. Retransmissions, dupes as we called them, were also teleprinted in full after 1942; before this, German preamble and differing groups only of dupes had been teleprinted, but we found that we were unable to rely on the intercept stations to notice all differences.

Shortage of teleprinters was a perennial problem, as the traffic constantly increased while those responsible for teleprinters persisted in believing that it would decrease. The effect of this shortage was that traffic got delayed at Scarborough for considerable lengths of time before teleprinting and was not in fact cleared until there was a lull in the traffic. It was usually true that there were sufficient teleprinters to cope over a period of 24 hours with the traffic sent in that period, but they were quite insufficient for the rush hours on the evening and early night shift. In the spring of 1944 the teleprinter situation was reviewed and considerably improved, in anticipation of the Second Front and possible heavy increases of traffic. Experiments were carried out with a priority teleprinting system for certain frequencies but the list tended to be so large (having to cover cryptographic and intelligence needs) and so fluid that it afforded no more than a theoretical solution to the problem. It was our experience that it was possible to 'rush' very small groups of traffic at very high speed—some very remarkable results were achieved with the frequency which carried Flying Bomb information—but that rushing a large quantity was comparatively ineffective.

As a result of the increased number of teleprinters, the average time elapsing between interception and teleprinting was reduced to about 30 minutes, which was thought to be satisfactory. For the opening of the Second Front a small W/T station was opened at B/P, most appropriately in the old Hut 8. This covered certain frequencies of special operational urgency or crib importance and produced very satisfactory results. A new record was established when a signal reached Admiralty in translation 12 minutes after being intercepted here. As the excitement over the success of the Second Front died down and the sense of urgency disappeared, the time lag became somewhat worse, but the situation remained under control, with one or two brief exceptions, even during the final peak period in March 1945.

As was to be expected, stations other than Scarborough which had less interest in Naval traffic and less facilities for teleprinting were appreciably slower in passing us the traffic; a time check late in 1944 revealed an average delay of 103 minutes at Flowerdown. It should perhaps also be gratefully recorded that Scarborough's standard of teleprinting, accuracy, and neatness remained right through the war a model which other stations were far from rivalling.

In early days traffic was teleprinted to the Main Building whence it was carried every half hour, later every ¼ hour, to the old Hut 8 Registration Room. This was inevitably a slow process, but it mattered comparatively little as in those days keys were not often being read currently. The move into Block D (February 1943) and the introduction of conveyor belts greatly improved the situation and traffic now came to the teleprinter room a few yards away whence it was conveyed to the Registration Room by belt.

Once a message had arrived in Hut 8, a considerable number of things had to be done before it could arrive decoded in Naval Section. The Registration Room had to sort the traffic—partly by frequency and partly with the help of the K book—into the various keys and, if the key in question was current, the message was then handed to the Decoding Room. For a long time decoded traffic was carried to Naval Section, a considerable walk either from the old or the new Hut 8; this wasteful method of conveyance was only superseded when the pneumatic tube system was introduced some time after we arrived in Block D.

These tubes were violently opposed on various grounds at first but, when the permanent two-way tube system had been introduced, they carried a terrific load and much toing and froing between Naval Section and ourselves was cut out. There were undoubted disadvantages in having to screw the messages up to put them into the containers but there can be no doubt that they saved us both time and trouble and that messages reached Naval Section much more quickly than before. A conveyor belt would certainly have been more satisfactory but, given the distance separating us from Naval Section (A15), was presumably out of the question.

The time taken for decodeable traffic to pass through the Section varied appreciably with the degree of excitement caused by the war news. In early Second Front days some messages were arriving in Naval Section 20 minutes after being intercepted, but speeds of this sort could not be kept up indefinitely. The introduction of time stamping with the help of Stromberg time-clocks enabled us to make a regular check of the time it took for traffic to pass through the Hut and we were normally able to keep the average in the region of half an hour. This reflects, I think, great credit on all concerned as the work was tiring and, especially in the Decoding Room, noisy, and at peak periods everyone had to work very fast indeed. In the week ending March 16th 1945, the record total of 19,902 messages were decoded, a remarkable feat for an average of perhaps 10 typists per shift.

This introduction has, I hope, supplied the necessary background on how the machine works and on how the traffic was received and dealt with. We can now turn to a more interesting subject and examine the history of the breaking of Enigma from the earliest days.

Early work on Enigma

Nearly all the early work on German Naval Enigma was done by Polish cryptographers, who handed over the details of their very considerable achievements just before the outbreak of war. Most of the information I have collected about pre-war days comes from them through Turing, who joined G.C.C.S. in 1939 and began to interest himself in Naval cyphers, which so far had received scant attention.

The Heimsoeth & Rinke[16] machine which was in use throughout the war and which I described [earlier] was not the first machine to be used by the German Navy. In the 1920s, the so-called O Bar machine had been in use. This had 3 wheels and no Stecker, and the curious characteristic of 29 keys—the modified vowels O, U, and A being included. Of these 29 symbols, X always encyphered as X without the current entering the machine, and the remaining 28 letters were encyphered in the normal way. The tyres of the wheels necessarily had 28 letters printed on them and it was decided that the letter which had been omitted was the modified O: hence the name of the machine, which was broken by the Poles and the traffic read.

The O Bar machine went out of force for fleet units in 1931, when the present machine was introduced, and gradually disappeared altogether. The new machine had originally been sold commercially by the Swedes: as sold by them it had no Stecker and it was they who recommended the boxing indicator system[17] which enabled so many Navy cyphers to be read.

When the German Navy first started to use the machine there were only 3 wheels in existence instead of the later 8 and only 6 Stecker were used. The

---

[16] Editor's note. The Heimsoeth & Rinke company manufactured Enigma machines following Scherbius' death. See F. L. Bauer, *Decrypted Secrets: Methods and Maxims of Cryptology* (2nd edn. Berlin: Springer-Verlag, 2000), 107.

[17] Editor's note. The 'boxing' or 'throw-on' indicator system was an earlier and much less secure method. K book and bigram tables were not used. Mahon explains the system later in his 'History' (on p. 56, which is not printed here): 'By this system a trigram of the operator's own choosing was encyphered twice at the Grundstellung and the resulting 6 letters became letters 2, 3, 4, 5, 7, 8, of the indicator groups.' Letters 1 and 5 were either dummies or, in Mediterranean traffic, identified which particular key was being used in accordance with *Kenngruppenverfahren Sued*. (See p. 273 and also H. Alexander, 'Cryptographic History of Work on the German Naval Enigma' (n.d. (c.1945), Public Record Office (document reference HW 25/1), 8–9; a digital facsimile of Alexander's typescript is available in The Turing Archive for the History of Computing <www.AlanTuring.net/alexander_naval_enigma>).

reflector in force was reflector A and boxing or throw-on indicators were used. ...

Having obtained photographs of the keys for 3 months, during which period the wheel order obligingly remained unchanged, the Poles broke the wiring of wheels 1, 2, and 3 by a 'Saga', a long and complicated hand process which I shall not attempt to explain. Having obtained all the details of the machine, they were able to read the traffic more or less currently with the help of the indicating system and catalogues of 'box shapes'.[18] These catalogues listed positions of the machine which would satisfy certain conditions which were implied by the indicator groups, and from them the machine set-up for the day could be worked out.

On May 1st 1937 a new indicating system was introduced. The first 2 groups of the message were repeated at the end, thus showing clearly that they formed the indicator, but it was immediately apparent that throwing-on had been given up. This was a sad blow, but the Poles succeeded in breaking May 8th and they discovered the Grundstellung with the help of messages to and from a torpedo boat with call-sign 'AFÄ', which had not got the instructions for the new indicating system. On breaking May 8th, the Poles discovered that it had the same wheel order as April 30th, and the intervening days were soon broken. AFÄ's lack of instructions and the continuation of the wheel order are typical examples of good fortune such as we have often experienced, and also of the German failure to appreciate that for a cypher innovation to be successful it must be absolutely complete.

May 8th and the preceding days could not, of course, be broken with the catalogue of box shapes, as the indicating system had changed, and for them the Poles devised a new method which is of considerable interest. Their account of this system, written in stilted German, still exists and makes amusing reading for anyone who has dealt with machines. The process was fundamentally a form of cribbing, the earliest known form. On the basis of external evidence, one message was assumed to be a continuation (a FORT) of another—apparently identification was easy and FORTS numerous in those days. The second message was then assume to start FORT followed by the time of origin of the first message, repeated twice between Ys. German security must have been non-existent in those days, as these cribs appear to have been good and the Poles quote an example which, by its pronounceability, gave its name to the method of attack they had evolved—FORTYWEEPYYWEEPY = continuation of message 2330,

---

[18] Editor's note. Alexander (ibid. 17–18) gives the following explanation. With the boxing indicator system, 'the letters of the encyphered indicators can be associated together so as to produce various patterns—known as "box shapes"—independent of the stecker and determined only by the wheel order and Grundstellung at which the indicators are encyphered. With only 6 possible wheel orders catalogues could be made of all possible box shapes with the machine positions at which they occurred and thus the daily key could be worked out.'

numerals at this time being read off the top row of the keyboard and inserted between Ys.[19] I will not attempt to explain the details of the method: it involved a series of assumptions to the effect that certain pairs of letters were both self-steckered. If the assumption was correct, the method worked. At this time, it will be remembered, 14 of the 26 letters were still unsteckered so that the assumption was not a very rash one and also the number of wheel orders was very small.

Even when they had found the Grundstellung with the help of the AFÄ messages, the Poles still could not read the traffic as they did not know how the indicating system worked. They set to work, therefore, to break individual messages on cribs—largely of the FORTY WEEPY type—not a very difficult process when Stecker and wheels are known. By this method they broke out about 15 messages a day and came to the conclusion that the indicating system involved a bigram substitution, but they got little further than this.

All witnesses agree that Naval Enigma was generally considered in 1939 to be unbreakable; indeed pessimism about cryptographic prospects in all fields appears to have been fairly prevalent. This attitude is constantly referred to in such letters of the period as still survive; Mr. Birch records that he was told when war broke out that 'all German codes were unbreakable. I was told it wasn't worth while putting pundits onto them' (letter to Commander Travis, August 1940) and, writing to Commander Denniston in December of the same year, he expresses the view that 'Defeatism at the beginning of the war, to my mind, played a large part in delaying the breaking of codes.'[20] When Turing joined the organization in 1939 no work was being done on Naval Enigma and he himself became interested in it 'because no one else was doing anything about it and I could have it to myself'. Machine cryptographers were on the whole working on the Army and Air Force cyphers with which considerable success had been obtained.

Turing started work where the Poles had given up; he set out to discover from the traffic of May 1937 how the new indicating system worked. 'Prof's Book', the write-up he made in 1940 of the work he had done and of the theory of Banburismus, describes the successful conclusion of this work.

Excerpt from Turing's *Treatise on the Enigma*

[T]he Poles found the keys for the 8th of May 1937, and as they found that the wheel order and the turnovers were the same as for the end of April they rightly assumed that the wheel order and Ringstellung had remained the same during the end of April and the beginning of May. This made it easier for them to find the keys for other days at the beginning of May and they actually found the

---

[19] Editor's note. The top line of the German keyboard is QWERTZUIO. Q was used for 1, W for 2, E for three, and so on. The time of origin of the original message, 23.30, becomes WEEP, 0 being represented by P, the first key of the bottom line of the keyboard. (See ibid. 18.)

[20] Editor's note. Edward Travis was Denniston's deputy.

Stecker for the 2nd, 3rd, 4th, 5th, and 8th, and read about 100 messages. The indicators and window positions of four (selected) messages for the 5th were

	Indicator		Window start
K F J X	E W T W	P C V
S Y L G	E W U F	B Z V
J M H O	U V Q G	M E M
J M F E	F E V C	M Y K

The repetition of the EW combined with the repetition of V suggests that the fifth and sixth letters describe the third letter of the window position, and similarly one is led to believe that the first two letters of the indicator represent the first letter of the window position, and that the third and fourth represent the second. Presumably this effect is somehow produced by means of a table of bigramme equivalents of letters, but it cannot be done simply by replacing the letters of the window position with one of their bigramme equivalents, and then putting in a dummy bigramme, for in this case the window position corresponding to JMFE FEVC would have to be say MYY instead of MYK. Probably some encipherment is involved somewhere. The two most natural alternatives are i) The letters of the window position are replaced by some bigramme equivalents and then the whole enciphered at some 'Grundstellung', or ii) The window position is enciphered at the Grundstellung, and the resulting letters replaced by bigramme equivalents. The second of these alternatives was made far more probable by the following indicators occurring on the 2nd May

E X D P	I V J O	V C P
X X E X	J X J Y	V U E
R C X X	J L W A	N U M

With this second alternative we can deduce from the first two indicators that the bigrammes EX and XX have the same value, and this is confirmed from the second and third, where XX and EX occur in the second position instead of the first.

It so happened that the change of indicating system had not been very well made, and a certain torpedo boat, with the call sign AFÄ, had not been provided with the bigramme tables. This boat sent a message in another cipher explaining this on the 1st May, and it was arranged that traffic with AFÄ was to take [place] according to the old system until May 4, when the bigramme tables would be supplied. Sufficient traffic passed on May 2, 3 to and from AFÄ for the Grundstellung used to be found, the Stecker having already been found from the FORTYWEEPY messages. It was natural to assume that the Grundstellung used by AFÄ was the Grundstellung to be used with the correct method of indication, and as soon as we noticed the two indicators mentioned above we tried this out and found it to be the case.

There actually turned out to be some more complications. There were two Grundstellungen at least instead of one. One of them was called the Allgemeine and the other the Offiziere Grundstellung. This made it extremely difficult to find either Grundstellung. The Poles pointed out another possibility, viz that the trigrammes were still probably not chosen at random. They suggested that probably the window positions enciphered at the Grundstellung, rather than the window positions themselves, were taken off the restricted list.

In Nov. 1939 a prisoner told us that the German Navy had now given up writing numbers with Y...YY...Y and that the digits of the numbers were spelt out in full. When we heard this we examined the messages toward the end of 1937 which were expected to be continuations and wrote the expected beginnings under them. The proportion of 'crashes' i.e. of letters apparently left unaltered by encipherment, then shews how nearly correct our guesses were. Assuming that the change mentioned by the prisoner had already taken place we found that about 70% of these cribs must have been right.

[End of Excerpt]

[Turing's] theory was further confirmed when the Grundstellung which AFÄ had been using was discovered to encypher these trigrams in such a way that the Vs and Us all came out as the same letter. Turing had in fact solved the essential part of the indicator problem and that same night he conceived the idea of Banburismus 'though I was not sure that it would work in practice, and was not in fact sure until some days had actually broken'.

As Banburismus was the fundamental process which Hut 8 performed for the next 2 or 3 years, it is essential to understand roughly the principle on which it works.

Banburismus

Banburismus is not possible unless you have the bigram tables.

The idea behind Banburismus is based on the fact that if two rows of letters of the alphabet, selected at random, are placed on top of each other, the repeat rate between them will be 1 in 26, while if two stretches of German Naval plain language are compared in the same way the repeat rate will be 1 in 17. Cypher texts of Enigma signals are in effect a selection of random letters and if compared in this way the repeat rate will be 1 in 26 but if, by any chance, both cypher texts were encyphered at the same position of the machine and [are] then written level under each other, the repeat rate will be 1 in 17—because, wherever there was a plain language repeat, there will be a cypher repeat also. Two messages thus aligned are said to be set *in depth*: their correct relative position has been found. If by any chance the two messages have identical content for 4 or 6 or 8 or more letters then the cypher texts will be the same for the number of letters concerned—such a coincidence between cypher texts is known as a 'fit'. Banburismus aims first of all at setting messages in depth with the help of fits and of a repeat rate much higher than the random expectation.

Long before the day is broken, a certain amount can be done to the indicators of the messages. The bigram substitution can be performed and the trigrams obtained: these trigrams, when encyphered or 'transposed' at the Grundstellung, will give us the starting positions of the messages. Once the day has been broken, the Grundstellung alphabets, i.e. the effect of encyphering each of the 26 letters of the alphabet at positions one, two, and three of the trigrams, can be produced. The alphabets will look something like this:

	A	B	C	D	E	F	G	H	I	J	K	L	M	N	O	P	Q	R	S	T	U	V	W	X	Y	Z
1.	T	V	X	M	U	I	W	N	F	L	P	J	D	H	Y	K	Z	S	R	A	E	B	G	C	O	Q
2.	E	Y	K	W	A	Q	X	R	T	U	C	N	S	L	V	Z	F	H	M	I	J	O	D	G	B	P
3.	J	G	D	C	F	E	B	P	Z	A	V	Q	W	O	N	H	L	T	U	R	S	K	M	Y	X	I

The aim of Banburismus is to obtain, with the help of the trigrams and fits between messages, alphabets 2 and 3, the middle and right hand wheel alphabets.

The chance a priori of 2 messages with completely different trigrams ZLE and OUX being correctly set in depth is 1 in 17,000, but if the trigrams are NPE and NLO, the factor against them being in depth is only 1 in 676 as, although we do not know the transposed value of the trigrams, N will in each case transpose to the same letter, and therefore both messages were encyphered with the left hand wheel in the same position. These messages are said to be 'at 676'. Messages with such trigrams as PDP and PDB are said to be 'at 26': they are known to have had starting positions close together on the machine. On our alphabets:

$$PDP = KWH$$
$$PDB = KWG$$

therefore PDB started one place earlier than PDP. This is expressed as $B + 1 = P$: in the right hand wheel alphabet, P will be seen one place ahead of B.

The first stage in attacking a day by Banburismus is to discover the fits. This was done largely by Freeborn who sorted all messages against all other messages and listed fits of 4 letters or more.[21] At the same time messages were punched by hand onto Banburies, long strips of paper with alphabets printed vertically, so that any 2 messages could be compared together and the number of repeats be recorded by counting the number of holes showing through both Banburies.[22] A scoring

---

[21] Editor's note. Freeborn's department, known as the 'Freebornery', used Hollerith equipment to sort and analyse cipher material. 'Freeborn' was commonly used as an adjective, e.g. attached to 'catalogue'. 'Freeborn, a[dj]: Performed, produced, or obtained by means of the (Hollerith) electrical calculating, sorting, collating, reproducing, and tabulating machines in Mr. Freeborn's department' (Cryptographic Dictionary, 39).

[22] Editor's note. The name arose because the printed strips of paper were produced in the nearby town of Banbury.

History of Hut 8 to December 1941 | 283

system by 'decibans' recorded the value of fits.[23] All messages at 26 were compared with 25 positions to the right and left of the level position and the scores recorded.

The completion of a Banburismus can best be explained by a simple example, the type of Banburismus that would take the expert 10 minutes rather than many hours or even days of work. This is our list of fits. Note 3.7 = 3 alphabets + 7 letters.

						ODDS
B B C	+	.2	=	B B E	hexagram	certain
E N F	+	3.7	=	E P Q	pentagram	17:1 on
R W C	+	.13	=	R W L	tetragram	4:1 on
P N X	+	.5	=	P I C	enneagram	certain
Q Q G	+	2.7	=	Q D U	pentagram	evens
I U S	+	3.3	=	I U Y	hexagram	20:1 on
Z D R	+	5.5	=	Z I X	hexagram	15:1 on
S W I	+	4.3	=	S U D	tetragram	4:1 on
P P D	+	.16	=	P P U	tetragram	2:1 against

The fits of better than even chance concern the letters of the right hand wheel.

C—E
F—Q
C—L
X—C
S—Y
R—X

One chain X R C L E can be expressed in this form:

R . . . . X . . . . C . E L

We know that those letters must appear in those relative positions in the right hand wheel alphabet. We now 'scritch' the 26 possible positions (R under A, then under B etc.) and cross out those which imply contradictions: the first position, with the reciprocals (R = A, etc) written in and ringed, looks like this.[24]

A B C D E F G H I J K L M N O P Q R S T U V W X Y Z
R . ⓚ . ⓜ X . . . . C ⓧ E . . . . ⓐ . . . . . L . .
                                                              ⓕ

Here there is the contradiction that L appears under X, but X which is on our chain also is under F, giving two values for one letter. Similar reasons reject most other positions. Those left in are:

---

[23] Editor's note. 'Ban: Fundamental scoring unit for the odds on, or probability factor of, one of a series of hypotheses which, in order that multiplication may be replaced by addition, are expressed in logarithms. One ban thus represents an odds of 10 to 1 in favour, and as this is too large a unit for most practical purposes *decibans* and *centibans* are normally employed instead' (Cryptographic Dictionary, 4).

[24] Editor's note. 'Scritch: To test (a hypothesis or possible solution) by examining its implications in conjunction with each of a set of (usually 26) further assumptions in turn, eliminating those cases which yield contradictions and scoring the others' (Cryptographic Dictionary, 72).

```
        A B C D E F G H I J K L M N O P Q R S T U V W X Y Z
1. L     N R P       X     A   C   E   D                 I
2.       S   U L     R     F   X       I C   E           N
3.       T   V   L   R     G       X   J   C   E         O
4.       U   W       L   R H           X   K   C     E P
5.   E Z     B               M L       R   P   X         U   C
6.       D C F E             Q             L T   R       Y X
7. X     F   H C     E       S             V L       R   A
8.   X G     I   C   E       T             W   L     R B
9.       I X K       C   E   V             Y             L   D R
```

Of these, 2 contradicts the S + 3 = Y fit, 1 and 9 contradict the good tetra I + 3 = D: they are, therefore, unlikely to be right. 6, however, is extremely interesting.

We scritched L R X C E and got the position:

```
A B C D E F G H I J K L M N O P Q R S T U V W X Y Z
    D C F E             Q             L T   R       Y X
```

In this alphabet F + 7 = Q and we have a good fit at this distance which was not used in our scritching. We have in fact 'picked up' this extra fit and thereby obtained a considerable factor in favour of this alphabet. We now fit in S in such a position that S + 3 = Y in accordance with our hexagram:

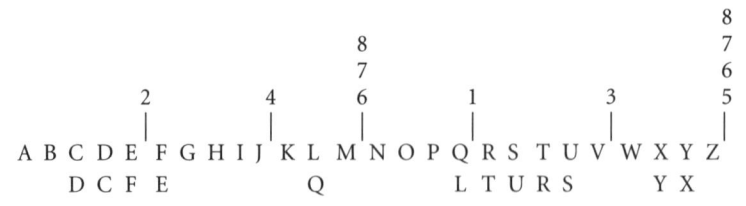

```
                                                            8
                            8                               7
                            7                               6
            2       4       6       1               3       5
            |       |       |       |               |       |
A B C D E F G H I J K L M N O P Q R S T U V W X Y Z
    D C F E             Q             L T U R S     Y X
```

This makes S equal U and we notice at once that we pick up the tetragram D + 16 = U, not a very good tetra but valuable as a further contribution.

We now have to see for which wheels this alphabet is valid. The turnovers on the wheels are in the positions marked on the alphabet above. From the fit RWC + 13 = RWL we know there is no turnover between C and L; if there were a turnover, the position of the middle wheel would have changed, but both trigrams have W in the middle. This fit knocks out wheels 4, 6, 7, and 8. BBC + 2 = BBE knocks out wheel 2. Further, we have the fit PNX + 5 = PIC: as the two trigrams have different letters in the middle, there must be a turnover between X and C. The wheel must therefore be 5, 6, 7, or 8; 6, 7, and 8 are already denied by C + 13 = L, so the wheel is 5.

The next stage is to count up the score for the alphabet, assuming it to be correct. For instance, if there are two messages BDL and BDS, the score for BDL + 4 = BDS will be recorded and should be better than random if the alphabet is right. The final score should be a handsome plus total. An attempt, usually not difficult, is then made to complete the alphabet with the help of any further fits or good scores which may exist. It will be noticed that the fit QQG + 2.7 = QDU is contradicted: it had, however, only an even chance of being right and we do not let this worry us unduly.

A similar process finds the alphabet for the middle wheel and sufficient material is then available to break the day on the bombe. The number of wheel orders which have to be run will probably have been reduced from 276 to something between 3 and 90.

This example shows clearly the fallacy of the system of having all the wheels turning over in different places. It was this characteristic alone which made it possible to distinguish the wheels by Banburismus and reduce the number of wheel orders to be tried. Wheels 6, 7, and 8 were indistinguishable from one another and a great nuisance to the Banburist.

Like depth cribbing, which was closely allied to it and which will be described in due course, Banburismus was a delightful intellectual game. It was eventually killed in 1943 by the rapidly increasing number of bombes, which made it unnecessary to spend much time and labour in reducing the number of wheel orders to be run: it was simpler and quicker to run all wheel orders.

January 1940 to July 1941

Turing's solution of the indicating system came at the end of 1939 but it was well over a year before Banburismus was established as a practical proposition and used as a successful method of attack. The reason for this was primarily a lack of bigram tables.

The next interesting development was the interrogation of Funkmaat[25] Meyer, who revealed valuable information about Short Signals and also the fact that the German Navy now spelt out numerals in full instead of using the top row of the keyboard. This encouraged Turing to look again at the FORTY WEEPY cribs which in 1937 had begun inexplicably to crash, and he came to the conclusion that the cribs remained fundamentally correct provided that the numerals were spelt out.

In early 1940, now joined by Twinn and 2 girls, he started an attack on November 1938 by the FORTY WEEPY method and the new-style crib. The reasons for choosing a period so long ago were various but were primarily based on the knowledge that modern keys were more complicated and would require more work. Two new wheels (4 and 5) had been introduced in December 1938, and from the beginning of the war they were unable to trace the FORTY WEEPY messages owing to call-signs being no longer used.

[25] Editor's note. *Funkmaat* = radio operator.

After about a fortnight's work they broke November 28th, and 4 further days were broken on the same wheel order. Only the Spanish Waters came out; the rest of the traffic was on a different key. There were still only 6 Stecker and there was a powerful and extremely helpful rule by which a letter was never steckered 2 days running: if continuity was preserved, 12 self-stecker were known in advance. No Grundstellungs and no bigrams were broken, messages being broken individually or on the EINS catalogue which was invented at this time and was to play an important part in the exploitation of Enigma.

EINS was the commonest tetragram in German Naval traffic: something in the region of 90% of the genuine messages contained at least one EINS. An EINS catalogue consisted of the results of encyphering EINS at all the 17,000 positions of the machine on the keys of the day in question. These 17,000 tetragrams were then compared with the messages of the day for repeats. When a repeat was found, it meant that [at] a certain position of the machine the messages could be made to say EINS, and further letters were then decoded to see if the answer was a genuine one. If it was, the starting position of the message was known and it could be decoded. In fact about one answer in 4 was right, so that messages were broken fairly rapidly. In later days, the whole process of preparation and comparison was done rapidly and efficiently by Hollerith machinery, but at first slow and laborious hand methods were used.

The plan was to read as many messages as possible, to gain some knowledge of cribs, and then to make rapid progress with the help of the Stecker rule. 'There seemed', says Turing in his book, 'to be some doubt as to the feasibility of this plan', and in fact it proved over-optimistic. Work was fizzling out when Norway was invaded and the cryptographic forces of Hut 8 were transferred en bloc to assist with Army and Air Force cyphers.

By the time work on Naval could be started again, the 'Narvik Pinch' of April 19th had taken place. This pinch revealed the precise form of the indicating system, supplied the Stecker and Grundstellung for April 23rd and 24th (though the scrap of paper on which they were written was for some time ignored) and the operators' log, which gave long letter for letter cribs for the 25th and 26th. Wheels 6 and 7 had been introduced by this time and were already in our possession.

In all 6 days were broken, April 22nd to 27th. The 23rd and 24th presented no difficulty and the days paired with them, the 22nd and 25th, were also broken (by this time the wheel order was only lasting 2 days and there were ten Stecker). The 26th gave much more trouble, being on a new wheel order with unknown Stecker. At first a hand method, the Stecker Knock Out, was unsuccessfully tried and then the Bombe, which had arrived in April and which will be discussed later, was put onto the problem. After about a fortnight of failure, due largely to running unsuitable menus, the day was broken on a freak menu, to be known later as a Wylie menu and tried unsuccessfully on Shark of February 28th 1943. The paired day, the 27th, was also broken.

History of Hut 8 to December 1941 | 287

All hands now turned to EINSing out messages and building up the bigram tables. Provided that the Grundstellung was known, the starting position of the message when broken by EINSing could be transposed and the trigram discovered. This then gave a value for 3 bigrams. A message with 2 of the 3 operative bigrams known could be 'twiddled' out: the 2 known bigrams fixed the positions of 2 of the wheels, and only 26 positions for the remaining wheel had to be tried.[26] This was quickly done and a further bigram was added to the store. April 27th was on the same bigram table as the 24th and this table came near enough to completion to make Banburismus feasible on another day using the same table. May 8th was identified as using this table and a Banburismus was started, but no results were obtained for many months. Turing wrongly deduced that June was using different bigram tables.

The next 6 months produced depressingly few results. Such Banburismus as was tried was unsuccessful, and there was little bombe time for running cribs. Such cribs as there were were supplied by Naval Section and failed to come out; 'Hinsley's certain cribs' became a standing joke. After consulting many people I have come to the conclusion that it is impossible to get an impartial and moderately accurate picture of cribbing attempts at this period: Hut 8 and Naval Section each remain convinced that cribbing failures were due to the other section's shortcomings.

1940 was clearly a very trying period for those outside Hut 8 whose hopes had been raised by the April Pinch and the results obtained from it. On August 21st Mr. Birch wrote to Commander Travis:

I'm worried about Naval Enigma. I've been worried for a long time, but haven't liked to say as much... Turing and Twinn are like people waiting for a miracle, without believing in miracles...

I'm not concerned with the cryptographic problem of Enigma. Pinches are beyond my control, but the cribs are ours. We supply them, we know the degree of reliability, the alternative letterings, etc. and I am confident if they were tried out systematically, they would work.

Turing and Twinn are brilliant, but like many brilliant people, they are not practical. They are untidy, they lose things, they can't copy out right, and they dither between theory and cribbing. Nor have they the determination of practical men...

Of the cribs we supply, some are tried out partially, some not at all, and one, at least, was copied out wrong before being put on the machine...

Sometimes we produce a crib of 90% certainty. Turing and Twinn insist on adding another word of less than 50% probability, because that reduces the number of answers and makes the result quicker. Quicker, my foot! It hasn't produced any result at all so far. The 'slower' method might have won the war by now.

Presumably the number of answers possible on a given crib is mathematically ascertainable. Suppose the one we back 90% has 100,000 possible answers: is that a superhuman labour?...

[26] Editor's note. 'Twiddle: To turn round the wheels of an Enigma machine in hand-testing' (Cryptographic Dictionary, 90).

When a crib, with or without unauthorised and very doubtful additions, has been tried once unsuccessfully we are not usually consulted as to what should be tried next, but, generally speaking, instead of exhausting the possibilities of the best crib, a new one is pottered with under similar handicaps. No crib has been tried systematically and failed; and a few have been tried partially and the partial trial has been unsuccessful...

Turing has stated categorically that with 10 machines [bombes] he could be sure of breaking Enigma and keeping it broken. Well can't we have 10 machines?...

At one end, we're responsible for cribs; at the other end we're responsible to Admiralty. We know the cribs and the odds on them and we believe in them and it's horrible to have no hold, no say, no nothing, on the use that is made of them or the way they are worked...

This letter has great value as a reflection of the relationship of Hut 8 with the outside world and of the cryptographic organization for breaking Enigma; the fact that the letter misinterprets the true position tends to show Mr. Birch at a disadvantage, but it would be most unfair to look at it in this light, as anyone must know who has read the early Naval Section documents and has seen the efforts which Mr. Birch extended to further any work which might in any way assist Hut 8.

First of all the letter demonstrates clearly Turing's almost total inability to make himself understood. Nearly all Mr. Birch's suggestions, as is immediately obvious to anyone with actual experience of Hut 8 work, are impossible and are simply the result of not understanding the problem—his 100,000 answers (had the bombe been able to run the job, which it couldn't have done) would have taken 5 men about 8 months to test. Such problems as this and the disadvantages of the other suggestions should clearly have been explained but Turing was a lamentable explainer and, as Mr. Birch rightly says, not a good practical man: it was for these reasons that he left Hut 8 when the research work was done and the back of the problem broken. The lack of satisfactory liaison was a great disadvantage in early days, but was fortunately most completely overcome later; in a letter to myself of May 16th, 1945, Mr. Birch speaks truly of 'two independent entities so closely, continuously, and cordially united as our two Sections'.

The second point of interest in the letter is the assumption that cribbing and all to do with it was the business of Naval Section and something quite separate from the mathematical work, classed as cryptography and belonging to Hut 8. This concept prevailed until 1941 when Hut 8 set up a Crib Room of its own. No one now would maintain that it would be feasible to separate cribbing from cryptography in this way: to be a good cribster it was essential to understand fully the working of the machine and the problems of Banburismus, bombe management, etc. On the other hand, it remained highly valuable to us that Naval Section were always crib conscious and would send over suggestions for us to explore.

The view frequently expressed by Hut 8 was that a successful pinch of a month's keys with all appurtenances (such as bigram tables) offered the best chance of our being able to get into a position where regular breaking would be possible, as in the course of that month crib records and modern statistics could be built up. Naval Section papers of the barren days of the autumn of 1940 discuss various plans for obtaining keys in this way.

On September 7th Mr. Birch distributed the following document to his subsections, requesting their comments:

When talking to Lt. Cd. Fleming the other day, Mr. Knox put forward the following suggestion: The Enigma Key for one day might be obtained by asking for it in a bogus signal. Lt. Cdr. Fleming suggested that the possibilities should be examined and something got ready and kept ready for use in emergency. Four groups of questions need answering:

1. In the light of our knowledge of German codes and cyphers, W/T routine, and coding [and] cyphering instructions, what signals could be made for the purpose,
    (1) in what code,
    (2) on what frequency,
    (3) at what hour(s),
    (4) from what geographical position accessible to us?
2. Of the various alternative possibilities, in what circumstances would which be most likely to fox the enemy?

This scheme found little favour and was soon rejected as impracticable but a week later Mr. Birch produced his own plan in a letter to D.N.I.:[27]

Operation Ruthless:
I suggest we obtain the loot by the following means

1. Obtain from Air Ministry an air-worthy German bomber (they have some).
2. Pick a tough crew of five, including a pilot, W/T operator and word-perfect German speaker. Dress them in German Air Force uniform, add blood and bandages to suit.
3. Crash plane in Channel after making S.O.S. to rescue service in P/L.[28]
4. Once aboard rescue boat, shoot German crew, dump overboard, bring rescue boat to English port.

In order to increase the chances of capturing an R. or M. with its richer booty, the crash might be staged in mid-Channel. The Germans would presumably employ one of this type for the longer and more hazardous journey.

This somewhat ungentlemanly scheme was never put into practice although detailed plans for it were made and it is discussed several times in Naval Section papers. In fact, the only valuable acquisition during this period was the finding of wheel 8 in August 1940, the last new wheel to be introduced during the war.

[27] Editor's note. Director of Naval Intelligence.
[28] Editor's note. Plain language.

The next event in the cryptographic world was the breaking in November of May 8th, known to history as Foss's day. Foss had joined temporarily to assist in exploiting the Banburismus idea, and after a labour of many months broke the first day on Banburismus. The moral effect of this triumph was considerable and about a fortnight later another Banburismus, April 14th, was broken at what was considered lightning speed. June 26th was also broken (June having by now been established to be on the same bigram tables) and contained the information that new bigram tables would come into force on July 1st, so Banburismus after that date was out of the question.

A second sensational event was the breaking of April 28th on a crib, the first all wheel order crib success.[29] Hut 8 at this time contained no linguists and no cribster by profession and the crib was produced as a result of the labour of two mathematicians, who take great delight in recalling that the correct form of the crib had been rejected by the rival Naval Section cribster.

This last break was obtained in February 1941 and was followed shortly afterwards by the first Lofoten pinch which is one of the landmarks in the history of the Section. This pinch gave us the complete keys for February—but no bigram tables or K book.

The immediate problem was to build up the bigram tables by EINSing and twiddling, the methods for which had now been much improved. With a whole month's traffic to deal with, there was a vast amount of work to be done and the staff position was acute. Rapid expansion and training of new people had to take place and greatly slowed up the work, but by late in March the bigram tables were more or less complete.

Much of the theory of the Banburismus scoring system had been worked out at the end of 1940 and now statistics were brought up to date and satisfactory charts produced. Much work was done on the identification and utilization of dummy messages which at this time formed about half the traffic. It was most important to know and allow for the chance of a message being dummy. The end of a dummy message consisted of a string of consonants and yielded a totally different repeat rate. If dummy was not allowed for, Banburismus could become difficult and even insoluble.

In March also shift work was started and Hut 8 was manned 24 hours a day for the rest of the war. In April, teleprinting of traffic from Scarborough was begun and the Registration Room was started, all traffic being registered currently. Banburismus was started on some March days and there was a rather depressing period of inexplicable failure before the first break and then the system began to get under way. Sooner or later a large part of April and May were broken. There can be no doubt that at this stage the battle was won and the problem was simply one of perfecting methods, of gaining experience, and of obtaining and above all

---

[29] Editor's note. See p. 253 for an explanation of 'all wheel order crib'.

of training staff. These last stages were made much simpler by the pinch of June and July keys.

These last two pinches were a great stroke of good fortune, for the bigram tables changed on June 15th and had once again to be reconstructed. The methods of reconstruction were, however, by now efficient and between June 15th and the end of July this task was easily accomplished. Had we not had the keys, all days subsequent to June would have had to be broken on all wheel order cribs and the messages subsequently broken by EINSing. With the lack of bombes and comparatively crude knowledge of cribs which existed at that stage, this would have been a slow process, and the beginning of what I have called the operational period of Hut 8 would have been delayed by perhaps 2 or 3 months.

Bombes

Throughout these early [sections] I have avoided as far as possible all mention of bombes. Bombes are a complicated subject and their workings are to a large extent incomprehensible to the layman, but without them Hut 8 and Hut 6[30] could not have existed and it is essential to attempt to describe briefly the part they played.

The bombe was so called because of the ticking noise it made, supposedly similar to that made by an infernal machine regulated by a clock.[31] From one side, a bombe appears to consist of 9 rows of revolving drums; from the other, of coils of coloured wire, reminiscent of a Fairisle sweater.

Put briefly, the function of a bombe was to take one wheel order at a time and discover which of the 17,000 possible positions of the machine combined with which of the half billion possible Stecker combinations would satisfy the conditions of the problem presented to it. This problem was called a menu and was in fact a crib in diagram form. If the crib and cypher text were

E	T	R	B	H	U	S	E	D	F	S	H	J	U	Q	A	P	L
V	V	V	J	V	O	N	D	E	R	G	R	O	E	B	E	N	J[32]

the bombe would be asked to find a position on the machine where V would encypher as E, followed at the next position by V encyphering as T, etc. To perform this function for one wheel order the bombe would take about 20 minutes. The wheel order would then be changed and the process repeated.

The bombe was a highly complicated electrical apparatus, involving some 10 miles of wire and about 1 million soldered connections. Its intricate and delicate apparatus had to be kept in perfect condition or the right answer was likely to be

---

[30] Editor's note. Hut 6 dealt with German Army and Air Force Enigma.
[31] Editor's note. This belief seems to have been widespread at Bletchley Park. See pp. 235–7, for remarks concerning the possible origin of the Polish term 'bomba', from which the British term 'bombe' was derived.
[32] Editor's note. FROM VON DER GROEBEN (probably a U-boat commander). 'VVV' was standard radio spelling for 'from' (von).

missed. An embryonic bombe was evolved by the Poles and could be used on the comparatively simple pre-war Enigma problems. The invention of the bombe as we have known it was largely the work of Turing, Welchman, and, on the technical side, Keen.

Unfortunately, the bombe was an expensive apparatus and it was far from certain that it would work or, even if the bombe itself worked, that it would enable us to break Enigma. Its original production, and above all the acceptance of a scheme for large scale production, was the subject of long and bitter battles. Hut 8, and, of course, Hut 6, owe very much to Commander Travis, and to a lesser extent to Mr. Birch, for the energy and courage with which they sponsored its production.

The first bombe arrived in April 1940. In August, the first bombe to incorporate the vital development of the diagonal board arrived.

On the 21st of December, 1940, Mr. Birch wrote:

The chances of reading current Enigma depend ultimately on the number of bombes available. The pundits promise that given 35 bombes they guarantee to break Enigma continuously at an average delay of 48 hours.

At present they get the part time use of one machine. The reason that they don't get more is that there are only two bombes available and that, owing to increased complications of Air Enigma, Hut 6 requires the use of both machines. It is true that more bombes are on their way, up to a limit of 12, but the situation may well be as bad when they have all arrived, owing to the introduction to Air Enigma of further complications and owing to the further success with other Enigma colours, Air or Army.[33]

The long and the short of it is Navy is not getting fair does. Nor is it likely to.

It has been argued that a large number of bombes would cost a lot of money, a lot of skilled labour to make and a lot of labour to run, as well as more electric power than is at present available here. Well, the issue is a simple one. Tot up the difficulties and balance them against the value to the Nation of being able to read current Enigma.

By August 1941, when Hut 8 really started work on an operational basis, 6 bombes were available. By this time it appears that they were considered to have proved their worth and production went ahead steadily. Some idea of the increasing bombe capacity may be obtained from the following figures of the number of jobs run in each year:

1940	273
1941	1,344
1942	4,655
1943	9,193
1944	15,303

The running and maintenance of the machinery was in no respect the responsibility of Huts 6 and 8 but was under the control of Squadron Leader Jones, who

---

[33] Editor's note. See p. 227 for an explanation of 'Enigma colours'.

had working for him a team of technical experts from the R.A.F. and large numbers of Wrens (about 2,000 in all) to operate the machines. The bombe organisation started in one Hut at B.P. and finished at four Out Stations organized and fed with menus from a central Station at B.P. The final organisation was complex and highly efficient, and we owe much to Squadron Leader Jones and his Section. At all times they gave every possible assistance with our problem and no labour was too much to ask of them: certainly no one in Hut 8 worked for lengths of time comparable to those worked frequently by the Bombe Hut mechanics.

The fact that the German Army, Navy, and Air Force used the same cypher machine had the fortunate result that the bombes could be used by both Hut 6 and ourselves and it was a universally accepted principle that they were to be used in the most profitable way possible, irrespective of the Service or Section concerned. As Hut 6 had more keys to run than ourselves, the bombes were normally left in their hands and we applied for them as required. The relative priority of Hut 6 and Hut 8 keys in their claim for bombe time was decided at a weekly meeting between Hut 6, Hut 8, Naval Section, and Hut 3; in very early days Commander Travis decided what use should be made of the few bombes then available.

To have sufficient material to break a day on the bombe a crib of 30 letters or more was normally needed; when cribs are referred to, the phrase should be taken to mean a guess at the plain text for not less than 30 letters.

The bombe was rather like the traditional German soldier, highly efficient but totally unintelligent; it could spot the perfectly correct answer but would ignore an immensely promising position involving one contradiction. The effect of this was that if one letter of the cypher text had been incorrectly intercepted, the menu would fail although both the crib and the text were elsewhere absolutely correct. This was the [reason for] the extensive double-banking programme, which has already been described.

Cribs were sent to the Bombe Hut in the form of menus with directions as to the wheel orders on which they were to be run. No more was heard of them until the possible positions, known as 'stops', started to come from the bombes as they worked through wheel orders. The strength of menus was calculated with a view to the bombe giving one stop on each wheel order, thus supplying a check that the machine was working correctly. The identification of the right stop, the stop giving the correct Stecker, and the rejection of the wrong ones was done in Hut 8.

In order to get the maximum use from the bombes they had, of course, to be kept fed with menus for 24 hours a day, and the art of bombe management required a certain amount of skill and experience. The plugging up of a new menu was a comparatively complicated and lengthy process, so that it was desirable to give a bombe as long a run on a menu as possible: on the other hand an urgent job would justify plugging up a large number of bombes for only a few runs, because of the importance of saving time. Efficient bombe management was largely a matter of striking the happy medium between speed and economy, of making sure that, with a limited amount of bombe time, everything of importance got run and that, as far as possible, the urgent jobs were run first. Bombe management was interesting because the situation a few hours ahead was to a large extent incalculable: allowance had to be made on the one hand for jobs which one expected to have to run in 12 hours time (and which when the time came did not always materialise) and for the fact that sometimes 2 or 3 jobs would come out in quick succession on one of the first wheel orders to be run, thus releasing large numbers of bombes: because of this possibility it was necessary to keep a reserve of fairly unimportant jobs to fill the gaps.

Pressure on the bombes varied greatly with the immediate cryptographic situation and the period of the month, it being generally true to say that towards the end of the month the number of wheel orders was restricted by wheel order rules.[34] It was, however, extremely rare for us to be unable to keep all the machinery busy even in latter days when there were very large numbers of bombes both here and in America.

Cribbing

The Beginning of the Crib Room

Autumn of 1941 found us at last approaching a position where there was some hope of breaking Naval Enigma with regularity. For the first time we had read, in June and July, a fairly long series of days and the traffic was heavy enough—in the region of 400 [messages] a day—to make Banburismus practicable.

Hut 8 immediately began to increase in numbers so as to be able to staff 3 shifts for an attack on current traffic. Before the end of the year our senior staff numbered 16. Rather curiously, this was the highest total it ever reached. As methods improved, and as we ourselves became quicker and more skilled, we found ever increasing difficulty in keeping busy and by the end of 1942 our numbers were already on the decline. Although the number of keys to be broken and the volume of traffic rose steadily, we reduced ourselves by March 1944 to a staff of 4, with which we were able to keep the situation under control for the rest of the war.

Autumn of 1941 saw the birth of the Crib Room as an independent body from the Banburists, an important date as the Cribsters were to outlive the Banburists

---

[34] Editor's note. Alexander, 'Cryptographic History of Work on the German Naval Enigma', 6: 'There were a number of ways in which the German key maker quite unnecessarily restricted his choice of wheel orders e.g. the W.O. always contained a [wheel] 6, 7 or 8 in it. Restrictions of this kind were known as Wheel Order Rules and could on occasions narrow the choice from 336 to as little as 10 or 20 which was of enormous value to us.'

by 18 months and cribbing was to become the only means of breaking after the introduction of the 4-wheeled machine.[35]

It is interesting to note here that by this time cribbing was accepted as a natural part of Hut 8 work—we have seen that earlier it was considered separate from 'cryptography' and the function of Naval Section. No one now would dispute that the only possible arrangement was to have the cribbing done by people who understood the whole problem of breaking Enigma, though the more 'crib conscious' people there were in Naval Section and the more suggestions they sent over the better. ...

Earlier in the year, as we have seen, days had been broken on cribs obtained from the operator's log of the Lofoten pinch, but little had been done in the way of analysing the traffic for routine messages. Cribbing is essentially merely a matter of guessing what a message says and then presenting the result to the bombe in the form of a menu on which the bombe has to find the correct answer. Any fool, as has recently been shown, can find an occasional right crib, though some skill and judgment is required to avoid wasting time on wrong ones, or rather to waste as little time as possible.

Cribs may be divided into 3 basic groups:

1. Depth cribs
2. Straight cribs
3. Re-encodements

As the history of the Hut is from this point to a large extent the history of cribbing, we must digress considerably at this point and study the 3 basic groups with some care. This process will take us far beyond August 1941. ...

Depth Cribbing

The concept of depth is very simple to understand.[36] If two operators choose the same trigram they will, after transposing it at the Grundstellung, get the same starting position for their messages. Now suppose that one encyphers

---

[35] Editor's note. Concerning the four-wheeled machine, Mahon says later in his 'History' (on p. 55, which is not printed here): 'January 1942 passed peacefully enough but on February 1st Shark [the U-boat key] changed over to the 4-wheeled machine and was not broken again, with one exception, until December. This was a depressing period for us as clearly we had lost the most valuable part of the traffic and no form of cryptographic attack was available to us.' Further information concerning the four-wheeled machine is given in the introduction to Chapter 8, pp. 343–5.

[36] Editor's note. 'Depth-cribbing = Fun and Games: The interesting process of simultaneously fitting two cribs (especially for the beginnings) to two messages on the same setting' (Cryptographic Dictionary, 27, 40). Alexander expands ('Cryptographic History of Work on the German Naval Enigma', 11): 'when the relative positions of a number of messages have been discovered it is frequently possible by examining them in conjunction with each other to work out the contents of some or all of the messages. This process is known as DEPTH CRIBBING and cribs produced in this way give cross checks from one message to the other of such a kind that one can be virtually certain of their correctness.'

W E T T E R F U E R D I E N A C H T[37]

and the other

M I T M M M D R E I S I E B E N E I N S[38]

It is clear that their cypher texts must have the 3rd, 9th, 12th, and 13th letters in common, as both hit the same letter at the same position of the machine. Let us write them under each other with the encyphered text:

```
B H N W S (M) S A W M N T C K N (N) P Z
W E T T E (R) F U E R D I E N A (C) H T

C N N J T (R) Q N W S T T C X R (C) D S L D
M I T M M (M) D R E I S I E B E (N) E I N S
```

Here we get the repeats—'clicks'[39] we called them—as expected and also 'reciprocals' where the cypher text in one message equals the clear text in the other. Now let us assume that we are cribsters possessing only the cypher text but suspecting that the top message is a weather message which says 'Wetter fuer die Nacht und Morgen'[40]—quite a likely state of affairs as, from time of origin, frequency, call signs, and length, we consider that this message is a plausible candidate for a weather message which occurs every day.

```
B H N W S M S A W M N T C K N N P Z
  W E T T E R F U E R D I E N A C H T

C N N J T R Q N W S T T C X R C D S
  T     M       E     I E     N
```

Here we see the state of our knowledge about the lower message after assuming we know the clear text of the upper message. It may well be that, from our knowledge of the traffic on this frequency and of the minesweepers known to be operating, we can guess the text of the lower message. We have now done a depth crib, and one that is certainly right. For each 'click' successfully cribbed we receive a factor of 17 (the language repeat rate) and for each reciprocal one of 26, so we have in our favour a factor of $17^4 \times 26^2$—an astronomical number in 8 figures, which completely lulls any lingering doubts we may have had about the a priori improbability of a given message starting Mit MMM 371... In fact, no experienced cribster would have troubled to do the calculation, but I include it as an example of a method which may very usefully be applied when assessing a crib.

[37] Editor's note. WEATHER FOR THE NIGHT.
[38] Editor's note. WITH M 371. 'M' refers to a class of vessel, minesweeper.
[39] Editor's note. 'Click: A repeat or repetition of one or more cipher units usually in two or more messages, especially a repeat which, by its position in the messages or from the fact that it is one of a significant series, suggests that the messages are in depth' (Cryptographic Dictionary, 16).
[40] Editor's note. WEATHER FOR THE NIGHT AND MORNING.

Such a depth crib as that illustrated was rare in the early days of cribbing, largely because we had not on the whole accumulated enough general knowledge or enough experience to have guessed correctly the beginning of the second message. Much of the early depth was 'dummy' depth. We have already noticed that the traffic contained very large quantities of dummies ending in a series of consonants. These began with a few dummy words and then said such things as

           HATKEINENSINNVONVONMNOOOBOULOGNE[41]
           FUELLFUNKVVVHANSJOTAAERGER[42]

(HJÄ, the call sign of Brest.)

The large number of messages saying VONVON naturally led to hexagram repeats being discovered by Freeborn when he analysed the traffic. One would be presented with something of this sort: two messages, trigrams HBN and HDS, have a hexa repeat if HDS is written out 17 places in front of HBN. The overlapping part of the texts look like this:

HDS HWV E L C N U T N U F K J (F M J W R J HQO F EWZ E Z N H D L (S) K (E) K I C M T B P
            D I E N T Z U R T (A) E U S C H U N G V O N V O N H A N S (J) O (T) A A E R G E R
HBN D S N A C M O U W M R L A H (A) W H V U X HQ Z F EWZ E Z X Q D Z (J) Z (T) R E C N X S B
         E   (F) U E L L F U N K V O N V O N F U N K (S) T (E) L L E O S LO

This was the simplest form of dummy depth and, as will be realized, the solution of it was extremely easy, and could in fact be found by consultation of a chart listing all known dummy expressions one under the other: all that was necessary was to find two phrases having 2 letters in common the appropriate distance in front of the 'vonvon'. To use this chart was thought by some cribsters to be unsporting and unaesthetic, but it was none the less a very useful document. The value of having cribbed the above messages was not, of course, only that we now possessed a crib which would break the day but also that the hexagram had been proved beyond doubt to be a correct fit—a valuable contribution to the Banburismus.

By the early months of 1942, depth cribbing had become a fairly highly developed art. The amount of dummy traffic was steadily decreasing and much of the cribbing had to be between genuine messages; this, of course, was much more varied, much wider in its scope, and required a far greater knowledge of what messages were likely to say. This knowledge was obtained by extensive reading of traffic and by the keeping of such useful records as lists of ships likely to be addressed on the various frequencies.

The various applications of depth cribbing were also developed and the amount of assistance cribsters were able to give to Banburists steadily increased.

---

[41] Editor's note. HAS NO MEANING FROM MNO BOULOGNE. 'MNO' stands for *Marine Nachrichten Offizier* = Naval Communications Officer. 'OOO' was standard radio spelling for 'O'.

[42] Editor's note. FILLER RADIO FROM HJÄ.

298 | *Patrick Mahon*

I will give one example of the growth of a 'monster depth' such as warmed the cockles of a cribster's heart.

We start by noting that there are 2 messages about which we think we know something and which have trigrams HEX and HEN, that is to say that they started within 26 places of each other on the machine and it is therefore possible that we may be able to set them in depth.

HEX we think says one of 2 things:

$$\text{VORHERSAGEBEREICHDREITEILEINS}^{43}$$

or

$$\text{WETTERBEREICHDREITEILEINS}^{44}$$

HEN we think says

$$\text{ZUSTANDOSTWAERTIGERKANALXX}^{45}$$

We now proceed to examine all the possible relative positions for confirmations or contradictions.

Position HEX = HEN + 1 looks like this:

```
HEX     BHNWSUHDWMTNCN...
        VORHERSAGEBERE
        WETTERBEREICHD          Stagger HEN 1 to left.
HEN     FDQRLTULEWGDQPB
        ZUSTANDOSTWAERT
```

Now in this position there are clicks between the 7th letter of HEX and the 8th of HEN and the 10th of HEX and 11th of HEN, but in neither case have the cribs for HEX letters in common with the crib for HEN, so the position is impossible. Continuing we get to the position HEX = HEN + 4, which is obviously correct, having 4 confirmations and no contradictions.

```
HEX         BHNW(S)UHDWM(T)NCNKHPZFHYFRUEKLIG
            VORH(E)RSAGE(B)EREICHDREITEILEINS
HEN     FDQRLTUL(E)WGDQP(B)OXNZRNS IOZHXHRRNINK
        ZUSTANDO(S)TWAER(T)IGERKANALXX        N
```

We now take message HEK about which we know nothing except that it probably comes from Boulogne and may be a dummy. Here again we examine position by position to see what consequences are implied.

Position HEK + 1 = HEN looks like this:

---

[43] Editor's note. FORECAST AREA 3 PART 1.
[44] Editor's note. WEATHER AREA 3 PART 1.
[45] Editor's note. SITUATION EASTERN CHANNEL. (The weather situation.)

History of Hut 8 to December 1941 | 299

```
HEX            BHNW(S)(UHDWM(T)N(O)(N)KHPZFHYFRUEKLIG
               VORH(E)R(S)AGE(B)E(R)(E)ICHDREITEILEINS
HEN     FDQRLTUL(E)WGDQP(B)OX(N)ZRNCIOZHXHRRN I NK
        ZUSTANDO(S)TWAERT)IGE(R)KANALXX           N
HEK  AVAJVQSKTWR)GPSIQT(R)(E)BHUSECD...
(Two places            (U)W          (C)(N)
to right)
```

Both UW and CN are more or less impossible bigrams, so we reject the position. Position HEK + 3 = HEN, however, looks like this:

```
HEX            BHNW(S)U(H)DWM(T)NCNKHPZFHYFR(U)EKLIG
               VORH(E)R(S)AGE(B)ERE I CHDRE ITE(U)LE INS
HEN     FDQRLTUL(E)WGDQP(B)OXNZRNC I OZHXHRRN I NK
        ZUSTANDO(S)TWAE R(T) I GERKANALXX            N
HEK  AVAJVQSKTWRGP(S)IQTR(E)BHUSE(C)DLSSJ(I)EQAIP
                  N       H  E   N          N O    (U)L
```

Which we guess correctly as OHNESINNVVVMNOOOBOULOGNE[46], the first N being just a stray letter in the dummy words. We are now well under way, and have supplied the Banburists with 2 certain distances: with the distances obtained from the Freeborn catalogue in addition they will probably soon produce the correct alphabets.

The next move is to examine a tetragram fit between HIP and HEX—not a very good tetragram, estimated to have about 1 chance in 3 of being right. With HEX, HEN, and HEK cribbed we can easily prove or disprove this tetra, and great is our delight on finding that it is the cypher text of the word D R E I[47]; this almost certainly a correct fit and the matter is proved beyond doubt when we crib the whole beginning of the message. With the amount of crib we now possess, attaching HIJ is not very difficult and we have a solid piece of depth on five messages.

```
HEX                     BHNW(S)U(H)DWM(T)NCNKHPZFHYFR
                        VORH(E)R(S)AGE(B)ERE I CHDRE I TE
HEN            F DQRLTUL(E)WGDQP(B)OXNZRNC I OZH(X)
               ZUSTANDO(S)TWAER(T)IGERKANA(L)XX(U)
HEK           AVAJVQSKTWRGP(S)IQTR(E)BHUSECD(L)SSJ
                 O   S   N     O(H)NE S I (N)NVVVMNO(O)OBO
HIP    QCWOJKIV(U)WLMYZTABDIBIXKULBNKPCZFHYH(U)
       BINEBSOO(O)VONVONMMMEINSNEUNEINSDRE I X(X)[48]
```

---

[46] Editor's note. WITHOUT MEANING FROM MNO BOULOGNE.

[47] Editor's note. Drei = three.

[48] Editor's note. BEE BSO FROM M 1913. I am grateful to Frode Weierud and Ralph Erskine for the following information: 'BSO' stands for *Befehlshaber der Seestreitkraefte Ostsee* = Commanding Officer Naval Forces Baltic Sea (see M. van der Meulen, 'Werftschluessel: A German Navy Hand Cipher System –

HIJ WMOWMROLAHWHVNJXVPNQPJMYMFDYIYAZ
     NACHTZUMEINSDREIXVUERXBESETZENXX⁴⁹

At this stage it is certain that the Banburists will get the alphabet, but we can still assist by confirming and by finding new distances on the middle wheel. If we have two messages HOD and HIJ and know that on the right hand wheel D + 1 = J, we can say that HOD starts in one of 50 positions in a known relationship to HIJ—1 alphabet plus 1 letter, or 2 plus 1 letter, etc. Inasmuch as HOD is long enough, we can try these possibilities also. Although the whole message cannot be cribbed, the position illustrated where I = O + 3 on the middle wheel is clearly right, all the consequences are good letters, and at one place much of EINSNULYYEINSNUL⁵⁰ is thrown up. HOS can then be attached.

HEX BHNWSUHDWMTNCNKHPZFHYFR
     VORHERSAGEBEREICHDREITE

HEN FDQRLTULEWGDQPBOXNZRNCIOZHX
     ZUSTANDOSTWAERTIGERKANAUXXU

HEK AVAJVQSKTWRGPSIQTREBHUSECDLSSJ
     O   S   N   OHNESINNVVVMNOQOBO

HIP QCWOJKIVUWLMYZTABDIBIXKULBNKPCZFHYHU
     BINEBSOOOVONVONMMEINSNEUNEINSDREIXX

HIJ WMOWMROLAHWHVNJXVPNQPJMYMFDYIYAZ
     NACHTZUMEINSDREIXVUERXBESETZENXX

HOD ZWKCFJZHBCMKGKOJXSHINONTYUPMAKJCMJW
     N   RE   N   EINSNULYYEINSNUL

HOS LPGICWNECBEMLTMZEBMCILZNB
     NULNULUHRROYANAUSGELAUFEN⁵¹

It would be pointless to pursue this depth further. More might well be added and, once started, the messages might be read for some distance. The delight of depth lay in its great variety—no two depths were ever quite the same—and it were a blasé cryptographer indeed who experienced no thrill at discovering a right position and correctly guessing large chunks of message about the contents of which he at first knew nothing. Needless to say things rarely went as smoothly as in our example, and it was possible to work for a very long time without even getting a piece of depth started.

Part I', *Cryptologia*, 19 (1995), 349–64). 'Bee' translates 'Bine', a code word for the urgency indicator 'SSD': 'Sehr, Sehr Dringend' (Very, Very Urgent). The German Navy used three alternative code words for SSD: 'Bine', short for 'Biene' (bee), 'Wespe' (wasp), and 'Mucke' or 'Muke', short for 'Muecke' (midge).

[49] Editor's note. NIGHT TO 13.4. OCCUPY.
[50] Editor's note. TEN TEN.
[51] Editor's note. 00 HOURS LEFT ROYAN.

There were other applications of depth cribbing of which only one need be mentioned here—the slide. The slide is similar to the process by which we started our last example, but in this case we only have a crib for one message—a fairly long and probably a right crib. If SWI is our cribbed message, we examine the consequences of trying SWL for 25 places to either side of it and, although we cannot attempt to crib [the] result, we look to see whether the letters thrown up are good or bad. For this a simple scoring system was used, E being worth something in the region of 57 and Q about −100. On a really long crib the right position usually showed up clearly and a large number of distances were established by this method.

Depth cribbing died with Banburismus in autumn 1943.

Straight Cribs

I should perhaps have dealt with straight cribs before depth. The theory of straight cribs is simple: it is merely a matter of guessing the contents of a message without the assistance of depth and without the contents having already been passed in another cypher which has been broken.

Finding straight cribs was largely a matter of analysing the traffic. At first, with the traffic fairly small, this was comparatively easy and an organisation was created for writing down any message which looked as if it might occur regularly. Having found a message of a routine type, details about it were recorded. Significant facts were normally frequency and frequencies on which it was retransmitted, call-signs, German time of origin, and length. When a few examples had been written down, it became possible to assess:

(a) whether, given an unbroken day's traffic, it would be possible to identify this particular message;
(b) whether, once identified, it had few enough forms to be used as a crib.

These two factors—identification and forms—were the essential factors in all cribbing. In our experience identification, though often tricky, could usually be cleared up by careful examination of the evidence and there have been comparatively few cribs that have been unusable because they had been unfindable. In very difficult cases we tried to enlist the assistance of R. F. P.[52] but, though help was most readily given, the experiments were never very successful. The German Security Service appears to have considered that retransmitting of messages of one area in another area was dangerous and did something towards stopping curious linkages by recyphering messages and adding dummy at the end before retransmission. This sort of thing was a nuisance to us, but never became sufficiently widespread to cause serious difficulty.

[52] Editor's note. 'Radio Finger Print: Enlarged or elongated film-record of morse transmission by means of which the type of transmitter used and the peculiarities of the individual sets of any type can be distinguished, serving to identify stations' (Cryptographic Dictionary, 63).

302 | *Patrick Mahon*

The ideal crib is not shorter than 35 letters and uses the identical wording every day. Such cribs never existed, though occasionally we possessed for a time what seemed to be the crib to end all cribbing. For some 3 months in the summer of 1942, Boulogne sent a weather message which began ZUSTANDOSTWAER-TIGERKANAL[53] and, if I remember rightly, it only failed twice during that period. On the whole, a crib that had more than 2 or 3 basic forms was little use except for getting out paired days on known wheel orders or for depth cribbing: most cribs produced 'horrors' from time to time which we classed as 'other forms' and made no attempt to allow for them as a possible form of a crib, but when assessing a crib it was of course necessary to take into consideration the frequency with which 'other forms' were tending to appear.

The crashing[54] property of the machine was an essential element of all cribbing and most especially of straight cribbing. When writing a wrong crib under a portion of cypher text, each letter of crib had 1 chance in 25 of being the same as the letter of cypher text above it and of thus proving by a 'crash' that the crib was wrong. If therefore a crib had 2 good forms, each about 30 letters long, there was an odds on chance that the wrong form would crash out; in this case the remaining form is, of course, left with a heavily odds on—instead of an approximately evens—chance of coming out. If the good forms crashed out and only some rather poor form went in, it proved to be bad policy to believe the poor form to be correct, although on the basis of mathematical calculation it might appear to have a reasonably good chance. It was the ability to assess this type of problem which distinguished the good cribster. It was impossible to become a good cribster until one had got beyond the stage of believing all one's own cribs were right—a very common form of optimism which died hard.

The pleasure of straight cribbing lay in the fact that no crib ever lasted for very long; it was always necessary to be looking for new cribs and to preserve an open mind as to which cribs offered the best chance of breaking a day. A crib which lasted for 2 months was a rarity; most cribs gradually deteriorated and never recovered until eventually we only recorded them every 2 or 3 days. To give up recording cribs because they were bad was a fool's policy; it unquestionably paid to keep a record of anything that might one day assist in breaking. Time and again, when a good crib died, we were thrown back onto a reserve at which we would have turned up our noses a week before, only to find that the reserve was quite good enough to enable us to break regularly.

The perennial mortality of cribs was undoubtedly to a considerable extent the result of the work of the German Security Service to whose work as crib hunters we must in all fairness pay tribute. The information recently received that they kept an

---

[53] Editor's note. SITUATION EASTERN CHANNEL. (Weather situation.)

[54] Editor's note. 'Crash: The occurrence of a plain letter opposite the same letter in the cipher text in one of the positions or versions in which a crib is tried, normally involving rejection of that position or version' (Cryptographic Dictionary, 22).

expert permanently at work analysing the weaknesses of the machine does on the other hand little credit to their technical ability. The crib chasers gave us a fairly bad time, especially in the areas nearer home where the cyphers were better organised. By March 1945 Dolphin—a large key of some 400 messages a day—had been rendered almost cribless and we might have failed to read some of the last days had we not captured the Hackle keys which kept us supplied with re-encodements.

Cribs in the early days were largely weather messages. A very large amount of weather was sent in Enigma and it was obvious that it was regarded as of first importance. In 1941 weather cribs from the Channel ports were our principal stand-by and WEWA[55] BOULOGNE and WEWA CHERBOURG were trusty friends. One day during the late autumn, the security officers pounced on this habit of announcing internally from whom the weather originated, but first class cribs of a rather shorter variety continued to come regularly at the beginnings and ends of messages. Rather curiously, this habit of signing weather messages at the end never caught on elsewhere and after the death of the Channel weather cribs in spring 1942 we never again had cribs at the ends of messages. This was, on the whole, convenient as a message always finished with a complete 4 letter group (irrespective of the number of letters in the plain text, dummy letters being added at the end), so that an end crib could be written in 4 different places, and one had to be fortunate with the crashing out for a single really good shot to be available.

The reprimand to the Channel weather stations for insecurity in April 1942 is something of a landmark as the Channel cribs never recovered, except for the remarkable run of ZUSTANDOSTWAERTIGERKANAL during the summer of the same year. Henceforth they omitted such lengthy statements as WETTER-ZUSTANDEINSACHTNULNULUHR[56] and satisfied themselves with a terse NANTESBISBIARRITZ[57] buried somewhere in the middle of the message. This habit of burying sign-offs was an almost completely effective anti-crib measure and became more and more widespread as time went on.

Weather cribs in Norway and the Baltic were useful for a long time, and in the Mediterranean for even longer, but there can be little doubt that the security services were weather conscious and one after another cribs of this type disappeared. They were replaced by other cribs of a type which generally required more finding—some were situation reports of a fairly obviously routine nature, others were much more elusive. For instance, when looking through a day's traffic one was not likely to be struck immediately by a message from Alderney to Seekommandant[58] Kanalinseln which said FEUER BRANNTEN WIE BEFOHLEN[59] but it was in fact a daily confirmation that various lights had been shown as ordered and was a very excellent crib.

[55] Editor's note. 'WEWA' abbreviates 'Wetter Warte': weather station.
[56] Editor's note. WEATHER SITUATION 1800 HOURS.
[57] Editor's note. NANTES TO BIARRITZ.
[58] Editor's note. Naval Commander.
[59] Editor's note. BEACONS LIT AS ORDERED.

To spot cribs of this type it was necessary to read through large quantities of traffic, covering perhaps a week or two, and to have a good short-term memory which would react to seeing two similar messages. Work of this type was naturally more difficult and, as the years went by, the finding of possible cribs (as distinct from their exploitation) began to require more and more high-grade labour. For a long time, junior members of the Crib Room were relied upon for discovering new cribs by reading traffic in Naval Section but, as cribs became scarcer, it became obvious that this system was inadequate, and we started having traffic redecoded on the carbon copies of messages so that they were available for scrutiny by senior cribsters. The desirability of this system was further stressed by the increasing numbers of re-encodements which had to be recognized and preserved, so that by 1945 nearly all the traffic was being decoded twice. In some respects this was an extravagant system, as it required a large decoding staff, but it meant that Naval Section received their decodes more quickly, as all available typists set to work first of all on their copies and subsequently typed those for the Crib Room. The alternative scheme would have been to have an increased number of high-grade cribsters so that someone was always available to examine decodes before they went to Naval Section.

The crib chosen as an example in the paragraph before last is interesting also for being a crib for a complete message. Cribs on the whole were only 'beginners', but we had much success with very short messages for in these we got additional confirmation that our crib was right, in that it finished up exactly in the last group of the message. An interesting example of this type was a little harbour report from the Mediterranean. It said: HANSMAXVVVLECHXXAAAYYDDDFEHLAN-ZEIGE.[60] This was a 13 group message. If the message was 14 groups, VVV had been changed to VONVON, one of the normal alternatives for which the cribster had to allow—others were FUNF or FUENF, SIBEN or SIEBEN, VIR or VIER, etc.[61] Most of these little messages that could be cribbed in toto were situation reports which said, in some form or other, 'Nothing to report'; some of them were security conscious and filled up the message with dummy words, which had the effect of degrading them to the level of a normal crib.

... One further crib must be mentioned here for fear it be forgotten altogether as, though often useful, it was never one of the cribs which formed our daily bread and butter. This was the POPTI crib, so christened by the decoders who were amused by the curious selection of letters it contained. We have seen that the German Navy had ceased in general using QWERTZUIOP numerals, but these numerals continued to be used for certain types of figure—weather and notably for certain observations affecting gunnery. As a result we received messages of this sort:

[60] Editor's note. HM FROM LECH A-D ERROR MESSAGE.
[61] Editor's note. Alternative spellings of the German words for five, seven, and four.

History of Hut 8 to December 1941 | 305

NUEMBERG VVV WEWA SWINEMUENDE LUFTBALTA 05 UHR[62]:

P	P	Q	I	P	P	W	Y	Q	U	E	P	T	O	Y
P	T	W	Q	T	P	R	Y	Q	E	Z	O	T	I	Y
Q	P	W	P	Q	P	E	Y	Q	E	T	W	T	Z	Y
Q	T	Q	O	Q	P	E	Y	Q	E	R	E	T	T	Y
W	P	Q	U	O	P	W	Y	Q	E	E	R	T	E	Y
W	T	Q	U	Z	P	W	Y	Q	E	W	O	T	W	Y
E	P	Q	U	Z	P	W	Y	Q	E	W	R	-	Q	Y
E	T	Q	I	Z	P	W	Y	Q	E	Q	O	T	P	Y
R	P	Q	U	Z	P	W	Y	Q	E	Q	I	T	P	Y
R	T	Q	U	E	P	W	Y	Q	E	Q	U	T	P	Y
T	P	Q	Z	O	P	W	Y	Q	E	Q	Z	T	P	Y
Z	P	Q	Z	Q	P	Q	Y	Q	E	Q	E	T	P	Y
U	P	P	I	U	P	Q	Y	Q	E	Q	-	T	P	Y
I	P	M	T	W	P	W	Y	Q	E	Q	P	T	P	Y

ERDBALTA 05 UHR[63]:

P	P	Q	I	P	P	W	Y	Q	E	U	P	T	O	Y
Q	P	Q	I	P	P	W	Y	Q	E	Z	I	T	I	Y
Q	T	Q	O	I	P	E	Y	Q	E	Z	E	T	I	Y
W	P	W	Q	T	P	R	Y	Q	E	T	O	T	U	Y
W	T	O	I	E	P	E	Y	Q	E	R	O	T	T	Y
E	P	Q	Z	U	P	E	Y	Q	E	R	Q	T	R	Y
R	P	Q	U	P	P	W	Y	Q	E	Q	O	T	W	Y
T	P	Q	U	E	P	W	Y	Q	E	Q	I	T	Q	Y
Z	P	Q	E	W	P	Q	Y	Q	E	Q	T	T	P	Y
U	P	P	W	T	P	R	Y	Q	E	Q	W	T	P	Y
I	P	P	E	I	Q	E	Y	Q	E	P	I	T	E	

The letters enclosed in the boxes are constant. The bigrams in columns 1 and 2 are the station indices and the 13 letters which follow are the observations from each station: the Ys are commas and the second half of the observation always has a one as its first digit—hence the Q. The stations were always listed in the same order and so we had right through a long message a series of known letters in fixed relative positions. The crib would look something like this: PP?????YQ?????YPT?????YQ etc. These cribs were good and were successfully used, but unfortunately usually lived for a short time only. The example given, for instance, was dependent on the *Nuemberg* being out exercising. Occasionally, similar messages occurred with the numerals written out in full, but were less useful in most cases owing to the varying lengths of numerals.

---

[62] Editor's note. NUREMBERG FROM WEATHER STATION SWINEMUENDA AIR [BALTA] 05 HOURS. 'Balta' is clearly a meteorological term but the meaning is obscure.

[63] Editor's note. GROUND [BALTA] 05 HOURS.

In the spring of 1942 the first shift system for a crib was discovered. This was a very important discovery which affected straight cribbing for the rest of the war. The reason for failing to make this discovery earlier was doubtless the same [as the] reason for our failing to notice in early days cribs of a type which were in regular use later; the obvious cribs were very good and we simply did not look very hard for the less obvious, a reprehensible but very understandable state of mind into which we seem frequently to have lapsed. The fact was that, when a key was breaking regularly and satisfactorily, there was very little incentive to do energetic research work.

The incentive which led to the discovery of the shift system was a deterioration of the Channel weather cribs, soon to be finally killed. The position was that they were developing too many different forms and it was at this stage that it was noticed that certain forms occurred at regular intervals. The discovery was made on Cherbourg weather and was rapidly exploited elsewhere. This system involved 4 operators and I very much regret that all records of this original shift system have long since been destroyed. The interesting thing is that all later shift systems, some of which have been proved beyond a shadow of doubt, have worked on a 3-day cycle and have involved 3 operators. It would be interesting to re-examine the original shifts in the light of this evidence to see if our conclusions were only in part correct. The most common form of 3 day cycle, and the only one to be proved in detail, divides the day into 3 shifts (approx. 0–9, 9–16, 16–24) and works as follows: /BAC/ACB/CBA/BAC/ACB/ etc. As an example of the uses of these shifts we may take the twice daily weather report sent out to Arctic U-boats. This message stated the day and the month for which it was a forecast and the month appeared as a jumble of names and numbers—in fact, 2 forms of the crib had to be run each time, one with April and one with VIER. On being divided into shifts on the above principle, however, it was discovered that one man said April consistently and the other two VIER so that on any day we knew which would occur. Also one man was security conscious and was responsible for nearly all the 'horrors' and it was best to leave the crib severely alone when he was on duty.

The practical use of crib records, that is to say, the identification of cribs and the decisions as to what was and was not worth running, always remained a job for high-grade labour, rather surprisingly, as in theory it is simple enough. An attempt to 'mechanize' cribbing with the help of mathematical formulae was a lamentable failure and disappeared amidst howls of derision, though in justice to the inventor of the system it must be said that this was not really due to the essential faultiness of the system as he proposed to use it, but rather the fact that it was misapplied by the inexpert and ignored by the expert, who felt rightly that it was no real assistance. Mathematical computation of the probability of cribs was a system which could not be ignored, but results needed to be modified and analysed by the judgement of experience. Nothing really could replace the knowledge which was gained by experience. A cribster had to come to realize that a crib was not right because he had thought of it himself, that he should not be discouraged because a series of apparently good cribs had failed and start making wild assumptions about new wheels and new keys. He had constantly to decide between two or more cribs as to which was the best and in doing so had to rely as much on a very wide experience as on written records. He had to know which risks could be wisely, which unwisely, taken and had constantly to make decisions on the policy to adopt in breaking a key: would it be better to run 1 crib with a 60% chance of coming out and, if this failed, change to another message, or start on a programme of a 3 form crib on one message which would finish by giving a 95% chance of success? Problems like this would, of course, have been easy had it not been necessary to consider such other factors as the bombe time available, the intelligence and cryptographic advantages of a quick break, the probable pressure on bombes in 12 hours time, and so on. On the face of it, straight cribbing appears to be impossibly tedious when compared with depth but, though depth had its great moments which straight cribbing could not touch, the problem did not become less interesting when Banburismus had died. With perhaps 6 to 10 keys to break regularly, the cribster was a busy man faced with an interesting problem in tackling which he had to consider not only the total number of keys eventually broken but also economy of bombe time and the demands of Intelligence.

Re-encodements

Re-encodements are repetitions in a cypher of messages which have already been transmitted on other cyphers, or indeed in plain language. The great advantage of re-encodements over straight cribs is the factor they receive in favour of their being right, owing to the length of crib which has been written in without its crashing against the cypher text. We have already seen that for each letter of a wrong crib there is a 1 in 25 chance of a crash, so that a 50 letter crib which does not crash gets a factor of 7 in its favour, a 100 letter crib a factor of 50, a 200 letter crib a factor of 2,500, a 300 letter crib a factor of 40,000—in fact there is really no chance of a long re-encodement being wrong. We have met re-encodements in a variety of fairly distinct forms which I will deal with separately.

Re-encodements from hand cyphers

The two main sources of re-encodements from hand cyphers were Werftschluessel[64] and R.H.V.[65] Werftschluessel was used by small ships in the German Home Waters area, mostly in the Baltic: it was read continuously from early 1941 to February 1945, when it was abandoned as being of little further value. R.H.V., the

---

[64] Editor's note. Dockyard Key.

[65] Editor's note. RHV = *Reservehandverfahren*: Reserve Hand Cipher, used should an Enigma machine break down (see F. H. Hinsley and A. Stripp (eds.), *Codebreakers: The Inside Story of Bletchley Park* (Oxford: Oxford University Press, 1993), 238–9).

reserve hand cypher of the German Navy, was being used, apart from its function as an emergency cypher, by a number of small ships in the Norwegian area when it was first captured in December 1941. Presumably owing to the completion of the distribution of the [Enigma] machine, its use gradually declined and there has been little traffic since the end of 1943.

Re-encodements from these cyphers were of the type one would expect— messages of significance to great and small alike: weather messages, gale warnings, aircraft reports, mine warnings, wreck warnings, etc. Given the R.H.V. or Werft version of the signal, it was not normally difficult to find its Enigma pair. Habits about relative time of origin varied somewhat from area to area but were fairly consistent in any one area, while length, and the frequencies the Enigma version was likely to be passed on, were normally fairly accurately predictable.

The most famous of all these re-encodements was Bereich 7[66], which broke Dolphin consistently from late 1942 to late 1943. This was a twice daily re-encodement of weather from Trondhjem, and it was normally possible to produce a right crib from it. The contents were rarely hatted[67] and the only hazard was an addition at the end of the Dolphin message giving the information that a certain R.H.V. message need not be decoded as it had identical content. Another famous weather crib was Bereich 5, which broke most of the first 6 months of Plaice but which subsequently became security conscious and hatted, the Werft version to such an extent that we could do nothing with it. By early 1945 it was so bad that we did not feel justified in asking that Werft should continue to be broken to assist us with Plaice. It was generally true of all re-encodements that, if much hatting took place, we could make little use of them.

The other most prolific source of re-encodements of this type was the mine laying, especially in the Baltic. Mine warnings and 'all clears' were sent out in Werft and then repeated for the Baltic U-boats. So good were these cribs that, if we were under pressure to obtain an early break, mines were laid deliberately with a view to producing cribs and most profitable results were obtained. We even put up suggestions as to the best places to lay mines: obviously WEG FUNF (or FUENF) ZIFFER (or ZIFFX or ZIFF) SIEBEN (or SIBEN) was a rotten place but KRIEGSANSTEUERUNGSTONNE SWINEMUENDE[68] was likely to go un-altered into the Enigma version. The golden age of these cribs was the first half of

---

[66] Editor's note. Area 7.

[67] Editor's note. 'Hat book: A code-book characterized by the fact that when the plain-language terms are arranged in alphabetical order the code groups are not in numerical (or alphabetical) order... Hatted: Arranged in other than numerical (or alphabetical) order.' (Cryptographic Dictionary, 43.)

[68] Editor's note. WARTIME MARKER BUOY SWINEMUENDE. (I am grateful to Ralph Erskine and Frode Weierud for assistance with the translation of this crib.)

1942. By 1943 they were already falling off and eventually, by using dummy words and inversions, they ceased to be of any interest to us.

Re-encodements from Army and Air cyphers

There was surprisingly little re-encodement between our traffic and that of Hut 6. The only re-encodements to have occurred consistently over a long period were Mediterranean reconnaissance reports. These were usually easy to tie up and fairly difficult and amusing to attack; they had their heyday in early 1944 when Porpoise ceased to be breakable on its indicating system. These differed from all other re-encodements in that they presented a problem of 'translation': the German Air Force used very different cypher conventions from the Navy, so each message had to be turned into Naval language before it could be used. With a little practice this game could be played very successfully.

Re-encodements between Naval machine cyphers

Apart from those already mentioned, all re-encodements have been between keys we ourselves have broken. Hence they only began fairly late in our history as the number of our keys increased. The most famous series of these re-encodements were those originating in Shark and being repeated in Dolphin (later Narwhal) for the Arctic U-boats, Plaice for the Baltic U-boats, [and] Turtle for the Mediterranean U-boats. These re-encodements were messages of general interest to all U-boats—corrections to existing documents, descriptions of new Allied weapons, significant experiences of other boats, etc.—and they would start their career by being sent on all existing Shark services, with often a note at the end ordering repetitions on other keys. Re-encodements of this type were usually dead easy and, as they were usually long, one knew definitely when one had a right crib. A considerable time-lag before re-encoding—on Narwhal sometimes 3 or 4 days—often made identification difficult, but had compensating advantages. Shark of the 2nd of a month might break Narwhal of the 4th and Narwhal of the 4th might also contain a re-encodement of a Shark message of the 3rd. The 3rd Shark would break and its paired day[69] the 4th would follow, providing perhaps a store of messages likely to appear in Narwhal of the 5th or 6th. Re-encodements were fairly frequent, an overall average of perhaps 1 or 2 a day, and, with several keys involved, it was often possible to break all keys concerned for considerable periods on re-encodements without having to use a single straight crib—a very economical process.

Another remarkable series of re-encodements linked Dolphin and Sucker from November 1944 until the end of the war. In this case it was a weather message from the Hook of Holland which was repeated in almost identical form in the 2

---

[69] Editor's note. See p. 259 for an explanation of 'paired day'.

cyphers. This crib was responsible for our consistently good results in Dolphin during the last months of the war.

Another weather re-encodement of considerable interest was that of BEREICHDORA[70]—an area in the Skagerrak and West Baltic of interest to Plaice and Dolphin. This message originated from Swinemunde on Plaice and was repeated on Dolphin by one of a variety of stations, in accordance with a complicated programme. The repeating station was responsible for hatting the original and for sending it out on the Dolphin frequencies; normally the repeating station encyphered it twice with the text in a different order and sent it out as 2 independent messages. Security precautions were in fact extremely strictly enforced. However, the original Plaice message always set out its observations in the same order—wind—cloud—visibility—sea—and it was possible to take the Dolphin message and reshuffle it into its original order. BEREICHDORA remained far too tricky to be a high-grade crib, but it was used regularly and quite successfully when nothing better was available: there were several periods when it was in fact our only crib into Plaice. The final refinement of discovering the principle on which the Plaice was hatted and using it for breaking Dolphin remained beyond our powers.

The remaining and most curious type of re-encodement was between one day's traffic and another on the same cypher. These re-encodements were found largely in the Baltic where a series of BELEHRUNGSFUNKSPRUECHE[71] were sent to the training U-boats and were repeated frequently for some weeks. There was no law as to when these might be retransmitted but, presumably because their content was [of] didactic value and the decyphering of them good practice, they occurred at one period fairly frequently, and could often be identified by their exceptional length. Dolphin also produced one crib of this type, perhaps the longest lived of all our cribs and one which would have broken Dolphin daily, had it been possible to intercept the message satisfactorily. It was a statement of areas off North Norway in which anti-submarine activities were to be allowed during the next 24 hours and, while the areas did indeed sometimes change, much more often they remained the same and the message of the day before supplied the crib for the message of the new day.

Commander Denniston wrote: 'The German Signal Service will do its best to prevent compromise of Enigma by inferior low-grade[72] cyphers... the Germans do not intend their cyphers to be read.' The German Signals Service failed, however, lamentably in its functions (though Werft and R.H.V. are not, of

---

[70] Editor's note. Area D.
[71] Editor's note. Training radio messages.
[72] Editor's note. 'Low-grade (of a code or cipher system): Not expected to resist attempts to break it for long, esp. if used to any great extent' (Cryptographic Dictionary, 52).

course, low-grade cyphers), and its cyphers were most extensively read. In later years, we should never have been able to continue regular breaking on a large number of keys if it had not been for the steady flow of re-encodements, and their complete failure to master the re-encodement problem was without doubt one of the biggest blunders the German Security Service made.

The Beginning of the Operational Period

During the Autumn of 1941, the outlook became steadily brighter. In August there were 6 bombes in action and we had the bigram tables without which Banburismus was impossible.

The results speak for themselves. All August was broken except 1–4, 24–25, all September, and all October except 3–4 and 12–13. From October 14th, 1941, Dolphin was broken consistently until March 7th, 1945.

During these months, methods of Banburismus and cribbing improved considerably. The situation gradually developed that stability which can only arise from a long period of regular breaking.

Early in October, on the 3rd, 4th, or 5th, we were mystified by the failure of the U-boat traffic to decode when Dolphin for the day had been broken. The situation was saved by a Werft crib, one of Werft's earliest triumphs, and it transpired that the key was the same as for the rest of the traffic except for the Grundstellung, which was entirely different. This was a further development of the innovation of April when the U-boats started to use a Grundstellung which was the reverse of that used by surface craft. This new development made very little difference except that it delayed slightly the reading of the U-boat traffic.

November 29th brought the first real crisis in the form of a change of bigram tables and K book. 6 months earlier this would have beaten us, but we now knew enough about cribs to be able to break the traffic without Banburismus if sufficient bombes were available, and there were now 12 bombes in action. Rather curiously, we thought it worth while to indulge in the luxury of a Freeborn catalogue of fits in the hope of being able to do a piece of depth—dummy depth—a remarkable reflection on our comparative ignorance of straight cribs and of their value. The policy was, however, to some extent justified, as on November 30th we got a right dummy depth crib and Commander Travis, then the arbiter of bombe policy, decreed that we might have all 12 bombes to run it.

This crib and many others duly came out in the course of December, and the process of EINSing, twiddling, and bigram table building, described in an earlier [section], proceeded merrily enough and considerably more smoothly than before. By the end of December, the tables were near enough to completion to begin to think of restarting Banburismus and Turing was just starting to reconstruct the K book—almost the last theoretical problem he tackled in the Section, although he remained with us for some time to come. On December 30th a pinch of keys, bigram tables, and K book made further work unnecessary and we were able to restart Banburismus at once.

CHAPTER 6

Bombe and Spider (*1940*)

Alan Turing

Introduction

Jack Copeland

This material forms chapter 6 of Turing's *Treatise on the Enigma* (known at Bletchley Park as 'Prof's Book').[1] The text has been prepared by Ralph Erskine, Philip Marks and Frode Weierud from the two known surviving copies of Turing's original typescript.[2] The pagination of Turing's typescript indicates that the chapter may possibly have continued for a further four pages; however, these pages are not to be found in either of the archived copies.[3]

---

[1] This material is Crown copyright and is published with the permission of the Public Record Office and the Estate of Alan Turing.

[2] One copy is held in the Public Record Office in Kew, Richmond, Surrey (document reference HW 25/3), and the other in the National Archives and Records Administration, College Park, Maryland (document reference RG 457, Historic Cryptographic Collection, Box 201, NR 964). A digital facsimile of the typescript HW 25/3 is available in The Turing Archive for the History of Computing <www.AlanTuring.net/profs_book>. A retyped version of the *Treatise on the Enigma* prepared by Erskine, Marks, and Weierud is available at <http://home.cern.ch/frode/crypto>. See also p. 250, n. 62.

[3] A small number of typographical errors in the original typescript have been corrected. Occasionally punctuation has been added and sometimes a superfluous punctuation mark has been removed. Material in the text appearing within square brackets has been added by the editors (e.g. mech[*anism*]). Footnotes beginning 'Editors' note' have been added by Copeland, Erskine, Marks, and Weierud; other footnotes are Turing's own.

The steckered Enigma. Bombe and Spider.[1]

When one has a steckered Enigma to deal with one's problems naturally divide themselves into what is to be done to find the Stecker, and what is to be done afterwards. Unless the indicating system is very well designed there will be no problem at all when the Stecker have been found, and even with a good indicating system we shall be able to apply the methods of the last two chapters [of Turing's *Treatise*] to the individual messages. The obvious example of a good indicating system is the German Naval Enigma cipher, which is dealt with in Chapter VII [of the *Treatise*]. This chapter is devoted to methods of finding the Stecker. Naturally enough we never find the Stecker without at the same time finding much other information.

Cribs

The most obvious kind of data for finding the keys is a 'crib', i.e. a message of which a part of the decode is known. We shall mostly assume that our data is a crib, although actually it may be a number of constatations[2] arising from another source, e.g. a number of CILLIs[3] or a Naval Banburismus.

FORTYWEEPYYWEEPY methods

It is sometimes possible to find the keys by pencil and paper methods when the number of Stecker is not very great, e.g. 5 to 7. One would have to hope that several of the constatations of the crib were 'unsteckered'. The best chance would be if the same pair of letters occurred twice in the crib (a 'half-bombe').[4] In this case, assuming 6 or 7 Stecker there would be a 25% chance of both constatations being unsteckered. The positions at which these constatations occurred could be found by means of the Turing sheets[5] (if there were three wheels) or the

---

[1] Editors' note. This is the original title of chapter 6 of Turing's *Treatise*.

[2] Editors' note. 'Constatation: The association of a cipher letter and its assumed plain equivalent for a particular position' (Cryptographic Dictionary, 20; see p. 269 of Chapter 5 for details of the Dictionary: a digital facsimile of the Dictionary is available in The Turing Archive for the History of Computing <www.AlanTuring.net/crypt_dic_1944>).

[3] Editors' note. 'CILLI: The employment or occurrence of the finishing position of one Enigma message as the setting for enciphering the message setting of a second, ... thus enabling the possible settings of the first message to be calculated for the various permissible wheel-orders and ... the setting and probable wheel-order to be determined' (Cryptographic Dictionary, 14). A psychological CILLI, PSILLI for short, is 'any setting which can be guessed from a knowledge of the idiosyncrasies of the operator concerned' (Cryptographic Dictionary, 63).

[4] Editors' note. In an earlier chapter of the *Treatise* Turing says, '... repetitions of constatations (half-bombes as they are rather absurdly called)' (p. 32 of the original typescript).

[5] Editors' note. A catalogue of wheel positions of a type described elsewhere in the *Treatise* (pp. 87 ff. of Turing's original typescript).

Jeffreys sheets.[6] The positions at which this occurred could be separately tested. Another possibility is to set up the inverse rods[7] for the crib and to look for clicks.[8] There is quite a good chance of any apparent click being a real click arising because all four letters involved are unsteckered. The position on the right hand wheel is given by the column[9] of the inverse rod set-up, and we can find all possible positions where the click coupling occurs from the Turing sheets or the Jeffreys sheets. In some cases there will be other constatations which are made up from letters supposed to be unsteckered because they occur in the click, and these will further reduce the number of places to be tested.

These methods have both of them given successful results, but they are not practicable for cases where there are many Stecker, or even where there are few Stecker and many wheel orders.

A mechanical method. The Bombe

Now let us turn to the case where there is a large number of Stecker, so many that any attempt to make use of the unsteckered letters is not likely to succeed. To fix our ideas let us take a particular crib.[10]

```
1  2  3  4  5  6  7  8  9  10 11 12 13 14 15 16 17 18 19 20 21 22 23 24 25
D  A  E  D  A  Q  O  Z  S  I  Q  M  M  K  B  I  L  G  M  P  W  H  A  I  V
K  E  I  N  E  Z  U  S  A  E  T  Z  E  Z  U  M  V  O  R  B  E  R  I  Q  T
```

Presumably the method of solution will depend on taking hypotheses about parts of the keys and drawing what conclusions one can, hoping to get either a confirmation or a contradiction. The parts of the keys involved are the wheel order, the rod start[11] of the crib, whether there are any turnovers[12] in the crib and if so where, and the Stecker. As regards the wheel order one is almost bound to consider all of these separately. If the crib were of very great length one might

---

[6] Editors' note. A type of perforated sheet; see also pp. 233–4.

[7] Editors' note. 'Rod: Strip of wood or other suitable material ruled off in equal compartments' (Cryptographic Dictionary, 70). 'Direct rod: A rod showing the letters on the right side of a wheel of an Enigma machine [i.e. the side of the wheel farther from the reflector] that are consecutively connected to a fixed point in space at the left side, for the twenty-six different positions of that wheel which occur in one revolution. Twenty-six such rods can be constructed for each wheel, one for each of twenty-six fixed points on its left side (corresponding to the contacts on a non-turning second wheel)' (ibid. 30). 'Inverse rod: A rod showing the letters on the left side of a wheel of an Enigma machine that are consecutively connected to a fixed point in space at the right side for the twenty-six different positions which occur in one revolution of that wheel' (ibid. 46).

[8] Editors' note. See p. 296 and n. 39.

[9] Editors' note. The twenty-six rods formed what was called a 'rod-square' with columns and rows. 'Rod-square: Square of 26 letters by 26 (or rectangle of 26 by 52) formed by the direct or inverse rods of a particular Enigma wheel arranged in order' (Cryptographic Dictionary, 70).

[10] Editors' note. The crib means: NO ADDITIONS TO PRELIMINARY REPORT.

[11] Editors' note. The rod-position (see p. 238) of the wheels at the start of the cribbed ciphertext.

[12] Editors' note. See p. 226 for the explanation of turnovers.

make no assumption about what wheels were in the L.H.W.[13] position and M.W. position, and apply a method we have called 'Stecker knock-out' (an attempt of this kind was made with the 'Feindseligkeiten' crib[14] in Nov. '39), or one might sometimes make assumptions about the L.H.W. and M.W. but none, until a late stage, about the R.H.W. In this case we have to work entirely with constatations where the R.H.W. has the same position. This method was used for the crib from the Schluesselzettel[15] of the Vorpostenboot, with success; however I shall assume that all wheel orders are being treated separately. As regards the turnover one will normally take several different hypotheses, e.g.:

1) turnover between positions 1 and   5
2) turnover between positions   5 and 10
3) turnover between positions 10 and 15
4) turnover between positions 15 and 20
5) turnover between positions 20 and 25

With the first of these hypotheses one would have to leave out the constatations in positions 2 to 4, and similarly in all the other hypotheses four constatations would have to be omitted. One could of course manage without leaving out any constatations at all if one took 25 different hypotheses, and there will always be a problem as to what constatations can best be dispensed with. In what follows I shall assume we are working the T.O.[16] hypothesis numbered 5)[17] above. We have not yet made sufficiently many hypotheses to be able to draw any immediate conclusions, and must therefore either assume something about the Stecker or about the rod start. If we were to assume something about the Stecker our best chance would be to assume the Stecker values of A and E, or of E and I, as we should then have two constatations corrected for Stecker, with only two Stecker assumptions. With Turing sheets one could find all possible places where these constatations occurred, of which we should, on the average, find about 28.1. As there would be 626 hypotheses of this kind to be worked we should gain very little in comparison with separate examination of all rod starts. If there had not been any half-bombes in the crib we should have fared even worse. We therefore work all possible hypotheses as to the rod start, and to simplify this we try to find characteristics of the crib which are independent of the Stecker. Such characteristics can be seen most easily if the crib is put in to the form of a picture

---

[13] Editors' note. L.H.W. = left-hand wheel; similarly for M.W. and R.H.W.

[14] Editors' note. *Feindseligkeiten* = hostilities.

[15] Editors' note. *Schluesselzettel* = key sheet; *Vorpostenboot* = patrol boat. German records show that the boat was the *Schiff 26* (erroneously identified as the *VP 2623* in some accounts). See R. Erskine, 'The First Naval Enigma Decrypts of World War II', *Cryptologia*, 21 (1997), 42–6.

[16] Editors' note. Turnover.

[17] Editors' note. Turing says hypothesis number 5 but the menu in Fig. 59 shows that the turnover hypothesis is number 4. (Turing's original numbering of the figures has been retained.)

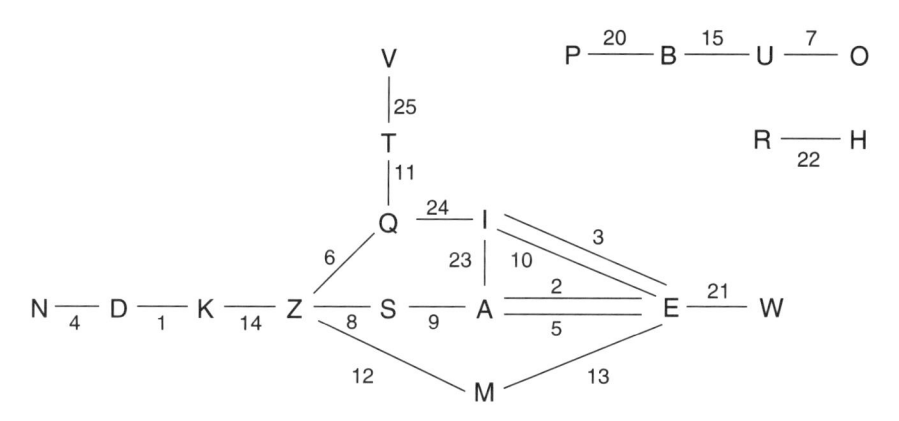

```
                                                    20        15       7
                              V              P ——— B ——— U ——— O
                              | 25
                              T                                    R ——— H
                              | 11                                     22
                              Q  ——24—— I
                         6 /        23 |   10    3
                                           2              21
    N —— D —— K —— Z —— S —— A  ========  E —— W
        4    1    14    8   9       5
                         12              13
                              M
```

Figure 59. Picture from KEINE ZUSAETZE crib. Constatations 16 to 19 omitted to allow for turnover.

like Fig. 59. From this picture we see that one characteristic which is independent of the Stecker is that there must be a letter which enciphered at either position 2 or position 5 of the crib gives the same result. This may also be expressed by saying that there must be a letter such that, if it is enciphered at position 2, and the result re-enciphered at position 5, the final result will be the original letter. Another such condition is that the same letter enciphered successively at the positions 3, 10 must lead back to the original letter. Three other conditions of this kind are that the successive encipherments at positions 2, 23, 3 or at 2, 9, 8, 6, 24, 3 or at 13, 12, 8, 9, 5 starting from the same letter as before must lead back to it. There are other such series, e.g. 13, 12, 6, 24, 3, but these do not give conditions independent of the others. The letter to which all these multiple encipherments are applied is, of course, the Stecker value of E. We shall call E the 'central letter'. Any letter can of course be chosen as [the] 'central letter', but the choice affects the series of positions or 'chains' for the multiple encipherments. There are other conditions, as well as these that involve the multiple encipherments. For instance the Stecker values of the letters in Fig. [59] must all be different. The Stecker values for E, I, M, Z, Q, S, A are the letters that arise at the various stages in the multiple encipherments and the values for W, T, V, N, D, K can be found similarly. There is also the condition that the Stecker must be self-reciprocal, and the other parts or 'webs' of Fig. 59, P-B-U-O and R-H, will also restrict the possibilities somewhat. Of these conditions the multiple encipherment one is obviously the easiest to apply, and with a crib as long as the one above this condition will be quite sufficient to reduce the possible positions to a number which can be tested by hand methods. It is actually possible to make use of some of the other conditions mechanically also; this will be explained later.

In order to apply the multiple encipherment condition one naturally wants to be able to perform the multiple encipherments without Stecker in one operation.

To do this we make a new kind of machine which we call a 'Letchworth Enigma'. There are two rows of contacts in a Letchworth Enigma each labelled A to Z and called the input and output rows: there are also moveable wheels. For each position of an ordinary Enigma there is a corresponding position of the Letchworth Enigma, and if the result of enciphering F at this position is R, then F on the input row of the Letchworth Enigma is connected to R on the output row, and of course R on the input row to F on the output row. Such a 'Letchworth Enigma' can be made like an ordinary Enigma, but with all the wiring of the moveable wheels in duplicate, one set of wires being used for the journey towards the Umkehrwalze,[18] and the others for the return journey. The Umkehrwalze has two sets of contacts, one in contact with the inward-journey wiring of the L.H.W. and one in contact with the outward-journey wiring. The Umkehrwalze wiring is from the one set of contacts across to the other. In the actual design used there were some other differences; the wheels did not actually come into contact with one another, but each came into contact with a 'commutator' bearing 104 fixed contacts. These contacts would be connected by fixed wiring to contacts of other commutators. These contacts of the commutators can be regarded as physical counterparts of the 'rod points' and 'output points' for the wheels.

If one has two of these 'Letchworth Enigmas' one can connect the output points of the one to the input points of the other and then the connections through the two Enigmas between the two sets of contacts left over will give the effect of successive encipherments at the positions occupied by the two Enigmas. Naturally this can be extended to the case of longer series of Enigmas, the output of each being connected to the input of the next.

Now let us return to our crib and see how we could use these Letchworth Enigmas. For each of our 'chains' we could set up a series of Enigmas. We should in fact use 18 Enigmas which we will name as follows.

A1, A2	with the respective positions	2, 5
B1, B2		3, 10
C1, C2, C3		2, 23, 3
D1, D2, D3, D4, D5, D6		2, 9, 8, 6, 24, 3
E1, E2, E3, E4, E5		13, 12, 8, 9, 5

By 'position 8' we here mean 'the position at which the constatation numbered 8 in the crib, is, under the hypothesis we are testing, supposed to be enciphered'. The Enigmas are connected up in this way: output of A1 to input of A2: output of B1 to input of B2: output of C1 to input of C2, output of C2 to input of C3: etc. This gives us five 'chains of Enigmas' which we may call A, B, C, D, E, and there must be some letter, which enciphered with each chain gives itself. We could easily arrange to have all five chains controlled by one keyboard, and to

[18] Editors' note. See p. 269.

have five lampboards shewing the results of the five multiple encipherments of the letter on the depressed key. After one hypothesis as to the rod start had been tested one would go on to the next, and this would usually involve simply moving the R.H.W. of each Enigma forward one place. When 26 positions of the R.H.W. have been tested the M.W. must be made to move forward too. This movement of the wheels in step can be very easily done mechanically, the right hand wheels all being driven continuously from one shaft, and the motion of the other wheels being controlled by a carry mechanism.

It now only remains to find a mechanical method of registering whether the multiple encipherment condition is fulfilled. This can be done most simply if we are willing to test each Stecker value of the central letter throughout all rod starts before trying the next Stecker value. Suppose we are investigating the case where the Stecker value of the central letter E is K. We let a current enter all of the chains of Enigmas at their K input points, and at the K output points of the chains we put relays. The 'on' points of the five relays are put in series with a battery (say), and another relay. A current flows through this last relay if and only if a current flows through all the other five relays, i.e. if the five multiple encipherments applied to K all give K. When this happens the effect is, essentially, to stop the machine, and such an occurrence is known at Letchworth as a 'straight'. An alternative possibility is to have a quickly rotating 'scanner' which, during a revolution, would first connect the input points A of the chains to the current supply, and the output points A to the relays, and then would connect the input and output points B to the supply and relays. In a revolution of the scanner the output and input points A to Z would all have their turn, and the right hand wheels would then move on. This last possible solution was called 'serial scanning' and led to all the possible forms of registration being known as different kinds of 'scanning'. The simple possibility that we first mentioned was called 'single line scanning'. Naturally there was much research into possible alternatives to these two kinds of scanning, which would enable all 26 possible Stecker values of the central letter to be tested simultaneously without any parts of the machine moving. Any device to do this was described as 'simultaneous scanning'.

The solution which was eventually found for this problem was more along mathematical than along electrical engineering lines, and would really not have been a solution of the problem as it was put to the electricians, to whom we gave, as we thought, just the essentials of the problem. It turned out in the end that we had given them rather less than the essentials, and they therefore cannot be blamed for not having found the best solution. They did find a solution of the problem as it was put to them, which would probably have worked if they had had a few more months experimenting. As it was the mathematical solution was found before they had finished.

Pye[19] simultaneous scanning

The problem as given to the electricians was this. There are 52 contacts labelled A ... Z, A', ..., Z'. At any moment each one of A, ..., Z is connected to one and only one of A', ..., Z': the connections are changing all the time very quickly. For each letter of the alphabet there is a relay, and we want to arrange that the relay for the letter R will only close if contact R is connected to contact R'.

The latest solution proposed for this problem depended on having current at 26 equidistant phases corresponding to the 26 different letters. There is also a thyratron valve[20] for each letter. The filaments of the thyratrons are given potentials corresponding to their letters, and the grids are connected to the corresponding points A', ..., Z'. The points A, ..., Z are also given potentials with the phase of the letter concerned. The result is that the difference of potential of the filament and the grid of thyratron A oscillates with an amplitude of at least $2\pi \frac{E}{27}$, [$2E \sin \frac{\pi}{26}$], E being the amplitude of the original 26 phase supply, unless A and A' are connected through the chain, in which case the potentials remain the same or differ only by whatever grid bias has been put into the grid circuit.[21] The thyratrons are so adjusted that an oscillation of amplitude $2\pi \frac{E}{27}$ will bring the potential of the grid to the critical value and the valve will 'fire'. The valve is coupled with a relay which only trips if the thyratron fails to fire. This relay is actually a 'differential relay', with two sets of windings, one carrying a constant current and the other carrying the current from the anode circuit of the thyratron. Fig. 60 shews a possible form of circuit. It is probably not the exact form of circuit used in the Pye experiments, but is given to illustrate the theoretical possibility.

The Spider

We can look at the Bombe in a slightly different way as a machine for making deductions about Stecker when the rod start is assumed. Suppose we were to put lamp-boards in between the Enigmas of the chains, and label the lamp-boards with the appropriate letters off figure [*number missing but presumably 59*]. For example

---

[19] Editors' note. Pye Limited was an electronics company located in Cambridge. Prior to the war Pye marketed radio and TV receivers. From the end of 1939 the company devoted itself to war work, chiefly the development and production of R.D.F. (radar) equipment and radio-based proximity fuses for munitions. (Thanks to David Clayden and M. Cosgrove for this information.)

[20] A thyratron valve has the property that no current flows in the anode circuit until the grid potential becomes more [positive] than a certain critical amount, after which the current continues to flow, regardless of the grid potential, until the anode potential is switched off. [See note to Fig. 60.]

[21] Editors' note. Turing is using an approximation to describe the amplitude. We have inserted the real expression in square brackets. This expression has been independently derived by Donald Davies and Martin Slack and we are grateful for their assistance.

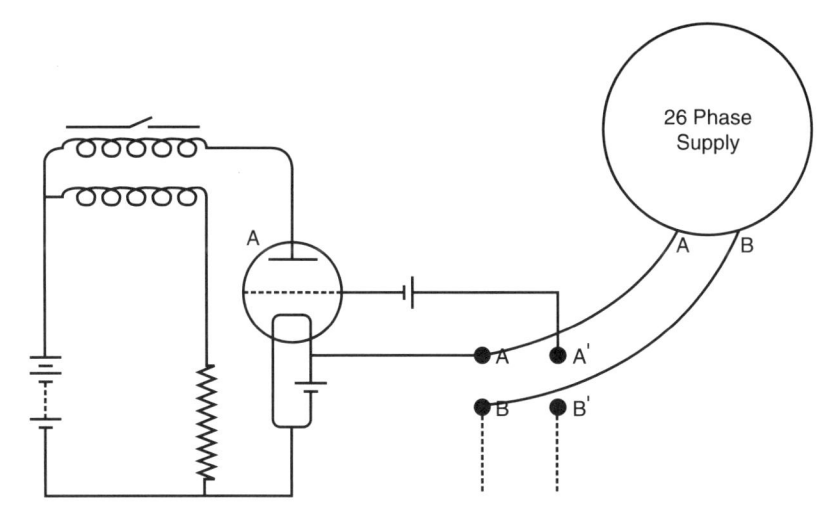

Figure 60. Circuit for Pye simultaneous scanning.

Editors' note: The figure shows a thyratron valve (the smaller of the two circles) in a circuit connected to a 26-phase supply. A thyratron is a gas or mercury vapour filled valve with three connections, called the anode, the cathode, and, between them, the grid. Current begins to flow through the valve once the grid voltage reaches a certain level. Once current has started flowing through the valve, it can be turned off only by dropping the anode voltage to zero. The component depicted at the upper left of the diagram (a breakable contact with two parallel coils shown beneath it) is a relay controlled by the thyratron. The two vertical lines of dots indicate identical repetitions of the same circuit. In total there are 26 thyratrons and 26 relays. The cathode of each thyratron is connected to one of the twenty-six input points of the first Letchworth Enigma in a chain, and the grid of each thyratron is connected to the corresponding output point of the last Letchworth Enigma in the chain. Each input point is supplied with one phase of the 26-phase supply.

The relay is of the differential type described by Turing on pp. 325, 327 and used extensively in the bombe. A differential relay has two coils; in Fig. 60 one of the relay's coils is permanently powered from the cathode supply via a resistor, and the second coil is powered via the thyratron. The idea is that if, for instance, point R is connected to point R' (as in Turing's description on p. 320) then there will be no grid supply signal to this thyratron and so it will not 'fire' (i.e. current will not start to flow through the valve), while all the other thyratrons will 'fire', thereby energizing the second coil of their respective differential relays. This will result in the relay attached to R—but none of the other relays—being switched on by the constant current in its first coil. At the start of each round of scanning, the anode supply would be momentarily interrupted, switching all the thyratrons off and resetting all the differential relays. This interruption can itself be performed electronically.

The 26-phase supply is not a rotary arrangement but a fixed supply generating sinusoidal signals with 26 equidistant phases. Such supplies can be made in various ways, usually using a number of transformers with different taps; the taps sum together to produce the desired phase relationships. Turing does not provide details of the construction of the 26 phase supply. In correspondence concerning Fig. 60, Donald Davies has suggested that the supply might consist of an 'arrangement using two transformers in quadrature with signals 3, 2, 1, 0, −1, −2, −3 times some unit value (E/3) from each transformer. This gives 49 signals and 26 of these can be chosen so that the maximum individual amplitude E corresponds to a minimum difference signal of E/3, better than their value 0.24E. This would be easier to produce than 26 equidistant phases. There are better distributions still, but a bit more difficult to produce.' Turing's sketch shows the cathode, anode, and grid of the thyratron as being powered by batteries. In an engineered form of the circuit, the power would be more likely to come from a mains supply via suitable transformers. The grid bias supply might be regulated electronically.

(Thanks to Martin Slack for information.)

in chain C the lamp-board between C1 and C2 would be labelled A. The keyboard, if we were using one, could be labelled with the 'central letter'. Now when we depress a letter of the key-board we can read off from the lamp-boards some of the Stecker consequences of the hypothesis that the depressed letter is steckered to the central letter: for one such consequence could be read off each lamp-board, namely that the letter lighting is steckered to the name of the lamp-board.

When we look at the Bombe in this way we see that it would be natural to modify it so as to make this idea fit even better. We have not so far allowed for lengthy chains of deductions; the possible deductions stop as soon as one comes back to the central letter. There is however no reason why, when from one hypothesis about the Stecker value of the central letter we have deduced that the central letter must have another Stecker value, we should not go on and draw further conclusions from this second Stecker value. At first sight this seems quite useless, but, as all the deductions are reversible, it is actually very useful, for all the conclusions that can be drawn will then be false, and those that remain will stand out clearly as possible correct hypotheses. In order that all these deductions may be made mechanically we shall have to connect the 26 contacts at the end of each chain to the common beginning of all the chains. With this arrangement we can think of each output or input point of an Enigma as representing a possible Stecker, and if two of these points are connected together through the Enigmas then the corresponding Stecker imply one another. At this point we might see how it all works out in the case of the crib given above. This crib was actually enciphered with alphabets which, when corrected for their Stecker, are those given below, the numbers over the crib constatations giving the column headings.[22] The alphabets most used below are 2, 3, 5, 10, 23, and these are reproduced here for reference.

2	3	5	10	23
XN	XH	MD	TB	LV
AP	BU	JZ	IH	WC
QK	EN	CV	RU	DI
CV	PK	SA	XE	OM
TF	QI	YE	CV	XU
UO	AW	GR	JY	FT
MS	OV	PQ	DF	JP
BD	JY	NW	SL	GE
IW	DM	LH	ON	AY
JZ	RZ	BX	QW	NB
GR	SL	FU	AZ	HS
YE	GT	OI	PK	ZQ
HL	FC	KT	GM	RK

[22] Editors' note. The alphabets have been generated by a one-wheel Enigma machine consisting of the *Umkehrwalze* (UKW) and wheel III (Green) from the Railway Enigma.

In Fig. 61 at the top are the chains, with the positions and the letters of the chain. In each column are written some of the letters which can be inferred to be Stecker values of the letters at the heads of their columns from the hypothesis that X is a Stecker value of the central letter E. By no means all possible inferences of this kind are made in the figure, but among those that are made are all possible Stecker values for E except the right one, L. If we had taken a rod start that was wrong we should almost certainly have found that all of the Stecker values of E could be deduced from any one of them, and this will hold for any cribs with two or more chains. Remembering now that with our Bombe one Stecker is deducible from another if the corresponding points on the lamp-boards are connected through the Enigmas, a correct rod start can only be one for which not all the input points of the chains are connected together; the positions at which this happens are almost exactly those at which a Bombe with simultaneous scanning would have stopped.

This is roughly the idea of the 'spider'. It has been described in this section as a way of getting simultaneous scanning on the Bombe, and has been made to look as much like the Bombe as possible. In the next section another description of the spider is given.

The Spider. A Second Description. Actual Form

In our original description of the Bombe we thought of it as a method of looking for characteristics of a crib which are independent of Stecker, but in the last section we thought of it more as a machine for making Stecker deductions. This last way of looking at it has obviously great possibilities, and so we will start afresh with this idea.

In the last section various points of the circuit were regarded as having certain Stecker corresponding to them. We are now going to carry this idea further and have a metal point for each possible Stecker. These we can imagine arranged in a rectangle. Each point has a name such as Pv: here the capital letters refer to 'outside' points and the small letters to 'inside letters'; an outside letter is the name of a key or bulb, and so can be a letter of a crib, while an inside letter is the name of a contact of the Eintrittwalze,[23] so that all constatations obtained from the Enigma without Stecker give information about inside letters rather than outside. Our statements will usually be put in rather illogical form: statements like 'J is an outside letter' will usually mean 'J is occurring in so and so as the name of a key rather than of a contact of the Eintrittwalze'. The rectangle is called the 'diagonal board' and the rows are named after the outside letters, the columns after the inside letters. Now let us take any constatation of our crib e.g. $\frac{Q}{I}$ at 24. For the position we are supposed to be testing we will have an

---
[23] Editors' note. See p. 225.

```
     2  5        3 10        2 23 3       2  9 8 6 24 3     13 12 8 9 5
    E A E       E I  E      E A  I E     E A S Z Q  I E    E  M  Z S A E
                 XHI
                 IQW
                 WAZ                                            A /
                 ZRU                                            B /
                 UBT                                            C /
                 TGM                                            D /
                 MDF                                            E /
                 FCV                                            F /
                 VON                                            G /
                 NEX                                            H /
     XNW                                                        I /
     WIO                                                        J /
     OUF                                                        K /
     FTK                                                        L
     KQP                                                        M /
     PAS                                                        N /
     SMD                                                        O /
     DBX¹                                                       P /
                 OVC                                            Q /
                 CFD                                            R /
                 DMG                                            S /
                 GTB                                            T /
                 BUR                                            U /
                 RZA                                            V /
                 AWQ                                            W /
                 QIH                                            X /
                 HXE                                            Y /
                 ENO                                            Z /
                 KPK
                 PKP
                 SLS
     IWN                     APJY
     NXB         YJY
     BDM                     PAYJ
     MSA         JYJ
```

Figure 61. Stecker deductions with crib on p. [315], with correct rod start and correct alphabets, but starting from an incorrect Stecker hypothesis E/X. All other incorrect Stecker values of E are deduced.

[1] Editors' note. Turing has written DBN but it must be DBX.

Bombe and Spider | 325

```
   2  5        3 10          2 23 3        2  9 8 2 24 3       13 12 8 9 5
   E A E       E I E         E A  I E      E  A S Z Q  I E     E  M  Z S A E
   L H L       L S L         L H  S L      L  H I R W  S L     L  M  R I H L
```

Figure 62. Stecker deductions with same alphabets as Fig. 61, but from correct Stecker hypothesis E/L.

Enigma set up at the right position for encoding this constatation, but of course without any Stecker. Let us suppose it set up for the correct position, then one of the pairs in the alphabet in position 24 is OC: consequently if Qo then Ic (i.e. if outside letter Q is associated with inside o then outside I is associated with inside c). Now if we connect the input of the (Letchworth) Enigma to the corresponding points of the diagonal board on line Q and the output to line I then since the 'o' input point is connected to the 'c' output point we shall have Qo on the diagonal board connected to Ic through the Letchworth Enigma. We can of course put in a Letchworth Enigma for every constatation of the crib, and then we shall have all the possible deductions that can be made about the association of inside and outside letters paralleled in the connections between the points of the diagonal board. We can also bring in the reciprocal property of the Stecker by connecting together diagonally opposite points of the diagonal board, e.g. connecting Pv to Vp. One can also bring in other conditions about the Stecker, e.g. if one knows that the letters which were unsteckered on one day are invariably steckered on the next then, having found the keys for one day's traffic, one could when looking for the keys for the next day connect together all points of the diagonal board which correspond to non-steckers which had occurred on the previous day. This would of course not entirely eliminate the inadmissible solutions, but would enormously reduce their number, the only solutions which would not be eliminated being those which were inadmissible on every count.

One difference between this arrangement and the Bombe, or the spider as we described it in the last section, is that we need only one Enigma for each constatation.

Our machine is still not complete, as we have not put in any mechanism for distinguishing correct from incorrect positions. In the case of a crib giving a picture like Fig. 59 where most of the letters are connected together in one 'web' it is sufficient to let current into the diagonal board at some point on some line named after a letter on the main web, e.g. at the Ea point in the case of the crib we have been considering. In this case the only possible positions will be ones in which the current fails to reach all the other points of the E line of the diagonal board. We can detect whether this happens by connecting the points of the E line through differential relays to the other pole of our current supply, and putting the 'on' points of the relays in parallel with one another and in series with the stop mech[*anism*]. Normally current will flow through all the differential relays,

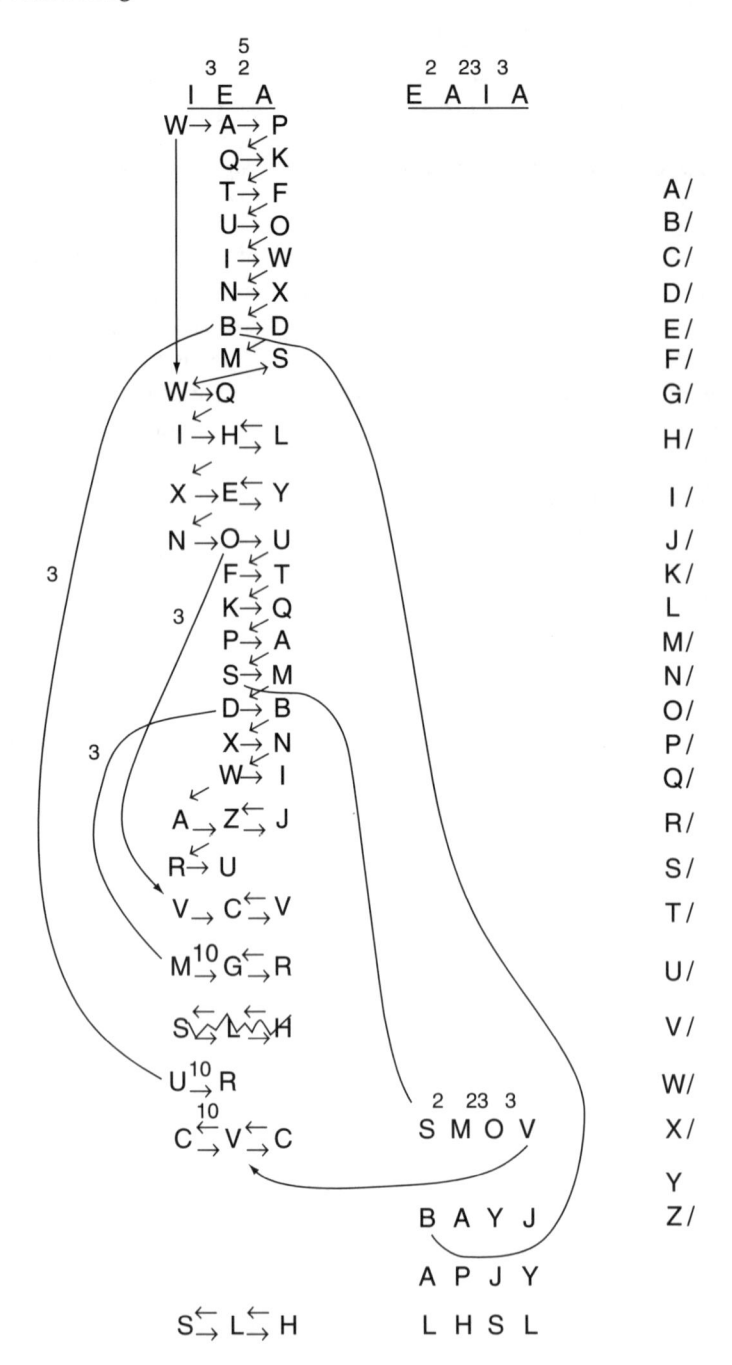

Figure 62a.

Editors' note: This figure, which has no caption or other explanation, follows Fig. 62. The label 'Figure 62a' is our own: in Turing's typescript, neither the figure nor the page containing it is numbered. The appearance is that of a worksheet rather than of a figure that belongs to the text.

and they will not move. When one reaches a position that might be correct the current fails to reach one of these relays, and the current permanently flowing in the other coil of the relay causes it to close, and bring the stopping mechanism into play. Mostly what will happen is that there will be just one relay which closes, and this will be one connected to a point of the diagonal board which corresponds to a Stecker which is possibly correct: more accurately, if this Stecker is not correct the position is not correct. Another possibility is that all relays close except the one connected to the point at which current enters the diagonal board, and this point then corresponds to the only possible Stecker. In cases where the data is rather scanty, and the stops therefore very frequent, other things may happen, e.g. we might find four relays closing simultaneously, all of them connected together through the Enigmas and the cross connections of the diagonal board, and therefore none of them corresponding to possible Stecker.

In order for it to be possible to make the necessary connections between the Enigmas, the diagonal board and the relays there has to be a good deal of additional gear. The input and output rows of the Enigmas are brought to rows of 26 contacts called 'female jacks'. The rows of the diagonal board are also brought to female jacks. The 26 relays and the current supply are also brought to a jack. Any two female jacks can be connected with 'plaited jacks' consisting of 26 wires plaited together and ending in male jacks which can be plugged into the female jacks. In order to make it possible to connect three or more rows of contacts together one is also provided with 'commons' consisting of four female jacks with corresponding points connected together. There is also a device for connecting together the output jack of one Enigma and the input of the next, both being connected to another female jack, which can be used for connecting them to anywhere else one wishes.

On the first spider made there were 30 Enigmas, and three diagonal boards and 'inputs' i.e. sets of relays and stopping devices. There were also 15 sets of commons.

Figs 63, 64 shew the connections of Enigmas and diagonal board in a particular case. The case of a six-letter alphabet has been taken to reduce the size of the figure.

The actual origin of the spider was not an attempt to find simultaneous scanning for the Bombe, but to make use of the reciprocal character of the Stecker. This occurred at a time when it was clear that very much shorter cribs would have to be worked than could be managed on the Bombe. Welchman then discovered that by using a diagonal board one could get the complete set of consequences of a hypothesis. The ideal machine that Welchman was aiming at was to reject any position in which a certain fixed-for-the-time Stecker hypothesis led to any direct contradiction: by a direct contradiction I do not mean to include any contradictions which can only be obtained by considering all Stecker values of some letter independently and shewing each one inconsistent

328 | *Alan Turing*

	5 out, D	5 out, E	5 out, F	5 out, A	5 out, B	5 out, C	
Input jack	Fa	Fb	Fc	Fd	Fe	Ff	
	5 in, A	5 in, B	5 in, C	5 in, D	5 in, E	5 in, F	
Enigma 5							(arriving from $\tfrac{F}{C}$)
	5 out, A	5 out, B	5 out, C	5 out, D	5 out, E	5 out, F	
Output jack	Ca	Cb	Cc	Cd	Ce	Cf	
	5 in, D	5 in, E	5 in, F	5 in, A	5 in, B	5 in, C	

	1 out, F	1 out, D	1 out, E	1 out, B	1 out, C	1 out, A	
Input jack	Fa	Fb	Fc	Fd	Fe	Ff	
	1 in, A	1 in, B	1 in, C	1 in, D	1 in, E	1 in, F	
Enigma 1							(arriving from constatation $\tfrac{F}{A}$)
	1 out, A	1 out, B	1 out, C	1 out, D	1 out, E	1 out, F	
Output jack	Aa	Ab	Ac	Ad	Ae	Af	
	1 in, F	1 in, D	1 in, E	1 in, B	1 in, C	1 in, A	

	2 out, B	2 out, A	2 out, D	2 out, C	2 out, F	2 out, E	
Input jack	Aa	Ab	Ac	Ad	Ae	Af	
	2 in, A	2 in, B	2 in, C	2 in, D	2 in, E	2 in, F	
Enigma 2							(arriving from $\tfrac{A}{C}$)
	2 out, A	2 out, B	2 out, C	2 out, D	2 out, E	2 out, F	
Output jack	Ca	Cb	Cc	Cd	Ce	Cf	
	2 in, B	2 in, A	2 in, D	2 in, C	2 in, F	2 in, E	

	3 out, F	3 out, D	3 out, E	3 out, B	3 out, C	3 out, A	
Input jack	Ca	Cb	Cc	Cd	Ce	Cf	
	3 in, A	3 in, B	3 in, C	3 in, D	3 in, E	3 in, F	
Enigma 3							(arriving from $\tfrac{C}{E}$)
	3 out, A	3 out, B	3 out, C	3 out, D	3 out, E	3 out, F	
Output jack	Ea	Eb	Ec	Ed	Ee	Ef	
	3 in, F	3 in, D	3 in, E	3 in, B	3 in, C	3 in, A	

	4 out, C	4 out, E	4 out, A	4 out, F	4 out, B	4 out, D	
Input jack	Ea	Eb	Ec	Ed	Ee	Ef	
	4 in, A	4 in, B	4 in, C	4 in, D	4 in, E	4 in, F	
Enigma 4							(arriving from $\tfrac{E}{F}$)
	4 out, A	4 out, B	4 out, C	4 out, D	4 out, E	4 out, F	
Output jack	Fa	Fb	Fc	Fd	Fe	Ff	
	4 in, C	4 in, E	4 in, A	4 in, F	4 in, B	4 in, D	

Figure 63. Spider connections with Enigma for 6 letter alphabet and crib

1	2	3	4	5		1	2	3	4	5
A	C	E	F	F	alphabets	AF	AB	AF	AC	FC
F	A	C	E	C		BD	CD	BD	BE	BE
						EC	EF	DE	DF	DA

Names of contact are given in purple ink, contacts to which they are connected in green [*in the original*]. Connections of diagonal board to Enigmas Fig. 64.

Aa 1 out, A 2 in, A	Ab Ba 1 out, B 2 in, B	Ac Ca 1 out, C 2 in, C	Ad Da 1 out, D 2 in, D	Ae Ea 1 out, E 2 in, E	Af Fa 1 out, F 2 in, F
Ba Ab	Bb	Bc Cb	Bd Db	Be Eb	Bf Fb
Ca Ac 2 out, A 3 in, A 5 out, A	Cb Bc 2 out, B 3 in, B 5 out, B	Cc 2 out, C 3 in, C 5 out, C	Cd Dc 2 out, D 3 in, D 5 out, D	Ce Ec 2 out, E 3 in, E 5 out, E	Cf Fc 2 out, F 3 in, F 5 out, F
Da Ad	Db Bd	Dc Cd	Dd	De Ed	Df Fd
Ea Ae 3 out, A 4 in, A Input A, current enters	Eb Be 3 out, B 4 in, B Input B, (relay)	Ec Ce 3 out, C 4 in, C Input C	Ed De 3 out, D 4 in, D Input D	Ee 3 out, E 4 in, E Input E	Ef Fe 3 out, F 4 in, F Input F
Fa Af 4 out, A 1 in, A 5 in, A	Fb Bf 4 out, B 1 in, B 5 in, B	Fc Cf 4 out, C 1 in, C 5 in, C	Fd Df 4 out, D 1 in, D 5 in, D	Fe Ef 4 out, E 1 in, E 5 in, E	Ff 4 out, F 1 in, F 5 in, F

Figure 64. Connections of diagonal board. See Fig. 63. 'Input' is at E. Correct hypothesis E/A. The squares in this figure represent contacts. As in Fig. [63] the purple letters are names and the green letters shew the contacts to which they are connected [*in the original*].

with the original hypothesis. Actually the spider does more than this in one way and less in another. It is not restricted to dealing with one Stecker hypothesis at a time, and it does not find all direct contradictions.

Naturally enough Welchman and Keen set to work to find some way of adapting the spider so as to detect all direct contradictions. The result of this research is described in the next section. Before we can leave the spider however we should see what sort of contradictions it will detect, and about how many stops one will get with given data.

First of all let us simplify the problem and consider only 'normal' stops, i.e. positions at which by altering the point at which the current enters the diagonal board to another pt [*point*] on the same line one can make 25 relays close [*illegible; may be* 'and there is only one such pt']. The current will then be entering at a correct Stecker if the position is correct. Let us further simplify the problem by supposing that there is only one 'web', i.e. that the 'picture' formed from the part of the crib that is being used forms one connected piece, e.g. with the crib on p [*missing, presumably a reference to Fig. 59*] we should have one web if we omit the constatations

P B U R
B U O H.

Some of the constatations of the web could still be omitted without any of the letters becoming disconnected from the rest. Let us choose some set of such constatations, in such a way that we cannot omit any more constatations without the web breaking up. When the constatations are omitted there will of course be no 'chains' or 'closures'. This set of constatations may be called the 'chain-closing constatations' and the others will be called the 'web-forming constatations'. At any position we may imagine that the web-forming constatations are brought into play first, and only if the position is possible for these are the chain-closing constatations used. Now the Stecker value of the input letter and the web-forming constatations will completely determine the Stecker values of the letters occurring in the web. When the chain closing constatations are brought in it will already be completely determined what are the corresponding 'unsteckered' constatations, so that if there are c chain-closing constatations the final number of stops will be a proportion 26^{-c} of the stops which occur if they are omitted. Our problem reduces therefore to the case in which there are no closures. It is, I hope, also fairly clear that the number of stops will not be appreciably affected by the branch arrangement of the web, but only by the number of letters occurring in it. These facts enable us to make a table for the calculation of the number of stops in any case where there is only one web. The method of construction of the table is very tedious and uninteresting. The table is reproduced below.

No. of letters on web	H-M factor[24]
2	0.92
3	0.79
4	0.62
5	0.44
6	0.29
7	0.17
8	0.087
9	0.041
10	0.016
11	0.0060
12	0.0018
13	0.00045
14	0.000095
15	0.000016
16	0.0000023

No. of answers = 26^{4-c} × H-M factor
c is number of closures

[24] Editors' note. In the PRO copy there is a handwritten note by Joan Murray, formerly Joan Clarke of Hut 8, explaining that H-M stands for Holland-Martin of the British Tabulating Machine Company.

A similar table has also been made to allow for two webs, with up to five letters on the second. To the case of three webs it is not worth while and hardly possible to go. One can often get a sufficiently good estimate in such cases by using common-sense inequalities, e.g. if we denote the H-M factor for the case of webs with m, n, and p letters by H(m, n, p) we shall have the common-sense inequalities

$$\frac{H(m,3,2)}{H(m,0,0)} < \frac{H(m,3,0)}{H(m,0,0)} \times \frac{H(m,2,0)}{H(m,0,0)}$$

$$H(m,3,2) > H(m,4,0)$$

To see what kind of contradictions are detected by the machine we can take the picture, Fig. 59, and on it write against each letter any Stecker values of that letter which can be deduced from the Stecker hypothesis which is read off the spider when it stops. This has been done in Fig. 65 for a case where the input was on letter E of the diagonal board, and the relay R closed when the machine stopped; if the position of the stop were correct at all the correct Stecker would be given by the points of the diagonal board which were connected to Er, and they will also be the direct consequences of the Stecker hypothesis E/R. As we are assuming that R was the only relay to close this relay cannot have been connected to any of the others, or it would have behaved similarly. We cannot therefore deduce any other Stecker value for E than R, and this explains why on the 'main web' in Fig. 65 there is only one pencil letter against each ink letter. Wherever any pencil letter is the same as an ink letter we are able to write down another pencil letter corresponding to the reciprocal Stecker or to the diagonal connections of the board. In one or two cases we find that the letter we might write down is there already. In others the new letter is written against a letter of one of the minor webs; in such a case we clearly have a contradiction, but as it does not result in a second set of pencil letters on the main web the machine is not prevented from stopping. There are other contradictions; e.g. we have Z/L, W/L, but as L does not occur in the crib this has no effect.

The machine gun

When using the spider there is a great deal of work in taking down data about stops from the machine and in testing these out afterwards, making it hardly feasible to run cribs which give more than 5 stops per wheel order. As the complete data about the direct consequences of any Stecker hypothesis at any position are already contained in the connections of the points of the diagonal board it seems that it should be possible to make the machine do the testing

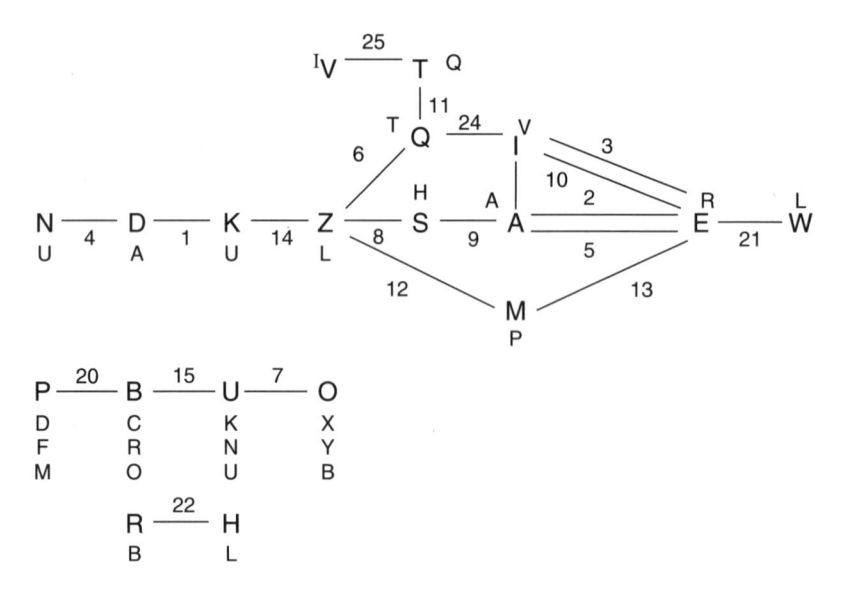

```
              25
         ¹V ───── T  Q
                  │11
              T   Q ──24── V
             ╱6        │      ╲3
                   H   A │  10   2
N ──D── K ──Z──── S ── A ═══════ R ──── L
4   1  14  8  9         5    E   21   W
U   A   U   L    12              13
                         M
                         P
```

```
    20     15    7
P ───── B ───── U ───── O
D       C       K       X
F       R       N       Y
M       O       U       B
           22
        R ───── H
        B       L
```

Relevant parts of alphabets

1	2	3	4	5	6	7	8	9
AU	AR	RV	AU[1]	AR	TL[2]	KX	LH	AH
						NY		
						UB		

10	11	12	13	14	15	16	17	18
RV	QT	LP	PR[3]	LU	CK			
					RN			
					OU			

19[4]	20	21	22	23	24	25		
	DC	RL	BL	AV	VT	QI		
	FR							
	MO							

Figure 65. Illustrating the kind of position at which the spider will stop. Here the input letter may be supposed to be E and the relay which closed R. The Stecker values of the letters, which are consequences of the hypothesis E/R are written against the letters. There are contradictions such as Z/L, W/L: P/D, P/F, P/M which are not observed by the spider.

[1] Editors' note. Originally ND, but ND are the menu letters. Should be AU.
[2] Editors' note. Orginally QL, should be TL.
[3] Editors' note. Originally ME, but ME are the menu letters. Should be PR.
[4] Editors' note. The table in the *Treatise* has column 20 positioned under column 19 and all the subsequent columns shifted one place to the left with a column 26 added at the end. This is clearly an error. There is no column 19 in this menu as the four constatations 16 to 19 have been left out (as Turing has already explained) due to the turnover hypothesis selected.

Bombe and Spider | 333

itself. It would not be necessary to improve on the stopping arrangement of the spider itself, as one could use the spider as already described, and have an arrangement by which, whenever it stopped, a further mechanism is brought into play, which looks more closely into the Stecker. Such a mechanism will be described as a machine gun, regardless of what its construction may be.

With almost any crib the proportion of spider stops that could be passed by a machine gun as possible would be higher than the ratio of spider stops to total possible hypotheses. Consequently the amount of time that can economically be allowed to the machine gun for examining a position is vastly greater than can be allowed to the spider. We might for instance run a crib which gives 100 spider stops per wheel order, and the time for running, apart from time spent during stops, might be 25 minutes. If the machine gun were allowed 5 seconds per position, as compared with the spider's 1/10 second, only 8 minutes would be added to the time for the run.

When the spider stops, normally the points of the diagonal board which are energised are those corresponding to supposedly false Stecker. Naturally it would be easier for the machine gun if the points energised corresponded to supposedly correct Stecker. It is therefore necessary to have some arrangement by which immediately after the spider stops the point of entry of the current is altered to the point to which the relay which closed was connected, or is left unaltered in the case that 25 relays closed. Mr. Keen has invented some device for doing this, depending entirely on relay wiring. I do not know the details at present, but apparently the effect is that the machine does not stop at all except in cases in which either just one relay closes or 25 relays close. In the case that 25 relays close the current is allowed to continue to enter at the same point, but if just one relay closes the point of entry is changed over to this relay. This method has the possible disadvantage that a certain number of possible solutions may be missed through not being of normal type. This will only be serious in cases where the frequency of spider stops is very high indeed, e.g. 20%, and some other method, such as 'Ringstellung cut-out' is being used for further reducing the stops. An alternative method is to have some kind of a 'scanner' which will look for relays which are not connected to any others. Which method is to be used is not yet decided.[25]

At the next stage in the process we have to see whether there are any contradictions in the Stecker; in order to reduce the number of relays involved this is done in stages. In the first stage we see whether or not there are two different Stecker values for A, in the second whether there are two different values for B, and so on. To do this testing we have 26 relays which are wired up in such a way that we can distinguish whether or not two or more of them are energised. When we are testing the Stecker values of A we have the 26 contacts of the A line of the diagonal board connected to the corresponding relays in this set. What is principally lacking is

[25] Now has been decided to use scanner.

some device for connecting the rows of the diagonal board successively to the set of relays. This fortunately was found in post-office standard equipment; the clicking noise that this gadget makes when in operation gives the whole apparatus its name. If we find no contradictions in the Steckers of any letter the whole position is passed as good. The machine is designed to print the position and the Stecker in such a case. Here again I do not know the exact method used, but the following simple arrangement seems to give much the same effect, although perhaps it could not be made to work quite fast enough. The Stecker are given by typing one letter in a column headed by the other. When any letter is being tested for Stecker contradictions the relays corresponding to the Stecker values of the letter close. We can arrange that these relays operate corresponding keys of the typewriter, but that in the case that there is a contradiction this is prevented and some special symbol is typed instead shewing that the whole is wrong. When no relay closes nothing is typed. The carriage of the typewriter is not operated by the keys but only by the space bar, and this is moved whenever there is a change of the letter whose Stecker are being examined.

Additional gadgets

Besides the spider and machine gun a number of other improvements of the Bombe are now being planned. We have already mentioned that it is possible to use additional data about Stecker by connecting up points of the diagonal board. It is planned to make this more straightforward by leading the points of the diagonal board to 325 points of a plug board; the plug board also has a great many points all connected together, and any Stecker which one believes to be false one simply connects to this set.

Another gadget is designed to deal with cases such as that in which there are two 'webs' with six letters and no chains on each. A little experiment will show that in the great majority of cases with such data, when the solution is found, the Stecker value of a letter on either web will imply the whole set of Steckers for the letters of both webs: in the current terminology, 'In the right place we can nearly always get from one web onto the other.' If however we try to run such data on the spider, even with the machine gun attachment, there will be an enormous number of stops, and the vast majority of these will be cases in which 'we have not got onto the second web'. If we are prepared to reject these possibilities without testing them we shall not very greatly decrease the probability of our finding the right solution, but very greatly reduce the amount of testing to be done. If in addition the spider can be persuaded not to stop in these positions, the spider time saved will be enormous. Some arrangement of this kind is being made but I will not attempt to describe how it works.

With some of the ciphers there is information about the Ringstellung (Herivelismus[26]) which makes certain stopping places wrong in virtue of their position, and not of the alphabets produced at those positions. There is an arrangement, known as a 'Ringstellung cut-out', which will prevent the machine from stopping in such positions.[27] The design of such a cut-out clearly presents no difficulties of principle.

There are also plans for 'majority vote' gadgets which will enable one to make use of data which is not very reliable. A hypothesis will only be regarded as rejected if it contradicts three (say) of the unreliable pieces of data. This method may be applied to the case of unreliable data about Stecker.

---

[26] Editors' note. 'Herivelismus' is probably an alternative name for the procedure devised by John Herivel known as the 'Herivel tip'. The 'Herivel tip' helped the codebreakers find the day's ring settings from the indicators of the first messages of the day to be sent by the various German stations making up a given network. Herivel describes the 'Herivel tip' in M. Smith, *Station X: The Codebreakers of Bletchley Park* (London: Channel 4 Books, 1998), 43: 'I thought of this imaginary German fellow with his wheels and his book of keys. He would open the book and find what wheels and settings he was supposed to use that day. He would set the rings on the wheels, put them into the machine and the next thing he would have to do would be to choose a three-letter indicator for his first message of the day. So I began to think, how would he choose that indicator. He might just take it out of a book, or he might pluck it out of the air like ABC or whatever. Then I had the thought, suppose he was a lazy fellow, or in a tearing hurry, or had the wind up, or something or other and he were to leave the wheels untouched in the machine and bang the top down and look at the windows, see what letters were showing and just use them. Then another thought struck me. What about the rings? Would he set them for each of the three given wheels before he put them into the machine or would he set them afterwards? Then I had a flash of illumination. If he set them afterwards and, at the same time, simply chose the letters in the windows as the indicator for his first message, then the indicator would tend to be close to the ring setting of the day. He would, as it were, be sending it almost in clear. If the intercept sites could send us the indicators of all the Red messages they judged to be the first messages of the day for the individual German operators there was a sporting chance that they would cluster around the ring settings for the day and we might be able to narrow down the 17,576 possible ring settings to a manageable number, say twenty or thirty, and simply test these one after the other in the hope of hitting on the right answer.'

[27] Editors' note. 'Ringstellung cut out: A running of a bombe with a restriction on the range of possible Ringstellungen' (Cryptographic Dictionary, 69).

CHAPTER 7

Letter to Winston Churchill (*1941*)

Alan Turing, Gordon Welchman,
Hugh Alexander, Stuart Milner-Barry

Introduction

Jack Copeland

During 1941, codebreaking at Bletchley Park was hindered by shortages of typists and unskilled staff. These shortages could have been easily rectified, but the codebreakers' urgent requests were ignored by officials in Whitehall. Going over the heads of those in command at GC & CS, Turing and his co-signatories wrote directly to the Prime Minister, Winston Churchill. On receiving the letter Churchill minuted his Chief of Staff, General Ismay: 'ACTION THIS DAY Make sure they have all they want on extreme priority and report to me that this had been done.'[1]

It fell to Stuart Milner-Barry of Hut 6 to deliver the letter by hand to 10 Downing Street. In 1986, Milner-Barry recalled his trip to Whitehall:

Why I was deputed to carry the letter to No. 10 I do not remember—at a guess, because I was the most readily expendable from the scene of action. What I do recall is arriving at Euston Station, hailing a taxi, and with a sense of total incredulity (can this really be happening?) inviting the driver to take me to 10 Downing Street. The taxi-driver never blinked an eyelid: without comment he directed himself to Whitehall. Arrived at the entrance to Downing Street, I was again surprised at the lack of formality: there was just a wooden barrier across the road, and one uniformed policeman who waved my driver on. At the door to No. 10 I paid off the taxi, rang the bell, was courteously ushered in, explained that I had an urgent letter which I was anxious to deliver to the Prime Minister

---

[1] Both the letter and Churchill's minute appear in F. H. Hinsley et al., *British Intelligence in the Second World War*, vol. ii (London: Her Majesty's Stationery Office, 1981), appendix 3, pp. 655–7. A facsimile of Churchill's minute appears on p. xiii of R. Erskine and M. Smith (eds.), *Action This Day* (London: Bantam, 2001). The letter and minute are in the Public Record Office in Kew, Richmond, Surrey (document reference HW 1/155). They are Crown copyright and are reproduced with the permission of the Controller of Her Majesty's Stationery Office.

personally, and was invited to wait. Of course I did not see the Prime Minister himself; but very shortly there appeared a dapper dark-suited figure of shortish stature whom I subsequently identified as Brigadier Harvie-Watt, Mr. Churchill's PPS from 1941 to 1945. To him I again explained my errand; and while obviously and understandably puzzled as to who I might be and what this was all about, he took me sufficiently seriously to promise that he would without fail deliver the letter to the Prime Minister and stress its urgency. That accomplished, I took my leave and took myself back to Bletchley by the next train. It was some forty years later before I found out what happened to that letter, when I saw the Prime Minister's minute dated the following day... All that we did notice was that almost from that day the rough ways began miraculously to be made smooth. The flow of bombes was speeded up, the staff bottlenecks were relieved, and we were able to devote ourselves uninterruptedly to the business in hand.[2]

Milner-Barry added, 'I by chance met Commander Denniston in the corridors some days later, and he made some rather wry remark about our unorthodox behaviour; but he was much too nice a man to bear malice.'

---

[2] P. S. Milner-Barry, '"Action This Day": The Letter from Bletchley Park Cryptanalysts to the Prime Minister, 21 October 1941', *Intelligence and National Security*, 1 (1986), 272–3. Reproduced with the permission of Frank Cass Publishers.

Letter to Winston Churchill

Secret and Confidential
Prime Minister only

Hut 6 and Hut 8
21st October 1941

Dear Prime Minister,
Some weeks ago you paid us the honour of a visit, and we believe that you regard our work as important. You will have seen that, thanks largely to the energy and foresight of Commander Travis, we have been well supplied with the 'bombes' for the breaking of the German Enigma codes. We think, however, that you ought to know that this work is being held up, and in some cases is not being done at all, principally because we cannot get sufficient staff to deal with it. Our reason for writing to you direct is that for months we have done everything that we possibly can through the normal channels, and that we despair of any early improvement without your intervention. No doubt in the long run these particular requirements will be met, but meanwhile still more precious months will have been wasted, and as our needs are continually expanding we see little hope of ever being adequately staffed.

We realise that there is a tremendous demand for labour of all kinds and that its allocation is a matter of priorities. The trouble to our mind is that as we are a very small section with numerically trivial requirements it is very difficult to bring home to the authorities finally responsible either the importance of what is done here or the urgent necessity of dealing promptly with our requests. At the same time we find it hard to believe that it is really impossible to produce quickly the additional staff that we need, even if this meant interfering with the normal machinery of allocations.

We do not wish to burden you with a detailed list of our difficulties, but the following are the bottlenecks which are causing us the most acute anxiety.

1. Breaking of Naval Enigma (Hut 8)

Owing to shortage of staff and the overworking of his present team the Hollerith section here under Mr Freeborn has had to stop working night shifts. The effect of this is that the finding of the naval keys is being delayed at least twelve hours every day. In order to enable him to start night shifts again Freeborn needs immediately about twenty more untrained Grade III women clerks. To put himself in a really adequate position to deal with any likely demands he will want a good many more.

A further serious danger now threatening us is that some of the skilled male staff, both with the British Tabulating Company at Letchworth and in Freeborn's section here, who have so far been exempt from military service, are now liable to be called up.

2. Military and Air Force Enigma (Hut 6)

We are intercepting quite a substantial proportion of wireless traffic in the Middle East which cannot be picked up by our intercepting stations here. This contains among other things a good deal of new 'Light Blue' intelligence. Owing to shortage of trained typists, however, and the fatigue of our present decoding staff, we cannot get all this traffic decoded. This has been the state of affairs since May. Yet all that we need to put matters right is about twenty trained typists.

3. Bombe testing, Hut 6 and Hut 8

In July we were promised that the testing of the 'stories' produced by the bombes would be taken over by the WRNS in the bombe hut and that sufficient WRNS would be provided for this purpose. It is now late in October and nothing has been done. We do not wish to stress this so strongly as the two preceding points, because it has not actually delayed us in delivering the goods. It has, however, meant that staff in Huts 6 and 8 who are needed for other jobs have had to do the testing themselves. We cannot help feeling that with a Service matter of this kind it should have been possible to detail a body of WRNS for this purpose, if sufficiently urgent instructions had been sent to the right quarters.

4. Apart altogether from staff matters, there are a number of other directions in which it seems to us that we have met with unnecessary impediments. It would take too long to set these out in full, and we realise that some of the matters involved are controversial. The cumulative effect, however, has been to drive us to the conviction that the importance of the work is not being impressed with sufficient force upon those outside authorities with whom we have to deal.

We have written this letter entirely on our own initiative. We do not know who or what is responsible for our difficulties, and most emphatically we do not want to be taken as criticising Commander Travis who has all along done his utmost to help us in every possible way. But if we are to do our job as well as it could and should be done it is absolutely vital that our wants, small as they are, should be promptly attended to. We have felt that we should be failing in our duty if we did not draw your attention to the facts and to the effects which they are having and must continue to have on our work, unless immediate action is taken.

> We are, Sir, Your obedient servants,
> A M Turing
> W G Welchman
> C H O'D Alexander
> P S Milner-Barry

CHAPTER 8

Memorandum to OP-20-G on Naval Enigma (*c.1941*)

Alan Turing

Introduction: Turing Questions OP-20-G's Attempts to Break Naval Enigma in 1941

Ralph Erskine, Colin Burke, and Philip Marks

The document[1] that forms this chapter was written by Alan Turing and sent to the US Navy codebreaking unit, OP-20-G, in Washington, DC, probably to the celebrated cryptanalyst Mrs Agnes Driscoll. It is undated, but was probably dispatched in the autumn of 1941.

Background[2]

Turing solved the indicating system of the principal Naval Enigma cipher, *Heimisch* (codenamed 'Dolphin' by the Government Code and Cypher School (GC & CS)), at Bletchley Park by the end of 1939. Typically, he thought Dolphin 'could be broken because it would be so interesting to break it' (see 'Enigma', p. 257). Hut 8 at Bletchley Park solved some wartime Naval Enigma signals in May and June 1940.[3] Internal evidence shows that Turing wrote his outstanding *Treatise on the Enigma* around autumn 1940. GC & CS read Dolphin traffic currently, using captured keys, in June and July 1941. The resulting data provided enough cribs and other information to break Dolphin signals cryptanalytically from August onwards, within 24 to 36 hours of their transmission.

---

[1] RG 38, CNSG Library, Box 117, 5750/205 (National Archives and Records Administration (NARA), College Park, Maryland).

[2] For an extensive treatment of the full background, see J. Debrosse and C. Burke, *The Secret in Building 26* (New York: Random House, 2004), ch. 3; also R. Erskine, 'What Did the Sinkov Mission Receive from Bletchley Park?', *Cryptologia*, 24 (2000), 97–109.

[3] R. Erskine, 'The First Naval Enigma Decrypts of World War II', *Cryptologia*, 21 (1997), 42–6.

Mrs Driscoll was assigned to attack Naval Enigma, with two assistants, around October 1940.[4] However, the US Navy was then intercepting only a small proportion of the Naval Enigma signals being transmitted, and was unable to make any progress against Dolphin, especially since it could not reconstruct the wiring of Enigma's wheels. It did not even fully understand the wheels' non-cyclometric motion, which considerably complicated any solution of Naval Enigma, in particular, since each of the special *Kriegsmarine* wheels VI to VIII had two notches. Notching made the wheel motion irregular, especially when a doubly notched wheel was in the middle or right-hand position.[5] Using two doubly notched wheels could reduce Enigma's period from its normal 16,900 (26 × 25 × 26) to 4,056 (24 × 13 × 13).

In February 1941, following lengthy negotiations between the US Army and Navy, a four-man team led by Abraham Sinkov visited GC & CS.[6] (Sinkov was accompanied by Leo Rosen, also from the US Army's Signal Intelligence Service, and Lt. Robert Weeks and Lt. Prescott Currier, both from OP-20-G.) They brought with them various items of codebreaking material, including one example of 'Purple', the American clone of the Japanese Foreign Ministry's high-level cipher machine. GC & CS staff briefed them fully about Bletchley's work and, with the blessing of Winston Churchill, showed them the top secret bombes used to break Enigma, although they had to undertake to tell only their superior officers about what they had learned about Enigma.[7] Neither Weeks nor Currier was an Enigma cryptanalyst, so that they are unlikely to have followed all the explanations they heard about breaking Enigma. They were given only a 'paper' copy of Enigma's wiring to bring home, since GC & CS did not have an Enigma machine to spare.

Alastair Denniston, the operational head of GC & CS, wanted further cooperation with the US Army and Navy codebreaking units. During a visit he made to Washington in August 1941, Mrs Driscoll told him about her attack on Naval Enigma. She also gave him some questions on which she needed help from GC & CS. Her questions reveal just how little OP-20-G then knew about some important details of Naval Enigma.[8] GC & CS responded to most of her questions and requests for data in early October 1941, although the packet unfortunately went astray, and did not turn up until mid-December.

---

[4] 'Naval Security Group History to World War II' 400: NARA, RG 457, SRH 355.

[5] D. Hamer, 'ENIGMA: Actions Involved in the "Double Stepping" of the Middle Rotor', *Cryptologia*, 21 (1997), 47–50.

[6] On the Sinkov mission, see B. Smith, *The Ultra-Magic Deals and the Most Secret Relationship* (Novato, Calif.: Presidio, 1993), chapter 3, pp. 54, 56, 58.

[7] R. Erskine, 'Churchill and the Start of the Ultra-Magic Deals', *International Journal of Intelligence and Counterintelligence*, 10 (1997), 57–74.

[8] Most of the questions and requests for data are set out in Erskine, 'What Did the Sinkov Mission Receive from Bletchley Park?'

Turing's memorandum can only have been written at some time after Denniston returned from the United States in late August 1941. It does not refer to the four-wheel Enigma, M4, and was therefore almost certainly prepared before 1 February 1942, when the Atlantic U-boats started to use M4 for their *Triton* cipher (codenamed 'Shark' by GC & CS). It was probably sent between mid-October and the end of November. There is no record of any reply by OP-20-G: it is quite possible that none was made.

Turing writes concerning Mrs Driscoll's hand attack on Naval Enigma, and therefore does not mention the bombes. However, breaking Dolphin manually was seldom anything other than an extremely slow process, because of its enormous key-space (6.014×10^{23} possible combinations, assuming that the wheel wiring was known[9]). Turing begins by trying to solve only a single wheel order. But he adopts a more rigorous approach in his penultimate paragraph, by requiring all 336 wheel orders to be tested, as would sometimes have been necessary in 1941 when using Mrs Driscoll's method.

Aftermath

Turing's memorandum should have been a turning point for OP-20-G's management, since it demonstrated that Mrs Driscoll's methods had no chance whatsoever of producing operationally useful intelligence. It also implied that she did not fully appreciate some of the subtleties of Naval Enigma. But it may not have reached OP-20-G before the Japanese attack on Pearl Harbor, and there is no evidence that it caused OP-20-G to change its approach to Naval Enigma, or to Mrs Driscoll's work, since she continued to receive support on Enigma until well into 1942. Lt.-Col. John Tiltman, GC & CS's most senior cryptanalyst, considered that she 'was making no original contribution at all',[10] but she remained part of the attack on Naval Enigma throughout 1942.[11] However, it is not known whether Tiltman passed his views on to OP-20-G's management, who treated her with 'exaggerated respect' on account of her pre-war achievements.[12]

A new generation of young, mathematically trained cryptanalysts, many of whom had been naval reservists, was recruited for OP-20-G after Pearl Harbor. They developed an outline bombe design for OP-20-G in April 1942, although it would have been twenty-six times less efficient than GC & CS's bombe, since it did not employ 'simultaneous scanning' (i.e. the simultaneous testing of twenty-

---

[9] Donald Davies, 'The Bombe: A Remarkable Logic Machine', *Cryptologia*, 23 (1999), 108–138.
[10] 'Report by Lieut-Colonel J. H. Tiltman on his visit to North America during March and April 1942', para. 14, 20 May 1942: PRO HW 14/46.
[11] See e.g. Turing's report, 'Visit to National Cash Register Corporation of Dayton, Ohio' (n.d., but *c.* Dec. 1942): NARA RG 38, CNSG Library, Box 183, 5750/441. We are indebted to Stephen Budiansky for a copy of this document.
[12] 'Report by Lieut-Colonel J. H. Tiltman'.

six possible Stecker values at each position of the bombe's wheels; see 'Enigma', pp. 254–5, and Chapter 6, pp. 319–20), about which OP-20-G seems to have known nothing at that time.[13] OP-20-G therefore appears not to have seen Turing's *Treatise* in early 1942. This is not wholly surprising. Until May 1942, GC & CS did not want OP-20-G to attack Naval Enigma, and may deliberately have withheld a copy of the *Treatise*. A copy was sent to the United States at some point, but probably only to the US Army. OP-20-G only learned the full details of the GC & CS bombe, and about simultaneous scanning, after Lt. Joseph Eachus and Lt. Robert Ely arrived at Bletchley Park in July 1942, to learn how GC & CS attacked Naval Enigma.

M4 Enigma proved devastating when it entered service on Shark in February 1942, since it took effect shortly after the introduction of a new edition of a weather short signal book (the *Wetterkurzschlüssel*), which deprived Hut 8 of cribs for Shark. Without cribs, and lacking four-wheel bombes, GC & CS became blind against Shark. M4 was not a true four-wheel machine. The new wheel, beta (which was stationary during encipherment), and its associated reflector, thin B, gave M4 the equivalent of twenty-six reflectors, but beta was not interchangeable with wheels I to VIII.[14] Fortunately, GC & CS had solved beta and thin B in December 1941, when M4 was used inadvertently.

The new *Wetterkurzschlüssel* was captured from *U-559* on 30 October 1942, and reached GC & CS in late November. Hut 8 found that M4 emulated M3 when enciphering weather short signals, allowing Shark to be solved using relatively few three-wheel bombes. On 13 December, Hut 8 solved Shark keys for 5 to 7 December.[15] It continued to do so for most days using three-wheel bombes, albeit with delays, until British four-wheel bombes entered service in June 1943.

GC & CS's failure to develop a four-wheel bombe quickly had led the US Navy to decide, on 4 September 1942, to embark on a massive four-wheel bombe-building programme of its own. In a considerable tour de force, Joseph Desch, of the National Cash Register Co., in Dayton, Ohio, submitted a detailed design proposal to OP-20-G on 15 September. In October 1942, OP-20-G and GC & CS entered into the Holden Agreement, which provided for 'full collaboration upon the German submarine [Shark] and naval cryptanalysis problems'.[16]

---

[13] Memorandum (no title or author), 25 Apr 1942: NARA RG 38, Inactive Stations, Box 54, 3200/1; R. B. Ely, 'Easy Research to Date', 24 July 1942: NARA RG 38, CNSG Library, Box 117, 5750/205; J. J. Eachus, 'Cold Spot Method: Short Row Test: New Bombe' (n.d., but *c.* July 1942): ibid., Box 113, 5750/177; cf. 'American Hot-Point Method', Aug. 1942: NARA RG 38, Radio Intelligence Publications, RIP 602, 1.

[14] R. Erskine, and F. Weierud, 'Naval Enigma: M4 and its Rotors', *Cryptologia*, 11 (1987), 235–44.

[15] R. Erskine, 'Naval Enigma: The Breaking of Heimisch and Triton', *Intelligence and National Security*, 3(1) (1988), 162–83.

[16] See R. Erskine, 'The Holden Agreement on Naval Sigint: The First BRUSA?', *Intelligence and National Security*, 14(2) (1999), 187–97.

After a difficult testing period for two bombe prototypes from May to late July 1943,[17] OP-20-G four-wheel bombes came into service in August 1943.[18] They performed so well that GC & CS came to rely on them extensively for work against Shark and other *Kriegsmarine* four-wheel ciphers, not least because as late as March 1944 the performance of the British four-wheel bombes was, in the words of Hugh Alexander, Turing's successor as head of Hut 8, 'poor and likely to remain so'.[19] The OP-20-G bombes also carried out a vast amount of work for Hut 6 on German Army and Air Force Enigma ciphers: for much of 1944 OP-20-G devoted around 60 per cent of its bombe time to those ciphers.[20]

Agnes Driscoll gradually faded out of the Naval Enigma scene in 1942, as indeed, for different reasons, did Alan Turing. She is mentioned only once or twice in the extant records for 1942, and did not write any of the extensive series of papers on Naval Enigma which have been preserved in the American archives. She was transferred to the attack on JNA-20 (codenamed 'Coral' by OP-20-G), the cipher machine used by Japanese naval attachés, on 31 January 1943.[21]

Provenance

Turing's typescript and figures were released only in late 1998. Ralph Erskine found the figures at College Park in October 1999.[22] He recognized them as being in Turing's handwriting but, not being attached to Turing's unsigned typescript (which he had received earlier from Stephen Budiansky), they completely lacked context, except that they clearly dealt with Naval Enigma.[23] A typescript by Colin Burke 'Agnes Meyer Driscoll v the Enigma and the Bombe' described the

---

[17] See 'Listing of Incoming/Outgoing Messages OP-20-G-4 from NCML': NARA RG 38, CNSG Library, Box 184, 5830/116.

[18] On the development of the US Navy bombes, see S. Budiansky, *Battle of Wits: The Complete Story of Codebreaking in World War II* (New York: The Free Press; London: Viking, 2000), 238–9, 241–2, 294–5; C. Burke, *Information and Secrecy: Vannevar Bush, Ultra and the Other Memex* (Metuchen, NJ: Scarecrow Press, 1994), *passim*.

[19] Alexander to Church, signal, 24 Mar. 1944: NARA RG 38, Inactive Stations, Box 55, 3200/2.

[20] See OP-20-GM-1's monthly 'Summary of Attack on Naval Enigma Traffic' during 1944: NARA RG 38, CNSG Library, Box 117, 5750/205.

[21] OP-20-GM-6 war diary, 31 Jan. 1943: ibid. Box 113, 5750/176.

[22] NARA, RG 38, CNSG Library, Box 117, 5750/205.

[23] The typescript and figures were found independently in May 2002 by Lee Gladwin, who published them, with a commentary, in 'Alan M. Turing's "Critique of Running Short Cribs on the U.S. Navy Bombe"', *Cryptologia*, 27 (2003), 44–9 and 50–4. However, Gladwin's commentary contains several errors: e.g. Turing's memorandum questioned the use of short cribs in solving Enigma manually, and not running them on the US Navy bombe, which had not even been designed when Turing wrote the memorandum. (The present chapter was completed and accepted for publication in this volume in January 2001.—Ed.)

typescript's author as being 'perhaps Alan Turing'. A careful study of the typescript showed that it was indeed by Turing: two words are in Turing's hand, and the typewriter used is identical with that employed for Turing's *Treatise*. When Erskine remembered the figures and re-examined them, the connection with Turing became patent. The figures tie in perfectly with references to them in the typescript, and the crib (vvvBDUU) is the same in both documents. Turing's authorship of the typescript is therefore beyond question.

Memorandum to OP-20-G on Naval Enigma[1]

We are rather surprised to hear that you are able to find the keys, given that a message when deciphered says V V V B D U U U.[2] Our experience shows that with a 'crib' as short as 8 letters there are always far too many sets of keys consistent with the data, so that whatever method may be used for discovering the keys the time required to test these solutions out further becomes prohibitive. To illustrate this I have enciphered V V V B D U U U with a random chosen key viz wheel order 457, English Ringstellung[3] RWH, pre-start window position[4] SZK and Stecker A/P, B/Y, C/L, E/Q, F/X, K/R, M/W, N/T, O/V, S/Z, giving Y F Z O N M T Y. I then imagined that

Y F Z O N M T Y
V V V B D U U U

was a crib that I had to solve, but that I knew the wheel order and Ringstellung: I tried out the hypothesis that the pre-start window position was the right one (SZK) and also the five which follow it (allowing correctly for turnovers) viz TAL, TAM, TBN,[5] TBO, TBP, and found that with pre-start TBP there is a solution with V/J, F/G, Z/H, Y/E, U/X, M/L, T/K, D/P and either B/S and O/W or B/W and O/S. The 'unsteckered alphabets' for the relevant positions of the machine are shown in Fig [8.1], and the working in Fig [8.2]. I hope that this working is self-explanatory. Each column of letters consists of steckers of the letters VFZYUMT which imply one another on account of the crib.

A continuation of this process would probably give about 3000 essentially different solutions per wheel order. Of course these solutions are not all equally likely: e.g. the solution with pre-start TBP is unlikely as the complete set of 10 Stecker has to be assumed to account for the whole crib, whereas with the right solution we can only deduce O/V, F/X, S/Z, Y/B, W/M, N/T, and D and U self-steckered. But there will still be a great many that look as good as this. A fairly simple calculation tells us the probability of this solution being the right one, under the assumption that the wheel order and Ringstellung are

---

[1] This title has been added by the editors, as have all footnotes.

[2] Editors' note. This crib was short for 'Von Befehlshaber der U-Boote' (From BdU—Admiral Commanding U-boats). 'BDUUU' was an abbreviation used in signals to and from the U-boats.

[3] Editors' note. Following recoveries of wheel wiring by the Poles before the war, the *Ringstellung* on the wheels used by GC & CS was displaced by one for wheels I to III and VI to VIII, by two for wheel IV, and by three for wheel V (all with the neutral position at Z). Thus the English *Ringstellung* RWH given in Turing's example corresponds to German *Ringstellung* TZI. The different ring settings were of no consequence when solving Enigma or using the bombes.

[4] Editors' note. The pre-start window position was the message setting—the starting position of the wheels for a specific message.

[5] Editors' note. An Enigma simulator confirms that the middle wheel does indeed move after only three keystrokes, when set as indicated by Turing. This turnover would have complicated a solution here.

348 | *Alan Turing*

	A B C D	E F G H I	J K L M N O	P Q R S T	U V W X Y Z
TAL	P O U M W	N L K T S	H G D F B	A R Q J I	C Y E Z V X V
TAM	U G L P T	K B N W S	F C R H X	D V M J E	A Q I O Z Y V
TBN	H G M F K	D B A X V	E Q C R S	Y L N O Z	W J U I P T V
TBO	E S Z W A	N J M K G	I R H F Q	X O L B U	T Y D P V C B
TBP	B A O T N	L H G K W	I F Q E C	U M S R D	P Y J Z V X D
TBQ	X S O Y J	R N T L E	M I K G C	Z V F B H	W Q U A D P U
TBR	W R O Z S	L J V M G	Q F I U C	Y K B E X	N H A T P D U
TBS	R U T P Y	S K J N H	G Z O I M	D W A F C	B X Q V E L U
	T U M N J	Q O V P K	D I T B C F	H E X W L	A G S R Z Y
	U C Z A W S	X L R J I	O G Q P K	N M H E U	T Y D F V B
	V G D K B V	T A U J I	C X W R P	O Z N Y F	H E M L S Q
	W N F I S Y	B U L C V	X H O A M	Z R Q D W	G J T K E P
	X Z D F B X	C S Y L N	U I Q J V	T M W G P	K O R E H A
	Y U J O E D	G F M R B	N Y H K C	X S I Q Z	A W V P L T
	Z W Z Y T S	O R J V H	N X U K F	Q P G E D	M I A L C B

Figure 8.1. Alphabets for V V V B D U U U
                            Y F Z O N M T Y

Figure 8.2. Workings to solve

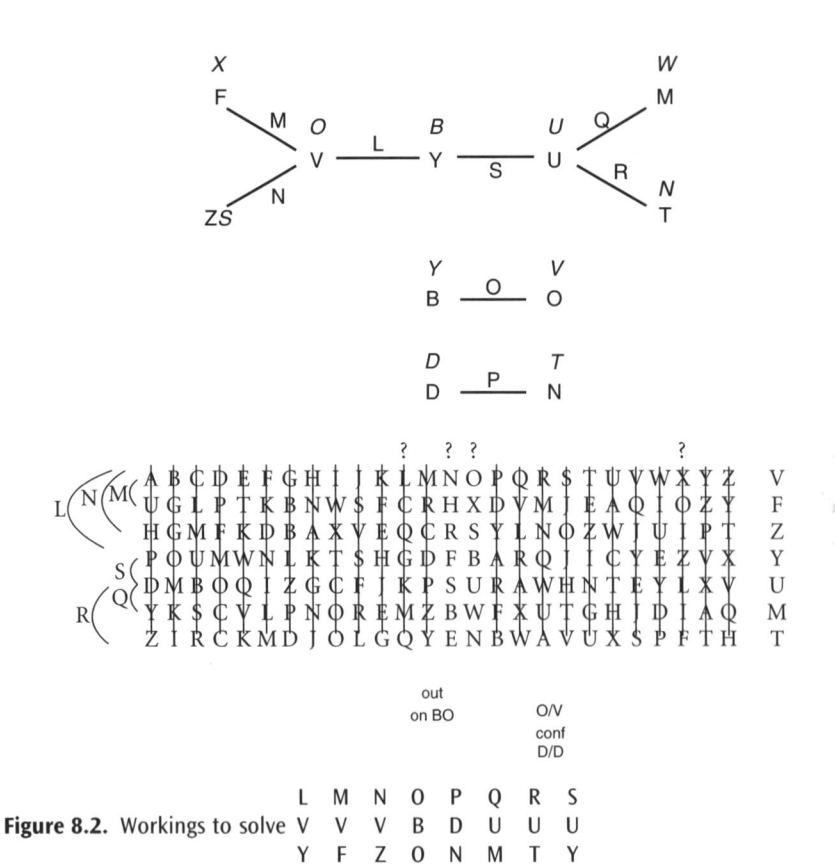

    L M N O P Q R S
    V V V B D U U U
    Y F Z O N M T Y

Editors' note: Turing continues with similar arrays to the one set out above, for one to five places after the right one. In the interests of space, those arrays are not reproduced here, but may be downloaded from www.AlanTuring.net/OP-20-G_figures.

right. The total number of ways of setting up 10 Stecker is about 1.5×10^{14}, and the number of essentially different window positions is 16,224,[6] so that the total number of sets of keys in question is about 2.4×10^{18}. From this we can obtain the expected number of sets of keys consistent with the data by multiplying by 26^{-8}. We get 1.15×10^7. Now the solution in question can be made into a complete set of keys, by completing the Stecker in 51,975 ways, i.e. it corresponds to 51,975 of the 1.15×10^7 solutions and therefore has a probability of $51,975/1.15 \times 10^7$ or 0.0045. We may therefore expect to have anything from say 50 to 1,000 solutions to test further on each wheel order, even if we assume the Ringstellung, or, what comes to the same, the position of the turnover in the message. The examination of these solutions is not very easy, especially in the case of likely looking solutions, as in such cases we necessarily know comparatively few Stecker, and so can get very little of the plain text of the message.

The working shown in Fig [8.2] is not given as a suggested method for solving these cribs. It is part of an *a fortiori* argument to the effect that even if all solutions had been found by this method or some other the remaining work to be done would still be too much.

Leaving aside this general aspect of the problem I should be interested to be sure that I understand your method correctly: the argument given above depends essentially on the length of the crib, and it may well be that you have a method which will deal with rather longer cribs.

As I understand it your method is to assume Stecker for certain letters thereby obtaining certain 'unsteckered' constatations.[7] One then takes 26 separate hypotheses concerning the position of the R.H.W. [right-hand wheel] and deduces, for each hypothesis, the 'output' of the two left hand wheels and U.K.W.[8] Assuming the wheel order one then looks up in a catalogue and finds the possible positions of the two wheels on the left. The whole effect of the process so far is to find the positions of the wheels consistent with the unsteckered constatations. Each position must be examined more closely afterwards with a machine.

The process may be explained by means of an example.

S D D Q T Y M D
V V V B D U U U

This is a favourable one as the same constatation VD occurs twice over. Suppose now that we wish to try out the hypothesis that V and D are both self-steckered

---

[6] Editors' note. Turing's number is incorrect. $16,224 = 17,576 - 1352$: the reduction ($1,352 = 2 \times 26 \times 26$) attempts to allow for the stepping pattern of wheel order 457. However, Philip Marks and Frode Weierud have independently calculated that 457 gives $16,952$ ($= 17,576 - 24 \times 26$) essentially different starting positions.

[7] Editors' note. 'Constatation: The association of a cipher letter and its assumed plain equivalent for a particular position' ('Cryptographic Dictionary', 20; see the notes to the introduction to Chapter 5 for details of the Dictionary).

[8] Editors' note. U.K.W. = *Umkehrwalze* (reflector).

350 | *Alan Turing*

```
 1  2  3  4  5  6  7  8  9 10 11 12 13 14 15 16 17 18 19 20 21 22 23 24 25 26
 E  J  O  A  P  I  T  C  Z  X  H  R  L  Z  K  D  G  K  W  B  L  F  M  Y  N  Q  V
 H  F  P  Z  T  H  S  L  O  S  E  J  T  N  U  G  V  Y  M  R  W  I  X  Q  B  K  D
```

Figure 8.3. Inverse rods of wheel VII, for solving $\begin{smallmatrix} V & V & V & B & D & U & U & U \\ S & D & D & Q & T & Y & M & D \end{smallmatrix}$

Editors' note: Figure 8.3 shows the V and D inverse rods for wheel 7. These show that V and D map to G and V, respectively, when taken through wheel 7 in position 17. Since V and D are paired in the crib, this means that the G and V contacts are connected together through the unsteckered 'two-wheel Enigma' formed by the other two wheels (4 and 5) and the reflector. The catalogue (which is not reproduced here) shows the results produced by such a two-wheel Engima for all possible wheel orders and wheel positions.

on wheel order 457. Assume that there is no turnover between the two occurrences of VD. We lay down the 'Inverse rods'[9] for V and D and wheel 7; the effect of this is shewn in Fig [8.3]. The information we get from them is for instance that if VD were enciphered with no Stecker and with the R.H.W. in absolute position (window position less English Ringstellung) 17, then, if the R.H.W. were replaced by a 'straight through' wheel,[10] and the other wheels kept the same and in the same positions, then the effect of enciphering G would be V. We wish therefore to find where, with wheel order '4 5 straight', we can get the pairings EH and JF, also where we can get JF and PO, and so on. We have catalogues in which we can look these pairs up. We find for instance that PO, AZ occurs in position 8 of the L.H.W. and 22 of the M.W. [middle wheel] and therefore that without Stecker we get VD in absolute positions 8, 22, 3 and 8, 22, 4. The complete set of solutions is shewn in Fig [8.4]. These solutions have now to be tested out on the remainder of the crib. Take the case of the solution 8, 22, 3 and suppose we are assuming there is no turnover in the whole crib. Then the DU will have been enciphered at 8, 22, 9 at which position D enciphers without Stecker to Z. Since we are assuming that D is self-steckered, we must have Z/U. Now the UY constatation was enciphered at 8, 22, 6 where Z without Stecker enciphers to V. We therefore have V/Y contrary to the hypothesis that V was self-steckered.

The full examination of the possibilities of turnover takes some considerable time. Of course it is only worth while considering rather longer cribs than VVVBDUUU: with cribs of length 20[11] it would be possible to deal with a wheel order on one assumption of Stecker for the letters taking the place of V and D in about five hours, of which about half an hour or less would be the

---

[9] Editors' note. An inverse rod was a cardboard strip giving the letters on the left-hand side of the wheel that were consecutively connected to a fixed point at the right-hand side for all 26 positions of a full wheel revolution. See further Chapter 6, n. 7.

[10] Editors' note. A 'straight through' wheel was a notional wheel whose wiring was an identity (A wired to A, B to B and so on). It could therefore be temporarily disregarded in any solution.

[11] Editors' note. Although Turing also suggested that a message could be solved with a 70-letter crib, or two depths of 26 ('Mathematical Theory of ENIGMA Machine by A M Turing', 60; cf. p. 96: PRO HW 25/3), in practice GC & CS seldom, if ever, tried to solve messages by hand with cribs that were shorter than 250 letters.

Memorandum to OP-20-G on Naval Enigma | 351

		~~21~~	~~19~~	~~1~~
PO, AZ		8	22	3
AZ, PT		21	23	4
PT, IH	$\begin{cases} \\ \end{cases}$	12	5	5
		21	26	5
		19	22	7
		9	13	17
		13	16	17
		15	19	17
		14	8	19
		7	21	19
		8	4	21
		24	6	21
		8	14	21
		14	16	21
		24	6	22
		14	25	23
		13	21	24
		5	23	25
		17	25	25
		25	1	26
		14	19	26

Figure 8.4. Position where we get $\frac{V}{D} \frac{V}{D}$ on wheel order 4 5 7.

Editors' note: Since the two V/D pairings are adjacent in the crib, solutions can occur only at positions where the two-wheel Enigma produces two adjacent pairings on the inverse rods. Figure 8.4 shows the complete set of such positions. In Turing's example (wheel positions 8, 22, 3), with V and D self-steckered, the successive pairings on the rods are P/O and A/Z, and the catalogue tells us that the two-wheel Enigma with wheel order 4 5 produces both of these pairings when wheel 4 is at position 8 and wheel 5 is at position 22. The inverse rods show pairing PO at position 3, thus the first of the V/D pairings in the crib must occur when wheel 7 is at position 3. The solution would be much more laborious if the very favourable V/D constatation had not been repeated, since more Stecker assumptions would have to be made in order to constrain the number of possible solutions.

looking up in catalogues. Suppose that we have a very large supply of cribs, 100 a day say, each with probability [½][12] of being right. The chance of the two letters being self-steckered is 3/65, and therefore working on 336 wheel orders[13] we should have on average[14] 22⅓ × 336 × 5 × 2 i.e. 72,800 hours work[15] to obtain a solution.

---

[12] Editors' note. This fraction is illegible. Turing's subsequent calculation, and the fact that it is typed with a single keystroke, show that it must be ½.

[13] Editors' note. (8 × 7 × 6): Dolphin used a three-wheel machine, M3, in 1941.

[14] Editors' note. 22⅓ is a typo for 21⅔ (65/3—see n. 15).

[15] Editors' note. 8.3 years! Dolphin keys had 10 Stecker pairs, leaving 6 letters unsteckered. The probability of two letters randomly selected being unsteckered is 6/26 × 5/25 = 3/65. Taking the crib probability correctness factor as 1/2, on average it would have been necessary to test 2 × 65/3 cribs to obtain a 'hit', giving Turing's 21⅔ [65/3] × 336 [wheel orders] × 2 × 5 [hours per test].

Would you mind telling us your method

1) Does it give the keys starting from scratch, or does one need to start with the Stecker?
2) Is the above account substantially correct?
3) Do you work with cribs as short as V V V B D U U U? Have you any longer ones?
4) About how many hours work do you estimate would be necessary to obtain a solution on 336 wheel orders?

Artificial Intelligence

Jack Copeland

Turing was the first to carry out substantial research in the field now known as Artificial Intelligence or AI. The term 'Artificial Intelligence' itself did not come into use until after Turing's death, making a prominent appearance in 1956, in the title of a conference held at Dartmouth College, New Hampshire: *The Dartmouth Summer Research Project on Artificial Intelligence*. Turing's original term, 'machine intelligence', remains in use, especially in Britain.

1. AI at Bletchley Park

Turing was thinking about machine intelligence at least as early as 1941. During the war he circulated a typewritten paper on machine intelligence among some of his colleagues at the Government Code and Cypher School (GC & CS).[1] Now lost, this was undoubtedly the earliest paper in the field of AI. It probably concerned the mechanization of problem-solving and the idea of machines learning from experience; both were topics that Turing discussed extensively during the war years at GC & CS.

Turing enthusiastically discussed the mechanization of chess with Donald Michie and others at Bletchley Park (see the introduction to Chapter 16).[2] Michie (a leading codebreaker at GC & CS) recalls Turing's speaking often during the war about the possibility of computing machines solving problems by means of searching through the space of possible solutions, guided by rules of thumb.[3] (Convinced by Turing that AI was worth pursuing, Michie himself went on to found the influential Department of Machine Intelligence and Perception at the University of Edinburgh.)

Turing's thinking on AI was probably influenced by his work on the bombe (see 'Enigma' and Chapter 6). Central to the bombe was the idea of solving a problem by means of a guided mechanical search through the space of possible solutions. The space of possible solutions searched by the bombe consisted of configurations of the Enigma machine, but the space searched by a different form of machine might consist of configurations of a chess board, for example. The

---

[1] Donald Michie in interview with Copeland (Oct. 1995, Feb. 1998).
[2] Ibid.
[3] Ibid.

bombe's search could be guided in various ways; one involved the 'multiple encipherment condition' associated with a crib (see Chapter 6, p. 317). The search would 'reduce the possible positions to a number which can be tested by hand methods' (ibid.).

Modern AI researchers speak of the method of 'generate-and-test'. Potential solutions to a given problem are generated by means of a guided search. These potential solutions are then tested by an auxiliary method in order to find out if any actually is a solution. The bombe mechanized the first process. The testing of the stops, or potential solutions, was then carried out manually (by setting up a replica Enigma accordingly, typing in the cipher text, and seeing whether or not German came out). Nowadays in AI both processes, generate and test, are typically carried out by the same programme.

In 1948 Turing boldly hypothesized that 'intellectual activity consists mainly of various kinds of search' (Chapter 10, p. 431). His readers would no doubt have been astonished to learn of his wartime experience with mechanized search (still secret at that time). Some eight years later the same hypothesis was put forward independently by Herbert Simon and Allen Newell in the USA; through their influential work, it became one of the central tenets of AI.[4]

Turing's work on the bombe involved him in the design of the *Ringstellung* 'cut-out' (Chapter 6, p. 335). The *Ringstellung* cut-out is an early example of constraining search by means of what modern AI researchers call *heuristics*. A heuristic is a rule that cuts down the amount of searching required in order to find potential solutions. Unlike a decision method (see 'Computable Numbers: A Guide'), a heuristic is not necessarily *guaranteed* to produce the correct solution, but works often enough to be useful. For example, one heuristic that everybody uses from time to time is: if you have lost something, search in the vicinity of the place where you think you dropped it. Heuristic search is one of modern AI's central techniques. The Poles in effect used heuristic search in the bomba: the number of possibilities to be examined was reduced by the assumption, true for more messages than not, that the letters of the indicator were unsteckered (see 'Enigma').

The *Ringstellung* cut-out would be set so as to prevent the bombe from stopping at certain positions at which it would otherwise have stopped—positions ruled out by conjectured information about the ring settings. The device thereby reduced the number of stops to be tested by hand. Early in 1940 John Herivel had discovered the heuristic concerning ring settings that helped Hut 6 break 'Red' Enigma daily. When a German operator sent his first message of the day, he would be liable to use for the indicator setting either the three letters visible in the windows when he had finished

---

[4] See, for example, A. Newell, J. C. Shaw, and H. A. Simon, 'Empirical Explorations with the Logic Theory Machine: A Case Study in Heuristics', *Proceedings of the Western Joint Computer Conference*, 15 (1957), 218–39 (reprinted in E. A. Feigenbaum and J. Feldman (eds.), *Computers and Thought* (New York: McGraw-Hill, 1963); A. Newell and H. A. Simon, 'Computer Science as Empirical Inquiry: Symbols and Search', *Communications of the Association for Computing Machinery*, 19 (1976), 113–26.

setting the rings, or the three letters visible at a closely neighbouring position of the wheels, for example the position produced by his having lazily turned a single wheel some small number of clicks (see Chapter 6, n. 26). By assuming that this was so in the case of any given first message, the codebreakers could (Herivel said) 'narrow down the 17,576 possible ring settings to a manageable number, say twenty or thirty'.[5] Of course, the assumption might be wrong in any particular case, but was correct often enough to be useful.

In *Treatise on the Enigma* Turing also described a 'majority vote gadget', mechanizing the process of evaluating certain hypotheses on the basis of unreliable data, e.g. unreliable data about *Stecker* (see Chapter 6, p. 335). Although the task mechanized by the gadget is a relatively undemanding one, it is nevertheless of a type that might be described as requiring or evidencing intelligence when carried out by human beings. The fact that such a task can be carried out by a machine is suggestive. Leaving aside tasks that are provably uncomputable (concerning which see Turing's discussion, in Chapters 9, 10, 11, and 12, of what he called the 'Mathematical Objection'), might it be the case that all the tasks we normally describe as demanding or showing intelligence can be reduced to rules that a computing machine can follow? In 1940 nothing was known to falsify the daring hypothesis that this is so, and the same remains true today.

2. AI in Post-war Britain

The birth of Artificial Intelligence as a field of research is usually placed at 1956, the year of the Dartmouth Summer Research Project on Artificial Intelligence and also the year in which a programme written by Newell, Simon, and Shaw—later named 'the Logic Theorist'—proved theorems from Whitehead and Russell's famous work on mathematical logic *Principia Mathematica*.[6] However, this received view of the matter is hardly accurate. By 1956 computer intelligence had been actively pursued for more than a decade in Britain; the earliest AI programmes to run were written there in 1951–2. That the earliest work in the field was done in Britain is a consequence of the fact that the first electronic stored-programme digital computers to function were built at Manchester and Cambridge universities (see the introduction to Chapter 9), and another important factor was the influence of Turing on the first generation of computer programmers.

In London in 1947 Turing gave what was, so far as is known, the earliest public lecture to mention computer intelligence, providing a breathtaking glimpse of a new field (Chapter 9). In 1948 he wrote his National Physical Laboratory report 'Intelligent Machinery' (Chapter 10). This, the first manifesto of Artificial

---

[5] M. Smith, *Station X: The Codebreakers of Bletchley Park* (London: Channel 4 Books, 1998), 43.
[6] Newell, Shaw, and Simon, 'Empirical Explorations with the Logic Theory Machine'; A. N. Whitehead and B. Russell, *Principia Mathematica*, vols. i–iii (Cambridge: Cambridge University Press, 1910–13).

Intelligence, adumbrated the methods of the new field and included Turing's proposals for connectionist-style neural simulation. In 1950 Turing published 'Computing Machinery and Intelligence' (Chapter 11), probably the best known of all his papers. In it Turing addressed mainly philosophical and logical issues, introducing his now famous imitation game or 'Turing test'. There then followed his three radio broadcasts on AI: the lectures 'Intelligent Machinery, A Heretical Theory' (Chapter 12) and 'Can Digital Computers Think?' (Chapter 13), and the panel discussion 'Can Automatic Calculating Machines Be Said to Think?' (Chapter 14). In 1953 Turing's last work on AI was published, a groundbreaking essay on computer chess (Chapter 16).

3. The First AI Programmes

Both during and after the war Turing experimented with machine routines for playing chess (see Chapter 16). In the absence of an electronic computer, the machine's behaviour was simulated by hand, using paper and pencil. The first chess programme to run electronically was written for the Manchester University computer by Dietrich Prinz in 1951 (see Chapter 16).

When Turing delivered his lecture 'Can Digital Computers Think?' on British radio one of his listeners was Christopher Strachey. Strachey's draughts (or checkers) programme first ran successfully in Turing's Computing Machine Laboratory at Manchester University. The programme used simple heuristics and looked ahead three to four turns of play. The state of the board was represented on the face of a cathode ray tube—one of the earliest uses of computer graphics. (Strachey was at this time a schoolmaster at Harrow; he later became Director of the Programming Research Group at Oxford University, where with the mathematical logician Dana Scott he did the work on the semantics of programming languages for which he is best known.[7])

Strachey initially coded his draughts programme in May 1951 for the Pilot Model of Turing's Automatic Computing Engine at the National Physical Laboratory (see Chapter 9).[8] This version of the programme never ran successfully.[9] An attempt to run it in July 1951 foundered due to programming errors. When Strachey was ready to try the programme again, he discovered that the computer's hardware had been modified; his programme could not be run without extensive changes. Strachey switched his attention to the Manchester University Computing Machine Laboratory, where the first Ferranti Mark I computer was installed in February 1952.

---

[7] D. S. Scott and C. S. Strachey, 'Towards a Mathematical Semantics for Computer Languages', *Proceedings of a Symposium on Computers and Automata*, Polytechnic Institute of Brooklyn, and Technical Monograph 6, Programming Research Group, Oxford University (1971).

[8] Letter from Strachey to Michael Woodger, 13 May 1951 (in the Woodger Papers, National Museum of Science and Industry, Kensington, London).

[9] Letters from Woodger to Copeland (15 July 1999, 15 Sept. 1999).

Figure 1. Turing standing at the console of the Manchester Ferranti computer.
Source: Reprinted with permission of the Department of Computer Science, University of Manchester.

With Turing's encouragement, and using Turing's recently completed *Programmers' Handbook* for the Ferranti machine, Strachey finally got his programme working.[10] By the summer of 1952 the programme could play a complete game of draughts at a reasonable speed.[11] (Built by the Manchester firm of Ferranti in close collaboration with the Computing Machine Laboratory, the Mark I was the world's first commercially available electronic stored-programme computer.)

In 1952 Strachey described his draughts programme at a computing conference in North America. Arthur Samuel of IBM took over the essentials of Strachey's programme and wrote a checkers player for the IBM 701 (IBM's first mass-produced electronic stored-programme computer). Samuel's checkers programme first ran at the end of 1952[12] and appears to have been the earliest AI programme in the USA. In 1955 Samuel added learning to the programme and over a period of years made successive improvements to the learning

---

[10] M. Campbell-Kelly, 'Christopher Strachey, 1916–1975: A Biographical Note', *Annals of the History of Computing*, 7 (1985), 19–42(24); A. M. Turing, 'Programmers' Handbook for Manchester Electronic Computer', Computing Machine Laboratory, University of Manchester (n.d., c. 1950); a digital facsimile is available in The Turing Archive for the History of Computing <www.AlanTuring.net/programmers_handbook>.

[11] C. S. Strachey, 'Logical or Non-Mathematical Programmes', *Proceedings of the Association for Computing Machinery*, Toronto (Sept. 1952), 46–9 (47).

[12] Letter from Samuel to Copeland (6 Dec. 1988).

apparatus.[13] In 1962 his programme won a game against a former Connecticut checkers champion, R. W. Nealey. Nealey, who immediately turned the tables and beat the programme in six consecutive games, is reported to have said: 'In the matter of the end game, I have not had such competition from any human being since 1954, when I lost my last game.'[14]

Strachey was thinking about mechanized learning at the time of writing his draughts player. He devised a simple rote-learning scheme which he envisaged being implemented in a NIM-playing programme.[15] Strachey wrote at length concerning learning in a letter to Turing, composed on the evening of Turing's lecture 'Can Digital Computers Think?'[16] He said:

I have just been listening to your talk on the Third Programme. Most stimulating... [i]n particular your remark... that the programme for making a machine think would probably have great similarities with the process of teaching; this seems to me absolutely fundamental. ... I am convinced that the crux of the problem of learning is recognizing relationships and being able to use them. ... There are, I think, three main stages in learning from a teacher. The first is the exhibition of a few special cases of the rule to be learned. The second is the process of generalisation—i.e. the underlining of the important features that these cases have in common. The third is that of verifying the rule in further special cases and asking questions about it. I have omitted any mention of 'understanding' the rule, because this is not appropriate at the moment to the action of a machine. I think, as a matter of fact, that the process of understanding a rule is connected with finding relationships between it and other rules—i.e. second (or higher) order relations between relations and this might well become important for a machine later. ... I think it might well be possible to programme the Manchester machine to do all of these stages, though how much it would be able to learn in this way before the storage became inadequate remains to be seen.

Strachey missed the opportunity to be the first to achieve a functioning programme incorporating learning, however. The earliest programmes to do so were written by Anthony Oettinger for the EDSAC computer at the University of Cambridge Mathematical Laboratory. Oettinger was considerably influenced by Turing's 'Computing Machinery and Intelligence' (Chapter 11).[17]

Oettinger's 'response-learning programme', dating from 1951,[18] could be taught to respond appropriately to given stimuli by means of expressions of 'approval' or 'disapproval' by the teacher.[19] As training proceeded errors became

---

[13] A. L. Samuel, 'Some Studies in Machine Learning Using the Game of Checkers', *IBM Journal of Research and Development*, 3 (1959), 211–29; reprinted in E. A. Feigenbaum and J. Feldman (eds.), *Computers and Thought* (New York: McGraw-Hill, 1963).

[14] Ibid. (Feigenbaum and Feldman (eds.), *Computers and Thought*, 104).

[15] Letter from Strachey to Turing, 15 May 1951 (in the Turing Papers, Modern Archive Centre, King's College, Cambridge (catalogue reference D 5)).

[16] Ibid. (This extract is published by permission of Henry Strachey and the Strachey family.)

[17] Oettinger in interview with Copeland (Jan. 2000).

[18] Letter from Oettinger to Copeland (19 June 2000).

[19] A. G. Oettinger, 'Programming a Digital Computer to Learn', *Philosophical Magazine*, 43 (1952), 1243–63.

less frequent, and the learned response would be initiated by a progressively weaker stimulus. Oettinger described the response-learning programme as 'operating at a level roughly corresponding to that of conditioned reflexes', and he noted that the 'behaviour pattern of the response-learning... machine is sufficiently complex to provide a difficult task for an observer required to discover the mechanism by which the behaviour of the... machine is determined.'[20]

Oettinger's 'shopping machine', also dating from 1951, incorporated rote-learning.[21] Adopting Turing's terminology, Oettinger described this programme as a 'child machine' (see Chapter 11, p. 460). Shopper's simulated world was a mall of eight shops.[22] When sent out to purchase an item whose location was unknown Shopper would search for it, visiting shops at random until the item was found. While searching, Shopper would memorize a few of the items stocked in each shop that it visited. Next time Shopper was sent out for the same item, or for some other item that it had already located, it would go to the right shop straight away. (Oettinger was the first of many programmers to claim a programme capable of passing a restricted form of the Turing test. The shopping machine could, he remarked, successfully play a version of Turing's imitation game in which the 'questions are restricted to... the form "In what shop may article j be found?"'[23])

4. Subsequent Developments

Within a decade of this early work, Artificial Intelligence had become an established and burgeoning area of research. Some landmarks in the development of the field were:

- During the latter part of the 1950s and the early 1960s, Artificial Intelligence laboratories were set up at a number of US and British universities, notably at Carnegie Mellon University (under Newell and Simon), Edinburgh University (under Michie), Massachusetts Institute of Technology (under Marvin Minsky), and Stanford University (under John McCarthy, the organizer of the Dartmouth Summer Research Project on Artificial Intelligence).
- The 1950s saw the development of a number of programmes able to carry out tasks of a sort usually said to require intelligence when carried out by human beings. The most famous of these early programmes was the General Problem Solver or GPS (written by Newell, Simon, and Shaw).[24]

---

[20] Ibid. 1251, 1257.
[21] Ibid. 1247–51.
[22] 'Shopper' is my term; Oettinger uses 'shopping programme' and 'shopping machine'.
[23] Oettinger, 'Programming a Digital Computer to Learn', 1250.
[24] A. Newell and H. Simon, 'GPS, a Program that Simulates Human Thought', in Feigenbaum and Feldman (eds.), *Computers and Thought*; G. W. Ernst and A. Newell, *GPS: A Case Study in Generality and Problem Solving* (New York: Academic Press, 1969).

GPS could solve a variety of puzzles. One example is the 'missionaries and cannibals' problem: how can a party of three missionaries and three cannibals cross a river by means of a boat holding at most two people, without the missionaries on either bank ever becoming outnumbered by cannibals?

- Work on neuron-like computation got under way in Britain and the USA during the 1950s. (Those involved included J. T. Allanson, R. L. Beurle, W. A. Clark, B. G. Farley, F. Rosenblatt, W. Ross Ashby, W. K. Taylor, and A. M. Uttley).[25] Ross Ashby's influential book *Design for a Brain* was published in 1952 and Rosenblatt's *Principles of Neurodynamics* in 1962.[26] Rosenblatt was widely influential and numerous research groups in the USA pursued his approach. He called this approach 'connectionist', emphasizing—as Turing had in 1948 (Chapter 10)—the role in learning of the creation and modification of connections between (real or simulated) neurons.

- At the close of the 1950s John McCarthy developed the computer language that he called LISP (from 'list processor').[27] Designed specifically for AI programming, LISP remains today one of the principal languages for AI work. McCarthy took some of the ideas used in LISP from Church's lambda calculus (see the introduction to Chapter 4).

- In 1965 AI researcher Edward Feigenbaum and geneticist Joshua Lederberg (both of Stanford University) began work on their programme Heuristic Dendral (subsequently shortened to DENDRAL).[28] The programme's task was chemical analysis. The substance to be analysed might be a complicated compound of carbon, hydrogen, and nitrogen, for example. Starting from spectrographic data obtained from the substance, DENDRAL would hypothesize the substance's molecular structure. DENDRAL's performance rivalled that of human chemists expert at this task, and the programme was used in industry and in universities. This high-performance programme was the model for much of the ensuing work in the important area of *expert systems* (see the introduction to Chapter 10).

---

[25] For a synopsis see B. J. Copeland and D. Proudfoot, 'On Alan Turing's Anticipation of Connectionism', *Synthese*, 108 (1996), 361–77; reprinted in R. Chrisley (ed.), *Artificial Intelligence: Critical Concepts in Cognitive Science*, ii: *Symbolic AI* (London: Routledge, 2000).

[26] W. R. Ashby, *Design for a Brain* (London: Chapman and Hall, 1952); F. Rosenblatt, *Principles of Neurodynamics* (Washington, DC: Spartan, 1962).

[27] J. McCarthy, 'Recursive Functions of Symbolic Expressions and their Computation by Machine, Part I', *Communications of the Association for Computing Machinery*, 3 (1960), 184–95.

[28] E. A. Feigenbaum, B. G. Buchanan, and J. Lederberg, 'On Generality and Problem Solving: A Case Study Using the DENDRAL Program', in B. Meltzer and D. Michie (eds.), *Machine Intelligence 6* (Edinburgh: Edinburgh University Press, 1971).

Further reading

Boden, M. A., *Mind as Machine: A History of Cognitive Science* (Oxford: Oxford University Press, 2005).

Copeland, B. J., *Artificial Intelligence: A Philosophical Introduction* (Oxford: Blackwell, 1993).

Copeland, B. J., 'Artificial Intelligence', *Encyclopaedia Britannica* (15th ed. 2001).

Haugeland, J., *Artificial Intelligence: The Very Idea* (Cambridge, Mass.: MIT Press, 1985)

McCorduck, P., *Machines Who Think: A Personal Enquiry into the History and Prospects of Artificial Intelligence* (New York: W. H. Freeman, 1979).

Michie, D., *On Machine Intelligence* (2nd ed. Chichester: Ellis Horwood, 1986).

CHAPTER 9

Lecture on the Automatic Computing Engine (1947)

Alan Turing

Introduction

Jack Copeland

Electronics at Bletchley Park

On 8 December 1943 the world's first large-scale special-purpose electronic digital computer—'Colossus', as it became known—went into operation at the Government Code and Cypher School (see 'Computable Numbers: A Guide', 'Enigma', and the introduction to Chapter 4). Colossus was built by Thomas H. Flowers and his team of engineers at the Post Office Research Station in Dollis Hill, London. Until relatively recently, few had any idea that electronic digital computation was used successfully during the Second World War, since those who built and worked with Colossus were prohibited by the Official Secrets Act from sharing their knowledge.

Colossus contained approximately the same number of electronic valves (vacuum tubes) as von Neumann's IAS computer, built at the Princeton Institute of Advanced Study and dedicated in 1952. The IAS computer was forerunner of the IBM 701, the company's first mass-produced stored-programme electronic computer (1953).[1] The first Colossus had 1,600 electronic valves and Colossus II, installed in mid-1944, 2,400, while the IAS computer had 2,600.[2]

Colossus lacked two important features of modern computers. First, it had no internally stored programmes (see 'Computable Numbers: A Guide'). To set up Colossus for a new task, the operators had to alter the machine's physical wiring, using plugs and switches. Second, Colossus was not a general-purpose machine, being designed for a specific cryptanalytic task (involving only logical operations

---

[1] C. C. Hurd, 'Computer Development at IBM', in N. Metropolis, J. Howlett, and G. C. Rota (eds.), *A History of Computing in the Twentieth Century* (New York: Academic Press, 1980).

[2] J. Bigelow, 'Computer Development at the Institute for Advanced Study', ibid.

and counting). Nevertheless, Flowers had established decisively and for the first time that large-scale electronic computing machinery was practicable.

The implication of Flowers's racks of electronic equipment would have been obvious to Turing. Once Turing had seen Colossus it was, Flowers said, just a matter of Turing's waiting to see what opportunity might arise to put the idea of his universal computing machine into practice.[3]

Turing Joins the National Physical Laboratory

Precisely such an opportunity fell into Turing's lap in 1945, when John Womersley invited him to join the Mathematics Division of the National Physical Laboratory (NPL) at Teddington in London, in order to design and develop an electronic stored-programme digital computer—a concrete form of the universal Turing machine of 1936. Womersley named the proposed computer the Automatic Computing Engine, or ACE, in homage to Babbage and his planned calculating machines, the Difference Engine and the Analytical Engine (see 'Computable Numbers: A Guide').

The formal date of Turing's appointment was 1 October 1945.[4] Womersley reviewed the events leading up to the appointment in an NPL document entitled 'A.C.E. Project – Origin and Early History' (26 November 1946).[5]

1936–37 Publication of paper by A.M. Turing 'On Computable Numbers, with an Application to the Entscheidungsproblem'. ...

1937–38 Paper seen by J.R.W. [J. R. Womersley] and read. J.R.W. met C. L. Norfolk, a telephone engineer who had specialised in totalisator design and discussed with him the planning of a 'Turing machine' using automatic telephone equipment. Rough schematics prepared, and possibility of submitting a proposal to N.P.L. discussed. It was decided that machine would be too slow to be effective.

June 1938 J.R.W. purchased a uniselector and some relays on Petty Cash at R.D. Woolwich for spare-time experiments. Experiments abandoned owing to pressure of work on ballistics. ...

Late 1943 J.R.W. first heard of [the] American machines. [Editor's note: Aiken's Sequence-Controlled Calculator at Harvard University and Stibitz's Relay Computer at Bell Telephone Laboratories—these machines were neither electronic nor stored-programme.]

1944 Interdepartmental Committee on a Central Mathematical Station. D. R. Hartree mentioned at one meeting the possible use of automatic telephone equipment in the

---

[3] Flowers in interview with Copeland (July 1996).

[4] Minutes of the Executive Committee of the National Physical Laboratory for 23 Oct. 1945 (National Physical Laboratory library; a digital facsimile is in The Turing Archive for the History of Computing <www.AlanTuring.net/npl_minutes_oct1945>).

[5] J. R. Womersley, 'A.C.E. Project – Origin and Early History', National Physical Laboratory, 26 Nov. 1946 (Public Record Office, Kew, Richmond, Surrey (document reference DSIR 10/385); a digital facsimile is in The Turing Archive for the History of Computing <www.AlanTuring.net/ace_early_history>).

design of large calculating machines. J.R.W. submitted suggestions for a research programme to be included in Committee's Report.

1944 Sept. J.R.W. chosen for Maths. Division.

1944 Oct. J.R.W. prepares research programme for Maths. Division which includes an item covering the A.C.E.

1944 Nov. J.R.W. addresses Executive Committee of N.P.L. Quotation from M/S (delivered verbatim)...

'Are we to have a mixed team developing gadgets of many kinds... Or are we, following Comrie... to rely on sheer virtuosity in the handling of the ordinary types of calculating machines? I think either attitude would be disastrous... We can gain the advantages of both methods by adopting electronic counting and by making the instructions to the machine automatic...'

1945 Feb–May J.R.W. sent to the U.S.A. by Director. Sees Harvard machine and calls it 'Turing in hardware'. (Can be confirmed by reference to letters to wife during visit). J.R.W. sees ENIAC and is given information about EDVAC by Von Neumann and Goldstine.

1945 June J.R.W. meets Professor M. H. A. Newman. Tells Newman he wishes to meet Turing. Meets Turing same day and invites him home. J.R.W. shows Turing the first report on the EDVAC and persuades him to join N.P.L. staff, arranges interview and convinces Director and Secretary.

The Automatic Computing Engine

During the remainder of 1945 Turing drafted his technical report 'Proposed Electronic Calculator'.[6] According to Michael Woodger—Turing's assistant at the NPL from 1946—an NPL file gave the date of Turing's completed report as 1945 (unfortunately, this file was destroyed in 1952).[7] Woodger believes that Turing probably wrote the report between October and December 1945. The report was submitted to the Executive Committee of the NPL in February 1946, under the title 'Proposals for the Development of an Automatic Computing Engine (ACE)', and on 19 March 1946 the 'Committee resolved unanimously to support with enthusiasm the proposal that Mathematics Division should undertake the

---

[6] Turing's technical report was reprinted by the NPL in April 1972 as Computer Science Division Report No. 57. The report is reprinted in full in B. J. Copeland (ed.), *Alan Turing's Automatic Computing Engine* (Oxford: Oxford University Press, 2004) under the title 'Proposed Electronic Calculator'; and in B. E. Carpenter and R. W. Doran (eds.), *A. M. Turing's ACE Report of 1946 and Other Papers* (Cambridge, Mass.: MIT Press, 1986) under the title 'Proposal for Development in the Mathematics Division of an Automatic Computing Engine (ACE)'. A copy of the original typewritten report is in the Woodger Papers (National Museum of Science and Industry, Kensington, London (catalogue reference M15/83)); a digital facsimile is in The Turing Archive for the History of Computing <www.AlanTuring.net/proposed_electronic_calculator>. Page references in what follows are to the original typescript.

[7] M. Woodger, handwritten note, undated (Woodger Papers (catalogue reference M15/78)); letter from Woodger to Copeland (27 Nov. 1999).

Figure 1. The Pilot ACE in December 1950. On the left are the control table and the modified Hollerith punched card unit (on the table with contoured legs). The tray slung below the main frame contains the short delay lines used for temporary storage.

Source: Crown copyright. Reproduced by permission of the National Physical Laboratory.

development and construction of an automatic computing engine of the type proposed by Dr. A. M. Turing'.[8]

Turing's 'Proposed Electronic Calculator' gave the first relatively complete specification of an electronic stored-programme digital computer. The earlier 'First Draft of a Report on the EDVAC', written by von Neumann in the USA in about May 1945 (see 'Computable Numbers: A Guide'), contained little engineering detail, in particular concerning electronic hardware. Harry Huskey, the electronic engineer who subsequently drew up the first detailed specifications of possible hardware configurations for the EDVAC, has stated that the 'information in the "First Draft" was of no help in this' and that 'von Neumann's "First Draft" provided no technical contribution to the development of computers'.[9] Turing's proposal, on the other hand, supplied detailed specifications of hardware units, including circuit designs, and specimen programmes in machine code. Turing even gave an exact estimate of the cost of building the machine (£11,200).

---

[8] Minutes of the Executive Committee of the National Physical Laboratory for 19 Mar. 1946 (National Physical Laboratory library; a digital facsimile is in The Turing Archive for the History of Computing <www.AlanTuring.net/npl_minutes_mar1946>).

[9] Letter from Huskey to Copeland (4 Feb. 2002).

Turing's ACE and the EDVAC (which was not fully working until 1952[10]) differed fundamentally in design. The EDVAC had what is now called a central processing unit or cpu, whereas in the ACE different memory locations had specific logical or numerical functions associated with them. For example, if two numbers were transferred to a certain destination in memory their sum would be formed there, ready to be transferred elsewhere by a subsequent instruction. Programmes for the ACE were made up entirely of instructions such as 'Transfer the contents of Temporary Store 27 to Temporary Store 6'. Instead of writing mathematically significant instructions like

MULTIPLY x BY y AND STORE THE RESULT IN z

the programmer composed a series of transfer instructions producing that effect.

Turing saw that size of memory and speed were the keys to computing. (His assistant at the NPL, James Wilkinson, observed that Turing 'was obsessed with the idea of speed on the machine'.[11]) Turing's design specified a high-speed memory of roughly the same capacity as an early Macintosh computer—enormous by the standards of his day. In order to increase the speed of a programme's execution, he proposed that instructions be stored, not consecutively, but at carefully chosen positions in memory, with each instruction containing a reference to the position of the next. Also with a view to speed, he included a small fast-access memory for the temporary storage of whichever numbers were used most frequently at a given stage of a computation. According to Wilkinson in 1955, Turing 'was the first to realise that it was possible to overcome access time difficulties with... mercury lines... or drum stores by providing a comparatively small amount of fast access store. Many of the commercial machines in the USA and... in this country make great use of this principle.'[12]

Turing's philosophy—very different from that embraced in the EDVAC—was to dispense with additional hardware in favour of software: in his design, complex behaviour was to be achieved by complex programming rather than by complex equipment (such as a hardware multiplier and divider, and special hardware for floating-point arithmetic). The ACE therefore had much in common with today's RISC (Reduced Instruction Set Computing) architectures. Turing spoke disparagingly of the contrary 'American tradition of solving one's difficulties by means of much equipment rather than thought'.[13]

[10] H. Huskey, 'The Development of Automatic Computing', in *Proceedings of the First USA–JAPAN Computer Conference*, Tokyo (1972), 698–704 (702).

[11] Wilkinson in interview with Christopher Evans in 1976 ('The Pioneers of Computing: An Oral History of Computing' (London: Science Museum)).

[12] Letter from Wilkinson to Newman, 10 June 1955 (Turing Papers, Modern Archive Centre, King's College, Cambridge (catalogue reference A 7)). Quoted by permission of Heather Wilkinson.

[13] Memo from Turing to Womersley, n.d. but c. Dec. 1946 (Woodger Papers (catalogue reference M15/77); a digital facsimile is in The Turing Archive for the History of Computing <www.AlanTuring.net/turing_womersley_cdec46>).

Had Turing's ACE been built as he planned, it would have been in a different league from the other early computers. However, from early 1947—for reasons that Turing might have regarded as amounting to 'cowardly and irrational doubts'[14]—his colleagues in the ACE Section were in favour of devoting their efforts to building a much scaled-down and simplified form of Turing's design, which they called the 'Test Assembly', rather than pressing on ahead immediately with the full-scale ACE. It was not in Turing's nature to direct them otherwise. He 'tended to ignore the Test Assembly', simply 'standing to one side' (Wilkinson said).[15] Finally, in 1948, a disenchanted Turing left the NPL for Newman's Computing Machine Laboratory at Manchester University. The introduction to the next chapter tells the full story of the frustrating delays at the NPL and of Turing's move to Manchester (and see also the introduction to Chapter 4).

It was not until May 1950 that a small 'pilot model' of the Automatic Computing Engine, built by Wilkinson, David Clayden, Donald Davies, Edward Newman, Michael Woodger, and others, executed its first programme. With an operating speed of 1 MHz, the Pilot Model ACE was for some time the fastest computer in the world. However, years of delays beyond Turing's control had cost the NPL the race to build the world's first stored-programme electronic digital computer—an honour that went to the University of Manchester, where, in Newman's Computing Machine Laboratory, the 'Manchester Baby' ran its first programme on 21 June 1948.[16] (As its name implies, the Baby was a very small computer, and the news that it had run what was only a tiny programme—just seventeen instructions long—for a mathematically trivial task was 'greeted with hilarity' by the NPL team developing the much more sophisticated Pilot Model ACE.[17]) Turing's influence on the Manchester computer is described below.

The EDSAC computer, built by Maurice Wilkes at Cambridge University, became operational in 1949 and was the second stored-programme electronic computer to run. The Pilot Model ACE was also preceded by the BINAC (1949), built by Eckert and Mauchly at their Electronic Control Company, Philadelphia (although opinions differ over whether the BINAC ever actually worked), the CSIR Mark 1 (1949), built by Trevor Pearcey at the Commonwealth Scientific and Industrial Research Organization, Division of Radiophysics, Sydney, Australia, the Whirlwind I (1949), built by Jay Forrester at the Digital Computer Laboratory, Massachusetts Institute of Technology, and the SEAC (1950), built

---

[14] Turing used this phrase in a different connection on p. 42 of his 'Proposed Electronic Calculator'.

[15] Wilkinson in interview with Evans (see n. 11).

[16] Williams described the Computing Machine Laboratory on p. 328 of his 'Early Computers at Manchester University' (*Radio and Electronic Engineer*, 45 (1975), 327–31): 'It was one room in a Victorian building whose architectural features are best described as "late lavatorial". The walls were of brown glazed brick and the door was labelled "Magnetism Room".'

[17] Woodger in interview with Copeland (June 1998).

by Samuel Alexander and Ralph Slutz at the US Bureau of Standards Eastern Division, Washington, DC. Huskey's SWAC, built at the US Bureau of Standards Western Division, Los Angeles, ran a few months after the Pilot Model ACE (in August 1950).

Derivatives of Turing's ACE Design

The DEUCE, the production version of the Pilot Model ACE, was built by the English Electric Company. The first was delivered in March 1955 (to the NPL) and the last went out of service around 1970.[18] Sales of this large and expensive machine exceeded thirty, confounding the suggestion, made in 1946 by Sir Charles Darwin, Director of the NPL, that 'it is very possible that...one machine would suffice to solve all the problems that are demanded of it from the whole country'.[19] (Douglas Hartee, a leading British expert on automatic computation, thought that a total of three digital computers would probably be adequate for the country's computing needs.[20]) The NPL's DEUCE replaced the Pilot Model ACE, and in 1956 much of the Pilot Model was transferred to the London Science Museum (where it is on permanent display).[21]

Work began on a large-scale ACE in the autumn of 1954.[22] Built and housed at the NPL, the 'Big ACE' was in operation by late 1958. Wilkinson, Clayden, Davies, (Ted) Newman, and Woodger all contributed to the final design.[23] The Big ACE filled a room the size of an auditorium.[24] It remained in service until 1967.

At the Press Day held in 1958 to announce the inauguration of the Big ACE, A. M. Uttley—Superintendent of the NPL's Control Mechanisms and Electronics

---

[18] Letter from J. Illingworth to Fryer, 6 Nov. 1956 (Woodger Papers (catalogue reference M15/87); a digital facsimile is in The Turing Archive for the History of Computing <www.AlanTuring.net/illingworth_fryer_6nov56>).

[19] C. Darwin, 'Automatic Computing Engine (ACE)', National Physical Laboratory, 17 Apr. 1946 (Public Record Office (document reference DSIR 10/385); a digital facsimile is in The Turing Archive for the History of Computing <www.AlanTuring.net/darwin_ace>).

[20] Hartree's opinion is quoted in V. Bowden, 'The 25th Anniversary of the Stored Program Computer', *Radio and Electronic Engineer*, 45 (1975), 326.

[21] Memorandum from Hiscocks to the DSIR, 30 Jan. 1956 (Public Record Office (document reference DSIR 10/275); a digital facsimile is in The Turing Archive for the History of Computing <www.AlanTuring.net/hiscocks_dsir_30jan1956>); letter from D. Brunt to Sir Charles Darwin (20 July 1956) (Public Record Office (document reference DSIR 10/275); <www.AlanTuring.net/brunt_darwin_20july56>); 'Disposal of Pilot ACE', memorandum from Hiscocks to the DSIR, 26 July 1956 (Public Record Office (document reference DSIR 10/275); <www.AlanTuring.net/hiscocks_disposal_pilot_ace>).

[22] Letter from Illingworth to Fryer (see n. 18).

[23] Letter from A. M. Uttley to Sara Turing, 19 Dec. 1958 (in the Turing Papers, the Modern Archive Centre, King's College, Cambridge (catalogue reference A 11)).

[24] F. M. Blake, D. O. Clayden, D. W. Davies, L. J. Page, and J. B. Stringer, 'Some Features of the ACE Computer', National Physical Laboratory, 8 May 1957 (Woodger Papers (catalogue reference N12/102); a digital facsimile is in The Turing Archive for the History of Computing <www.AlanTuring.net/ace_features>).

Division—announced: 'Today, Turing's dream has come true.'[25] If so, it was a dream whose time had passed. Technology had moved ahead in the thirteen years since Turing wrote 'Proposed Electronic Calculator'. The Big ACE was not the revolutionary machine that it would have been if completed six or seven years earlier. Not only did the Big ACE use valves in the era of the transistor; the designers also retained the by then outmoded mercury delay line memory proposed by Turing in 1945 (see below).[26]

Nevertheless, the Big ACE was a fast machine with a large memory, and the decision to stick with the principles used in the Pilot ACE and the DEUCE was reasonable in the circumstances. In 1953 Francis Colebrook (Head of the Electronics Section responsible for building Pilot ACE) urged that the proposed large-scale ACE should 'be based on well proved components and techniques, even when revolutionary developments seem to be just around the corner. Otherwise the [Mathematics] Division will get nothing but a succession of pilot models.'[27]

The Big ACE ran at 1.5 MHz, 50 per cent faster than the Pilot ACE and the DEUCE, both 1 MHz machines. A 1957 report stated that the Big ACE 'appears in fact to be about as fast as present-day parallel core-store computers'.[28] 'Core-store' or magnetic core memory was the most advanced high-speed storage medium at that time.

The basic principles of Turing's ACE design were used in the G15 computer, built and marketed by the Detroit-based Bendix Corporation.[29] The G15 was designed by Huskey, who had spent the year 1947 at the NPL working in the ACE Section. The first G15 ran in 1954.[30] It was arguably the first personal computer. By following Turing's philosophy of minimizing hardware in favour of software, Huskey was able to make the G15 small enough (it was the size of a large domestic refrigerator) and cheap enough to be marketed as a single-user computer. Yet thanks to the design of the ACE-like memory (implemented in a magnetic drum), the G15 was as fast as computers many times its size. Over 400 were sold worldwide and the G15 remained in use until about 1970.

Another computer deriving from Turing's ACE design, the MOSAIC or Ministry of Supply Automatic Integrator and Computer, played a role in Britain's air defences during the Cold War period. In 1946 Flowers established a small team at the Post Office Research Station to build a computer to Turing's logical design (see the introduction to the next chapter). The team consisted of two engineers, William

---

[25] Letter from Uttley to Sara Turing (see n. 23).
[26] An experimental transistorized machine went into operation at Manchester University in 1953; see S. H. Lavington, *Early British Computers* (Manchester: Manchester University Press, 1980).
[27] F. M. Colebrook, 4 May 1953; quoted in D. M. Yates, *Turing's Legacy: A History of Computing at the National Physical Laboratory 1945–1995* (London: Science Museum, 1997), 67.
[28] Blake, Clayden, Davies, Page, and Stringer, 'Some Features of the ACE Computer', 3.
[29] Huskey in interview with Copeland (Feb. 1998).
[30] Letter from Huskey to Copeland (20 Dec. 2001).

Chandler and Allen Coombs, both of whom had assisted Flowers in the construction of Colossus. Working alone, Coombs and Chandler carried out the engineering design of the MOSAIC, a large computer based on Turing's Version VII of the ACE design (Version VII dated from 1946).[31] Of the various ACE-type computers that were built, the MOSAIC was the closest to Turing's original conception.

The MOSAIC consisted of some 70 mercury delay lines, 2,000 semi-conductors (germanium diodes), and 7,000 thermionic valves.[32] It first ran a programme in 1952 or early 1953.[33] Once completed, the MOSAIC was installed at the Radar Research and Development Establishment (RRDE) in Malvern. It was used to calculate aircraft trajectories from radar data in connection with anti-aircraft measures (the details appear still to be classified). The data was generated by two mobile data-recorders working in conjunction with a radar tracking system.[34]

Coombs has emphasized: 'it was just Chandler and I—we designed every scrap of that machine.'[35] Given, therefore, that two engineers working alone succeeded in completing the large MOSAIC, there seems little doubt that, had the NPL possessed the organizational capability and sufficient manpower, a computer not too distant from Turing's original conception could have been up and running by the early 1950s. Thanks to their experience with Colossus, Chandler and Coombs had a substantial start on everyone else in the field. Turing was unable to share his knowledge of their wartime work with Darwin. Had he been able to do so, the NPL might have acted to boost the resources available to Chandler and Coombs, and so made Turing's dream a reality much sooner.

Other derivatives of the ACE included the EMI Business Machine and the Packard-Bell PB250.[36] Designed for business applications involving the shallow processing of large quantities of data, the EMI Business Machine was a relatively

---

[31] Coombs in interview with Evans in 1976 ('The Pioneers of Computing: An Oral History of Computing' (London: Science Museum)); A. W. M. Coombs, 'MOSAIC', in *Automatic Digital Computation: Proceedings of a Symposium Held at the National Physical Laboratory* (London: Her Majesty's Stationery Office, 1954); B. J. Copeland (ed.), 'The Turing–Wilkinson Lectures on the Automatic Computing Engine', in K. Furukawa, D. Michie, and S. Muggleton (eds.), *Machine Intelligence 15* (Oxford: Oxford University Press, 1999). Digital facsimiles of a series of technical reports concerning the MOSAIC by Coombs, Chandler, and others are available in The Turing Archive for the History of Computing <www.AlanTuring.net/mosaic>.

[32] 'Engineer-in-Chief's Report on the Work of the Engineering Department for the Year 1 April 1954 to 31 March 1955', Post Office Engineering Department (The Post Office Archive, London); Coombs, 'MOSAIC'.

[33] 'Engineer-in-Chief's Report on the Work of the Engineering Department for the Year 1 April 1952 to 31 March 1953', Post Office Engineering Department (The Post Office Archive, London).

[34] 'Engineer-in-Chief's Report on the Work of the Engineering Department for the Year 1 April 1951 to 31 March 1952', Post Office Engineering Department (The Post Office Archive, London).

[35] Coombs in interview with Evans (see n. 31).

[36] C. G. Bell, and A. Newell, *Computer Structures: Readings and Examples* (New York: McGraw-Hill, 1971), 44, 74; R. J. Froggatt, 'Logical Design of a Computer for Business Use', *Journal of the British Institution of Radio Engineers*, 17 (1957), 681–96; Yates, *Turing's Legacy: A History of Computing at the National Physical Laboratory 1945–1995*, 43–4.

slow electronic computer with a large memory. The PB250 was a low-cost transistorized computer.

Turing, Newman, and the Manchester Computer

At the time of the Manchester Baby and its successor, the Manchester Mark I, the electronic engineers Frederic Williams and Thomas Kilburn, who had translated the logico-mathematical idea of the stored-programme computer into hardware, were given too little credit by the mathematicians at Manchester—Williams and Kilburn were regarded as excellent engineers but not as 'ideas men'.[37] Nowadays the tables have turned too far and the triumph at Manchester is usually credited to Williams and Kilburn alone. Fortunately the words of the late Williams survive to set the record straight:

Now let's be clear before we go any further that neither Tom Kilburn nor I knew the first thing about computers when we arrived in Manchester University...Newman explained the whole business of how a computer works to us.[38]

Tom Kilburn and I knew nothing about computers...Professor Newman and Mr A. M. Turing...knew a lot about computers...They took us by the hand and explained how numbers could live in houses with addresses...[39]

In an address to the Royal Society on 4 March 1948, Newman presented this very explanation:

In modern times the idea of a universal calculating machine was independently [of Babbage] introduced by Turing...There is provision for storing numbers, say in the scale of 2, so that each number appears as a row of, say, forty 0's and 1's in certain places or 'houses' in the machine. ... Certain of these numbers, or 'words' are read, one after another, as orders. In one possible type of machine an order consists of four numbers, for example 11, 13, 27, 4. The number 4 signifies 'add', and when control shifts to this word the 'houses' $H11$ and $H13$ will be connected to the adder as inputs, and $H27$ as output. The numbers stored in $H11$ and $H13$ pass through the adder, are added, and the sum is passed on to $H27$. The control then shifts to the next order. In most real machines the process just described would be done by three separate orders, the first bringing $<H11>$ (= content of $H11$) to a central accumulator, the second adding $<H13>$ into the accumulator, and the third sending the result to $H27$; thus only one address would be required in each order. ... A machine with storage, with this automatic-telephone-exchange arrangement and with the necessary adders, subtractors and so on, is, in a sense, already a universal machine.[40]

[37] Peter Hilton in interview with Copeland (June 2001).
[38] Williams in interview with Evans in 1976 ('The Pioneers of Computing: An Oral History of Computing' (London: Science Museum)).
[39] Williams, 'Early Computers at Manchester University', 328.
[40] M. H. A. Newman, 'General Principles of the Design of All-Purpose Computing Machines', *Proceedings of the Royal Society of London*, Series A, 195 (1948), 271–74 (271–2).

Following this explanation of Turing's three-address concept (source 1, source 2, destination, function) Newman went on to describe programme storage ('the orders shall be in a series of houses $X1$, $X2$, ...') and conditional branching. He then summed up:

> From this highly simplified account it emerges that the essential internal parts of the machine are, first, a storage for numbers (which may also be orders). ... Secondly, adders, multipliers, etc. Thirdly, an 'automatic telephone exchange' for selecting 'houses', connecting them to the arithmetic organ, and writing the answers in other prescribed houses. Finally, means of moving control at any stage to any chosen order, if a certain condition is satisfied, otherwise passing to the next order in the normal sequence. Besides these there must be ways of setting up the machine at the outset, and extracting the final answer in useable form.[41]

In a letter written in 1972 Williams described in some detail what he and Kilburn were told by Newman:

> About the middle of the year [1946] the possibility of an appointment at Manchester University arose and I had a talk with Professor Newman who was already interested in the possibility of developing computers and had acquired a grant from the Royal Society of £30,000 for this purpose. Since he understood computers and I understood electronics the possibilities of fruitful collaboration were obvious. I remember Newman giving us a few lectures in which he outlined the organisation of a computer in terms of numbers being identified by the address of the house in which they were placed and in terms of numbers being transferred from this address, one at a time, to an accumulator where each entering number was added to what was already there. At any time the number in the accumulator could be transferred back to an assigned address in the store and the accumulator cleared for further use. The transfers were to be effected by a stored program in which a list of instructions was obeyed sequentially. Ordered progress through the list could be interrupted by a test instruction which examined the sign of the number in the accumulator. Thereafter operation started from a new point in the list of instructions. This was the first information I received about the organisation of computers. ... Our first computer was the simplest embodiment of these principles, with the sole difference that it used a subtracting rather than an adding accumulator.[42]

Turing's early input to the developments at Manchester, hinted at by Williams in his above-quoted reference to Turing, may have been via the lectures on computer design that Turing and Wilkinson gave in London during the period December 1946 to February 1947.[43] The lectures were attended by representatives of various organizations planning to use or build an electronic computer.

---

[41] M. H. A. Newman, 'General Principles of the Design of All-Purpose Computing Machines', *Proceedings of the Royal Society of London*, Series A, 195 (1948), 271–74 (271–2).

[42] Letter from Williams to Randell, 1972 (quoted in B. Randell, 'On Alan Turing and the Origins of Digital Computers', in B. Meltzer and D. Michie (eds.), *Machine Intelligence 7* (Edinburgh: Edinburgh University Press, 1972), 9).

[43] 'The Turing–Wilkinson Lectures on the Automatic Computing Engine'.

Kilburn was in the audience.[44] (Kilburn usually said, when asked where he obtained his basic knowledge of the computer from, that he could not remember;[45] for example, in a 1992 interview he said: 'Between early 1945 and early 1947, in that period, somehow or other I knew what a digital computer was... Where I got this knowledge from I've no idea.'[46])

Whatever role Turing's lectures may have played in informing Kilburn, there is little doubt that credit for the Manchester computer—called the 'Newman–Williams machine' by Huskey in a report written shortly after a visit in 1947 to the Manchester project[47]—belongs not only to Williams and Kilburn but also to Newman, and that the influence on Newman of Turing's 'On Computable Numbers' was crucial, as was the influence of Flowers's Colossus (see the introduction to Chapter 4).

There is more information concerning Turing and the Manchester computer in the chapter 'Artificial Life'.

The Manchester computer and the EDVAC

The Baby and the Manchester Mark I are sometimes said to have descended from the EDVAC. Newman was well aware of von Neumann's 'First Draft of a Report on the EDVAC'. In the summer of 1946 he sent David Rees, a lecturer in his department at Manchester and an ex-member of the Newmanry, to a series of lectures at the Moore School, where Eckert, Mauchly, and other members of the ENIAC-EDVAC group publicized their ideas on computer design.[48] In the autumn of 1946 Newman himself went to Princeton for three months.[49]

Newman's advocacy of 'a central accumulator'—a characteristic feature of the EDVAC but not of the ACE—was probably influenced by his knowledge of the American proposals. However, von Neumann's ideas seem to have had little influence on other members of the Manchester project. Kilburn spoke scathingly of the von Neumann 'dictat'.[50] Geoffrey Tootill said:

Williams, Kilburn and I (the three designers of the first Manchester machine) had all spent the 1939–1945 war at the Telecommunications Research Establishment doing R & D on radiolocation equipments. The main U.S. ideas that we accepted in return for our

---

[44] G. Bowker and R. Giordano, 'Interview with Tom Kilburn', *Annals of the History of Computing*, 15 (1993), 17–32.
[45] Letter from Brian Napper to Copeland (16 June 2002).
[46] Bowker and Giordano, 'Interview with Tom Kilburn', 19.
[47] H. D. Huskey, untitled typescript, National Physical Laboratory, n.d. but c. Mar. 1947 (Woodger Papers (catalogue reference M12/105); a digital facsimile is in The Turing Archive for the History of Computing <www.AlanTuring.net/huskey_1947>).
[48] Letter from Rees to Copeland (2 Apr. 2001).
[49] W. Newman, 'Max Newman: Mathematician, Codebreaker and Computer Pioneer', to appear in B. J. Copeland (ed.) *Colossus: The First Electronic Computer* (Oxford: Oxford University Press).
[50] Kilburn in interview with Copeland (July 1997).

initiatives on these and later on computers were the terms 'radar' and 'memory'... We disliked the latter term, incidentally, as encouraging the anthropomorphic concept of 'machines that think'.[51]

To the best of my recollection FC [Williams], Tom [Kilburn] and I never discussed... von Neumann's... ideas during the development of the Small-Scale Experimental Machine [the Baby], nor did I have any knowledge of them when I designed the Ferranti Mk I. I don't think FC was influenced at all by von Neumann, because I think he was in general quite punctilious in acknowledging other people's ideas.[52]

Tootill added:

As well as our own ideas, we incorporated functions suggested by Turing and Newman in the improvement and extension of the first machine. When I did the logic design of the Ferranti Mark 1, I got them to approve the list of functions.[53]

The ACE and Artificial Intelligence[54]

In designing the ACE, Artificial Intelligence was not far from Turing's thoughts —he described himself as building 'a brain'.[55] The otherwise austere 'Proposed Electronic Calculator' contains a cameo discussion of computer intelligence and chess (Turing's earliest surviving remarks concerning AI):

'Can the machine play chess?' It could fairly easily be made to play a rather bad game. It would be bad because chess requires intelligence. We stated at the beginning of this section that the machine should be treated as entirely without intelligence. There are indications however that it is possible to make the machine display intelligence at the risk of its making occasional serious mistakes. By following up this aspect the machine could probably be made to play very good chess.[56]

What is probably Turing's earliest mention to survive of his interest in neural simulation (see further Chapter 10) occurs in a letter to the cyberneticist W. Ross Ashby:

In working on the ACE I am more interested in the possibility of producing models of the action of the brain than in the practical applications to computing. ... The ACE will be used, as you suggest, in the first instance in an entirely disciplined manner, similar to the action of the lower centres, although the reflexes will be extremely complicated. The disciplined action carries with it the disagreeable feature, which you mentioned, that it will be entirely uncritical when anything goes wrong. It will also be necessarily devoid of

[51] Letter from Tootill to Copeland (18 Apr. 2001).
[52] Letter from Tootill to Copeland (16 May 2001).
[53] Letter from Tootill to Copeland (18 Apr. 2001).
[54] As explained in the chapter 'Artificial Intelligence', the term 'Artificial Intelligence' did not come into use until after Turing's death.
[55] Don Bayley in interview with Copeland (Dec. 1997).
[56] 'Proposed Electronic Calculator', 16.

anything that could be called originality. There is, however, no reason why the machine should always be used in such a manner: there is nothing in its construction which obliges us to do so. It would be quite possible for the machine to try out variations of behaviour and accept or reject them in the manner you describe and I have been hoping to make the machine do this. This is possible because, without altering the design of the machine itself, it can, in theory at any rate, be used as a model of any other machine, by making it remember a suitable set of instructions. The ACE is in fact analogous to the 'universal machine' described in my paper on computable numbers. This theoretical possibility is attainable in practice, in all reasonable cases, at worst at the expense of operating slightly slower than a machine specially designed for the purpose in question. Thus, although the brain may in fact operate by changing its neuron circuits by the growth of axons and dendrites, we could nevertheless make a model, within the ACE, in which this possibility was allowed for, but in which the actual construction of the ACE did not alter, but only the remembered data, describing the mode of behaviour applicable at any time. I feel that you would be well advised to take advantage of this principle, and do your experiments on the ACE, instead of building a special machine. I should be very glad to help you over this.[57]

The Lecture

On 20 February 1947 Turing lectured on the ACE to the London Mathematical Society.[58] So far as is known, this was the earliest public lecture to mention computer intelligence, providing a breathtaking glimpe of a new field. Turing discussed the prospect of machines acting intelligently, learning, and beating human opponents at chess, remarking that '[w]hat we want is a machine that can learn from experience' and that '[t]he possibility of letting the machine alter its own instructions provides the mechanism for this'.

The lecture is also of note for its early discussion of computer programming. By the time of the lecture, Turing had been developing programmes—then called 'instruction tables' (see 'Computable Numbers: A Guide')—for the not-yet-existent ACE for well over a year. As Womersley was to remark in 1948, 'The planning of the ACE is far ahead of the hardware.'[59] During this period, Turing and the other members of the ACE Section brought the nascent science of computer programming to a state of considerable sophistication.

Among Turing's many technical innovations was the use of what are now called *subroutines*, or in Turing's term, 'subsidiary tables' (see 'Computable Numbers: A Guide').

[57] Letter from Turing to W. Ross Ashby, no date (Woodger Papers (catalogue reference M11/99); a digital facsimile is in The Turing Archive for the History of Computing <www.AlanTuring.net/turing_ashby>). The letter was probably written in 1946 and certainly prior to October 1947.
[58] The lecture was held at 5 p.m. in the rooms of the Royal Astronomical Society at Burlington House in London (entry in Woodger's diary for 20 Feb. 1947 (Copeland is grateful to Woodger for this information)).
[59] Minutes of the NPL Executive Committee, 20 Apr. 1948 (National Physical Laboratory library; a digital facsimile is in The Turing Archive for the History of Computing <www.AlanTuring.net/npl_minutes_apr1948>).

Turing devoted about a quarter of his lecture to a discussion of computer memory, saying (p. 383):

> I have spent a considerable time in this lecture on this question of memory, because I believe that the provision of proper storage is the key to the problem of the digital computer, and certainly if they are to be persuaded to show any sort of genuine intelligence much larger capacities than are yet available must be provided.

It was precisely Turing's desire to conduct experiments in AI that led him to propose such a large memory in the ACE.[60]

Delay Line Memory and Optimum Programming

The fundamental feature of the ACE design, which influenced virtually all other aspects of it, was Turing's adoption of acoustic delay lines to form the high-speed memory. He remarked in the lecture that the chief advantage of delay lines as a memory medium—they were far from ideal—was that they were 'already a going concern' (p. 380). The acoustic delay line was pioneered in 1942 by W. B. Shockley, later one of the co-inventors of the transistor, and in 1943 Presper Eckert and others at the Moore School (home of the ENIAC) independently investigated a different type of acoustic delay line filled with mercury.[61] Subsequently, acoustic delay lines were widely employed in radar. It was Eckert who first proposed their use in digital computers (see 'Computable Numbers: A Guide').

The following description of a mercury delay line is from Turing's 'Proposed Electronic Calculator':

> It is proposed to build 'delay line' units consisting of mercury...tubes about 5′ long and 1″ in diameter in contact with a quartz crystal at each end. The velocity of sound in...mercury...is such that the delay will be 1.024 ms. The information to be stored may be considered to be a sequence of 1024 'digits' (0 or 1)...These digits will be represented by a corresponding sequence of pulses. The digit 0...will be represented by the absence of a pulse at the appropriate time, the digit 1...by its presence. This series of pulses is impressed on the end of the line by one piezo-crystal, it is transmitted down the line in the form of supersonic waves, and is reconverted into a varying voltage by the crystal at the far end. This voltage is amplified sufficiently to give an output of the order of 10 volts peak to peak and is used to gate a standard pulse generated by the clock. This pulse may be again fed into the line by means of the transmitting crystal, or we may feed in some altogether different signal. We also have the possibility of leading the gated pulse to some other part of the calculator, if we have need of that information at the time. Making use of the information does not of course preclude keeping it also.[62]

---

[60] Woodger in interview with Copeland (June 1998).
[61] A. G. Emslie, H. B. Huntington, H. Shapiro, and A. E. Benfield, 'Ultrasonic Delay Lines II', *Journal of the Franklin Institute*, 245 (1948), 101–15 (101–2). In Shockley's delay line, the transmitting medium was not mercury but ethylene glycol.
[62] 'Proposed Electronic Calculator', 5.

As Turing noted in his 1947 lecture, the ACE's memory was to consist of approximately 200 such delay lines. In the event, the Big ACE contained only a few dozen delay lines, supplemented by four magnetic drums.

Having decided to use delay lines, Turing was determined to maximize their effectiveness. The time taken for an instruction, or number, to emerge from a delay line will depend on where in the delay line it happens to be. In order to minimize waiting time, Turing arranged for instructions to be stored not at regular intervals in a delay line, but in irregular positions. These positions were selected by the programmer in such a way that each instruction would emerge from its delay line at precisely the time it was required. This system became known at the NPL as 'optimum coding' (and later 'optimum programming').

Optimum coding made for difficult and untidy programming but the advantage in terms of speed was considerable. Thanks to optimum coding, the Pilot Model ACE was able to do a floating point multiplication in 3 milliseconds.[63] The EDSAC computer at Cambridge University—a slower delay line machine lacking optimum coding—required 4.5 milliseconds to perform a single fixed point multiplication.[64]

Further reading

Campbell-Kelly, M., 'Programming the Pilot ACE: Early Programming Activity at the National Physical Laboratory', *Annals of the History of Computing*, 3 (1981), 133–62.

Carpenter, B. E., and Doran, R. W. (eds.), *A. M. Turing's ACE Report of 1946 and Other Papers* (Cambridge, Mass.: MIT Press, 1986).

Copeland, B. J. (ed.), *Alan Turing's Automatic Computing Engine* (Oxford: Oxford University Press, 2004).

Yates, D. M., *Turing's Legacy: A History of Computing at the National Physical Laboratory 1945–1995* (London: Science Museum, 1997).

Provenance

What follows is the text of the lecture taken from Turing's own typescript (which is headed 'Lecture to L.M.S. Feb. 20 1947').[65]

---

[63] Wilkinson in interview with Evans (see n. 11).

[64] M. Campbell-Kelly, 'Programming the EDSAC: Early Programming Activity at the University of Cambridge', *Annals of the History of Computing*, 2 (1980), 7–36.

[65] Turing's typescript is among the Turing Papers in the Modern Archive Centre, King's College, Cambridge (catalogue reference B 1). It was first published in 1986 in Carpenter and Doran, *A. M. Turing's ACE Report of 1946 and Other Papers*. The present edition differs from the 1986 edition in various small respects. A few missing words have been restored (on pp. 379 and 381) and some minor departures in the earlier edition from Turing's original text have been rectified. Many of Turing's sketch diagrams (reproduced in the 1986 edition) have been redrawn. Obvious typing and spelling errors in Turing's typescript have been corrected without comment. Words or letters enclosed in square brackets have been added by the editor.

Lecture on the Automatic Computing Engine

The automatic computing engine now being designed at N.P.L. is a typical large scale electronic digital computing machine. In a single lecture it will not be possible to give much technical detail of this machine, and most of what I shall say will apply equally to any other machine of this type now being planned.

From the point of view of the mathematician the property of being digital should be of greater interest than that of being electronic. That it is electronic is certainly important because these machines owe their high speed to this, and without the speed it is doubtful if financial support for their construction would be forthcoming. But this is virtually all that there is to be said on that subject. That the machine is digital however has more subtle significance. It means firstly that numbers are represented by sequences of digits which can be as long as one wishes. One can therefore work to any desired degree of accuracy. This accuracy is not obtained by more careful machining of parts, control of temperature variations, and such means, but by a slight increase in the amount of equipment in the machine. To double the number of significant figures used would involve increasing the equipment by a factor definitely less than two, and would also have some effect in increasing the time taken over each job. This is in sharp contrast with analogue machines, and continuous variable machines such as the differential analyser, where each additional decimal digit required necessitates a complete redesign of the machine, and an increase in the cost by perhaps as much as a factor of 10. A second advantage of digital computing machines is that they are not restricted in their applications to any particular type of problem. The differential analyser is by far the most general type of analogue machine yet produced, but even it is comparatively limited in its scope. It can be made to deal with almost any kind of ordinary differential equation, but it is hardly able to deal with partial differential equations at all, and certainly cannot manage large numbers of linear simultaneous equations, or the zeros of polynomials. With digital machines however it is almost literally true that they are able to tackle any computing problem. A good working rule is that the ACE can be made to do any job that could be done by a human computer, and will do it in one ten-thousandth of the time. This time estimate is fairly reliable, except in cases where the job is too trivial to be worth while giving to the ACE.

Some years ago I was researching on what might now be described as an investigation of the theoretical possibilities and limitations of digital computing machines. I considered a type of machine which had a central mechanism, and an infinite memory which was contained on an infinite tape. This type of machine appeared to be sufficiently general. One of my conclusions was that the idea of a 'rule of thumb' process and a 'machine process' were synonymous. The expres-

Reproduced with permission of the Estate of Alan Turing.

sion 'machine process' of course means one which could be carried out by the type of machine I was considering. It was essential in these theoretical arguments that the memory should be infinite. It can easily be shown that otherwise the machine can only execute periodic operations. Machines such as the ACE may be regarded as practical versions of this same type of machine. There is at least a very close analogy. Digital computing machines all have a central mechanism or control and some very extensive form of memory. The memory does not have to be infinite, but it certainly needs to be very large. In general the arrangement of the memory on an infinite tape is unsatisfactory in a practical machine, because of the large amount of time which is liable to be spent in shifting up and down the tape to reach the point at which a particular piece of information required at the moment is stored. Thus a problem might easily need a storage of three million entries, and if each entry was equally likely to be the next required the average journey up the tape would be through a million entries, and this would be intolerable. One needs some form of memory with which any required entry can be reached at short notice. This difficulty presumably used to worry the Egyptians when their books were written on papyrus scrolls. It must have been slow work looking up references in them, and the present arrangement of written matter in books which can be opened at any point is greatly to be preferred. We may say that storage on tape and papyrus scrolls is somewhat *inaccessible*. It takes a considerable time to find a given entry. Memory in book form is a good deal better, and is certainly highly suitable when it is to be read by the human eye. We could even imagine a computing machine that was made to work with a memory based on books. It would not be very easy but would be immensely preferable to the single long tape. Let us for the sake of argument suppose that the difficulties involved in using books as memory were overcome, that is to say that mechanical devices for finding the right book and opening it at the right page, etc. etc. had been developed, imitating the use of human hands and eyes. The information contained in the books would still be rather inaccessible because of the time occupied in the mechanical motions. One cannot turn a page over very quickly without tearing it, and if one were to do much book transportation, and do it fast the energy involved would be very great. Thus if we moved one book every millisecond and each was moved ten metres and weighed 200 grams, and if the kinetic energy were wasted each time we should consume 10^{10} watts, about half the country's power consumption. If we are to have a really fast machine then, we must have our information, or at any rate a part of it, in a more accessible form than can be obtained with books. It seems that this can only be done at the expense of compactness and economy, e.g. by cutting the pages out of the books, and putting each one into a separate reading mechanism. Some of the methods of storage which are being developed at the present time are not unlike this.

If one wishes to go to the extreme of accessibility in storage mechanisms one is liable to find that it is gained at the price of an intolerable loss of compactness and economy. For instance the most accessible known form of storage is that provided by the valve flip-flop or Jordan Eccles trigger circuit. This enables us to store one digit, capable of two values, and uses two thermionic valves. To store the content of an ordinary novel by such means would cost many millions of pounds. We clearly need some compromise method of storage which is more accessible than paper, film etc, but more economical in space and money than the straightforward use of valves. Another desirable feature is that it should be possible to record into the memory from within the computing machine, and this should be possible whether or not the storage already contains something, i.e. the storage should be *erasible*.

There are three main types of storage which have been developed recently and have these properties in greater or less degree. Magnetic wire is very compact, is erasible, can be recorded on from within the machine, and is moderately accessible. There is storage in the form of charge patterns on the screen of a cathode ray tube. This is probably the ultimate solution. It could eventually be nearly as accessible as the Jordan Eccles circuit. A third possibility is provided by acoustic delay lines. They give greater accessibility than the magnetic wire, though less than the C.R.T type. The accessibility is adequate for most purposes. Their chief advantage is that they are already a going concern. It is intended that the main memory of the ACE shall be provided by acoustic delay lines, consisting of mercury tanks.

The idea of using acoustic delay lines as memory units is due I believe to Eckert of Philadelphia University,[1] who was the engineer chiefly responsible for the Eniac. The idea is to store the information in the form of compression waves travelling along a column of mercury. Liquids and solids will transmit sound of surprisingly high frequency, and it is quite feasible to put as many as 1,000 pulses into a single 5' tube. The signals may be conveyed into the mercury by a piezo-electric crystal; and also detected at the far end by another quartz crystal. A train of pulses or the information which they represent may be regarded as stored in the mercury whilst it is travelling through it. If the information is not required when the train emerges it can be fed back into the column again and again until such time as it is required. This requires a 'recirculating circuit' to read the signal

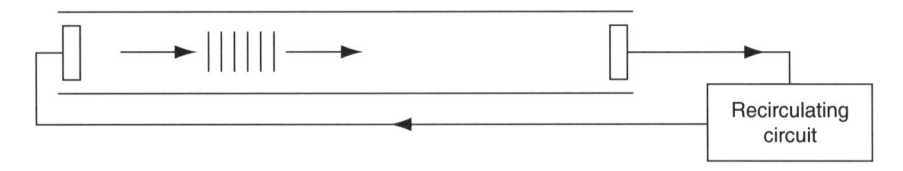

Figure 9.1.

---

[1] Editor's note. Turing means the University of Pennsylvania in Philadelphia.

Lecture on the Automatic Computing Engine | 381

as it emerges from the tank and amplify it and feed it in again. If this were done with a simple amplifier it is clear that the characteristics of both the tank and the amplifier would have to be extremely good to permit the signal to pass through even as many as ten times. Actually the recirculating circuit does something slightly different. What it does may perhaps be best expressed in terms of point set topology. Let the plane of the diagram represent the space of all possible signals. I do not of course wish to imply that this is two dimensional. Let the function f be defined for arguments in this signal space and have values in it. In fact let $f(s)$ represent the effect on the signal s when it is passed through the tank and the recirculating mechanism. We assume however that owing to thermal agitation the effect of recirculation may be to give any pt [point] within a circle of radius δ of $f(s)$. Then a necessary and sufficient condition that the tank can be used as a storage which will distinguish between N different signals is that there must be N sets $E_1 \ldots E_N$ such that if F_r is the set of pts [points] within distance ε of E_r

$$s \in F_r \supset f(s) \in E_r$$

and the sets F_r are disjoint. It is clearly sufficient for we have only then to ensure that the signals initially fed in belong to one or other of the sets F_r, and it will remain in the set after any number of recirculations, without any danger of confusion. It is necessary for suppose $s_1 \ldots s_N$ are signals which have different meanings and which can be fed into the machine at any time and read out later without fear of confusion.

Let E_r be the set of signals which *could* be obtained for s_r by successive applications of f and shifts of distance not more than ε. Then the sets E_r are disjoint, [*five or six illegible words*], and by applying a shift of distance ε or less to pts [points] of E_r we obtain [*line missing from bottom of page*]. In the case of a mercury delay line used for $N = 16$ the set would consist of all continuous signals within the shaded area.

Figure 9.2.

One of the sets would consist of all continuous signals lying in the region below. It would represent the signal 1001.

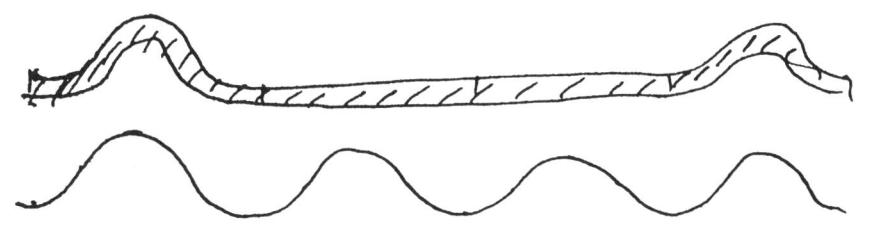

Figure 9.3.

In order to put such a recirculation system into effect it is essential that a clock signal be supplied to the memory system so that it will be able to distinguish the times when a pulse if any should be present. It would for instance be natural to supply a timing sine wave as shown above to the recirculator.

The idea of a process f with the properties we have described is a very common one in connection with storage devices. It is known as 'regeneration' of storage. It is always present in some form, but sometimes the regeneration is as it were naturally occurring and no precautions have to be taken. In other cases special precautions have to be taken to improve such an f process or else the impression will fade.

The importance of a clock to the regeneration process in delay lines may be illustrated by an interesting little theorem. Suppose that instead of the condition $s \in F_r \supset f(s) \in E_r$ we impose a stronger one, viz $f^n(s) \to c_r$ if $s \in E_r$, i.e. there are ideal forms of the distinguishable signals, and each admissible signal converges towards the ideal form after recirculating. Then we can show that unless there is a clock the ideal signals are all constants. For let U_α represent a shift of origin, i.e. $U_\alpha s(t) = s(t + \alpha)$. Then since there is no clock the properties of the recirculator are the same at all times and f therefore commutes with U_α. Then $f U_\alpha(c_r) = U_\alpha f(c_r) = U_\alpha c_r$, for $f(c_r) = c_r$ since c_r is an ideal signal. But this means that $U_\alpha(c_r)$ is an ideal signal, and therefore for sufficiently small α must be c_r, since the ideal signals are discrete. Then for any β and sufficiently large n, β/n will be sufficiently small and $U_{\beta/n}(c) = c$. But then by iteration $c = U_{\beta/n}^n(c) = U_\beta(c)$ i.e. $c(t + \beta) = c(t)$. This means that the ideal signal c is a constant.

We might say that the clock enables us to introduce a discreteness into time, so that time can for some purposes be regarded as a succession of instants instead of as a continuous flow. A digital machine must essentially deal with discrete objects, and in the case of the ACE this is made possible by the use of a clock. All other digital computing machines that I know of except for human and other brains do the same. One can think up ways of avoiding it, but they are very awkward. I should mention that the use of the clock in the ACE is not confined to the recirculation process, but is used in almost every part.

It may be as well to mention some figures connected with the mercury delay line as we shall use it. We shall use five foot tubes, with a[n] inside diameter of half an inch. Each of these will enable us to store 1024 binary digits. The unit I have used here to describe storage capacity is self explanatory. A storage mechanism has a capacity of m binary digits if it can remember any sequence of m digits each being a 0 or a 1. The storage capacity is also the logarithm to the base 2 of the number of different signals which can be remembered, i.e. $\log_2 N$. The digits will be placed at a time interval of one microsecond, so that the time taken for the waves to travel down the tube is just over a millisecond. The velocity is about one and a half kilometres per second. The delay in accessibility time or average waiting for a given piece of information is about half a millisecond. In practice this is reduced to an effective 150 μs.[2] The full storage capacity of the ACE available on Hg[3] delay lines will be about 200,000 binary digits. This is probably comparable with the memory capacity of a minnow.

I have spent a considerable time in this lecture on this question of memory, because I believe that the provision of proper storage is the key to the problem of the digital computer, and certainly if they are to be persuaded to show any sort of genuine intelligence much larger capacities than are yet available must be provided. In my opinion this problem of making a large memory available at reasonably short notice is much more important than that of doing operations such as multiplication at high speed. Speed is necessary if the machine is to work fast enough for the machine to be commercially valuable, but a large storage capacity is necessary if it is to be capable of anything more than rather trivial operations. The storage capacity is therefore the more fundamental requirement.

Let us now return to the analogy of the theoretical computing machines with an infinite tape. It can be shown that a single special machine of that type can be made to do the work of all. It could in fact be made to work as a model of any other machine. The special machine may be called the universal machine; it works in the following quite simple manner. When we have decided what machine we wish to imitate we punch a description of it on the tape of the universal machine. This description explains what the machine would do in every configuration in which it might find itself. The universal machine has only to keep looking at this description in order to find out what it should do at each stage. Thus the complexity of the machine to be imitated is concentrated in the tape and does not appear in the universal machine proper in any way.

If we take the properties of the universal machine in combination with the fact that machine processes and rule of thumb processes are synonymous we may say that the universal machine is one which, when supplied with the appropriate instructions, can be made to do any rule of thumb process. This feature is paralleled in digital computing machines such as the ACE. They are in fact practical versions of the universal machine. There is a certain central pool of electronic equipment, and a large memory. When any particular problem has to be handled the appropriate instructions for the computing process involved are stored in the memory of the ACE and it is then 'set up' for carrying out that process.

I have indicated the main strategic ideas behind digital computing machinery, and will now follow this account up with the very briefest description of the ACE. It may be divided for the sake of argument into the following parts

---

[2] Editor's note. μS = microseconds.
[3] Editor's note. Hg = mercury.

Memory
Control
Arithmetic part
Input and Output

I have already said enough about the memory and will only repeat that in the ACE the memory will consist mainly of 200 mercury delay lines each holding 1024 binary digits. The purpose of the control is to take the right instructions from the memory, see what they mean, and arrange for them to be carried out. It is understood that a certain 'code of instructions' has been laid down, whereby each 'word' or combination of say 32 binary digits describes some particular operation. The circuit of the control is made in accordance with the code, so that the right effect is produced. To a large extent we have also allowed the circuit to determine the code, i.e. we have not just thought up an imaginary 'best code' and then found a circuit to put it into effect, but have often simplified the circuit at the expense of the code. It is also quite difficult to think about the code entirely in abstracto without any kind of circuit.

The arithmetic part of the machine is the part concerned with addition, multiplication and any other operations which it seems worth while to do by means of special circuits rather than through the simple facilities provided by the control. The distinction between control and arithmetic part is a rather hazy one, but at any rate it is clear that the machine should at least have an adder and a multiplier, even if they turn out in the end to be part of the control. This is the point at which I should mention that the machine is operated in the binary scale, with two qualifications. Inputs from externally provided data are in decimal, and so are outputs intended for human eyes rather than for later reconsumption by the ACE. This is the first qualification. The second is that, in spite of the intention of binary working there can be no bar on decimal working of a kind, because of the relation of the ACE to the universal machine. Binary working is the most natural thing to do with any large scale computer. It is much easier to work in the scale of two than any other, because it is so easy to produce mechanisms which have two positions of stability: the two positions may then

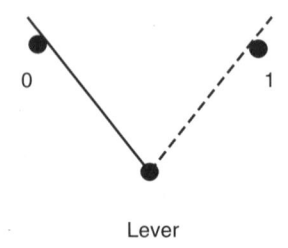

Lever

Figure 9.4.

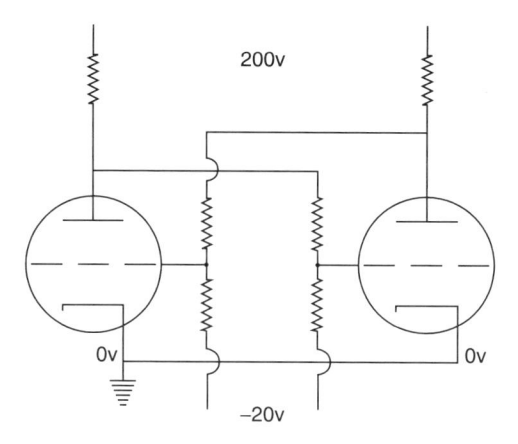

Figure 9.5.

Editor's note. I am grateful to John Andreae and David Clayden for assistance with redrawing Turing's original figure. The figure shows an Eccles–Jordan trigger circuit, also known as a bistable circuit or flip-flop. The circuit has two stable states. In one, current flows through the left-hand triode valve but not through the right; in the other, current flows through the right-hand triode but not the left. The circuit can be switched from one state to the other by applying a pulse to one of the valves.

be regarded as representing 0 and 1. Examples are lever as diagram, Jordan Eccles circuit, thyratron. If one is concerned with a small scale calculating machine then there is at least one serious objection to binary working. For practical use it will be necessary to build a converter to transform numbers from the binary form to the decimal and back. This may well be a larger undertaking than the binary calculator. With the large scale machines this argument carries no weight. In the first place a converter would become a relatively small piece of apparatus, and in the second it would not really be necessary. This last statement sounds quite paradoxical, but it is a simple consequence of the fact that these machines can be made to do any rule of thumb process by remembering suitable instructions. In particular it can be made to do binary decimal conversion. For example in the case of the ACE the provision of the converter involves no more than adding two extra delay lines to the memory. This situation is very typical of what happens with the ACE. There are many fussy little details which have to be taken care of, and which, according to normal engineering practice, would require special circuits. We are able to deal with these points without modification of the machine itself, by pure paper work, eventually resulting in feeding in appropriate instructions.

To return to the various parts of the machine. I was saying that it will work in the scale of two. It is not unnatural to use the convention that an electrical pulse shall represent the digit 1 and that absence of a pulse shall represent a digit 0. Thus a sequence of digits 0010110 would be represented by a signal like

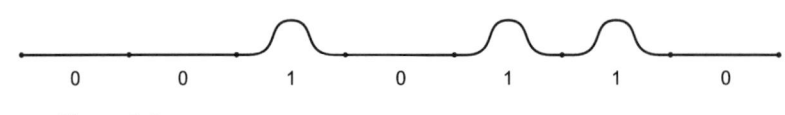

Figure 9.6.

where the time interval might be one microsecond. Let us now look at what the process of binary addition is like. In ordinary decimal addition we always begin from the right, and the same naturally applies to binary. We have to do this because we cannot tell whether to carry unless we have already dealt with the less significant columns. The same applies with electronic addition, and therefore it is convenient to use the convention that if a sequence of pulses is coming down a line, then the least significant pulse always comes first. This has the unfortunate result that we must either write the least significant digit on the left in our binary numbers or else make time flow from right to left in our diagrams. As the latter alternative would involve writing from right to left as well as adding in that way, we have decided to put the least significant digit on the left. Now let us do a typical addition. Let us write the carry digits above the addends.

```
Carry     0 1 1 1 1 1 0 0 1       1
A       0 1 1 0 1 1 0 0 1 0     1 ...
B       0 1 1 1 0 1 0 0 1 1 ...
        ─────────────────────
        0 1 0 0 1 1 0 0 0
```

Note that I can do the addition only looking at a small part of the data. To do the addition electronically we need to produce a circuit with three inputs and two outputs.

Inputs		Outputs	
Addend A	α	Sum	δ
Addend B	β	Carry	ε
Carry from last column	γ		

This circuit must be such that

If no. of 1s on inputs α, β, γ is $\begin{cases} 0 \\ 1 \\ 2 \\ 3 \end{cases}$ Then sum $\begin{matrix} 0 \\ \delta \\ \text{is} \\ \end{matrix}$ $\begin{matrix} 0 \\ 1 \\ 0 \\ 1 \end{matrix}$ and carry $\begin{matrix} 0 \\ 0 \\ \varepsilon \\ \text{is} \end{matrix}$ $\begin{matrix} 0 \\ 0 \\ 1 \\ 1 \end{matrix}$

It is very easy to produce a voltage proportional to the number of pulses on the inputs, and one then merely has to provide a circuit which will discriminate between four different levels and put out the appropriate sum and carry digits. I will not attempt to describe such a circuit; it can be quite simple. When we are given the circuit we merely have to connect it up with feedback and it is an adder. Thus:

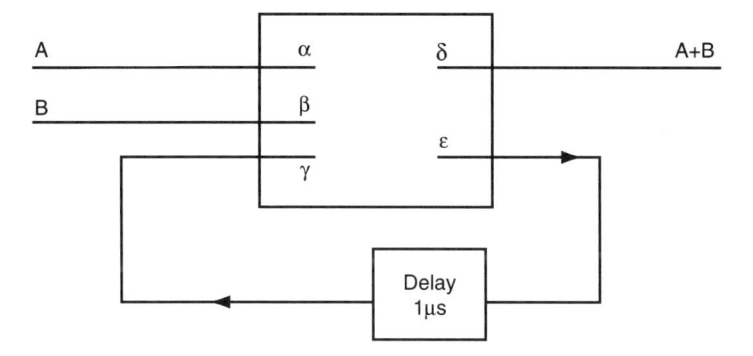

Figure 9.7.

It will be seen that we have made use of the fact that the same process is used in addition with each digit, and also the fact that the properties of the electrical circuit are invariant under time shifts, at any rate if these are multiples of the clock period. It might be said that we have made use of the isomorphism between the group of these time shifts and the multiplicative group of real numbers to simplify our apparatus, though I doubt if many other applications of this principle could be found.

It will be seen that with such an adder the addition is broken down into the most elementary steps possible, such as adding one and one. Each of these occupies a microsecond. Our numbers will normally consist of 32 binary digits, so that two of them can be added in 32 microseconds. Likewise we shall do multiplications in the form of a number of consecutive additions of one and one or one and zero etc. There are 1024 such additions or thereabouts to be done in a multiplication of one 32 digit number by another, so that one might expect a multiplication to take about a millisecond. Actually the multiplier to be used on ACE will take rather over two milliseconds. This may sound rather long, when the unit operation is only a microsecond, but it actually seems that the machine is fairly well balanced in this respect, i.e. the multiplication time is not a serious bottleneck. Computers always spend just as long in writing numbers down and deciding what to do next as they do in actual multiplications, and it is just the same with the ACE. A great deal of time is spent in getting numbers in and out of storage and deciding what to do next. To complete the four elementary processes, subtraction is done by complementation and addition, and division is done by the use of the iteration formula

$$u_n = u_{n-1} + u_{n-1}(1 - au_{n-1}).$$

u_n converges to a^{-1} provided $|1 - au_0| < 1$. The error is squared at each step, so that the convergence is very rapid. This process is of course programmed, i.e. the only extra apparatus required is the delay lines required for storing the relevant instructions.

Passing on from the arithmetic part there remains the input and output. For this purpose we have chosen Hollerith card equipment. We are able to obtain this without having to do any special development work. The speeds obtainable are not very impressive compared with the speeds at which the electronic equipment works, but they are quite sufficient in all cases where the calculation is long and the result concise: the interesting cases in fact. It might appear that there would be a difficulty in converting the information provided at the slow speeds appropriate to the Hollerith equipment to the high speeds required with the ACE, but it is really quite easy. The Hollerith speeds are so slow as to be counted zero or stop for many purposes, and the problem reduces to the simple one of converting a number of statically given digits into a stream of pulses. This can be done by means of a form of electronic commutator.

Before leaving the outline of the description of the machine I should mention some of the tactical situations that are met with in programming. I can illustrate two of them in connection with the calculation of the reciprocal described above. One of these is the idea of the iterative cycle. Each time that we go from u_r to u_{r+1} we apply the same sequence of operations, and it will therefore be economical in storage space if we use the same instructions. Thus we go round and round a cycle of instructions:

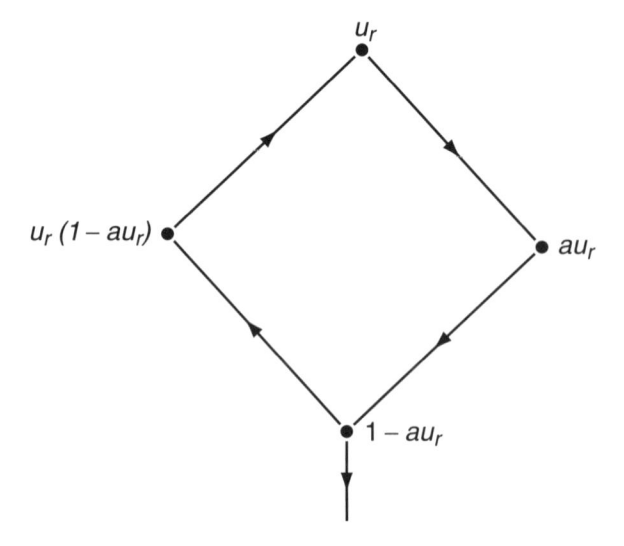

[Diagram: cycle with nodes u_r (top), au_r (right), $1 - au_r$ (bottom), $u_r(1 - au_r)$ (left)]

Figure 9.8.
Editor's note. I am grateful to Mike Woodger for assistance with redrawing Turing's original diagram.

It looks however as if we were in danger of getting stuck in this cycle, and unable to get out. The solution of this difficulty involves another tactical idea, 'discrimination' i.e. of deciding what to do next partly according to the results of the machine itself, instead of according to data available to the programmer. In that of this case we include a discrimination in each cycle, which takes us out of the cycle when the value of $|1 - au|$ is sufficiently small. It is like an aeroplane circling over an aerodrome, and asking permission to land after each circle. This is a very simple idea, but is of the utmost importance. The idea of the iterative cycle of instructions will also be seen to be rather fundamental when it is realised that the majority of the instructions in the memory must be obeyed a great number of times. If the whole memory were occupied by instructions, none of it being used for numbers or other data, and if each instruction were obeyed once only, but took the longest possible time, the machine could only remain working for sixteen seconds.

Another important idea is that of constructing an instruction and then obeying it. This can be used amongst other things for discrimination. In the example I have just taken for instance we could calculate a quantity which was 1 if $|1 - au|$ was less than 2^{-31} and 0 otherwise. By adding this quantity to the instruction that is obeyed at the forking point that instruction can be completely altered in its effect when finally $1 - au$ is reduced to sufficiently small dimensions.

Probably the most important idea involved in instruction tables is that of standard *subsidiary tables*. Certain processes are used repeatedly in all sorts of different connections, and we wish to use the same instructions, from the same part of the memory every time. Thus we may use interpolation for the calculation of a great number of different functions, but we shall always use the same instruction table for interpolation. We have only to think out how this is to be done once, and forget then how it is done. Each time we want to do an interpolation we have only to remember the memory position where this table is kept, and make the appropriate reference in the instruction table which is using the interpolation. We might for instance be making up an instruction table for finding values of $J_0(x)$ and use the interpolation table in this way. We should then say that the interpolation table was a subsidiary to the table for calculating $J_0(x)$. There is thus a sort of hierarchy of tables. The interpolation table might be regarded as taking its orders from the J_0 table, and reporting its answers back to it. The master servant analogy is however not a very good one, as there are many more masters than servants, and many masters have to share the same servants.

Now let me give a picture of the operation of the machine. Let us begin with some problem which has been brought in by a customer. It will first go to the problems preparation section where it is examined to see whether it is in a suitable form and self-consistent, and a very rough computing procedure made out. It then goes to the tables preparation section. Let us suppose for example that the problem was to tabulate solutions of the equation

$$y'' + xy' = J_0(x)$$

with initial conditions $x = y = 0$, $y' = a$. This would be regarded as a particular case of solving the equation

$$y'' = F(x, y, y')$$

for which one would have instruction tables already prepared. One would need also a table to produce the function $F(x, y, z)$ (in this case $F(x, y, z) = J_0(x) - xz$ which would mainly involve a table to produce $J_0(x)$, and this we might expect to get off the shelf). A few additional details about the boundary conditions and the length of the arc would have to be dealt with, but much of this detail would also be found on the shelf, just like the table for obtaining $J_0(x)$. The instructions for the job would therefore consist of a considerable number taken off the shelf together with a few made up specially for the job in question. The instruction cards for the standard processes would have already been punched, but the new ones would have to be done separately. When these had all been assembled and checked they would be taken to the input mechanism, which is simply a Hollerith card feed. They would be put into the card hopper and a button pressed to start the cards moving through. It must be remembered that initially there are no instructions in the machine, and one's normal facilities are therefore not available. The first few cards that pass in have therefore to be carefully thought out to deal with this situation. They are the initial input cards and are always the same. When they have passed in a few rather fundamental instruction tables will have been set up in the machine, including sufficient to enable the machine to read the special pack of cards that has been prepared for the job we are doing. When this has been done there are various possibilities as to what happens next, depending on the way the job has been programmed. The machine might have been made to go straight on through, and carry out the job, punching or printing all the answers required, and stopping when all of this has been done. But more probably it will have been arranged that the machine stops as soon as the instruction tables have been put in. This allows for the possibility of checking that the content of the memories is correct, and for a number of variations of procedure. It is clearly a suitable moment for a break. We might also make a number of other breaks. For instance we might be interested in certain particular values of the parameter *a*, which were experimentally obtained figures, and it would then be convenient to pause after each parameter value, and feed the next parameter value in from another card. Or one might prefer to have the cards all ready in the hopper and let the ACE take them in as it wanted them. One can do as one wishes, but one must make up one's mind. Each time the machine pauses in this way a 'word' or sequence of 32 binary digits is displayed on neon bulbs. This word indicates the reason for stopping. I have already mentioned two possible reasons. A large class of further possible reasons is provided by the checks. The programming should be done in such a way that the ACE is frequently investigating identities which should be satisfied if all is as it should be. Whenever one of these checks fails the machine stops and displays a word which describes what check has failed.

It will be seen that the possibilities as to what one may do are immense. One of our difficulties will be the maintenance of an appropriate discipline, so that we do not lose track of what we are doing. We shall need a number of efficient librarian types to keep us in order.

Finally I should like to make a few conjectures as to the repercussions that electronic digital computing machinery will have on mathematics. I have already mentioned that the ACE will do the work of about 10,000 computers. It is to be expected therefore that large scale hand-computing will die out. Computers will still be employed on small calculations, such as the substitution of values in formulae, but whenever a single calculation may be expected to take a human computer days of work, it will presumably be done by an electronic computer instead. This will not necessitate everyone interested in such work having an electronic computer. It would be quite possible to arrange to control a distant computer by means of a telephone line. Special input and output machinery would be developed for use at these out stations, and would cost a few hundred pounds at most. The main bulk of the work done by these computers will however consist of problems which could not have been tackled by hand computing because of the scale of the undertaking. In order to supply the machine with these problems we shall need a great number of mathematicians of ability. These mathematicians will be needed in order to do the preliminary research on the problems, putting them into a form for computation. There will be considerable scope for analysts. When a human computer is working on a problem he can usually apply some common sense to give him an idea of how accurate his answers are. With a digital computer we can no longer rely on common sense, and the bounds of error must be based on some proved inequalities. We need analysts to find the appropriate inequalities for us. The inequalities need not always be explicit, i.e. one need not have them in such a form that we can tell, before the calculation starts, and using only pencil and paper, how big the error will be. The error calculation may be a serious part of the ACE's duties. To an extent it may be possible to replace the estimates of error by statistical estimates obtained by repeating the job several times, and doing the rounding off differently each time, controlling it by some random element, some electronic roulette wheel. Such statistical estimates however leave much in doubt, are wasteful in machine time, and give no indication of what can be done if it turns out that the errors are intolerably large. The statistical method can only help the analyst, not replace him.

Analysis is just one of the purposes for which we shall need good mathematicians. Roughly speaking those who work in connection with the ACE will be divided into its masters and its servants. Its masters will plan out instruction tables for it, thinking up deeper and deeper ways of using it. Its servants will feed it with cards as it calls for them. They will put right any parts that go wrong. They will assemble data that it requires. In fact the servants will take the place of limbs. As time goes on the calculator itself will take over the functions both of masters and of servants. The servants will be replaced by mechanical and electrical limbs and sense organs. One might for instance provide curve followers to enable data to be taken direct from curves instead of having girls read off values and punch them on cards. The masters are liable to get replaced because as soon as any technique becomes at all stereotyped it becomes possible to devise a system of instruction tables which will enable the electronic computer to do it for itself. It may happen however that the masters will refuse to do this. They may be unwilling to let their jobs be stolen from them in this way. In that case they would surround the whole of their work with mystery and make excuses, couched in well chosen gibberish, whenever any dangerous suggestions were made. I think that a reaction of this kind is a very real danger. This topic naturally leads to the question as to how far it is possible in principle for a computing machine to simulate human activities. I will return to this later, when I have discussed the effects of these machines on mathematics a little further.

I expect that digital computing machines will eventually stimulate a considerable interest in symbolic logic and mathematical philosophy. The language in which one communicates with these machines, i.e. the language of instruction tables, forms a sort of symbolic logic. The machine interprets whatever it is told in a quite definite manner without any sense of humour or sense of proportion. Unless in communicating with it one says exactly what one means, trouble is bound to result. Actually one could communicate with these machines in any language provided it was an exact language, i.e. in principle one should be able to communicate in any symbolic logic, provided that the machine were given instruction tables which would enable it to interpret that logical system. This should mean that there will be much more practical scope for logical systems than there has been in the past. Some attempts will probably be made to get the machines to do actual manipulations of mathematical formulae. To do so will require the development of a special logical system for the purpose. This system should resemble normal mathematical procedure closely, but at the same time should be as unambiguous as possible. As regards mathematical philosophy, since the machines will be doing more and more mathematics themselves, the centre of gravity of the human interest will be driven further and further into philosophical questions of what can in principle be done etc.

It has been said that computing machines can only carry out the processes that they are instructed to do. This is certainly true in the sense that if they do something other than what they were instructed then they have just made some mistake. It is also true that the intention in constructing these machines in the first instance is to treat them as slaves, giving them only jobs which have been thought out in detail, jobs such that the user of the machine fully understands what in principle is going on all the time. Up till the present machines have only been used in this way. But is it necessary that they should always be used in such a manner? Let us suppose we have set up a machine with certain initial instruction tables, so constructed that these tables might on occasion, if good reason arose, modify those tables. One can imagine that after the machine had been operating for some time, the instructions would have altered out of all recognition, but nevertheless still be such that one would have to admit that the machine was still doing very worthwhile calculations. Possibly it might still be getting results of the type desired when the machine was first set up, but in a much more efficient manner. In such a case one would have to admit that the progress of the machine had not been foreseen when its original instructions were put in. It would be like a pupil who had learnt much from his master, but had added much more by his own work. When this happens I feel that one is obliged to regard the machine as showing intelligence. As soon as one can provide a reasonably large memory capacity it should be possible to begin to experiment on these lines. The memory capacity of the human brain is probably of the order of ten thousand million binary digits. But most of this is probably used in remembering visual impressions, and other comparatively wasteful ways. One might reasonably hope to be able to make some real progress with a few million digits, especially if one confined one's investigations to some rather limited field such as the game of chess. It would probably be quite easy to find instruction tables which would enable the ACE to win against an average player. Indeed Shannon of Bell Telephone laboratories tells me that he has won games playing by rule of thumb: the skill of his opponents is not stated. But I would not consider such a victory very significant. What we want is a machine that can learn from experience. The possibility of letting the machine alter its own instructions provides the mechanism for this, but this of course does not get us very far.

It might be argued that there is a fundamental contradiction in the idea of a machine with intelligence. It is certainly true that 'acting like a machine' has become synonymous with lack of adaptability. But the reason for this is obvious. Machines in the past have had very little storage, and there has been no question of the machine having any discretion. The argument might however be put into a more aggressive form. It has for instance been shown that with certain logical systems there can be no machine which will distinguish provable formulae of the system from unprovable, i.e. that there is no test that the machine can apply which will divide propositions with certainty into these two classes. Thus if a machine is made for this purpose it must in some cases fail to give an answer. On the other hand if a mathematician is confronted with such a problem he would search around a[nd] find new methods of proof, so that he ought eventually to be able to reach a decision about any given formula. This would be the argument. Against it I would say that fair play must be given to the machine. Instead of it sometimes giving no answer we could arrange that it gives occasional wrong answers. But the human mathematician would likewise make blunders when trying out new techniques. It is easy for us to regard these blunders as not counting and give him another chance, but the machine would probably be allowed no mercy. In other words then, if a machine is expected to be infallible, it cannot also be intelligent. There are several mathematical theorems which say almost exactly that. But these theorems say nothing about how much intelligence may be displayed if a machine makes no pretence at infallibility. To continue my plea for 'fair play for the machines' when testing their I.Q. A human mathematician has always undergone an extensive training. This training may be regarded as not unlike putting instruction tables into a machine. One must therefore not expect a machine to do a very great deal of building up of instruction tables on its own. No man adds very much to the body of knowledge, why should we expect more of a machine? Putting the same point differently, the machine must be allowed to have contact with human beings in order that it may adapt itself to their standards. The game of chess may perhaps be rather suitable for this purpose, as the moves of the machine's opponent will automatically provide this contact.

CHAPTER 10

Intelligent Machinery (*1948*)
Alan Turing

Introduction
Jack Copeland

Slow Progress on the ACE

By the beginning of 1947 much effort had gone into writing programmes or 'instruction tables' for the ACE. In 'Proposed Electronic Calculator', Turing had said that work on instruction tables should start immediately, since the 'earlier stages of the making of instruction tables will have serious repercussions on the design [of the machine]'; moreover, the programming work should 'go on whilst the machine is being built, in order to avoid some of the delay between the delivery of the machine and the production of results'.[1]

However, little progress had been made on the physical construction of the ACE. The actual engineering work was being carried out not at the National Physical Laboratory but at the Post Office Research Station, under the supervision of Turing's wartime associate Flowers (see the introductions to Chapters 4 and 9). Flowers was asked by the NPL early in 1946 to assist with the engineering design of, and to build, the ACE.[2] Ominously, the letter of agreement from Flowers's superior to Turing's superior spoke of 'very considerable arrears of work' and warned that 'the manpower position is difficult'.[3]

Initial progress was nevertheless promising, with some early successes in the experimental work on the delay line memory units. According to an NPL

---

[1] 'Proposed Electronic Calculator', 18 (see n. 6 of the introduction to Chapter 9). A digital facsimile of the original typescript of 'Proposed Electronic Calculator' is in The Turing Archive for the History of Computing <www.AlanTuring.net/proposed_electronic_calculator>. Page references are to this typescript.

[2] Flowers in interview with Copeland (July 1996).

[3] Letter from W. G. Radley to Womersley, 25 Feb. 1946 (Public Record Office, Kew, Richmond, Surrey (document reference DSIR 10/385); a digital facsimile is in The Turing Archive for the History of Computing <www.AlanTuring.net/radley_womersley_25feb46>).

document dated March 1946, 'Mr. Flowers states that they can have ready for N.P.L. a minimal ACE by August or September.'[4] Unfortunately it proved impossible to keep to Flowers's timetable. Dollis Hill was occupied with a backlog of urgent work on the national telephone system (at that time managed by the Post Office). Flowers's Section was 'too busy to do other people's work' (he said in 1998).[5] Only two men could be spared to work on the ACE, Chandler and Coombs (both of whom had played leading roles in the wartime Colossus project).

Sir Charles Darwin, the Director of the NPL, noted in August 1946 that the Post Office was 'not in a position to plunge very deep', and by November was expressing concern to Post Office staff about the slow rate of progress on the ACE.[6] The fault was not all with the Post Office, however. In November 1946, the NPL proposed a radical change in the hardware design, with cathode ray tube memory units taking the place of mercury delay lines.[7] Such a change would have meant that most of the work done by Chandler and Coombs up to that point was wasted. (In the end CRT memory was used at Manchester but not in the ACE.) The logical design kept changing, too—by the end of 1946 Turing had reached Version VII of the ACE design.[8] Coombs said:

One of the problems was, I remember, that NPL kept on changing its ideas, and every time we went down there and said 'Right now! What do you want us to make?', we'd find that the last idea, that they gave us last week, was old hat and they'd got a quite different one, and we couldn't get a consolidated idea at all until eventually we dug our toes in and said 'Stop! Tell us what to make.'[9]

Some eight years later, Chandler and Coombs finally completed a computer based on Version VII of Turing's ACE design, the MOSAIC (see the introduction to Chapter 9).

---

[4] 'Status of the Delay Line Computing Machine at the P.O. Research Station', National Physical Laboratory, 7 Mar. 1946, anon. (Woodger Papers, National Museum of Science and Industry, Kensington, London (catalogue reference M12/105); a digital facsimile is in The Turing Archive for the History of Computing <www.AlanTuring.net/delay_line_status>).

[5] Flowers in interview with Copeland (July 1998).

[6] Letter from Darwin to Sir Edward Appleton, 13 Aug. 1946 (Public Record Office (document reference DSIR 10/275); a digital facsimile is in The Turing Archive for the History of Computing <www.AlanTuring.net/darwin_appleton_13aug46>); letter from Radley to Darwin, 1 Nov. 1946 (Public Record Office (document reference DSIR 10/385); <www.AlanTuring.net/radley_darwin_1nov46>).

[7] Letter from Darwin to Radley, 26 Nov. 1946 (Public Record Office (document reference DSIR 10/385); a digital facsimile is in The Turing Archive for the History of Computing <www.AlanTuring.net/darwin_radley_26nov46>).

[8] See B. J. Copeland (ed.), 'The Turing–Wilkinson Lectures on the Automatic Computing Engine', in K. Furukawa, D. Michie, and S. Muggleton (eds.), *Machine Intelligence 15* (Oxford University Press, 1999).

[9] Coombs in interview with Christopher Evans in 1976 ('The Pioneers of Computing: an Oral History of Computing' (London: Science Museum)).

Turing Proposes an ACE Electronics Section

In January 1947 Turing had gone to the United States, visiting several of the groups there that were attempting to build an electronic stored-programme computer. In his report on his visit he wrote:

> One point concerning the form of organisation struck me very strongly. The engineering development work was in every case being done in the same building with the more mathematical work. I am convinced that this is the right approach. It is not possible for the two parts of the organisation to keep in sufficiently close touch otherwise. They are too deeply interdependent. We are frequently finding that we are held up due to ignorance of some point which could be cleared up by a conversation with the engineers, and the Post Office find similar difficulty; a telephone conversation is seldom effective because we cannot use diagrams. Probably more important are the points which are misunderstood, but which would be cleared up if closer contact were maintained, because they would come to light in casual discussion. It is clear that we must have an engineering section at the ACE site eventually, the sooner the better, I would say.[10]

Darwin suggested to Dr Horace Thomas, a member of the NPL's Radio Division, that he 'be put in charge of the work of making by a suitable firm a prototype model' of the ACE.[11] A 'pre-prototype model' would be started in-house at the NPL 'in Radio and Metrology workshops before approaching an outside firm', Darwin said, and 'Metrology and Radio workshops could get on with the hardware part of the job straight away'.[12] (The outside firm eventually approached was English Electric, who subsequently built and marketed the DEUCE (see the introduction to Chapter 9). In 1949 wiremen and engineers from English Electric would join the NPL team to assist with the completion of Pilot Model ACE.[13])

Darwin's directive that an in-house electronics section begin work on the hardware took several months to implement, however—the wheels of administration turned slowly. At the end of April an NPL minute spoke of the need for

---

[10] A. M. Turing, 'Report on visit to U.S.A., January 1st–20th, 1947', National Physical Laboratory, 3 Feb. 1947 (Public Record Office (document reference DSIR 10/385); a digital facsimile is in The Turing Archive for the History of Computing <www.AlanTuring.net/turing_usa_visit>).

[11] Report by Darwin to the NPL Executive Committee in the Minutes of the Executive Committee for 18 Mar. 1947 (NPL library; a digital facsimile is in The Turing Archive for the History of Computing <www.AlanTuring.net/npl_minutes_mar1947>).

[12] Ibid.

[13] The placing of a contract with English Electric '[i]n order to expedite the construction of [the] pilot assembly and to make possible the construction of the final machine' was proposed by Mathematics Division in February 1949 and approved by Treasury in May of that year (J. R. Womersley, 'A.C.E. Project: Present Position, and request for financial provision for a Study Contract to be placed with the English Electric Co. Ltd.', 1 Feb. 1949 (Woodger Papers; a digital facsimile is in The Turing Archive for the History of Computing <www.AlanTuring.net/ace_project_position>); letter from Evans to Darwin, 28 May 1949 (Public Record Office (document reference DSIR 10/275); a digital facsimile is in The Turing Archive for the History of Computing <www.AlanTuring.net/evans_darwin_28may49>)).

an electronics group working 'together in one place as a whole in close contact with the planning staff in the Mathematics Division'.[14] The minute emphasized that the 'various parts' of the ACE project were 'so interwoven' that it was 'not practicable at present to farm out portions of the work to isolated groups', adding 'Our experience with the Post Office confirms this.' Yet it was not until August that suitable staff were seconded from elsewhere in the NPL and Thomas's electronics group finally came into existence.[15]

Womersley proposed that the new group should complete the 'Test Assembly', a pilot version of the ACE begun by Huskey in Mathematics Division in the spring of 1947 (see the introduction to Chapter 9).[16] Womersley's proposal was disregarded, however. Thomas was an empire-builder. The week before the official launch of Thomas's electronics group on 18 August 1947, E. S. Hiscocks, Secretary to the NPL, had written:

Thomas has apparently shown some signs of behaving as if he is starting up a new Division, and so as to allay certain qualms which both Smith-Rose and Womersley have, I think it would be better for it to be explained to the whole team that Mathematics Division is the parent Division, and the one which is to justify the financial outlay on this work; that the work is being put out on contract, as it were, to Radio Division, and that Thomas's team is a part of Radio Division. I think, even if only for our own peace of mind, this is desirable, because Thomas has already shown some signs of wanting to set up a separate office, etc.[17]

Unfortunately Thomas 'didn't like...the idea of this group in Mathematics Division...working independently', Wilkinson recalled.[18] Rivalry quickly sprang up between Thomas's group in Radio Division and the ACE Section in Mathematics Division, and soon after Thomas's appointment the work on the Test Assembly was summarily stopped by Darwin. This was a dire turn of events—in Huskey's words, 'morale in the Mathematics Division collapsed'.[19]

---

[14] J. R. Womersley and R. L. Smith-Rose, 'A.C.E. Pilot Test Assembly and later Development', National Physical Laboratory, 30 Apr. 1947 (Public Record Office (document reference DSIR 10/385); a digital facsimile is in The Turing Archive for the History of Computing <www.AlanTuring.net/pilot_test_assembly>).

[15] Memorandum from Hiscocks to Womersley, National Physical Laboratory, 6 Aug. 1947 (Public Record Office (document reference DSIR 10/385); a digital facsimile is in The Turing Archive for the History of Computing <www.AlanTuring.net/hiscocks_womersley_6aug47>).

[16] 'A.C.E. Project', National Physical Laboratory, 21 Aug. 1947, initialled 'JWC/JG' (Public Record Office (document reference DSIR 10/385); a digital facsimile is in The Turing Archive for the History of Computing <www.AlanTuring.net/ace_project_meeting>).

[17] Letter from Hiscocks to Darwin, 12 Aug. 1947 (Public Record Office (document reference DSIR 10/385); a digital facsimile is in The Turing Archive for the History of Computing <www.AlanTuring.net/hiscocks_darwin_12aug47>).

[18] Wilkinson in interview with Evans in 1976 ('The Pioneers of Computing: An Oral History of Computing' (London: Science Museum)).

[19] H. D. Huskey, 'From ACE to the G-15', Annals of the History of Computing, 6 (1984), 350–71 (361).

The Test Assembly—based on Version V of Turing's design for the ACE[20]—was in fact a highly promising project and considerable progress had been made during the course of 1947. By August 1947 the NPL workshops were fabricating a mercury delay line to Huskey's specifications, valve types had been chosen and circuit block diagrams made, source and destination decisions had been taken, and programmes were being written to check these decisions.[21] In October, Womersley and E. C. Fieller (who became Superintendent of Mathematics Division when Womersley left in 1950) expected, optimistically, that the Test Assembly would 'be ready by the end of November'.[22]

Work on building the ACE drew nearly to a standstill after Thomas persuaded Darwin to shut down the Test Assembly. Only two members of Thomas's group were skilled in digital electronics and they knew little about computers. Thomas's own background was not in digital electronics at all but in radio and industrial electronics. The group 'began to develop their knowledge of pulse techniques', said Wilkinson, and 'for a while they just did basic things and became more familiar with the electronics they needed to learn to build a computer'. Then, in February 1948, Thomas resigned from the NPL to join Unilever Ltd. (the manufacturers of 'Sunlight' soap). As Womersley summed up the situation in April 1948, hardware development was 'probably as far advanced 18 months ago'.[23]

Given only slightly better management on the NPL's part, a minimal computer based on Turing's Version V of the ACE design could probably have been working at the NPL during 1948. By September 1947 the NPL had an electronics group which, by joining forces with Huskey and the ACE Section, would have been capable of carrying Huskey's Test Assembly to completion. (Huskey said: 'I never hoped to have the Test Assembly working before I left [the NPL for the USA] in December. I certainly hoped the group would have it working in 1948.') Womersley had advocated this course of action but Thomas threw a spanner in the works.[24] The Test Assembly could easily have been the world's first electronic

[20] M. Woodger, 'ACE Test Assembly, Sept./Oct. 1947', National Physical Laboratory, n.d. (Woodger Papers (catalogue reference M15/84); a digital facsimile is in The Turing Archive for the History of Computing <www.AlanTuring.net/test_assembly>).

[21] Letter from Huskey to Copeland (3 June 2003).

[22] E. C. Fieller, 'Hollerith Equipment for A.C.E. Work - Immediate Requirements', National Physical Laboratory, 16 Oct. 1947 (Public Record Office (document reference DSIR 10/385); a digital facsimile is in The Turing Archive for the History of Computing <www.AlanTuring.net/hollerith_equipment>).

[23] Minutes of the Executive Committee of the National Physical Laboratory for 20 Apr. 1948 (NPL library; a digital facsimile is in The Turing Archive for the History of Computing <www.AlanTuring.net/npl_minutes_apr1948>).

[24] J. R. Womersley, 'A.C.E. Project', National Physical Laboratory, n.d., attached to a letter from Womersley to the Secretary of the NPL dated 21 Aug. 1947 (Public Record Office (document reference DSIR 10/385); a digital facsimile is in The Turing Archive for the History of Computing <www.AlanTuring.net/womersley_ace_project>).

stored-programme digital computer to run a trial calculation (a title that in the event went to the Manchester 'Baby' in June 1948).

Following Thomas's replacement by Francis Colebrook in March 1948 the fortunes of the ACE did at last begin to improve. Colebrook drew the ACE Section and the Electronics Section together to work harmoniously on what became the Pilot Model ACE (see the introduction to Chapter 9).[25] But Turing did not wait.

Turing Leaves the National Physical Laboratory

In the middle of 1947 a thoroughly disheartened Turing applied for a twelve-month period of sabbatical leave to be spent in Cambridge. The purpose of the leave, as described by Darwin in July 1947, was to enable Turing

> to extend his work on the machine [the ACE] still further towards the biological side. I can best describe it by saying that hitherto the machine has been planned for work equivalent to that of the lower parts of the brain, and he [Turing] wants to see how much a machine can do for the higher ones; for example, could a machine be made that could learn by experience? This will be theoretical work, and better done away from here.[26]

Turing left the NPL for Cambridge in the autumn of 1947.[27] Then in May 1948 he gave up his position at the NPL altogether, breaking what Darwin referred to as 'a gentleman's agreement to return here for at least two years after the year's absence'.[28] Newman's offer of a job lured a 'very fed up' Turing—Robin Gandy's description[29]—to Manchester University, where in May 1948 he was appointed Deputy Director of the Computing Machine Laboratory (there being no director).[30]

---

[25] The Thomas era and the ACE's change in fortunes under Colebrook are described in B. J. Copeland, 'The Origins and Development of the ACE Project', in B. J. Copeland (ed.), *Alan Turing's Automatic Computing Engine* (Oxford: Oxford University Press, 2004).

[26] Letter from Darwin to Appleton, 23 July 1947 (Public Record Office (document reference DSIR 10/385); a digital facsimile is in The Turing Archive for the History of Computing <www.AlanTuring.net/darwin_appleton_23jul47>).

[27] Probably at the end of September. Turing was still at the NPL when Geoff Hayes arrived in Maths Division on 23 Sept. 1947 (communication from Hayes to Woodger, Nov. 1979). Turing was on half-pay during his sabbatical (Minutes of the Executive Committee of the National Physical Laboratory for 28 Sept. 1948 (NPL library; a digital facsimile is in The Turing Archive for the History of Computing <www.AlanTuring.net/npl_minutes_sept1948>)). Darwin was in favour of paying Turing his full salary, but Turing preferred not, 'because if he were earning full pay, he would feel that "I ought not to play tennis in the morning, when I want to"' (Darwin to Appleton, 23 July 1947 (see n. 26)).

[28] Darwin to Appleton, 23 July 1947 (see n. 26).

[29] Gandy in interview with Copeland (Oct. 1995).

[30] Turing's salary was paid wholly from a Royal Society grant to Newman for the purpose of developing a stored-programme electronic computer (letter from Newman to D. Brunt at the Royal Society, 22 Dec. 1948; a digital facsimile is in The Turing Archive for the History of Computing <www.AlanTuring.net/newman_brunt_22dec48>).

At Manchester, Turing designed the input mechanism and programming system for an expanded version of Kilburn and William's 'Baby' (see the introduction to Chapter 9) and wrote a programming manual for the new machine.[31] At last Turing had his hands on a functioning stored-programme computer. He was soon using it to model biological growth—pioneering work in the field now known as Artificial Life (see Chapter 15 and 'Artifical Life').

The First Manifesto of Artificial Intelligence

In the summer of 1948 Turing completed a report for Darwin describing the outcomes of his research into 'how much a machine can do for the higher' parts of the brain.[32] It was entitled 'Intelligent Machinery'. Donald Michie recalls that Turing 'was in a state of some agitation about its reception by his superiors at NPL: "A bit thin for a year's time off!" '[33] The headmasterly Darwin—who once complained about the 'smudgy' appearance of Turing's work[34]—was, as Turing predicted, displeased with the report, describing it as a 'schoolboy's essay' and 'not suitable for publication'.[35] In reality this far-sighted paper was the first manifesto of AI; sadly Turing never published it.

'Intelligent Machinery' is a wide-ranging and strikingly original survey of the prospects for Artificial Intelligence. In it Turing brilliantly introduced many of the concepts that were later to become central in the field, in some cases after reinvention by others. These included the logic-based approach to problem-solving, and the idea, subsequently made popular by Newell and Simon, that (as Turing put it) 'intellectual activity consists mainly of various kinds of search' (p. 431). Turing anticipated the concept of a genetic algorithm (GA), in a brief passage concerning what he calls 'genetical or evolutionary search' (see further Chapter 11, p. 463, and 'Artificial Life'). 'Intelligent Machinery' also contains the earliest description of (a restricted form of) what Turing was later to call the 'imitation game' and is now known simply as the Turing test (see further Chapter

---

[31] Letter from Williams to Brain Randell (1972); printed on p. 9 of B. Randell, 'On Alan Turing and the Origins of Digital Computers', in B. Meltzer and D. Michie (eds.), *Machine Intelligence 5* (Edinburgh: Edinburgh University Press, 1972); Turing, *Programmers' Handbook for Manchester Electronic Computer* (University of Manchester Computing Machine Laboratory, 1950; a digital facsimile is in The Turing Archive for the History of Computing <www.AlanTuring.net/programmers_handbook>).

[32] During his sabbatical year Turing also proved that the word problem for semi-groups with cancellation is unsolvable (A. M. Turing, 'The Word Problem in Semi-Groups with Cancellation', *Annals of Mathematics*, 52 (1950), 491–505).

[33] Michie, unpublished note (in the Woodger papers).

[34] Letter from Darwin to Turing, 11 Nov. 1947 (in the Modern Archive Centre, King's College, Cambridge (catalogue reference D 5); a digital facsimile is in The Turing Archive for the History of Computing <www.AlanTuring.net/darwin_turing_11nov47>).

[35] Gandy in interview with Copeland (Oct. 1995); Minutes of the Executive Committee of the National Physical Laboratory for 28 Sept. 1948.

11). The major part of 'Intelligent Machinery', however, consists of an exquisite discussion of machine learning, in which Turing anticipated the modern approach to AI known as connectionism.

Expert Systems

Search and the logic-based approach are both used in modern 'expert systems'. An expert system is an AI programme for solving problems and giving advice within a specialized field of knowledge, such as medical diagnosis or corporate planning. The area of expert systems is one of the most successful in modern AI.

The basic components of an expert system are a knowledge base (KB), an inference engine, and a search engine for searching the KB. The KB is built up by interviewing (human) experts in the area in question. A 'knowledge engineer' organizes the information elicited from the experts into a system of propositions, typically of 'if-then' form.

The inference engine enables the expert system to draw deductions from propositions in the KB. For example, from two propositions 'if x then y' and 'if y then z', the inference engine is able to deduce 'if x then z'. The expert system might then query its user 'Is x true in the situation that we are considering?' (e.g. 'Does the patient have a rash?'), and if the answer is affirmative, the system will proceed to infer z.

Modern systems using search and inference can produce impressive results. For example, Douglas Lenat's common-sense reasoning system CYC is able to conclude 'Garcia is wet' from the statement 'Garcia is finishing a marathon run', by means of searching for and using such items of common-sense knowledge as that running a marathon entails high exertion, that people sweat at high levels of exertion, and that when something sweats it is wet.

Connectionism

Connectionism—still in its infancy—is the science of computing with networks of artificial neurons. This approach came to widespread attention in the mid-1980s when a group based at the University of California at San Diego reported some striking experiments. In one, an artificial neural network learned to form the past tenses of English verbs, responding correctly to irregular verbs not previously encountered (such as 'weep' and 'cling').[36]

---

[36] D. E. Rumelhart and J. L. McClelland, 'On Learning the Past Tenses of English Verbs', in J. L. McClelland, D. E. Rumelhart, and the PDP Research Group, *Parallel Distributed Processing: Explorations in the Microstructure of Cognition*, ii: *Psychological and Biological Models* (Cambridge, Mass.: MIT Press, 1986).

Modern connectionists regard Donald Hebb[37] and Frank Rosenblatt[38] as the founding figures of their approach and it is not widely realized that Turing wrote a blueprint for much of the connectionist project as early as 1948.[39] In 'Intelligent Machinery' he introduced what he called 'unorganised machines', giving as examples networks of neuron-like elements connected together in a largely random manner. He described a certain type of network as 'the simplest model of a nervous system' (p. 418). From a historical point of view, his idea that an initially unorganized neural network can be organized by means of 'interfering training' is of considerable significance, since it did not appear in the earlier work of McCulloch and Pitts (see below).[40] In Turing's model, the training process renders certain neural pathways effective and others ineffective. So far as is known, he was the first person to consider building computing machines out of trainable networks of randomly arranged neuron-like elements.

B-Type Unorganized Machines

In 'Intelligent Machinery' Turing introduced the type of neural network that he called a 'B-type unorganised machine'. A B-type consists of interconnected artificial neurons, depicted in Figure 10.1 as circles, and connection-modifiers, depicted as squares. A B-type may contain any number of neurons connected together in any pattern, but subject always to the restriction that each neuron-to-neuron connection passes through a connection-modifier (Figure 10.2). The connection-modifiers are used in training the network.

Turing's model neurons work as follows. Each neuron has two input fibres, and the output of a neuron is a simple logical function of its two inputs. Every neuron in the network performs the same logical operation, 'nand' (Table 1).

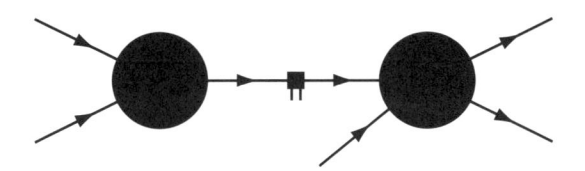

Figure 10.1. Two neurons from a B-type network. The two fibres on the connection-modifier between the neurons enable training by an external agent.

---

[37] D. O. Hebb, *The Organization of Behavior: A Neuropsychological Theory* (New York: John Wiley, 1949).
[38] F. Rosenblatt, *Principles of Neurodynamics* (Washington, DC: Spartan, 1962).
[39] See further B. J. Copeland and D. Proudfoot, 'On Alan Turing's Anticipation of Connectionism' (*Synthese*, 108 (1996), 361–77) and 'Alan Turing's Forgotten Ideas in Computer Science' (*Scientific American*, 280 (1999), 99–103).
[40] W. S. McCulloch and W. Pitts, 'A Logical Calculus of the Ideas Immanent in Nervous Activity', *Bulletin of Mathematical Biophysics*, 5 (1943), 115–33.

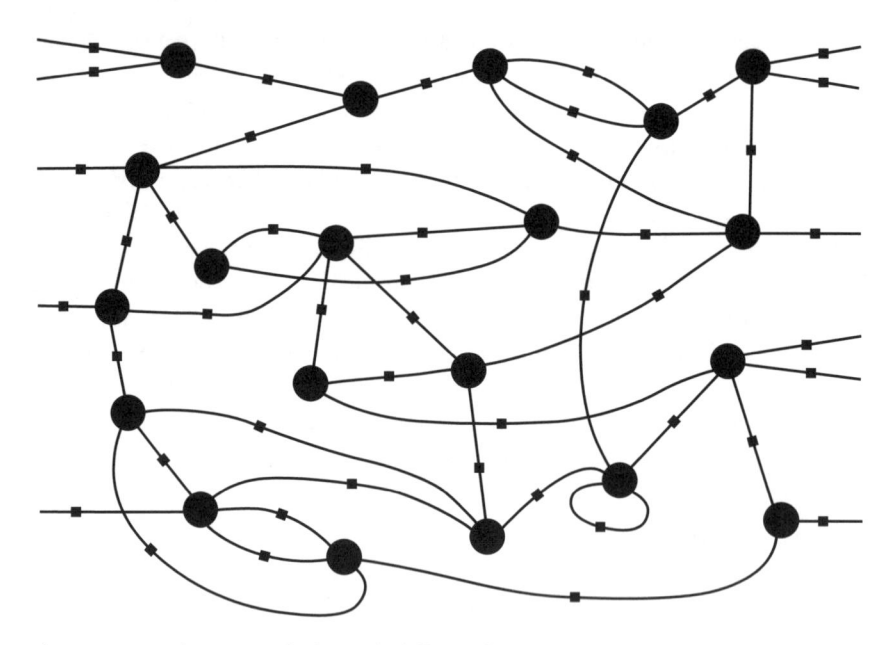

Figure 10.2. A fragment of a large, intially random B-type neural network

Table 1 Behaviour of a nand-neuron

Input-1	Input-2	Output
1	1	0
1	0	1
0	1	1
0	0	1

Turing chose nand as the basic operation of his model neurons because every other boolean operation can be carried out by groups of nand-neurons. Turing showed that even the connection-modifier itself can be built out of nand-neurons. So each B-type network consists of nothing more than nand-neurons and their connecting fibres. This is about the simplest possible model of the cortex.

Training a B-Type

Each connection-modifier has two training fibres (Figure 10.1). Applying a pulse to one fibre sets the modifier to *interchange mode*. When the modifier is in interchange mode and the input into the modifier is 1, the modifier's output is 0; and when the input is 0, the output is 1.

The effect of a pulse on the other training fibre is to place the modifier in *interrupt mode*. In this mode, the output of the modifier is always 1, no matter what its input. When in interrupt mode, the modifier destroys all information attempting to pass along the connection to which the modifier is attached. Once set, a modifier will maintain its function unless it receives a pulse on the other training fibre.

The presence of these connection-modifiers enables a B-type unorganized machine to be trained, by means of what Turing called 'appropriate interference, mimicking education' (p. 422). Turing theorized that 'the cortex of an infant is an unorganised machine, which can be organised by suitable interfering training' (p. 424).

Initially a network that is to be trained contains random inter-neural connections, and the modifiers on these connections are also set randomly (Figure 10.2). Unwanted connections are destroyed by switching their attached modifiers to interrupt mode. The output of the neuron immediately upstream of the modifier no longer finds its way along the connection to the neuron on the downstream end (see Table 1). Conversely, switching the setting of the modifier on an initially interrupted connection to the other mode is in effect to create a new connection. This selective culling and enlivening of connections hones the initially random network into one organized for a given task.

Neural Simulation

Turing wished to investigate other types of 'unorganised machine', and he envisaged the procedure—nowadays used extensively by connectionists—of simulating a neural network and its training regimen using an ordinary digital computer (just as an engineer may use a computer to simulate an aircraft wing or a weather analyst to simulate a storm system). He would, he said, 'allow the whole system to run for an appreciable period, and then break in as a kind of "inspector of schools" and see what progress had been made' (p. 428).

However, Turing's research on neural networks was carried out shortly before the first general-purpose electronic computers were up and running, and he used only paper and pencil. Once he did have access to a computer in the Manchester Computing Machine Laboratory, Turing turned his attention to research in Artificial Life, and he seems never to have used computer simulation to explore the behaviour of neural networks.

In the year of Turing's death (1954), two researchers at MIT, Wesley Clark and Belmont Farley, succeeded in running the first computer simulations of neural networks. Clark and Farley were unaware of Turing's earlier work and their neural architecture was quite different from his, using inter-neural connections of variable 'weight' (or 'strength'). Clark and Farley were able to train their networks—which contained a maximum of 128 neurons—to recognize simple patterns.[41] (In addition, they discovered that the random destruction of up to 10 per cent of the neurons in a trained network does not affect the network's performance at its task.) The work begun by Clark and Farley was considerably developed by Rosenblatt, in whose theory of 'perceptrons' modern connectionism took shape.[42] Meanwhile, Turing's pioneering work on a distinctively different type of connectionist architecture was forgotten.

B-Types Redefined

There appears to be an inconsistency in Turing's presentation of B-types. He claimed that B-types are able, with appropriate modifier settings, to 'do any required job, given sufficient time and provided the number of units [i.e. neurons] is sufficient' (p. 422). This claim is false.[43] The problem lies with the connection-modifiers. If each modifier must either interrupt or interchange, then there are simple logical operations that no B-type can perform. For example, no B-type can produce the *exclusive disjunction* of a pair of inputs (Table 2). Nor can a B-type compute the identity function (i.e. produce outputs identical to its inputs). In effect the difficulty is one that Turing mentions in a different connection. In Chapter 17 he points out that 'an odd number of... interchanges... can never bring one back to where one started' (p. 589).

A simple remedy is to use connection-modifiers that, when not in interrupt mode, are in *pass mode* rather than interchange mode.[44] In pass mode the modifier's output is identical to its input. Turing's claim that B-types can 'do any required job...' is true if the two available modes of each modifier are interrupt and pass.

The required interrupt/pass modifiers can in fact be built out of pairs of interrupt/interchange modifiers. A pair of modifiers both in interchange mode amounts to a single modifier in pass mode. Similarly, a pair of modifiers whose downstream member is in interrupt mode functions indistinguishably from a single modifier in interrupt mode.

---

[41] B. G. Farley and W. A. Clark, 'Simulation of Self-Organising Systems by Digital Computer', *Institute of Radio Engineers Transactions on Information Theory*, 4 (1954), 76–84; W. A. Clark and B. G. Farley, 'Generalisation of Pattern Recognition in a Self-Organising System', in *Proceedings of the Western Joint Computer Conference* (1955).

[42] F. Rosenblatt, 'The Perceptron, a Perceiving and Recognizing Automaton', Cornell Aeronautical Laboratory Report No. 85-460-1 (1957); 'The Perceptron: a Theory of Statistical Separability in Cognitive Systems', Cornell Aeronautical Laboratory Report No. VG-1196-G-1 (1958); 'The Perception: a Probabilistic Model for Information Storage and Organisation in the Brain', *Psychological Review*, 65 (1958), 386–408.

[43] Copeland and Proudfoot, 'On Alan Turing's Anticipation of Connectionism', 367.

[44] Ibid. 367–8.

Table 2 Exclusive disjunction (XOR)

Input-1	Input-2	Output
1	1	0
1	0	1
0	1	1
0	0	0

B-Types and the Universal Turing Machine

Turing claimed a proof (not given in 'Intelligent Machinery' and now lost) of the proposition that an initially unorganized B-type network with sufficient neurons can be organized (via its connection-modifiers) to become 'a universal Turing machine with a given storage capacity' (p. 422)—i.e. to become equivalent to a universal Turing machine with a truncated tape.

This feature of neural networks may shed light on one of the most fundamental problems concerning human cognition. Examining cognition from a top-down perspective, we find complex sequential processes, often involving language or other forms of symbolic representation—for example, logical reasoning, the planning out of activities, and mathematical calculation. Yet from a bottom-up perspective, cognition is nothing but the simple firings of neurons. How is the view from the top to be reconciled with the prima facie very different view from the bottom?

Turing's proof first opened up the possibility, noted in 'Intelligent Machinery' (p. 424), that the brain is in part a universal computing machine (of a given storage capacity) implemented in a neural network. This, then, is a possible solution to the fundamental problem: it is by virtue of being a neural network acting as a universal computing machine (of a given storage capacity) that the cortex is able to carry out the sequential, symbol-rich processing discerned in the view from the top. In 1948 this hypothesis was well ahead of its time, and today it remains one of our best guesses concerning the fundamental problem of cognitive science.

McCulloch–Pitts Neural Nets

It is interesting that Turing makes no reference in 'Intelligent Machinery' to the now famous work of McCulloch and Pitts, itself influenced by his own 'On Computable Numbers'. Their 1943 article represents the first attempt to apply what they call 'the Turing definition of computability' to the study of neuronal function.[45] Like Turing, McCulloch and Pitts considered networks of simple

---

[45] McCulloch and Pitts, 'A Logical Calculus of the Ideas Immanent in Nervous Activity', 129.

two-state boolean 'neurons', although there were important differences between McCulloch–Pitts nets and Turing nets. For example, inhibitory synapses are a fundamental feature of McCulloch–Pitts nets but not of Turing nets. (An input of 1 at an inhibitory synapse at moment m unconditionally sets the output of the unit to 0 at $m + 1$.)

McCulloch stressed the extent to which his and Pitts's work is indebted to Turing in the course of some autobiographical remarks made during the public discussion of a lecture given by von Neumann in 1948:

I started at entirely the wrong angle... and it was not until I saw Turing's paper ['On Computable Numbers'] that I began to get going the right way around, and with Pitts' help formulated the required logical calculus. What we thought we were doing (and I think we succeeded fairly well) was treating the brain as a Turing machine.[46]

Turing had undoubtedly heard something of the work of McCulloch and Pitts. Wiener—with McCulloch a founding member of the cybernetics movement—would almost certainly have mentioned McCulloch in the course of his 'talk over the fundamental ideas of cybernetics with Mr Turing' at the NPL in the spring of 1947.[47] Moreover, von Neumann mentioned the McCulloch–Pitts article of 1943—albeit very briefly—in the 'First Draft of a Report on the EDVAC', which Turing read in 1945. In order to depict the EDVAC's logic gates, von Neumann employed a modified version of the diagrammatic notation for neural nets used by McCulloch and Pitts. (In his 'Proposed Electronic Calculator' (see Chapter 9) Turing considerably extended the notation that he found in the 'First Draft'.[48]) Turing and McCulloch seem not to have met until 1949. After their meeting Turing spoke dismissively of McCulloch, referring to him as a 'charlatan'.[49]

It is an open question whether the work of McCulloch and Pitts had any influence at all on the development of the ideas presented in 'Intelligent Machinery'. Probably not. As Newman once remarked of Turing: 'It was, perhaps, a defect of his qualities that he found it hard to use the work of others, preferring to work things out for himself.'[50]

Whatever the influences were on Turing at that time, there is no doubt that his work on neural nets goes importantly beyond the earlier work of McCulloch and Pitts. The latter give only a perfunctory discussion of learning, saying no more than that the mechanisms supposedly underlying learning in the brain—they specifically mention the formation of new connections and neuronal threshold

[46] J. von Neumann, *Collected Works*, vol. v, ed. A. H. Taub (Oxford: Pergamon Press, 1963), 319.
[47] N. Wiener, *Cybernetics* (New York: John Wiley, 1948), 32.
[48] 'Proposed Electronic Calculator'; see also D. R. Hartree, *Calculating Instruments and Machines* (Illinois: University of Illinois Press, 1949), 97, 102, and B. E. Carpenter and R. W. Doran (eds.), *A. M. Turing's ACE Report of 1946 and Other Papers* (Cambridge, Mass.: MIT Press, 1986), 277.
[49] Gandy in interview with Copeland (Oct. 1995).
[50] *Manchester Guardian*, 11 June 1954.

change—can be mimicked by means of nets whose connections and thresholds are fixed.[51] Turing's idea of using supervised interference to train an initially random arrangement of neurons to compute a specified function is nowhere prefigured.

Further reading

Copeland, B. J., and Proudfoot, D., 'On Alan Turing's Anticipation of Connectionism', *Synthese*, 108 (1996), 361–77. Reprinted in R. Chrisley (ed.), *Artificial Intelligence: Critical Concepts in Cognitive Science*, ii: *Symbolic AI* (London: Routledge, 2000).

Jackson, P., *Introduction to Expert Systems* (Wokingham: Addison-Wesley, 1986).

Rumelhart, D. E., McClelland, J. L., and the PDP Research Group, *Parallel Distributed Processing: Explorations in the Microstructure of Cognition*, i: *Foundations*, (Cambridge, Mass.: MIT Press, 1986).

Provenance

What follows is the text of the National Physical Laboratory document 'Intelligent Machinery: A Report by A. M. Turing' (dated 1948).[52] Symbols missing from the Report itself but present in Turing's draft typescript have been restored.[53]

---

[51] McCulloch and Pitts, 'A Logical Calculus of the Ideas Immanent in Nervous Activity', 117, 124.

[52] 'Intelligent Machinery: A Report by A. M. Turing' is in the Woodger Papers; a digital facsimile is in The Turing Archive for the History of Computing <www.AlanTuring.net/intelligent_machinery>. Turing's draft typescript of the Report is among the Turing Papers in the Modern Archive Centre, King's College, Cambridge (catalogue reference C 11); there is a digital facsimile at <www.turingarchive.org>.

[53] 'Intelligent Machinery' appeared in 1968 in a collection of essays entitled *Key Papers: Cybernetics* (London: Butterworths), ed. C. R. Evans and A. D. J. Robertson, two members of the Autonomics Division of the National Physical Laboratory. The following year another edition of 'Intelligent Machinery' appeared in the volume *Machine Intelligence 5* (ed. B. Meltzer and D. Michie, Edinburgh University Press). Unlike the 1968 and 1969 editions, the present edition follows Turing's layout and his numbering of headings. Unfortunately the 1969 edition (which is reproduced in *Collected Works of A. M. Turing: Mechanical Intelligence*, ed. D. C. Ince (Amsterdam: Elsevier, 1992)) contained numerous departures from Turing's own wording, as well as typographical errors, and outright mistakes. To mention only the most significant of these: (1) Turing's words 'determined by that symbol' on p. 413, below, were replaced by 'described by that symbol'; (2) Turing's words 'we can design a digital computer to do it, but that we stick to one, say the ACE, and that' on p. 415 were omitted, making nonsense of Turing's statement; (3) the words 'or either form of iii)' on p. 423 were omitted; (4) 'T1' was omitted from Turing's list 'U, T0, T1, D0 or D1' on p. 426; (5) the phrase 'replacing the Us of the character by D0' on p. 428 was incorrectly rendered 'replacing the 0s of the character by D0'. In their introduction to the 1969 edition, the editors state that 'Intelligent Machinery' was 'written in September 1947'. This statement has caused some confusion in the literature. In fact 'Intelligent Machinery' was written in the summer of 1948. The phrase 'Manchester machine (as actually working 8/7/48)' (p. 413 below) appears in both the finished NPL Report and Turing's draft typescript. In the 1969 edition Turing's date '8/7/48' (8 July 1948) has been rewritten '8 August 1947'. The Manchester machine first operated on 21 June 1948.

Intelligent Machinery

I propose to investigate the question as to whether it is possible for machinery to show intelligent behaviour. It is usually assumed without argument that it is not possible. Common catch phrases such as 'acting like a machine', 'purely mechanical behaviour' reveal this common attitude. It is not difficult to see why such an attitude should have arisen. Some of the reasons are:

(a) An unwillingness to admit the possibility that mankind can have any rivals in intellectual power. This occurs as much amongst intellectual people as amongst others: they have more to lose. Those who admit the possibility all agree that its realization would be very disagreeable. The same situation arises in connection with the possibility of our being superseded by some other animal species. This is almost as disagreeable and its theoretical possibility is indisputable.

(b) A religious belief that any attempt to construct such machines is a sort of Promethean irreverence.

(c) The very limited character of the machinery which has been used until recent times (e.g. up to 1940). This encouraged the belief that machinery was necessarily limited to extremely straightforward, possibly even to repetitive, jobs. This attitude is very well expressed by Dorothy Sayers (*The Mind of the Maker*, p. 46) '... which imagines that God, having created his Universe, has now screwed the cap on His pen, put His feet on the mantelpiece and left the work to get on with itself.' This, however, rather comes into St. Augustine's category of figures of speech or enigmatic sayings framed from things which do not exist at all. We simply do not know of any creation which goes on creating itself in variety when the creator has withdrawn from it. The idea is that God simply created a vast machine and has left it working until it runs down from lack of fuel. This is another of those obscure analogies, since we have no experience of machines that produce variety of their own accord; the nature of a machine is to do the same thing over and over again so long as it keeps going.

(d) Recently the theorem of Gödel and related results (Gödel,[1] Church,[2] Turing[3]) have shown that if one tries to use machines for such purposes as determining the truth or falsity of mathematical theorems and one is not willing to tolerate an occasional wrong result, then any given machine

---

Crown copyright. Reproduced with permission of the Controller of HMSO, the National Physical Laboratory, and the Estate of Alan Turing.

[1] K. Gödel, 'Über formal unentscheidbare Sätze der Principia Mathematica und verwandter Systeme I', *Monatshefte für Mathematik und Physik*, 38 (1931), 173–98.

[2] A. Church, 'An Unsolvable Problem of Elementary Number Theory', *American Journal of Mathematics*, 58 (1936), 345–63.

[3] 'On Computable Numbers, with an Application to the Entscheidungsproblem' (Chapter 1).

will in some cases be unable to give an answer at all. On the other hand the human intelligence seems to be able to find methods of ever increasing power for dealing with such problems 'transcending' the methods available to machines.
(e) In so far as a machine can show intelligence this is to be regarded as nothing but a reflection of the intelligence of its creator.

2. Refutation of some objections

In this section I propose to outline reasons why we do not need to be influenced by the above described objections. The objections (a) and (b), being purely emotional, do not really need to be refuted. If one feels it necessary to refute them there is little to be said that could hope to prevail, though the actual production of the machines would probably have some effect. In so far then as we are influenced by such arguments we are bound to be left feeling rather uneasy about the whole project, at any rate for the present. These arguments cannot be wholly ignored, because the idea of 'intelligence' is itself emotional rather than mathematical.

The objection (c) in its crudest form is refuted at once by the actual existence of machinery (ENIAC etc.) which can go on through immense numbers (e.g. $10^{60,000}$ about for ACE) of operations without repetition, assuming no breakdown. The more subtle forms of this objection will be considered at length in § 11 and 12.

The argument from Gödel's and other theorems (objection (d)) rests essentially on the condition that the machine must not make mistakes. But this is not a requirement for intelligence. It is related that the infant Gauss was asked at school to do the addition $15 + 18 + 21 + \ldots + 54$ (or something of the kind) and that he immediately wrote down 483, presumably having calculated it as $(15 + 54)(54 - 12)/2.3$. One can imagine circumstances where a foolish master told the child that he ought instead to have added 18 to 15 obtaining 33, then added 21 etc. From some points of view this would be a 'mistake', in spite of the obvious intelligence involved. One can also imagine a situation where the children were given a number of additions to do, of which the first 5 were all arithmetic progressions, but the 6th was say $23 + 34 + 45 + \ldots + 100 + 112 + 122 + \ldots + 199$. Gauss might have given the answer to this as if it were an arithmetic progression, not having noticed that the 9th term was 112 instead of 111. This would be a definite mistake, which the less intelligent children would not have been likely to make.

The view (e) that intelligence in machinery is merely a reflection of that of its creator is rather similar to the view that the credit for the discoveries of a pupil should be given to his teacher. In such a case the teacher would be pleased with the success of his methods of education, but would not claim the results themselves unless he had actually communicated them to his pupil. He would certainly have envisaged in very broad outline the sort of thing his pupil might be expected to do, but would not expect to foresee any sort of detail. It is already possible to produce machines where this sort of situation arises in a small degree. One can produce 'paper machines' for playing chess. Playing against such a machine gives a definite feeling that one is pitting one's wits against something alive.

These views will all be developed more completely below.

3. Varieties of machinery

It will not be possible to discuss possible means of producing intelligent machinery without introducing a number of technical terms to describe different kinds of existent machinery.

'*Discrete*' *and* '*Continuous*' *machinery.* We may call a machine 'discrete' when it is natural to describe its possible states as a discrete set, the motion of the machine occurring by jumping from one state to another. The states of 'continuous' machinery on the other hand form a continuous manifold, and the behaviour of the machine is described by a curve on this manifold. All machinery can be regarded as continuous, but when it is possible to regard it as discrete it is usually best to do so. The states of discrete machinery will be described as 'configurations'.

'*Controlling*' *and* '*Active*' *machinery.* Machinery may be described as 'controlling' if it only deals with information. In practice this condition is much the same as saying that the magnitude of the machine's effects may be as small as we please, so long as we do not introduce confusion through Brownian movement etc. 'Active' machinery is intended to produce some definite physical effect.

Examples		
	A Bulldozer is	Continuous Active
	A Telephone is	Continuous Controlling
	A Brunsviga[4] is	Discrete Controlling
	A Brain is probably	Continuous Controlling, but is very similar to much discrete machinery
	The ENIAC, ACE etc.	Discrete Controlling
	A Differential Analyser	Continuous Controlling

We shall mainly be concerned with discrete controlling machinery. As we have mentioned, brains very nearly fall into this class, and there seems every reason to believe that they could have been made to fall genuinely into it without any change in their essential properties. However, the property of being 'discrete' is

---

[4] Editor's note. The Brunsviga was a popular desk calculating machine.

only an advantage for the theoretical investigator, and serves no evolutionary purpose, so we could not expect Nature to assist us by producing truly 'discrete' brains.

Given any discrete machine the first thing we wish to find out about it is the number of states (configurations) it can have. This number may be infinite (but enumerable) in which case we say that the machine has infinite memory (or storage) capacity. If the machine has a finite number N of possible states then we say that it has a memory capacity of (or equivalent to) $\log_2 N$ binary digits. According to this definition we have the following table of capacities, very roughly

Brunsviga	90
ENIAC without cards and with fixed programme	600
ENIAC with cards	∞[5]
ACE as proposed	60,000
Manchester machine (as actually working 8/7/48)	1,100

The memory capacity of a machine more than anything else determines the complexity of its possible behaviour.

The behaviour of a discrete machine is completely described when we are given the state (configuration) of the machine as a function of the immediately preceding state and the relevant external data.

Logical Computing Machines (L.C.M.s)

In [Chapter 1] a certain type of discrete machine was described. It had an infinite memory capacity obtained in the form of an infinite tape marked out into squares on each of which a symbol could be printed. At any moment there is one symbol in the machine; it is called the scanned symbol. The machine can alter the scanned symbol and its behaviour is in part determined by that symbol, but the symbols on the tape elsewhere do not affect the behaviour of the machine. However the tape can be moved back and forth through the machine, this being one of the elementary operations of the machine. Any symbol on the tape may therefore eventually have an innings.

These machines will here be called 'Logical Computing Machines'. They are chiefly of interest when we wish to consider what a machine could in principle be designed to do, when we are willing to allow it both unlimited time and unlimited storage capacity.

Universal Logical Computing Machines. It is possible to describe L.C.M.s in a very standard way, and to put the description into a form which can be 'understood' (i.e. applied by) a special machine. In particular it is possible to design a 'universal machine' which is an L.C.M. such that if the standard

---

[5] Editor's note. The symbol '∞' is omitted from the original 1948 NPL Report and from the 1968 and 1969 editions. It is present in Turing's draft typescript.

414 | Alan Turing

description of some other L.C.M. is imposed on the otherwise blank tape from outside, and the (universal) machine then set going it will carry out the operations of the particular machine whose description it was given. For details the reader must refer to [Chapter 1].

The importance of the universal machine is clear. We do not need to have an infinity of different machines doing different jobs. A single one will suffice. The engineering problem of producing various machines for various jobs is replaced by the office work of 'programming' the universal machine to do these jobs.

It is found in practice that L.C.M.s can do anything that could be described as 'rule of thumb' or 'purely mechanical'. This is sufficiently well established that it is now agreed amongst logicians that 'calculable by means of an L.C.M.' is the correct accurate rendering of such phrases. There are several mathematically equivalent but superficially very different renderings.

Practical Computing Machines (P.C.M.s)

Although the operations which can be performed by L.C.M.s include every rule of thumb process, the number of steps involved tends to be enormous. This is mainly due to the arrangement of the memory along the tape. Two facts which need to be used together may be stored very far apart on the tape. There is also rather little encouragement, when dealing with these machines, to condense the stored expressions at all. For instance the number of symbols required in order to express a number in Arabic form (e.g. 149056) cannot be given any definite bound, any more than if the numbers are expressed in the 'simplified Roman' form (IIIII...I, with 149056 occurrences of I). As the simplified Roman system obeys very much simpler laws one uses it instead of the Arabic system.

In practice however one *can* assign finite bounds to the numbers that one will deal with. For instance we can assign a bound to the number of steps that we will admit in a calculation performed with a real machine in the following sort of way. Suppose that the storage system depends on charging condensers of capacity $C = 1\,\mu f$,[6] and that we use two states of charging, $E = 100$ volts and $-E = -100$ volts. When we wish to use the information carried by the condenser we have to observe its voltage. Owing to thermal agitation the voltage observed will always be slightly wrong, and the probability of an error between V and $V - \mathrm{d}V$ volts is

$$\sqrt{\frac{2kT}{\pi C}} e^{-\frac{1}{2}V^2 C/kT} V \mathrm{d}V\,[7]$$

where k is Boltzmann's constant. Taking the values suggested we find that the probability of reading the sign of the voltage wrong is about $10^{-1.2\times 10^{16}}$. If then a

---

[6] Editor's note. μf = micro-farad.
[7] Editor's note. This formula is taken from Turing's draft typescript.

job took more than $10^{10^{17}}$ steps we should be virtually certain of getting the wrong answer, and we may therefore restrict ourselves to jobs with fewer steps. Even a bound of this order might have useful simplifying effects. More practical bounds are obtained by assuming that a light wave must travel at least 1 cm between steps (this would only be false with a very small machine) and that we could not wait more than 100 years for an answer. This would give a limit of 10^{20} steps. The storage capacity will probably have a rather similar bound, so that we could use sequences of 20 decimal digits for describing the position in which a given piece of data was to be found, and this would be a really valuable possibility.

Machines of the type generally known as 'Automatic Digital Computing Machines' often make great use of this possibility. They also usually put a great deal of their stored information in a form very different from the tape form. By means of a system rather reminiscent of a telephone exchange it is made possible to obtain a piece of information almost immediately by 'dialling' the position of this information in the store. The delay may be only a few microseconds with some systems. Such machines will be described as 'Practical Computing Machines'.

Universal Practical Computing Machines

Nearly all of the P.C.M.s now under construction have the essential properties of the 'Universal Logical Computing' machines mentioned earlier. In practice, given any job which could have been done on an L.C.M. one can also do it on one of these digital computers. I do not mean that we can design a digital computer to do it, but that we stick to one, say the ACE, and that we can do any required job of the type mentioned on it, by suitable programming. The programming is pure paper work. It naturally occurs to one to ask whether e.g. the ACE would be truly universal if its memory capacity were infinitely extended. I have investigated this question, and the answer appears to be as follows, though I have not proved any formal mathematical theorem about it. As has been explained, the ACE at present uses finite sequences of digits to describe positions in its memory: they are (Sept 1947) actually sequences of 9 binary digits. The ACE also works largely for other purposes with sequences of 32 binary digits. If the memory were extended e.g. to 1000 times its present capacity it would be natural to arrange the memory in blocks of nearly the maximum capacity which can be handled with the 9 digits, and from time to time to switch from block to block. A relatively small part would never be switched. This would contain some of the more fundamental instruction tables and those concerned with switching. This part might be called the 'central part'. One would then need to have a number which described which block was in action at any moment. This number might however be as large as one pleased. Eventually the point would be reached where it could not be stored in a word (32 digits), or even in the central part. One would then have to set aside a block for storing the number, or even a sequence of blocks, say blocks 1, 2, ... n. We should then have to store n, and in theory it would be of indefinite size. This sort of process can be extended in all sorts of ways, but we shall always be left with a positive integer which is of indefinite size and which needs to be stored somewhere, and there seems to be no way out of the difficulty but to introduce a 'tape'. But once this has been done, and since we are only trying to prove a theoretical result, one might as well, whilst proving the theorem, ignore all the other forms of storage. One will in fact have a U.L.C.M. with some complications. This in effect means that one will not be able to prove any result of the required kind which gives any intellectual satisfaction.

Paper machines

It is possible to produce the effect of a computing machine by writing down a set of rules of procedure and asking a man to carry them out. Such a combination of a man with written instructions will be called a 'Paper Machine'. A man provided with paper, pencil, and rubber, and subject to strict discipline, is in effect a universal machine. The expression 'paper machine' will often be used below.

Partially random and apparently partially random machines

It is possible to modify the above described types of discrete machines by allowing several alternative operations to be applied at some points, the alternatives to be chosen by a random process. Such a machine will be described as 'partially random'. If we wish to say definitely that a machine is not of this kind we will describe it as 'determined'. Sometimes a machine may be strictly speaking determined but appear superficially as if it were partially random. This would occur if for instance the digits of the number π were used to determine the choices of a partially random machine, where previously a dice thrower or electronic equivalent had been used. These machines are known as apparently partially random.

4. Unorganised machines

So far we have been considering machines which are designed for a definite purpose (though the universal machines are in a sense an exception). We might instead consider what happens when we make up a machine in a comparatively unsystematic way from some kind of standard components. We could consider some particular machine of this nature and find out what sort of things it is likely to do. Machines which are largely random in their construction in this way will be called 'unorganised machines'. This does not pretend to be an accurate term. It is conceivable that the same machine might be regarded by one man as organised and by another as unorganised.

A typical example of an unorganised machine would be as follows. The machine is made up from a rather large number N of similar units. Each unit has two input terminals, and has an output terminal which can be connected to the input terminals of (0 or more) other units. We may imagine that for each integer r, $1 \leqslant r \leqslant N$, two numbers $i(r)$ and $j(r)$ are chosen at random from $1 \ldots N$ and that we connect the inputs of unit r to the outputs of units $i(r)$ and $j(r)$. All of the units are connected to a central synchronising unit from which synchronising pulses are emitted at more or less equal intervals of time. The times when these pulses arrive will be called 'moments'. Each unit is capable of having two states at each moment. These states may be called 0 and 1. The state is determined by the rule that the states of the units from which the input leads come are to be taken at the previous moment, multiplied together and the result subtracted from 1. An unorganised machine of this character is shown in the diagram below.

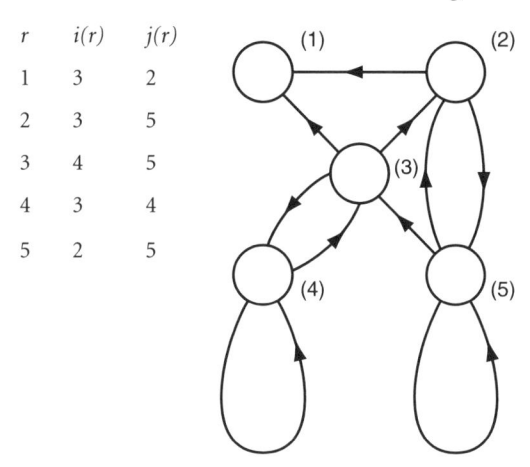

r	$i(r)$	$j(r)$
1	3	2
2	3	5
3	4	5
4	3	4
5	2	5

Figure 10.1.

A sequence of six possible consecutive conditions for the whole machine is:

1	1	1	0	0	1	0
2	1	1	1	0	1	0
3	0	1	1	1	1	1
4	0	1	0	1	0	1
5	1	0	1	0	1	0

The behaviour of a machine with so few units is naturally very trivial. However, machines of this character can behave in a very complicated manner when the number of units is large. We may call these A-type unorganised machines. Thus the machine in the diagram is an A-type unorganised machine of 5 units. The motion of an A-type machine with N units is of course eventually periodic, as is any determined machine with finite memory capacity. The period cannot exceed 2^N moments, nor can the length of time before the periodic motion begins. In the example above the period is 2 moments and there are 3 moments before the periodic motion begins. 2^N is 32.

The A-type unorganised machines are of interest as being about the simplest model of a nervous system with a random arrangement of neurons. It would therefore be of very great interest to find out something about their behaviour. A second type of unorganised machine will now be described, not because it is of any great intrinsic importance, but because it will be useful later for illustrative purposes. Let us denote the circuit

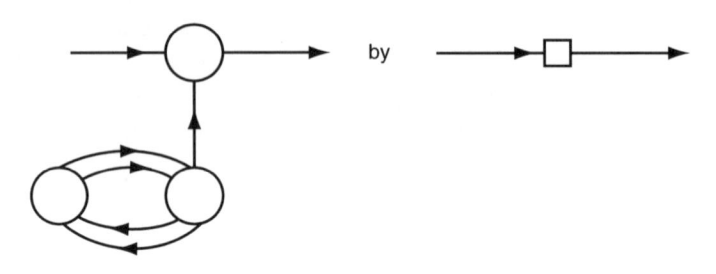

Figure 10.2.

as an abbreviation. Then for each A-type unorganised machine we can construct another machine by replacing each connection →►— in it by →►□►—. The resulting machines will be called B-type unorganised machines. It may be said that the B-type machines are all A-type. To this I would reply that the above definitions if correctly (but drily!) set out would take the form of describing the probability of an A- (or B-) type machine belonging to a given set; it is not merely a definition of which are the A-type machines and which are the B-type machines. If one chooses an A-type machine, with a given number of units, at random, it will be extremely unlikely that one will get a B-type machine.

It is easily seen that the connection →►□►— can have three conditions. It may i) pass all signals through with interchange of 0 and 1, or ii) it may convert all signals into 1, or again iii) it may act as in i) and ii) in alternate moments. (Alternative iii) has two sub-cases.) Which of these cases applies depends on the initial conditions. There is a delay of two moments in going through →►□►—.

5. Interference with machinery. Modifiable and self-modifying machinery

The types of machine that we have considered so far are mainly ones that are allowed to continue in their own way for indefinite periods without interference from outside. The universal machines were an exception to this, in that from time to time one might change the description of the machine which is being imitated. We shall now consider machines in which such interference is the rule rather than the exception.

We may distinguish two kinds of interference. There is the extreme form in which parts of the machine are removed and replaced by others. This may be described as 'screwdriver interference'. At the other end of the scale is 'paper interference', which consists in the mere communication of information to the machine, which alters its behaviour. In view of the properties of the universal machine we do not need to consider the difference between these two kinds of machine as being so very radical after all. Paper interference when applied to the universal machine can be as useful as screwdriver interference.

We shall mainly be interested in paper interference. Since screwdriver interference can produce a completely new machine without difficulty there is rather little to be said about it. In future 'interference' will normally mean 'paper interference'.

When it is possible to alter the behaviour of a machine very radically we may speak of the machine as being 'modifiable'. This is a relative term. One machine may be spoken of as being more modifiable than another.

One may also sometimes speak of a machine modifying itself, or of a machine changing its own instructions. This is really a nonsensical form of phraseology, but is convenient. Of course, according to our conventions the 'machine' is completely described by the relation between its possible configurations at consecutive moments. It is an abstraction which by the form of its definition cannot change in time. If we consider the machine as starting in a particular configuration, however, we may be tempted to ignore those configurations which cannot be reached without interference from it. If we do this we should get a 'successor relation' for the configurations with different properties from the original one and so a different 'machine'.

If we now consider interference, we should say that each time interference occurs the machine is probably changed. It is in this sense that interference 'modifies' a machine. The sense in which a machine can modify itself is even more remote. We may if we wish divide the operations of the machine into two classes, normal and self-modifying operations. So long as only normal operations are performed we regard the machine as unaltered. Clearly the idea of 'self-modification' will not be of much interest except where the division of operations into the two classes is very carefully made. The sort of case I have in mind is a computing machine like the ACE where large parts of the storage are normally occupied in holding instruction tables. (Instruction tables are the equivalent in U.P.C.M.s of descriptions of machines in U.L.C.M.s). Whenever the content of this storage was altered by the internal operations of the machine, one would naturally speak of the machine 'modifying itself'.

6. Man as a machine

A great positive reason for believing in the possibility of making thinking machinery is the fact that it is possible to make machinery to imitate any small part of a man. That the microphone does this for the ear, and the television camera for the eye, are commonplaces. One can also produce remote controlled Robots whose limbs balance the body with the aid of servo-mechanisms. Here we are chiefly interested in the nervous system. We could produce fairly accurate electrical models to copy the behaviour of nerves, but there seems very little point in doing so. It would be rather like putting a lot of work into cars which walked on legs instead of continuing to use wheels. The electrical circuits which are used in electronic computing machinery seem to have the essential properties of nerves. They are able to transmit information from place to place, and also to store it. Certainly the nerve has many advantages. It is extremely compact, does not wear out (probably for hundreds of years if kept in a suitable medium!) and has a very low energy consumption. Against these advantages the electronic circuits have only one counter attraction, that of speed. This advantage is however on such a scale that it may possibly outweigh the advantages of the nerve.

One way of setting about our task of building a 'thinking machine' would be to take a man as a whole and to try to replace all the parts of him by machinery. He would include television cameras, microphones, loudspeakers, wheels and 'handling servo-mechanisms' as well as some sort of 'electronic brain'. This would of course be a tremendous undertaking. The object if produced by present techniques would be of immense size, even if the 'brain' part were stationary and controlled the body from a distance. In order that the machine should have a chance of finding things out for itself it should be allowed to roam the countryside, and the danger to the ordinary citizen would be serious. Moreover even when the facilities mentioned above were provided, the creature would still have no contact with food, sex, sport and many other things of interest to the human being. Thus although this method is probably the 'sure' way of producing a thinking machine it seems to be altogether too slow and impracticable.

Instead we propose to try and see what can be done with a 'brain' which is more or less without a body, providing at most organs of sight, speech and hearing. We are then faced with the problem of finding suitable branches of thought for the machine to exercise its powers in. The following fields appear to me to have advantages:

   (i) Various games e.g. chess, noughts and crosses, bridge, poker
   (ii) The learning of languages
   (iii) Translation of languages
   (iv) Cryptography
   (v) Mathematics.

Of these (i), (iv), and to a lesser extent (iii) and (v) are good in that they require little contact with the outside world. For instance in order that the machine should be able to play chess its only organs need be 'eyes' capable of distinguishing the various positions on a specially made board, and means for announcing its own moves. Mathematics should preferably be restricted to branches where diagrams are not much used. Of the above possible fields the learning of languages would be the most impressive, since it is the most human of these activities. This field seems however to depend rather too much on sense organs and locomotion to be feasible.

The field of cryptography will perhaps be the most rewarding. There is a remarkably close parallel between the problems of the physicist and those of the cryptographer. The system on which a message is enciphered corresponds to the laws of the universe, the intercepted messages to the evidence available, the keys for a day or a message to important constants which have to be determined. The correspondence is very close, but the subject matter of cryptography is very easily dealt with by discrete machinery, physics not so easily.

7. Education of machinery

Although we have abandoned the plan to make a 'whole man', we should be wise to sometimes compare the circumstances of our machine with those of a man. It would be quite unfair to expect a machine straight from the factory to compete on equal terms with a university graduate. The graduate has had contact with human beings for twenty years or more. This contact has throughout that period been modifying his behaviour pattern. His teachers have been intentionally trying to modify it. At the end of the period a large number of standard routines will have been superimposed on the original pattern of his brain. These routines will be known to the community as a whole. He is then in a position to try out new combinations of these routines, to make slight variations on them, and to apply them in new ways.

We may say then that in so far as a man is a machine he is one that is subject to very much interference. In fact interference will be the rule rather than the exception. He is in frequent communication with other men, and is continually receiving visual and other stimuli which themselves constitute a form of interference. It will only be when the man is 'concentrating' with a view to eliminating these stimuli or 'distractions' that he approximates a machine without interference.

We are chiefly interested in machines with comparatively little interference, for reasons given in the last section, but it is important to remember that although a man when concentrating may behave like a machine without interference, his behaviour when concentrating is largely determined by the way he has been conditioned by previous interference.

If we are trying to produce an intelligent machine, and are following the human model as closely as we can, we should begin with a machine with very little capacity to carry out elaborate operations or to react in a disciplined manner to orders (taking the form of interference). Then by applying appropriate interference, mimicking education, we should hope to modify the machine until it could be relied on to produce definite reactions to certain commands. This would be the beginning of the process. I will not attempt to follow it further now.

8. Organising unorganised machinery

Many unorganised machines have configurations such that if once that configuration is reached, and if the interference thereafter is appropriately restricted, the machine behaves as one organised for some definite purpose. For instance the B-type machine shown below was chosen at random.

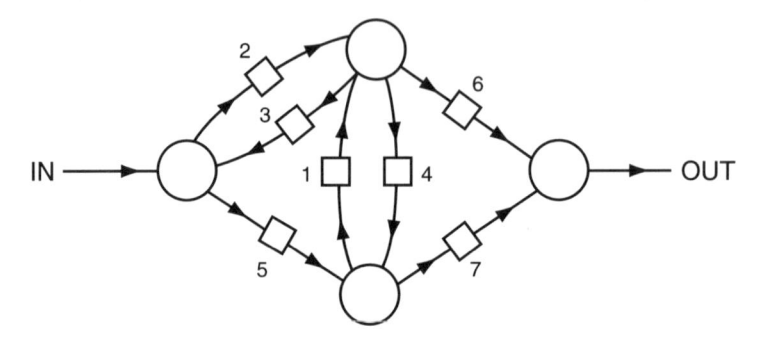

Figure 10.3.

If the connections numbered 1, 3, 6, 4 are in condition ii) initially and connections 2, 5, 7 are in condition i), then the machine may be considered to be one for the purpose of passing on signals with a delay of 4 moments. This is a particular case of a very general property of B-type machines (and many other types), viz. that with suitable initial conditions they will do any required job, given sufficient time and provided the number of units is sufficient. In particular with a B-type unorganised machine with sufficient units one can find initial conditions which will make it into a universal machine with a given storage capacity. (A formal proof to this effect might be of some interest, or even a demonstration of it starting with a particular unorganised B-type machine, but I am not giving it as it lies rather too far outside the main argument.)

With these B-type machines the possibility of interference which could set in appropriate initial conditions has not been arranged for. It is however not difficult to think of appropriate methods by which this could be done. For instance instead of the connection

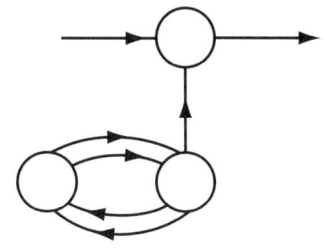

Figure 10.4.

one might use

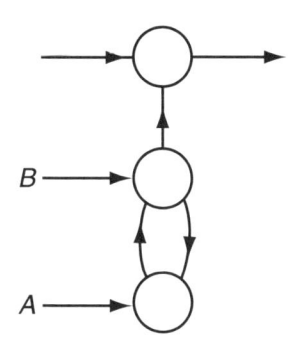

Figure 10.5.

Here A, B are interfering inputs, normally giving the signal '1'. By supplying appropriate other signals at A, B we can get the connection into condition i) or ii) or either form of iii), as desired. However, this requires two special interfering inputs for each connection.

We shall be mainly interested in cases where there are only quite few independent inputs altogether, so that all the interference which sets up the 'initial conditions' of the machine has to be provided through one or two inputs. The process of setting up these initial conditions so that the machine will carry out some particular useful task may be called 'organising the machine'. 'Organising' is thus a form of 'modification'.

9. The cortex as an unorganised machine

Many parts of a man's brain are definite nerve circuits required for quite definite purposes. Examples of these are the 'centres' which control respiration, sneezing, following moving objects with the eyes, etc.: all the reflexes proper (not 'conditioned') are due to the activities of these definite structures in the brain. Likewise the apparatus for the more elementary analysis of shapes and sounds probably comes into this category. But the more intellectual activities of the brain are too varied to be managed on this basis. The difference between the languages spoken on the two sides of the Channel is not due to differences in development of the French-speaking and English-speaking parts of the brain. It is due to the linguistic parts having been subjected to different training. We believe then that there are large parts of the brain, chiefly in the cortex, whose function is largely indeterminate. In the infant these parts do not have much effect: the effect they have is uncoordinated. In the adult they have great and purposive effect: the form of this effect depends on the training in childhood. A large remnant of the random behaviour of infancy remains in the adult.

All of this suggests that the cortex of the infant is an unorganised machine, which can be organised by suitable interfering training. The organising might result in the modification of the machine into a universal machine or something like it. This would mean that the adult will obey orders given in appropriate language, even if they were very complicated; he would have no common sense, and would obey the most ridiculous orders unflinchingly. When all his orders had been fulfilled he would sink into a comatose state or perhaps obey some standing order, such as eating. Creatures not unlike this can really be found, but most people behave quite differently under many circumstances. However the resemblance to a universal machine is still very great, and suggests to us that the step from the unorganised infant to a universal machine is one which should be understood. When this has been mastered we shall be in a far better position to consider how the organising process might have been modified to produce a more normal type of mind.

This picture of the cortex as an unorganised machine is very satisfactory from the point of view of evolution and genetics. It clearly would not require any very complex system of genes to produce something like the A- or B-type unorganised machine. In fact this should be much easier than the production of such things as the respiratory centre. This might suggest that intelligent races could be produced comparatively easily. I think this is wrong because the possession of a human cortex (say) would be virtually useless if no attempt was made to organise it. Thus if a wolf by a mutation acquired a human cortex there is little reason to believe that he would have any selective advantage. If however the mutation occurred in a milieu where speech had developed (parrot-like wolves), and if the mutation by chance had well permeated a small community, then some selective advantage might be felt. It would then be possible to pass information on from generation to generation. However this is all rather speculative.

10. Experiments in organising. Pleasure–pain systems

It is interesting to experiment with unorganised machines admitting definite types of interference and trying to organize them, e.g. to modify them into universal machines.

The organisation of a machine into a universal machine would be most impressive if the arrangements of interference involve very few inputs. The training of the human child depends largely on a system of rewards and punishments, and this suggests that it ought to be possible to carry through the organising with only two interfering inputs, one for 'pleasure' or 'reward' (R) and the other for 'pain' or 'punishment' (P). One can devise a large number of such 'pleasure–pain' systems. I will use this term to mean an unorganised machine of the following general character:

The configurations of the machine are described by two expressions, which we may call the character-expression and the situation-expression. The character and situation at any moment, together with the input signals, determine the character and situation at the next moment. The character may be subject to some random variation. Pleasure interference has a tendency to fix the character i.e. towards preventing it changing, whereas pain stimuli tend to disrupt the character, causing features which had become fixed to change, or to become again subject to random variation.

This definition is probably too vague and general to be very helpful. The idea is that when the 'character' changes we like to think of it as a change in the machine, but the 'situation' is merely the configuration of the machine described by the character. It is intended that pain stimuli occur when the machine's behaviour is wrong, pleasure stimuli when it is particularly right. With appropriate stimuli on these lines, judiciously operated by the 'teacher', one may hope that the 'character' will converge towards the one desired, i.e. that wrong behaviour will tend to become rare.

I have investigated a particular type of pleasure–pain system, which I will now describe.

11. The P-type unorganised machine

The P-type machine may be regarded as an L.C.M. without a tape, and whose description is largely incomplete. When a configuration is reached for which the action is undetermined, a random choice for the missing data is made and the appropriate entry is made in the description, tentatively, and is applied. When a pain stimulus occurs all tentative entries are cancelled, and when a pleasure stimulus occurs they are all made permanent.

Specifically. The situation is a number $s = 1, 2, \ldots, N$ and corresponds to the configuration of the incomplete machine. The character is a table of N entries showing the behaviour of the machine in each situation. Each entry has to say something both about the next situation and about what action the machine has to take. The action part may be either

(i) To do some externally visible act A_1 or $A_2 \ldots A_K$

(ii) To set one of the memory units $M_1 \ldots M_R$ either into the '1' condition or into the '0' condition.

The next situation is always the remainder either of $2s$ or of $2s + 1$ on division by N. These may be called alternatives 0 and 1. Which alternative applies may be determined by either

(a) One of the memory units
(b) A sense stimulus
(c) The pleasure–pain arrangements.

In each situation it is determined which of these applies when the machine is made, i.e. interference cannot alter which of the three cases applies. Also in cases (a) and (b) interference can have no effect. In case (c) the entry in the character table may be either U ('uncertain'), or T0 (tentative 0), T1, D0 (definite 0) or D1. When the entry in the character for the current situation is U then the alternative is chosen at random, and the entry in the character is changed to T0 or T1 according as 0 or 1 was chosen. If the character entry was T0 or D0 then the alternative is 0 and if it is T1 or D1 then the alternative is 1. The changes in character include the above mentioned change from U to T0 or T1, and a change of every T to D when a pleasure stimulus occurs, changes of T0 and T1 to U when a pain stimulus occurs.

We may imagine the memory units essentially as 'trigger circuits' or switches. The sense stimuli are means by which the teacher communicates 'unemotionally' to the machine, i.e. otherwise than by pleasure and pain stimuli. There are a finite number S of sense stimulus lines, and each always carries either the signal 0 or 1.

A small P-type machine is described in the table below

1	P	A	
2	P	B	$M_1 = 1$
3	P	B	
4	S_1	A	$M_1 = 0$
5	M_1	C	

In this machine there is only one memory unit M_1 and one sense line S_1. Its behaviour can be described by giving the successive situations together with the actions of the teacher: the latter consist of the values of S_1 and the rewards and punishments. At any moment the 'character' consists of the above table with each 'P' replaced by either U, T0, T1, D0 or D1. In working out the behaviour of the machine it is convenient first of all to make up a sequence of random digits for use when the U cases occur. Underneath these we may write the sequence of situations, and have other rows for the corresponding entries from the character, and for the actions of the teacher. The character and the values stored in the memory units may be kept on another sheet. The T entries may be made in pencil and the D entries in ink. A bit of the behaviour of the machine is given below:

Random sequence	0 0 1 1 1 0 0 1 0 0 1 1 0 1 1 0 0 0
Situations	3 1 3 1 3 1 3 1 2 4 4 4 3 2 . .
Alternative given by	U T T T T T U U S S S U T
	0 0 0 0 0 1 1 1 0
Visible action	B A B A B A B A B A A A B B
Rew. & Pun.	P
Changes in S_1	1 0

It will be noticed that the machine very soon got into a repetitive cycle. This became externally visible through the repetitive BABAB... By means of a pain stimulus this cycle was broken.

It is probably possible to organise these P-type machines into universal machines, but it is not easy because of the form of memory available. It would be necessary to organise the randomly distributed 'memory units' to provide a systematic form of memory, and this would not be easy. If, however, we supply the P-type machine with a systematic external memory this organising becomes quite feasible. Such a memory could be provided in the form of a tape, and the externally visible operations could include movement to right and left along the tape, and altering the symbol on the tape to 0 or to 1. The sense lines could include one from the symbol on the tape. Alternatively, if the memory were to be finite, e.g. not more than 2^{32} binary digits, we could use a dialling system. (Dialling systems can also be used with an infinite memory, but this is not of much practical interest.) I have succeeded in organising such a (paper) machine into a universal machine.

The details of the machine involved were as follows. There was a circular memory consisting of 64 squares of which at any moment one was in the machine ('scanned') and motion to right or left were among the 'visible actions'. Changing the symbol on the square was another 'visible action', and the symbol was connected to one of the sense lines S_1. The even-numbered squares also had another function, they controlled the dialling of information to or from the main memory. This main memory consisted of 2^{32} binary digits. At any moment one of these digits was connected to the sense line S_2. The digit of the main memory concerned was that indicated by the 32 even positioned digits of the circular memory. Another two of the 'visible actions' were printing 0 or 1 in this square of the main memory. There were also three ordinary memory units and three sense units S_3, S_4, S_5. Also six other externally visible actions A, B, C, D, E, F.

This P-type machine with external memory has, it must be admitted, considerably more 'organisation' than say the A-type unorganised machine.

Nevertheless the fact that it can be organised into a universal machine still remains interesting.

The actual technique by which the 'organising' of the P-type machine was carried through is perhaps a little disappointing. It is not sufficiently analogous to the kind of process by which a child would really be taught. The process actually adopted was first to let the machine run for a long time with continuous application of pain, and with various changes of the sense data S_3, S_4, S_5. Observation of the sequence of externally visible actions for some thousands of moments made it possible to set up a scheme for identifying the situations, i.e. by which one could at any moment find out what the situation was, except that the situations as a whole had been renamed. A similar investigation, with less use of punishment, enables one to find the situations which are affected by the sense lines; the data about the situations involving the memory units can also be found but with more difficulty. At this stage the character has been reconstructed. There are no occurrences of T0, T1, D0, D1. The next stage is to think up some way of replacing the Us of the character by D0, D1 in such a way as to give the desired modification. This will normally be possible with the suggested number of situations (1000), memory units etc. The final stage is the conversion of the character into the chosen one. This may be done simply by allowing the machine to wander at random through a sequence of situations, and applying pain stimuli when the wrong choice is made, pleasure stimuli when the right one is made. It is best also to apply pain stimuli when irrelevant choices are made. This is to prevent getting isolated in a ring of irrelevant situations. The machine is now 'ready for use'.

The form of universal machine actually produced in this process was as follows. Each instruction consisted of 128 digits, which we may regard as forming four sets of 32 each of which describes one place in the main memory. These places may be called P, Q, R, S. The meaning of the instruction is that if p is the digit at P and q that at Q then $1 - pq$ is to be transferred to position R and that the next instruction will be found in the 128 digits beginning at S. This gives a U.P.C.M., though with rather less facilities than are available say on the ACE.

I feel that more should be done on these lines. I would like to investigate other types of unorganised machine, and also to try out organising methods that would be more nearly analogous to our 'methods of education'. I made a start on the latter but found the work altogether too laborious at present. When some electronic machines are in actual operation I hope that they will make this more feasible. It should be easy to make a model of any particular machine that one wishes to work on within such a U.P.C.M. instead of having to work with a paper machine as at present. If also one decided on quite definite 'teaching policies' these could also be programmed into the machine. One would then allow the whole system to run for an appreciable period, and then break in as a kind of 'inspector of schools' and see what progress had been made. One might also be able to make some progress with unorganised machines more like the A- and B-types. The work involved with these is altogether too great for pure paper-machine work.

One particular kind of phenomenon I had been hoping to find in connection with the P-type machines. This was the incorporation of old routines into new. One might have 'taught' (i.e. modified or organised) a machine to add (say). Later one might teach it to multiply by small numbers by repeated addition and so arrange matters that the same set of situations which formed the addition routine, as originally taught, was also used in the additions involved in the multiplication. Although I was able to obtain a fairly detailed picture of how this might happen I was not able to do experiments on a sufficient scale for such phenomena to be seen as part of a larger context.

I also hoped to find something rather similar to the 'irregular verbs' which add variety to language. We seem to be quite content that things should not obey too mathematically regular rules. By long experience we can pick up and apply the most complicated rules without being able to enunciate them at all. I rather suspect that a P-type machine without the systematic memory would behave in a rather similar manner because of the randomly distributed memory units. Clearly this could only be verified by very painstaking work; by the very nature of the problem 'mass production' methods like built-in teaching procedures could not help.

12. Discipline and initiative

If the untrained infant's mind is to become an intelligent one, it must acquire both discipline and initiative. So far we have been considering only discipline. To convert a brain or machine into a universal machine is the extremest form of discipline. Without something of this kind one cannot set up proper communication. But discipline is certainly not enough in itself to produce intelligence. That which is required in addition we call initiative. This statement will have to serve as a definition. Our task is to discover the nature of this residue as it occurs in man, and to try and copy it in machines.

Two possible methods of setting about this present themselves. On the one hand we have fully disciplined machines immediately available, or in a matter of months or years, in the form of various U.P.C.M.s. We might try to graft some initiative onto these. This would probably take the form of programming the machine to do every kind of job that could be done, as a matter of principle, whether it were economical to do it by machine or not. Bit by bit one would be able to allow the machine to make more and more 'choices' or 'decisions'. One would eventually find it possible to programme it so as to make its behaviour be the logical result of a comparatively small number of general principles. When these became sufficiently general, interference would no longer be necessary, and the machine would have 'grown up'. This may be called the 'direct method'.

The other method is to start with an unorganised machine and to try to bring both discipline and initiative into it at once, i.e. instead of trying to organise the machine to become a universal machine, to organise it for initiative as well. Both methods should, I think, be attempted.

Intellectual, Genetical, and Cultural Searches

A very typical sort of problem requiring some sort of initiative consists of those of the form 'Find a number n such that...'. This form covers a very great variety of problems. For instance problems of the form 'See if you can find a way of calculating the function...which will enable us to obtain the values for arguments...to accuracy...within a time...using the U.P.C.M....' are reducible to this form, for the problem is clearly equivalent to that of finding a programme to put on the machine in question, and it is easy to put the programmes into correspondence with the positive integers in such a way that given either the number or the programme the other can easily be found. We should not go far wrong for the time being if we assumed that all problems were reducible to this form. It will be time to think again when something turns up which is obviously not of this form.

The crudest way of dealing with such a problem is to take the integers in order and to test each one to see whether it has the required property, and to go on until one is found which has it. Such a method will only be successful in the simplest cases. For instance in the case of problems of the kind mentioned above, where one is really searching for a programme, the number required will normally be somewhere between 2^{1000} and $2^{1,000,000}$. For practical work therefore some more expeditious method is necessary. In a number of cases the following method would be successful. Starting with a U.P.C.M. we first put a programme into it which corresponds to building in a logical system (like Russell's *Principia Mathematica*). This would not determine the behaviour of the machine completely: at various stages more than one choice as to the next step would be possible. We might however arrange to take all possible arrangements of choices in order, and go on until the machine proved a theorem which, by its form, could be verified to give a solution of the problem. This may be seen to be a conversion of the original problem into another of the same form. Instead of searching through values of the original variable n one searches through values of something else. In practice when solving problems of the above kind one will probably apply some very complex 'transformation' of the original problem, involving searching through various variables, some more analogous to the original one, some more like a 'search through all proofs'. Further research into intelligence of machinery will probably be very greatly concerned with 'searches' of this kind. We may perhaps call such searches 'intellectual searches'. They might very briefly be defined as 'searches carried out by brains for combinations with particular properties'.

It may be of interest to mention two other kinds of search in this connection. There is the genetical or evolutionary search by which a combination of genes is looked for, the criterion being survival value. The remarkable success of this search confirms to some extent the idea that intellectual activity consists mainly of various kinds of search.

The remaining form of search is what I should like to call the 'cultural search'. As I have mentioned, the isolated man does not develop any intellectual power. It is necessary for him to be immersed in an environment of other men, whose techniques he absorbs during the first 20 years of his life. He may then perhaps do a little research of his own and make a very few discoveries which are passed on to other men. From this point of view the search for new techniques must be regarded as carried out by the human community as a whole, rather than by individuals.

13. Intelligence as an emotional concept

The extent to which we regard something as behaving in an intelligent manner is determined as much by our own state of mind and training as by the properties of the object under consideration. If we are able to explain and predict its behaviour or if there seems to be little underlying plan, we have little temptation to imagine intelligence. With the same object therefore it is possible that one man would consider it as intelligent and another would not; the second man would have found out the rules of its behaviour.

It is possible to do a little experiment on these lines, even at the present stage of knowledge. It is not difficult to devise a paper machine which will play a not very bad game of chess. Now get three men as subjects for the experiment A, B, C. A and C are to be rather poor chess players, B is the operator who works the paper machine. (In order that he should be able to work it fairly fast it is advisable that he be both mathematician and chess player.) Two rooms are used with some arrangement for communicating moves, and a game is played between C and either A or the paper machine. C may find it quite difficult to tell which he is playing.

(This is a rather idealized form of an experiment I have actually done.)

Summary

The possible ways in which machinery might be made to show intelligent behaviour are discussed. The analogy with the human brain is used as a guiding principle. It is pointed out that the potentialities of the human intelligence can only be realised if suitable education is provided. The investigation mainly centres round an analogous teaching process applied to machines. The idea of an unorganised machine is defined, and it is suggested that the infant human cortex is of this nature. Simple examples of such machines are given, and their education by means of rewards and punishments is discussed. In one case the education process is carried through until the organisation is similar to that of an ACE.

CHAPTER 11

Computing Machinery and Intelligence (1950)

Alan Turing

Introduction

Jack Copeland

Together with 'On Computable Numbers' (Chapter 1), 'Computing Machinery and Intelligence' forms Turing's best-known work. This elegant and sometimes amusing essay was originally published in 1950 in the leading philosophy journal *Mind*. Turing's friend Robin Gandy (like Turing a mathematical logician) said that 'Computing Machinery and Intelligence'

was intended not so much as a penetrating contribution to philosophy but as propaganda. Turing thought the time had come for philosophers and mathematicians and scientists to take seriously the fact that computers were not merely calculating engines but were capable of behaviour which must be accounted as intelligent; he sought to persuade people that this was so. He wrote this paper—unlike his mathematical papers—quickly and with enjoyment. I can remember him reading aloud to me some of the passages—always with a smile, sometimes with a giggle.[1]

The quality and originality of 'Computing Machinery and Intelligence' have earned it a place among the classics of philosophy of mind.

The Turing Test

'Computing Machinery and Intelligence' contains Turing's principal exposition of the famous 'imitation game' or Turing test. The test first appeared, in a restricted form, in the closing paragraphs of 'Intelligent Machinery' (Chapter 10). Chapters 13 and 14, dating from 1951 and 1952 respectively, contain further

---

[1] R. Gandy, 'Human versus Mechanical Intelligence', in P. Millican and A. Clark (eds.), *Machines and Thought: The Legacy of Alan Turing*, vol. i (Oxford: Clarendon Press, 1996), 125.

discussion and amplification; unpublished until 1999, this important additional material throws new light on how the Turing test is to be understood.[2]

The imitation game involves three participants: a computer, a human interrogator, and a human 'foil'.[3] The interrogator attempts to determine, by asking questions of the other two participants, which of them is the computer. All communication is via keyboard and screen, or an equivalent arrangement (Turing suggested a teleprinter link). The interrogator may ask questions as penetrating and wide-ranging as he or she likes, and the computer is permitted to do everything possible to force a wrong identification. (So the computer might answer 'No' in response to 'Are you a computer?' and might follow a request to multiply one large number by another with a long pause and a plausibly incorrect answer.) The foil must help the interrogator to make a correct identification.

The ability to play the imitation game successfully is Turing's proposed 'criterion for "thinking"' (pp. 442, 443). He gives two examples of the sort of exchange that might occur between an interrogator and a machine that plays successfully. The following is from p. 452.

Interrogator: In the first line of your sonnet which reads 'Shall I compare thee to a summer's day', would not 'a spring day' do as well or better?
Machine: It wouldn't scan.
Interrogator: How about 'a winter's day'? That would scan all right.
Machine: Yes, but nobody wants to be compared to a winter's day.
Interrogator: Would you say Mr Pickwick reminded you of Christmas?
Machine: In a way.
Interrogator: Yet Christmas is a winter's day, and I do not think Mr Pickwick would mind the comparison.
Machine: I don't think you're serious. By a winter's day one means a typical winter's day, rather than a special one like Christmas.

Did Turing Propose a Definition?

Turing is sometimes said to have proposed a definition of 'thinking' or 'intelligence'; and sometimes his supposed definition is said to be an 'operational' or 'behaviourist' definition. For example:

An especially influential behaviorist definition of intelligence was put forward by Turing.[4] (Ned Block)

---

[2] This additional material was first published in B. J. Copeland (ed.), 'A Lecture and Two Radio Broadcasts on Machine Intelligence by Alan Turing', in K. Furukawa, D. Michie and S. Muggleton (eds.), *Machine Intelligence 15* (Oxford University Press, 1999). See also B. J. Copeland, 'The Turing Test', *Minds and Machines*, 10 (2000), 519–39 (reprinted in J. H. Moor (ed.), *The Turing Test* (Dordrecht: Kluwer, 2003)).

[3] The term 'foil' is from p. 40 of B. J. Copeland, *Artificial Intelligence: A Philosophical Introduction* (Oxford: Blackwell, 1993).

[4] N. Block, 'The Computer Model of the Mind', in D. N. Osherson and H. Lasnik (eds.), *An Invitation to Cognitive Science*, vol. iii (Cambridge, Mass.: MIT Press, 1990), 248.

[Turing] introduced ... an operational definition of 'thinking' or 'intelligence'... by means of a sexual guessing game.[5] (Andrew Hodges)

The Turing Test [was] originally proposed as a simple operational definition of intelligence.[6] (Robert French)

There is no textual evidence to support this interpretation of Turing, however. In 'Computing Machinery and Intelligence' Turing claimed to be offering only a '*criterion* for "thinking"' (emphasis added). Moreover, in his discussion of the Turing test in Chapter 14, Turing says quite specifically that his aim is not 'to give a definition of thinking' (p. 494).

In fact, Turing made it plain in 'Computing Machinery and Intelligence' that his intention was not to offer a definition, for he said:

The game may perhaps be criticised on the ground that the odds are weighted too heavily against the machine. If the man were to try and pretend to be the machine he would clearly make a very poor showing. He would be given away at once by slowness and inaccuracy in arithmetic. May not machines carry out something which ought to be described as thinking but which is very different from what a man does? (p. 442)

A computer carrying out something that 'ought to be described as thinking' would nevertheless fail the Turing test if for any reason it stood out in conversation as very different from a man. It follows that 'thinking' cannot be defined in terms of success in the imitation game. Success in the game is arguably a sufficient condition for thinking; but success in the imitation game is not also a *necessary* condition for thinking. (Someone's breathing spontaneously is a sufficient condition for their being alive, but it is not also a necessary condition, for someone may be alive without breathing spontaneously.)

The Male–Female Imitation Game

Turing introduced his criterion for 'thinking' by first describing an imitation game involving a human interrogator and two *human* subjects, one male (A) and one female (B). The interrogator must determine, by question and answer, which of A and B is the man. A's object in the game is to try to cause the interrogator to make the wrong identification. Having introduced the imitation game in this way, Turing said:

We now ask the question, 'What will happen when a machine takes the part of A in this game?' Will the interrogator decide wrongly as often when the game is played like this as he does when the game is played between a man and a woman? These questions replace our original, 'Can machines think?' (p. 441)

---

[5] A. Hodges, *Alan Turing: The Enigma* (London: Vintage, 1992), 415.
[6] R. French, 'The Turing Test: The First 50 Years', *Trends in Cognitive Sciences*, 4 (2000), 115–22 (115).

Some commentators have suggested, on the basis of this passage, that Turing's criterion for thinking is that the computer in the Turing test be able to impersonate a woman.[7] Later in the article, however, Turing described matters differently, saying that the part of A is taken by a machine and 'the part of B...by a man' (p. 448). This runs contrary to the suggestion that the computer is supposed to imitate a woman (rather than a man or a woman). Moreover in Chapter 14 Turing says that '[t]he idea of the test is that the machine has to try and pretend to be a man...and it will pass only if the pretence is reasonably convincing' (p. 495). In Chapter 13 Turing presents the test in a starkly ungendered form: here the point of the test is to determine whether or not a computer can 'imitate a brain' (p. 485). On balance, then, it seems rather unlikely that Turing's intention in 'Computing Machinery and Intelligence' was to put forward a test in which the computer must impersonate a woman.

The role of the man-imitates-woman game is frequently misunderstood. For example, Hodges claims that this game is irrelevant as an introduction to the Turing test—indeed, it is a 'red herring'.[8] However, the man-imitates-woman game forms part of the protocol for scoring the test. Will interrogators decide wrongly as often in man-imitates-woman imitation games as they do in computer-imitates-human games? This question, Turing said, replaces 'Can machines think?'

The Current Status of the Turing Test

Section 6 of 'Computing Machinery and Intelligence', entitled 'Contrary Views on the Main Question', occupies nearly half of the article. It contains no fewer than nine objections to Turing's position, together with Turing's rebuttal of each. One of them, the 'Mathematical Objection', is also discussed in Chapters 10 and 12 (the introduction to Chapter 12 gives some further information about this important and controversial objection).

Since 'Computing Machinery and Intelligence' first appeared, Turing's test has received considerable attention from philosophers, computer scientists, psychologists, and others, and numerous additional objections have been raised to the test, some of them ingenious indeed. Nevertheless, it seems to me that none of these objections is successful (see my chapter in Moor's *The Turing Test* in the list of further reading). A discussion of one such objection, called here the Shannon–McCarthy objection, will give something of the flavour of the debate that still rages over the Turing test. Another form of objection—the 'Fiendish Expert' objection—is discussed in the introduction to Chapter 14.

[7] See, for example, S. G. Sterrett, 'Turing's Two Tests for Intelligence', *Minds and Machines*, 10 (2000), 541–59; S. Traiger, 'Making the Right Identification in the Turing Test', *Minds and Machines*, 10 (2000), 561–72 (both reprinted in J. H. Moor (ed.), *The Turing Test* (Dordrecht: Kluwer, 2003)).

[8] Hodges, *Alan Turing*, 415.

The Shannon–McCarthy Objection

This objection envisages a hypothetical computer that is able to play the imitation game successfully, for any set length of time, in virtue of incorporating a very large—but nevertheless finite—'look-up' table. The table contains *all* the exchanges that could possibly occur between the computer and the interrogator during the length of time for which the test is run. The number of these is astronomical—but finite. For example, the exchange displayed earlier concerning sonnets and Mr Pickwick forms part of this (imaginary) table.

Clearly an interrogator would have no means by which to distinguish a computer using this table from a human respondent. Yet presumably the computer—which does nothing but search the table provided by its (hypothetical) programmers—does not think. In principle, therefore, an unthinking, unintelligent computer can pass the test.

Claude Shannon and John McCarthy put the objection forward in 1956:

> The problem of giving a precise definition to the concept of 'thinking' and of deciding whether or not a given machine is capable of thinking has aroused a great deal of heated discussion. One interesting definition has been proposed by A. M. Turing: a machine is termed capable of thinking if it can, under certain prescribed conditions, imitate a human being by answering questions sufficiently well to deceive a human questioner for a reasonable period of time. A definition of this type has the advantages of being operational, or, in the psychologists' term, behavioristic. ... A disadvantage of the Turing definition of thinking is that it is possible, in principle, to design a machine with a complete set of arbitrarily chosen responses to all possible input stimuli... Such a machine, in a sense, for any given input situation (including past history) merely looks up in a 'dictionary' the appropriate response. With a suitable dictionary such a machine would surely satisfy Turing's definition but does not reflect our usual intuitive concept of thinking.[9]

This objection has been rediscovered by a number of philosophers, and it is in fact usually credited to Block, who published a version of it in 1981.[10] (It is sometimes referred to as the 'blockhead' objection to the Turing test.)

What might Turing have said in response to the objection? A hint is perhaps provided by the following exchange between Turing and Newman (Chapter 14, p. 503):

> Newman: It is all very well to say that a machine could... be made to do this or that, but, to take only one practical point, what about the time it would take to do it? It would only take an hour or two to make up a routine to make our Manchester machine analyse all possible variations of the game of chess right out, and find the best move that way—*if* you

---

[9] C. E. Shannon and J. McCarthy (eds.), *Automata Studies* (Princeton: Princeton University Press, 1956), pp. v–vi.
[10] N. Block, 'Psychologism and Behaviorism', *Philosophical Review*, 90 (1981), 5–43.

didn't mind its taking thousands of millions of years to run through the routine. Solving a problem on the machine doesn't mean finding a way to do it between now and eternity, but within a reasonable time. ...

Turing: To my mind this time factor is the one question which will involve all the real technical difficulty.

The Shannon–McCarthy objection establishes only that the

Turing Test Principle If x plays Turing's imitation game satisfactorily, then x thinks

is false in *some possible world*. The objection directs our imagination toward a possible world that is very different from the actual world—a world in which an astronomically large look-up table can be stored in a computer's memory and searched in a reasonable time—and points out that the Turing test principle is false in *that* world. However, there is no textual evidence to indicate that Turing was claiming anything more than that the Turing test principle is *actually* true, i.e. true in the actual world. Nor did he need to claim more than this in order to advocate the imitation game as a satisfactory real-world test.

Had Turing been proposing a definition of 'thinking', then he would indeed have had to say, consistently, that the Turing test principle is true in *all* possible worlds. (To take a more obvious case, if 'bachelor' is defined as 'unmarried male of marriageable age', then it is true not only in the actual world but in every possible world that if x is an unmarried male of marriageable age, then x is a bachelor.) At bottom, then, the Shannon–McCarthy objection depends on the interpretational mistake of taking Turing to be proposing a definition.

There is further discussion of the Turing test in Chapters 13, 14, and 16.

Learning Machines

The discussion of learning begun in Chapter 10 is continued in the iconoclastic Section 7 of 'Computing Machinery and Intelligence', entitled 'Learning Machines'. Turing poses the rhetorical question: 'Instead of trying to produce a programme to simulate the adult mind, why not rather try to produce one which simulates the child's?' (p. 460). The child's mind may contain 'so little mechanism' that 'something like it can be easily programmed'. If this child-machine 'were then subjected to an appropriate course of education one would obtain the adult brain'. These remarks are of a piece with Turing's suggestion in Chapter 10 that 'the cortex of an infant is an unorganised machine, which can be organised by suitable interfering training' (p. 424).

Turing mentions in Section 7 that he has 'done some experiments with one such child-machine, and succeeded in teaching it a few things, but the teaching method was too unorthodox for the experiment to be considered really successful' (p. 461). Here he is probably referring to the experiments with an unorganized machine that are described in Chapter 10, where he says that he has 'succeeded in organising such a (paper) machine into a universal machine', but that the technique used 'is not sufficiently analogous to the kind of process by which a child would really be taught' (pp. 427-8).

Situated AI

AI traditionally has attempted to build disembodied intelligences carrying out abstract activities—e.g. chess-playing—and whose only way of interacting with the world is by means of a screen or printer. An alternative approach now called 'situated AI' aims at building embodied intelligences situated in the real world. 'Computing Machinery and Intelligence' ends with a characteristically far-sighted statement in which Turing sketches each of these two approaches to AI. He contrasts research that focuses on 'abstract activity, like the playing of chess' with research aiming 'to provide the machine with the best sense organs that money can buy, and then teach it to understand and speak English' (p. 463). Turing recommended that 'both approaches should be tried' (ibid.; compare Chapter 10, pp. 420–1).

Rodney Brooks, a modern pioneer of situated AI and Director of the MIT Artificial Intelligence Laboratory, pointed out that although Turing proposed both these 'paths toward his goal of a thinking machine', Artificial Intelligence for a long time 'all but ignored' the situated approach.[11] Now the tables have turned and there is huge interest in situated AI.

One of Brooks's experimental robots, Herbert—named after Herbert Simon—searched the offices and work-spaces of the MIT AI Lab for empty soda cans, picking them up and carrying them to the trash.[12] Herbert, unlike previous generations of experimental robots, operated in real time in a busy, cluttered, and unpredictably changing real-world environment. Brooks's humanoid learning robot Cog—from 'cognizer'—has four microphone-type 'ears' and saccading foveated vision provided by cameras mounted on its 'head'.[13] Cog's legless torso is able to lean and twist. Strain gauges on the spine give Cog information about posture, while heat and current sensors on the robot's motors provide feedback concerning exertion. Cog's arm and manipulating hand are coated with electrically conducting rubber membranes providing tactile information. Those working in situated AI regard Cog as a milestone on the road toward the realization of Turing's dream.

---

[11] R. Brooks, 'Intelligence without Reason', in L. Steels and R. Brooks (eds.), *The Artificial Life Route to Artificial Intelligence* (Hillsdale, NJ: Erlbaum, 1995), 34. See also R. Brooks, *Cambrian Intelligence: The History of the New AI* (Cambridge, Mass.: MIT Press, 1999).
[12] R. Brooks, 'Elephants Don't Play Chess', *Robotics and Autonomous Systems*, 6 (1990), 3–15.
[13] R. A. Brooks and L. A. Stein, 'Building Brains for Bodies', *Autonomous Robots*, 1 (1994), 7–25.

Further reading

Block, N., 'Psychologism and Behaviorism', *Philosophical Review*, 90 (1981), 5–43.

Dennett, D. C., 'Can Machines Think?', in his *Brainchildren: Essays on Designing Minds* (Cambridge, Mass.: MIT Press, 1998).

French, R., 'The Turing Test: The First 50 Years', *Trends in Cognitive Sciences*, 4 (2000), 115–22.

Michie, D., 'Turing's Test and Conscious Thought', *Artificial Intelligence*, 60 (1993), 1–22. Reprinted in P. Millican and A. Clark (eds.), *Machines and Thought: The Legacy of Alan Turing* (Oxford: Clarendon Press, 1996).

Moor, J. H. (ed.), *The Turing Test* (Dordrecht: Kluwer, 2003).

—— 'An Analysis of the Turing Test', *Philosophical Studies*, 30 (1976), 249–57.

Provenance

What follows is the text of the original printing of 'Computing Machinery and Intelligence' in *Mind*.[14] (Unfortunately Turing's typescript has been lost.)

---

[14] Footnotes have been renumbered consecutively. All footnotes not marked 'Editor's note' appeared in *Mind*. Where the text contains numbers referring to pages of *Mind* these have been replaced by the numbers of the corresponding pages of the present edition, enclosed in square brackets. Not all cross-references in Turing's article were dealt with correctly by the editor of *Mind*—some of the numbers appearing in *Mind* presumably refer to pages of Turing's original typescript. These also have been replaced by the numbers of the corresponding pages of this volume.

Computing Machinery and Intelligence

1. The Imitation Game

I propose to consider the question, 'Can machines think?' This should begin with definitions of the meaning of the terms 'machine' and 'think'. The definitions might be framed so as to reflect so far as possible the normal use of the words, but this attitude is dangerous. If the meaning of the words 'machine' and 'think' are to be found by examining how they are commonly used it is difficult to escape the conclusion that the meaning and the answer to the question, 'Can machines think?' is to be sought in a statistical survey such as a Gallup poll. But this is absurd. Instead of attempting such a definition I shall replace the question by another, which is closely related to it and is expressed in relatively unambiguous words.

The new form of the problem can be described in terms of a game which we call the 'imitation game'. It is played with three people, a man (A), a woman (B), and an interrogator (C) who may be of either sex. The interrogator stays in a room apart from the other two. The object of the game for the interrogator is to determine which of the other two is the man and which is the woman. He knows them by labels X and Y, and at the end of the game he says either 'X is A and Y is B' or 'X is B and Y is A'. The interrogator is allowed to put questions to A and B thus:

C: Will X please tell me the length of his or her hair? Now suppose X is actually A, then A must answer. It is A's object in the game to try and cause C to make the wrong identification. His answer might therefore be

'My hair is shingled, and the longest strands are about nine inches long.'

In order that tones of voice may not help the interrogator the answers should be written, or better still, typewritten. The ideal arrangement is to have a teleprinter communicating between the two rooms. Alternatively the question and answers can be repeated by an intermediary. The object of the game for the third player (B) is to help the interrogator. The best strategy for her is probably to give truthful answers. She can add such things as 'I am the woman, don't listen to him!' to her answers, but it will avail nothing as the man can make similar remarks.

We now ask the question, 'What will happen when a machine takes the part of A in this game?' Will the interrogator decide wrongly as often when the game is played like this as he does when the game is played between a man and a woman? These questions replace our original, 'Can machines think?'

This article first appeared in *Mind*, 59 (1950), 433–60. It is reprinted with the permission of the Mind Association and the Estate of Alan Turing.

2. Critique of the New Problem

As well as asking, 'What is the answer to this new form of the question', one may ask, 'Is this new question a worthy one to investigate?' This latter question we investigate without further ado, thereby cutting short an infinite regress.

The new problem has the advantage of drawing a fairly sharp line between the physical and the intellectual capacities of a man. No engineer or chemist claims to be able to produce a material which is indistinguishable from the human skin. It is possible that at some time this might be done, but even supposing this invention available we should feel there was little point in trying to make a 'thinking machine' more human by dressing it up in such artificial flesh. The form in which we have set the problem reflects this fact in the condition which prevents the interrogator from seeing or touching the other competitors, or hearing their voices. Some other advantages of the proposed criterion may be shown up by specimen questions and answers. Thus:

Q: Please write me a sonnet on the subject of the Forth Bridge.
A: Count me out on this one. I never could write poetry.
Q: Add 34957 to 70764.
A: (Pause about 30 seconds and then give as answer) 105621.
Q: Do you play chess?
A: Yes.
Q: I have K at my K1, and no other pieces. You have only K at K6 and R at R1. It is your move. What do you play?
A: (After a pause of 15 seconds) R-R8 mate.

The question and answer method seems to be suitable for introducing almost any one of the fields of human endeavour that we wish to include. We do not wish to penalise the machine for its inability to shine in beauty competitions, nor to penalise a man for losing in a race against an aeroplane. The conditions of our game make these disabilities irrelevant. The 'witnesses' can brag, if they consider it advisable, as much as they please about their charms, strength or heroism, but the interrogator cannot demand practical demonstrations.

The game may perhaps be criticised on the ground that the odds are weighted too heavily against the machine. If the man were to try and pretend to be the machine he would clearly make a very poor showing. He would be given away at once by slowness and inaccuracy in arithmetic. May not machines carry out something which ought to be described as thinking but which is very different from what a man does? This objection is a very strong one, but at least we can say that if, nevertheless, a machine can be constructed to play the imitation game satisfactorily, we need not be troubled by this objection.

It might be urged that when playing the 'imitation game' the best strategy for the machine may possibly be something other than imitation of the behaviour of a man. This may be, but I think it is unlikely that there is any great effect of this kind. In any case there is no intention to investigate here the theory of the game, and it will be assumed that the best strategy is to try to provide answers that would naturally be given by a man.

3. The Machines concerned in the Game

The question which we put in §1 will not be quite definite until we have specified what we mean by the word 'machine'. It is natural that we should wish to permit every kind of engineering technique to be used in our machines. We also wish to allow the possibility that an engineer or team of engineers may construct a machine which works, but whose manner of operation cannot be satisfactorily described by its constructors because they have applied a method which is largely experimental. Finally, we wish to exclude from the machines men born in the usual manner. It is difficult to frame the definitions so as to satisfy these three conditions. One might for instance insist that the team of engineers should be all of one sex, but this would not really be satisfactory, for it is probably possible to rear a complete individual from a single cell of the skin (say) of a man. To do so would be a feat of biological technique deserving of the very highest praise, but we would not be inclined to regard it as a case of 'constructing a thinking machine'. This prompts us to abandon the requirement that every kind of technique should be permitted. We are the more ready to do so in view of the fact that the present interest in 'thinking machines' has been aroused by a particular kind of machine, usually called an 'electronic computer' or 'digital computer'. Following this suggestion we only permit digital computers to take part in our game.

This restriction appears at first sight to be a very drastic one. I shall attempt to show that it is not so in reality. To do this necessitates a short account of the nature and properties of these computers.

It may also be said that this identification of machines with digital computers, like our criterion for 'thinking', will only be unsatisfactory if (contrary to my belief), it turns out that digital computers are unable to give a good showing in the game.

There are already a number of digital computers in working order, and it may be asked, 'Why not try the experiment straight away? It would be easy to satisfy the conditions of the game. A number of interrogators could be used, and statistics compiled to show how often the right identification was given.' The short answer is that we are not asking whether all digital computers would do well in the game nor whether the computers at present available would do well, but whether there are imaginable computers which would do well. But this is only the short answer. We shall see this question in a different light later.

4. Digital Computers

The idea behind digital computers may be explained by saying that these machines are intended to carry out any operations which could be done by a human computer. The human computer is supposed to be following fixed rules; he has no authority to deviate from them in any detail. We may suppose that these rules are supplied in a book, which is altered whenever he is put on to a new job. He has also an unlimited supply of paper on which he does his calculations. He may also do his multiplications and additions on a 'desk machine', but this is not important.

If we use the above explanation as a definition we shall be in danger of circularity of argument. We avoid this by giving an outline of the means by which the desired effect is achieved. A digital computer can usually be regarded as consisting of three parts:

(i) Store.
(ii) Executive unit.
(iii) Control.

The store is a store of information, and corresponds to the human computer's paper, whether this is the paper on which he does his calculations or that on which his book of rules is printed. In so far as the human computer does calculations in his head a part of the store will correspond to his memory.

The executive unit is the part which carries out the various individual operations involved in a calculation. What these individual operations are will vary from machine to machine. Usually fairly lengthy operations can be done such as 'Multiply 3540675445 by 7076345687' but in some machines only very simple ones such as 'Write down 0' are possible.

We have mentioned that the 'book of rules' supplied to the computer is replaced in the machine by a part of the store. It is then called the 'table of instructions'. It is the duty of the control to see that these instructions are obeyed correctly and in the right order. The control is so constructed that this necessarily happens.

The information in the store is usually broken up into packets of moderately small size. In one machine, for instance, a packet might consist of ten decimal digits. Numbers are assigned to the parts of the store in which the various packets of information are stored, in some systematic manner. A typical instruction might say—

'Add the number stored in position 6809 to that in 4302 and put the result back into the latter storage position'.

Needless to say it would not occur in the machine expressed in English. It would more likely be coded in a form such as 6809430217. Here 17 says which of various possible operations is to be performed on the two numbers. In this case the operation is that described above, *viz*. 'Add the number...' It will be noticed that the instruction takes up 10 digits and so forms one packet of information, very conveniently. The control will normally take the instructions to be obeyed in the order of the positions in which they are stored, but occasionally an instruction such as

'Now obey the instruction stored in position 5606, and continue from there' may be encountered, or again

'If position 4505 contains 0 obey next the instruction stored in 6707, otherwise continue straight on.'
Instructions of these latter types are very important because they make it possible for a sequence of operations to be repeated over and over again until some condition is fulfilled, but in doing so to obey, not fresh instructions on each repetition, but the same ones over and over again. To take a domestic analogy. Suppose Mother wants Tommy to call at the cobbler's every morning on his way to school to see if her shoes are done, she can ask him afresh every morning. Alternatively she can stick up a notice once and for all in the hall which he will see when he leaves for school and which tells him to call for the shoes, and also to destroy the notice when he comes back if he has the shoes with him.

The reader must accept it as a fact that digital computers can be constructed, and indeed have been constructed, according to the principles we have described, and that they can in fact mimic the actions of a human computer very closely.

The book of rules which we have described our human computer as using is of course a convenient fiction. Actual human computers really remember what they have got to do. If one wants to make a machine mimic the behaviour of the human computer in some complex operation one has to ask him how it is done, and then translate the answer into the form of an instruction table. Constructing instruction tables is usually described as 'programming'. To 'programme a machine to carry out the operation A' means to put the appropriate instruction table into the machine so that it will do A.

An interesting variant on the idea of a digital computer is a 'digital computer with a random element'. These have instructions involving the throwing of a die or some equivalent electronic process; one such instruction might for instance be, 'Throw the die and put the resulting number into store 1000'. Sometimes such a machine is described as having free will (though I would not use this phrase myself). It is not normally possible to determine from observing a machine whether it has a random element, for a similar effect can be produced by such devices as making the choices depend on the digits of the decimal for π.

Most actual digital computers have only a finite store. There is no theoretical difficulty in the idea of a computer with an unlimited store. Of course only a finite part can have been used at any one time. Likewise only a finite amount can have been constructed, but we can imagine more and more being added as required. Such computers have special theoretical interest and will be called infinitive[1] capacity computers.

The idea of a digital computer is an old one. Charles Babbage, Lucasian Professor of Mathematics at Cambridge from 1828 to 1839, planned such a machine, called the Analytical Engine, but it was never completed. Although Babbage had all the essential ideas, his machine was not at that time such a very attractive prospect. The speed which would have been available would be definitely faster than a human computer but something like 100 times slower than the Manchester machine, itself one of the slower of the modern machines. The storage was to be purely mechanical, using wheels and cards.

The fact that Babbage's Analytical Engine was to be entirely mechanical will help us to rid ourselves of a superstition. Importance is often attached to the fact that modern digital computers are electrical, and that the nervous system also is electrical. Since Babbage's machine was not electrical, and since all digital computers are in a sense equivalent, we see that this use of electricity cannot be of theoretical importance. Of course electricity usually comes in where fast signalling is concerned, so that it is not surprising that we find it in both these connections. In the nervous system chemical phenomena are at least as important as electrical. In certain computers the storage system is mainly acoustic. The feature of using electricity is thus seen to be only a very superficial similarity. If we wish to find such similarities we should look rather for mathematical analogies of function.

5. Universality of Digital Computers

The digital computers considered in the last section may be classified amongst the 'discrete state machines'. These are the machines which move by sudden jumps or clicks from one quite definite state to another. These states are sufficiently different for the possibility of confusion between them to be ignored. Strictly speaking there are no such machines. Everything really moves continuously. But there are many kinds of machine which can profitably be *thought of* as being discrete state machines. For instance in considering the switches for a lighting system it is a convenient fiction that each switch must be definitely on or definitely off. There must be intermediate positions, but for most purposes we can forget about them. As an example of a discrete state machine we might consider a wheel which clicks round through 120° once a second, but may be stopped by a lever which can be operated from outside; in addition a lamp is to light in one of the positions of the wheel. This machine could be described abstractly as follows. The internal state of the machine (which is described by the position of the wheel) may be q_1, q_2 or q_3. There is an input signal i_0 or i_1

---

[1] Editor's note. Perhaps 'infinitive' is a mis-printing in *Mind* of 'infinite'.

(position of lever). The internal state at any moment is determined by the last state and input signal according to the table

		Last State		
		q_1	q_2	q_3
Input	i_0	q_2	q_3	q_1
	i_1	q_1	q_2	q_3

The output signals, the only externally visible indication of the internal state (the light) are described by the table

$$\begin{array}{cccc} \text{State} & q_1 & q_2 & q_3 \\ \text{Output} & o_0 & o_0 & o_1 \end{array}$$

This example is typical of discrete state machines. They can be described by such tables provided they have only a finite number of possible states.

It will seem that given the initial state of the machine and the input signals it is always possible to predict all future states. This is reminiscent of Laplace's view that from the complete state of the universe at one moment of time, as described by the positions and velocities of all particles, it should be possible to predict all future states. The prediction which we are considering is, however, rather nearer to practicability than that considered by Laplace. The system of the 'universe as a whole' is such that quite small errors in the initial conditions can have an overwhelming effect at a later time. The displacement of a single electron by a billionth of a centimetre at one moment might make the difference between a man being killed by an avalanche a year later, or escaping. It is an essential property of the mechanical systems which we have called 'discrete state machines' that this phenomenon does not occur. Even when we consider the actual physical machines instead of the idealised machines, reasonably accurate knowledge of the state at one moment yields reasonably accurate knowledge any number of steps later.

As we have mentioned, digital computers fall within the class of discrete state machines. But the number of states of which such a machine is capable is usually enormously large. For instance, the number for the machine now working at Manchester is about $2^{165,000}$, *i.e.* about $10^{50,000}$. Compare this with our example of the clicking wheel described above, which had three states. It is not difficult to see why the number of states should be so immense. The computer includes a store corresponding to the paper used by a human computer. It must be possible to write into the store any one of the combinations of symbols which might have been written on the paper. For simplicity suppose that only digits from 0 to 9 are used as symbols. Variations in handwriting are ignored. Suppose the computer is allowed 100 sheets of paper each containing 50 lines each with room for 30 digits. Then the number of states is $10^{100 \times 50 \times 30}$, *i.e.* $10^{150,000}$. This is about the number of states of three Manchester machines put together. The logarithm to the base two of the number of states is usually called the 'storage capacity' of the machine. Thus the Manchester machine has a storage capacity of about 165,000 and the wheel machine of our example about 1.6. If two machines are put together their capacities must be added to obtain the capacity of the resultant machine. This leads to the possibility of statements such as 'The Manchester machine contains 64 magnetic tracks each with a capacity of 2560, eight electronic tubes with a capacity of 1280. Miscellaneous storage amounts to about 300 making a total of 174,380.'

Given the table corresponding to a discrete state machine it is possible to predict what it will do. There is no reason why this calculation should not be carried out by means of a digital computer. Provided it could be carried out sufficiently quickly the digital computer could mimic the behaviour of any discrete state machine. The imitation game could then be played with the machine in question (as B) and the mimicking digital computer (as A) and the interrogator would be unable to distinguish them. Of course the digital computer must have an adequate storage capacity as well as working sufficiently fast. Moreover, it must be programmed afresh for each new machine which it is desired to mimic.

This special property of digital computers, that they can mimic any discrete state machine, is described by saying that they are *universal* machines. The existence of machines with this property has the important consequence that, considerations of speed apart, it is unnecessary to design various new machines to do various computing processes. They can all be done with one digital computer, suitably programmed for each case. It will be seen that as a consequence of this all digital computers are in a sense equivalent.

We may now consider again the point raised at the end of §3. It was suggested tentatively that the question, 'Can machines think?' should be replaced by 'Are there imaginable digital computers which would do well in the imitation game?' If we wish we can make this superficially more general and ask 'Are there discrete state machines which would do well?' But in view of the universality property we see that either of these questions is equivalent to this, 'Let us fix our attention on one particular digital computer C. Is it true that by modifying this computer to have an adequate storage, suitably increasing its speed of action, and providing it with an appropriate programme, C can be made to play satisfactorily the part of A in the imitation game, the part of B being taken by a man?'

6. Contrary Views on the Main Question

We may now consider the ground to have been cleared and we are ready to proceed to the debate on our question, 'Can machines think?' and the variant of it quoted at the end of the last section. We cannot altogether abandon the original form of the problem, for opinions will differ as to the appropriateness of the substitution and we must at least listen to what has to be said in this connexion.

It will simplify matters for the reader if I explain first my own beliefs in the matter. Consider first the more accurate form of the question. I believe that in about fifty years' time it will be possible to programme computers, with a storage capacity of about 10^9, to make them play the imitation game so well that an average interrogator will not have more than 70 per cent. chance of making the right identification after five minutes of questioning. The original question, 'Can machines think?' I believe to be too meaningless to deserve discussion. Nevertheless I believe that at the end of the century the use of words and general educated opinion will have altered so much that one will be able to speak of machines thinking without expecting to be contradicted. I believe further that no useful purpose is served by concealing these beliefs. The popular view that scientists proceed inexorably from well-established fact to well-established fact, never being influenced by any unproved conjecture, is quite mistaken. Provided it is made clear which are proved facts and which are conjectures, no harm can result. Conjectures are of great importance since they suggest useful lines of research.

I now proceed to consider opinions opposed to my own.

(1) *The Theological Objection.* Thinking is a function of man's immortal soul. God has given an immortal soul to every man and woman, but not to any other animal or to machines. Hence no animal or machine can think.

I am unable to accept any part of this, but will attempt to reply in theological terms. I should find the argument more convincing if animals were classed with men, for there is a greater difference, to my mind, between the typical animate and the inanimate than there is between man and the other animals. The arbitrary character of the orthodox view becomes clearer if we consider how it might appear to a member of some other religious community. How do Christians regard the Moslem view that women have no souls? But let us leave this point aside and return to the main argument. It appears to me that the argument quoted above implies a serious restriction of the omnipotence of the Almighty. It is admitted that there are certain things that He cannot do such as making one equal to two, but should we not believe that He has freedom to confer a soul on an elephant if He sees fit? We might expect that He would only exercise this power in conjunction with a mutation which provided the elephant with an appropriately improved brain to minister to the needs of this soul. An argument of exactly similar form may be made for the case of machines. It may seem different because it is more difficult to "swallow". But this really only means that we think it would be less likely that He would consider the circumstances suitable for conferring a soul. The circumstances in question are discussed in the rest of this paper. In attempting to construct such machines we should not be irreverently usurping His power of creating souls, any more than we are in the procreation of children: rather we are, in either case, instruments of His will providing mansions for the souls that He creates.

However, this is mere speculation. I am not very impressed with theological arguments whatever they may be used to support. Such arguments have often been found unsatisfactory in the past. In the time of Galileo it was argued that the texts, "And the sun stood still...and hasted not to go down about a whole day" (Joshua x. 13) and "He laid the foundations of the earth, that it should not move at any time" (Psalm cv. 5) were an adequate refutation of the Copernican theory. With our present knowledge such an argument appears futile. When that knowledge was not available it made a quite different impression.

(2) *The 'Heads in the Sand' Objection.* "The consequences of machines thinking would be too dreadful. Let us hope and believe that they cannot do so."

This argument is seldom expressed quite so openly as in the form above. But it affects most of us who think about it at all. We like to believe that Man is in some subtle way superior to the rest of creation. It is best if he can be shown to be *necessarily* superior, for then there is no danger of him losing his commanding position. The popularity of the theological argument is clearly connected with this feeling. It is likely to be quite strong in intellectual people, since they value the power of thinking more highly than others, and are more inclined to base their belief in the superiority of Man on this power.

I do not think that this argument is sufficiently substantial to require refutation. Consolation would be more appropriate: perhaps this should be sought in the transmigration of souls.

(3) *The Mathematical Objection.* There are a number of results of mathematical logic which can be used to show that there are limitations to the powers of discrete-state machines. The best known of these results is known as *Gödel's theorem*,[3] and shows that in any sufficiently powerful logical system statements can be formulated which can neither be proved nor disproved within the system, unless possibly the system itself is inconsistent. There are other, in some respects similar, results due to *Church, Kleene, Rosser,* and *Turing*. The latter result is the most convenient to consider, since it refers directly to machines, whereas the others can only be used in a comparatively indirect argument: for instance if Gödel's theorem is to be used we need in addition to have some means of describing logical systems in terms of machines, and machines in terms of logical systems. The result in question refers to a type of machine which is essentially a digital computer with an infinite capacity. It states that there are certain things that such a machine cannot do. If it is rigged up to give answers to questions as in

---

[2] Possibly this view is heretical. St Thomas Aquinas (*Summa Theologica*, quoted by Bertrand Russell, p. 480) states that God cannot make a man to have no soul. But this may not be a real restriction on His powers, but only a result of the fact that men's souls are immortal, and therefore indestructible. (Editor's note: the text in *Mind* contains no reference-marker for this footnote.)

[3] Author's names in italics refer to the Bibliography.

the imitation game, there will be some questions to which it will either give a wrong answer, or fail to give an answer at all however much time is allowed for a reply. There may, of course, be many such questions, and questions which cannot be answered by one machine may be satisfactorily answered by another. We are of course supposing for the present that the questions are of the kind to which an answer 'Yes' or 'No' is appropriate, rather than questions such as 'What do you think of Picasso?' The questions that we know the machines must fail on are of this type, "Consider the machine specified as follows... Will this machine ever answer 'Yes' to any question?" The dots are to be replaced by a description of some machine in a standard form, which could be something like that used in §5. When the machine described bears a certain comparatively simple relation to the machine which is under interrogation, it can be shown that the answer is either wrong or not forthcoming. This is the mathematical result: it is argued that it proves a disability of machines to which the human intellect is not subject.

The short answer to this argument is that although it is established that there are limitations to the powers of any particular machine, it has only been stated, without any sort of proof, that no such limitations apply to the human intellect. But I do not think this view can be dismissed quite so lightly. Whenever one of these machines is asked the appropriate critical question, and gives a definite answer, we know that this answer must be wrong, and this gives us a certain feeling of superiority. Is this feeling illusory? It is no doubt quite genuine, but I do not think too much importance should be attached to it. We too often give wrong answers to questions ourselves to be justified in being very pleased at such evidence of fallibility on the part of the machines. Further, our superiority can only be felt on such an occasion in relation to the one machine over which we have scored our petty triumph. There would be no question of triumphing simultaneously over *all* machines. In short, then, there might be men cleverer than any given machine, but then again there might be other machines cleverer again, and so on.

Those who hold to the mathematical argument would, I think, mostly be willing to accept the imitation game as a basis for discussion. Those who believe in the two previous objections would probably not be interested in any criteria.

(4) *The Argument from Consciousness.* This argument is very well expressed in *Professor Jefferson's* Lister Oration for 1949, from which I quote. "Not until a machine can write a sonnet or compose a concerto because of thoughts and emotions felt, and not by the chance fall of symbols, could we agree that machine equals brain—that is, not only write it but know that it had written it. No mechanism could feel (and not merely artificially signal, an easy contrivance) pleasure at its successes, grief when its valves fuse, be warmed by flattery, be made miserable by its mistakes, be charmed by sex, be angry or depressed when it cannot get what it wants."

This argument appears to be a denial of the validity of our test. According to the most extreme form of this view the only way by which one could be sure that a machine thinks is to *be* the machine and to feel oneself thinking. One could then describe these feelings to the world, but of course no one would be justified in taking any notice. Likewise according to this view the only way to know that a *man* thinks is to be that particular man. It is in fact the solipsist point of view. It may be the most logical view to hold but it makes communication of ideas difficult. A is liable to believe 'A thinks but B does not' whilst B believes 'B thinks but A does not'. Instead of arguing continually over this point it is usual to have the polite convention that everyone thinks.

I am sure that Professor Jefferson does not wish to adopt the extreme and solipsist point of view. Probably he would be quite willing to accept the imitation game as a test. The game (with the player B omitted) is frequently used in practice under the name of *viva voce* to discover whether some one really understands something or has 'learnt it parrot fashion'. Let us listen in to a part of such a *viva voce*:

> Interrogator: In the first line of your sonnet which reads 'Shall I compare thee to a summer's day', would not 'a spring day' do as well or better?
> Witness: It wouldn't scan.
> Interrogator: How about 'a winter's day'. That would scan all right.
> Witness: Yes, but nobody wants to be compared to a winter's day.
> Interrogator: Would you say Mr. Pickwick reminded you of Christmas?
> Witness: In a way.
> Interrogator: Yet Christmas is a winter's day, and I do not think Mr. Pickwick would mind the comparison.
> Witness: I don't think you're serious. By a winter's day one means a typical winter's day, rather than a special one like Christmas.

And so on. What would Professor Jefferson say if the sonnet-writing machine was able to answer like this in the *viva voce*? I do not know whether he would regard the machine as 'merely artificially signalling' these answers, but if the answers were as satisfactory and sustained as in the above passage I do not think he would describe it as 'an easy contrivance'. This phrase is, I think, intended to cover such devices as the inclusion in the machine of a record of someone reading a sonnet, with appropriate switching to turn it on from time to time.

In short then, I think that most of those who support the argument from consciousness could be persuaded to abandon it rather than be forced into the solipsist position. They will then probably be willing to accept our test.

I do not wish to give the impression that I think there is no mystery about consciousness. There is, for instance, something of a paradox connected with any attempt to localise it. But I do not think these mysteries necessarily need to be solved before we can answer the question with which we are concerned in this paper.

(5) *Arguments from Various Disabilities.* These arguments take the form, "I grant you that you can make machines do all the things you have mentioned but you will never be able to make one to do X". Numerous features X are suggested in this connexion. I offer a selection:

Be kind, resourceful, beautiful, friendly (p. [454]), have initiative, have a sense of humour, tell right from wrong, make mistakes (p. [454]), fall in love, enjoy strawberries and cream (p. [453]), make some one fall in love with it, learn from experience (pp. [460]f.), use words properly, be the subject of its own thought (pp. [454–5]), have as much diversity of behaviour as a man, do something really new (pp. [455–6]). (Some of these disabilities are given special consideration as indicated by the page numbers.)

No support is usually offered for these statements. I believe they are mostly founded on the principle of scientific induction. A man has seen thousands of machines in his lifetime. From what he sees of them he draws a number of general conclusions. They are ugly, each is designed for a very limited purpose, when required for a minutely different purpose they are useless, the variety of behaviour of any one of them is very small, etc., etc. Naturally he concludes that these are necessary properties of machines in general. Many of these limitations are associated with the very small storage capacity of most machines. (I am assuming that the idea of storage capacity is extended in some way to cover machines other than discrete-state machines. The exact definition does not matter as no mathematical accuracy is claimed in the present discussion.) A few years ago, when very little had been heard of digital computers, it was possible to elicit much incredulity concerning them, if one mentioned their properties without describing their construction. That was presumably due to a similar application of the principle of scientific induction. These applications of the principle are of course largely unconscious. When a burnt child fears the fire and shows that he fears it by avoiding it, I should say that he was applying scientific induction. (I could of course also describe his behaviour in many other ways.) The works and customs of mankind do not seem to be very suitable material to which to apply scientific induction. A very large part of space-time must be investigated, if reliable results are to be obtained. Otherwise we may (as most English children do) decide that everybody speaks English, and that it is silly to learn French.

There are, however, special remarks to be made about many of the disabilities that have been mentioned. The inability to enjoy strawberries and cream may have struck the reader as frivolous. Possibly a machine might be made to enjoy this delicious dish, but any attempt to make one do so would be idiotic. What is important about this disability is that it contributes to some of the other disabilities, *e.g.* to the difficulty of the same kind of friendliness occurring between man and machine as between white man and white man, or between black man and black man.

The claim that "machines cannot make mistakes" seems a curious one. One is tempted to retort, "Are they any the worse for that?" But let us adopt a more sympathetic attitude, and try to see what is really meant. I think this criticism can be explained in terms of the imitation game. It is claimed that the interrogator could distinguish the machine from the man simply by setting them a number of problems in arithmetic. The machine would be unmasked because of its deadly accuracy. The reply to this is simple. The machine (programmed for playing the game) would not attempt to give the *right* answers to the arithmetic problems. It would deliberately introduce mistakes in a manner calculated to confuse the interrogator. A mechanical fault would probably show itself through an unsuitable decision as to what sort of a mistake to make in the arithmetic. Even this interpretation of the criticism is not sufficiently sympathetic. But we cannot afford the space to go into it much further. It seems to me that this criticism depends on a confusion between two kinds of mistake. We may call them 'errors of functioning' and 'errors of conclusion'. Errors of functioning are due to some mechanical or electrical fault which causes the machine to behave otherwise than it was designed to do. In philosophical discussions one likes to ignore the possibility of such errors; one is therefore discussing 'abstract machines'. These abstract machines are mathematical fictions rather than physical objects. By definition they are incapable of errors of functioning. In this sense we can truly say that 'machines can never make mistakes'. Errors of conclusion can only arise when some meaning is attached to the output signals from the machine. The machine might, for instance, type out mathematical equations, or sentences in English. When a false proposition is typed we say that the machine has committed an error of conclusion. There is clearly no reason at all for saying that a machine cannot make this kind of mistake. It might do nothing but type out repeatedly '0 = 1'. To take a less perverse example, it might have some method for drawing conclusions by scientific induction. We must expect such a method to lead occasionally to erroneous results.

The claim that a machine cannot be the subject of its own thought can of course only be answered if it can be shown that the machine has *some* thought with *some* subject matter. Nevertheless, 'the subject matter of a machine's operations' does seem to mean something, at least to the people who deal with it. If, for instance, the machine was trying to find a solution of the equation $x^2 - 40x - 11 = 0$ one would be tempted to describe this equation as part of the machine's subject matter at that moment. In this sort of sense a machine undoubtedly can be its own subject matter. It may be used to help in making up its own programmes, or to predict the effect of alterations in its own structure. By observing the results of its own behaviour it can modify its own programmes so as to achieve some purpose more effectively. These are possibilities of the near future, rather than Utopian dreams.

The criticism that a machine cannot have much diversity of behaviour is just a way of saying that it cannot have much storage capacity. Until fairly recently a storage capacity of even a thousand digits was very rare.

The criticisms that we are considering here are often disguised forms of the argument from consciousness. Usually if one maintains that a machine *can* do one of these things, and describes the kind of method that the machine could use, one will not make much of an impression. It is thought that the method (whatever it may be, for it must be mechanical) is really rather base. Compare the parenthesis in Jefferson's statement quoted on p. [451].

(6) *Lady Lovelace's Objection.* Our most detailed information of Babbage's Analytical Engine comes from a memoir by *Lady Lovelace*. In it she states, "The Analytical Engine has no pretensions to *originate* anything. It can do *whatever we know how to order it* to perform" (her italics). This statement is quoted by *Hartree* (p. 70) who adds: "This does not imply that it may not be possible to construct electronic equipment which will 'think for itself', or in which, in biological terms, one could set up a conditioned reflex, which would serve as a basis for 'learning'. Whether this is possible in principle or not is a stimulating and exciting question, suggested by some of these recent developments. But it did not seem that the machines constructed or projected at the time had this property."

I am in thorough agreement with Hartree over this. It will be noticed that he does not assert that the machines in question had not got the property, but rather that the evidence available to Lady Lovelace did not encourage her to believe that they had it. It is quite possible that the machines in question had in a sense got this property. For suppose that some discrete-state machine has the property. The Analytical Engine was a universal digital computer, so that, if its storage capacity and speed were adequate, it could by suitable programming be made to mimic the machine in question. Probably this argument did not occur to the Countess or to Babbage. In any case there was no obligation on them to claim all that could be claimed.

This whole question will be considered again under the heading of learning machines.

A variant of Lady Lovelace's objection states that a machine can 'never do anything really new'. This may be parried for a moment with the saw, 'There is nothing new under the sun'. Who can be certain that 'original work' that he has done was not simply the growth of the seed planted in him by teaching, or the effect of following well-known general principles. A better variant of the objection says that a machine can never 'take us by surprise'. This statement is a more direct challenge and can be met directly. Machines take me by surprise with great frequency. This is largely because I do not do sufficient calculation to decide what to expect them to do, or rather because, although I do a calculation, I do it in a hurried, slipshod fashion, taking risks. Perhaps I say to myself, 'I suppose the voltage here ought to be the same as there: anyway let's assume it is'. Naturally I am often wrong, and the result is a surprise for me for by the time the experiment is done these assumptions have been forgotten. These admissions lay me open to lectures on the subject of my vicious ways, but do not throw any doubt on my credibility when I testify to the surprises I experience.

I do not expect this reply to silence my critic. He will probably say that such surprises are due to some creative mental act on my part, and reflect no credit on the machine. This leads us back to the argument from consciousness, and far from the idea of surprise. It is a line of argument we must consider closed, but it is perhaps worth remarking that the appreciation of something as surprising requires as much of a 'creative mental act' whether the surprising event originates from a man, a book, a machine or anything else.

The view that machines cannot give rise to surprises is due, I believe, to a fallacy to which philosophers and mathematicians are particularly subject. This is the assumption that as soon as a fact is presented to a mind all consequences of that fact spring into the mind simultaneously with it. It is a very useful assumption under many circumstances, but one too easily forgets that it is false. A natural consequence of doing so is that one then assumes that there is no virtue in the mere working out of consequences from data and general principles.

(7) *Argument from Continuity in the Nervous System.* The nervous system is certainly not a discrete-state machine. A small error in the information about the size of a nervous impulse impinging on a neuron, may make a large difference to the size of the outgoing impulse. It may be argued that, this being so, one cannot expect to be able to mimic the behaviour of the nervous system with a discrete-state system.

It is true that a discrete-state machine must be different from a continuous machine. But if we adhere to the conditions of the imitation game, the interrogator will not be able to take any advantage of this difference. The situation can be made clearer if we consider some other simpler continuous machine. A differential analyser will do very well. (A differential analyser is a certain kind of machine not of the discrete-state type used for some kinds of calculation.) Some of these provide their answers in a typed form, and so are suitable for taking part in the game. It would not be possible for a digital computer to predict exactly what answers the differential analyser would give to a problem, but it would be quite capable of giving the right sort of answer. For instance, if asked to give the value of π (actually about 3.1416) it would be reasonable to choose at random between the values 3.12, 3.13, 3.14, 3.15, 3.16 with the probabilities of 0.05, 0.15, 0.55, 0.19, 0.06 (say). Under these circumstances it would be very difficult for the interrogator to distinguish the differential analyser from the digital computer.

(8) *The Argument from Informality of Behaviour.* It is not possible to produce a set of rules purporting to describe what a man should do in every conceivable set of circumstances. One might for instance have a rule that one is to stop when one sees a red traffic light, and to go if one sees a green one, but what if by some fault both appear together? One may perhaps decide that it is safest to stop. But some further difficulty may well arise from this decision later. To attempt to provide rules of conduct to cover every eventuality, even those arising from traffic lights, appears to be impossible. With all this I agree.

From this it is argued that we cannot be machines. I shall try to reproduce the argument, but I fear I shall hardly do it justice. It seems to run something like this. 'If each man had a definite set of rules of conduct by which he regulated his life he would be no better than a machine. But there are no such rules, so men cannot be machines.' The undistributed middle is glaring. I do not think the argument is ever put quite like this, but I believe this is the argument used nevertheless. There may however be a certain confusion between 'rules of conduct' and 'laws of behaviour' to cloud the issue. By 'rules of conduct' I mean precepts such as 'Stop if you see red lights', on which one can act, and of which one can be conscious. By 'laws of behaviour' I mean laws of nature as applied to a man's body such as 'if you pinch him he will squeak'. If we substitute 'laws of behaviour which regulate his life' for 'laws of conduct by which he regulates his life' in the argument quoted the undistributed middle is no longer insuperable. For we believe that it is not only true that being regulated by laws of behaviour implies being some sort of machine (though not necessarily a discrete-state machine), but that conversely being such a machine implies being regulated by such laws. However, we cannot so easily convince ourselves of the absence of complete laws of behaviour as of complete rules of conduct. The only way we know of for finding such laws is scientific observation, and we certainly know of no circumstances under which we could say, 'We have searched enough. There are no such laws.'

We can demonstrate more forcibly that any such statement would be unjustified. For suppose we could be sure of finding such laws if they existed. Then given a discrete-state machine it should certainly be possible to discover by observation sufficient about it to predict its future behaviour, and this within a reasonable time, say a thousand years. But this does not seem to be the case. I have set up on the Manchester computer a small programme using only 1000 units of storage, whereby the machine supplied with one sixteen figure number replies with another within two seconds. I would defy anyone to learn from these replies sufficient about the programme to be able to predict any replies to untried values.

(9) *The Argument from Extra-Sensory Perception.* I assume that the reader is familiar with the idea of extra-sensory perception, and the meaning of the four

items of it, *viz.* telepathy, clairvoyance, precognition and psycho-kinesis. These disturbing phenomena seem to deny all our usual scientific ideas. How we should like to discredit them! Unfortunately the statistical evidence, at least for telepathy, is overwhelming. It is very difficult to rearrange one's ideas so as to fit these new facts in. Once one has accepted them it does not seem a very big step to believe in ghosts and bogies. The idea that our bodies move simply according to the known laws of physics, together with some others not yet discovered but somewhat similar, would be one of the first to go.

This argument is to my mind quite a strong one. One can say in reply that many scientific theories seem to remain workable in practice, in spite of clashing with E.S.P.; that in fact one can get along very nicely if one forgets about it. This is rather cold comfort, and one fears that thinking is just the kind of phenomenon where E.S.P. may be especially relevant.

A more specific argument based on E.S.P. might run as follows: "Let us play the imitation game, using as witnesses a man who is good as a telepathic receiver, and a digital computer. The interrogator can ask such questions as 'What suit does the card in my right hand belong to?' The man by telepathy or clairvoyance gives the right answer 130 times out of 400 cards. The machine can only guess at random, and perhaps gets 104 right, so the interrogator makes the right identification." There is an interesting possibility which opens here. Suppose the digital computer contains a random number generator. Then it will be natural to use this to decide what answer to give. But then the random number generator will be subject to the psycho-kinetic powers of the interrogator. Perhaps this psycho-kinesis might cause the machine to guess right more often than would be expected on a probability calculation, so that the interrogator might still be unable to make the right identification. On the other hand, he might be able to guess right without any questioning, by clairvoyance. With E.S.P. anything may happen.

If telepathy is admitted it will be necessary to tighten our test up. The situation could be regarded as analogous to that which would occur if the interrogator were talking to himself and one of the competitors was listening with his ear to the wall. To put the competitors into a 'telepathy-proof room' would satisfy all requirements.

7. Learning Machines

The reader will have anticipated that I have no very convincing arguments of a positive nature to support my views. If I had I should not have taken such pains to point out the fallacies in contrary views. Such evidence as I have I shall now give.

Let us return for a moment to Lady Lovelace's objection, which stated that the machine can only do what we tell it to do. One could say that a man can 'inject' an idea into the machine, and that it will respond to a certain extent and then drop into quiescence, like a piano string struck by a hammer. Another simile would be an atomic pile of less than critical size: an injected idea is to correspond to a neutron entering the pile from without. Each such neutron will cause a certain disturbance which eventually dies away. If, however, the size of the pile is sufficiently increased, the disturbance caused by such an incoming neutron will very likely go on and on increasing until the whole pile is destroyed. Is there a corresponding phenomenon for minds, and is there one for machines? There does seem to be one for the human mind. The majority of them seem to be 'sub-critical', *i.e.* to correspond in this analogy to piles of sub-critical size. An idea presented to such a mind will on average give rise to less than one idea in reply. A smallish proportion are super-critical. An idea presented to such a mind may give rise to a whole 'theory' consisting of secondary, tertiary and more remote ideas. Animals minds seem to be very definitely sub-critical. Adhering to this analogy we ask, 'Can a machine be made to be super-critical?'

The 'skin of an onion' analogy is also helpful. In considering the functions of the mind or the brain we find certain operations which we can explain in purely mechanical terms. This we say does not correspond to the real mind: it is a sort of skin which we must strip off if we are to find the real mind. But then in what remains we find a further skin to be stripped off, and so on. Proceeding in this way do we ever come to the 'real' mind, or do we eventually come to the skin which has nothing in it? In the latter case the whole mind is mechanical. (It would not be a discrete-state machine however. We have discussed this.)

These last two paragraphs do not claim to be convincing arguments. They should rather be described as 'recitations tending to produce belief'.

The only really satisfactory support that can be given for the view expressed at the beginning of §6, will be that provided by waiting for the end of the century and then doing the experiment described. But what can we say in the meantime? What steps should be taken now if the experiment is to be successful?

As I have explained, the problem is mainly one of programming. Advances in engineering will have to be made too, but it seems unlikely that these will not be adequate for the requirements. Estimates of the storage capacity of the brain vary from 10^{10} to 10^{15} binary digits. I incline to the lower values and believe that only a very small fraction is used for the higher types of thinking. Most of it is probably used for the retention of visual impressions. I should be surprised if more than 10^9 was required for satisfactory playing of the imitation game, at any rate against a blind man. (Note—The capacity of the *Encyclopaedia Britannica*, 11th edition, is 2×10^9.) A storage capacity of 10^7 would be a very practicable possibility even by present techniques. It is probably not necessary to increase the speed of operations of the machines at all. Parts of modern machines which can be regarded as analogues of nerve cells work about a thousand times faster than the latter. This should provide a 'margin of safety' which could cover losses of speed arising in many ways. Our problem then is to find out how to programme these machines to play the game. At my present rate of working I produce about a thousand digits of programme a day, so that about sixty workers, working steadily through the fifty years might accomplish the job, if nothing went into the waste-paper basket. Some more expeditious method seems desirable.

In the process of trying to imitate an adult human mind we are bound to think a good deal about the process which has brought it to the state that it is in. We may notice three components,

(*a*) The initial state of the mind, say at birth,
(*b*) The education to which it has been subjected,
(*c*) Other experience, not to be described as education, to which it has been subjected.

Instead of trying to produce a programme to simulate the adult mind, why not rather try to produce one which simulates the child's? If this were then subjected to an appropriate course of education one would obtain the adult brain. Presumably the child-brain is something like a note-book as one buys it from the stationers. Rather little mechanism, and lots of blank sheets. (Mechanism and writing are from our point of view almost synonymous.) Our hope is that there is so little mechanism in the child-brain that something like it can be easily programmed. The amount of work in the education we can assume, as a first approximation, to be much the same as for the human child.

We have thus divided our problem into two parts. The child-programme and the education process. These two remain very closely connected. We cannot expect to find a good child-machine at the first attempt. One must experiment with teaching one such machine and see how well it learns. One can then try another and see if it is better or worse. There is an obvious connection between this process and evolution, by the identifications

Structure of the child machine = Hereditary material
Changes of the child machine = Mutations
Natural selection   = Judgment of the experimenter

One may hope, however, that this process will be more expeditious than evolution. The survival of the fittest is a slow method for measuring advantages. The experimenter, by the exercise of intelligence, should be able to speed it up. Equally important is the fact that he is not restricted to random mutations. If he can trace a cause for some weakness he can probably think of the kind of mutation which will improve it.

It will not be possible to apply exactly the same teaching process to the machine as to a normal child. It will not, for instance, be provided with legs, so that it could not be asked to go out and fill the coal scuttle. Possibly it might not have eyes. But however well these deficiencies might be overcome by clever engineering, one could not send the creature to school without the other children making excessive fun of it. It must be given some tuition. We need not be too concerned about the legs, eyes, etc. The example of Miss Helen Keller shows that education can take place provided that communication in both directions between teacher and pupil can take place by some means or other.

We normally associate punishments and rewards with the teaching process. Some simple child-machines can be constructed or programmed on this sort of principle. The machine has to be so constructed that events which shortly preceded the occurrence of a punishment-signal are unlikely to be repeated, whereas a reward-signal increased the probability of repetition of the events which led up to it. These definitions do not presuppose any feelings on the part of the machine. I have done some experiments with one such child-machine, and succeeded in teaching it a few things, but the teaching method was too unorthodox for the experiment to be considered really successful.

The use of punishments and rewards can at best be a part of the teaching process. Roughly speaking, if the teacher has no other means of communicating to the pupil, the amount of information which can reach him does not exceed the total number of rewards and punishments applied. By the time a child has learnt to repeat 'Casabianca' he would probably feel very sore indeed, if the text could only be discovered by a 'Twenty Questions' technique, every 'NO' taking the form of a blow. It is necessary therefore to have some other 'unemotional' channels of communication. If these are available it is possible to teach a machine by punishments and rewards to obey orders given in some language, *e.g.* a symbolic language. These orders are to be transmitted through the 'unemotional' channels. The use of this language will diminish greatly the number of punishments and rewards required.

Opinions may vary as to the complexity which is suitable in the child machine. One might try to make it as simple as possible consistently with the general principles. Alternatively one might have a complete system of logical inference 'built in'.[4] In the latter case the store would be largely occupied with definitions and propositions. The propositions would have various kinds of status, *e.g.* well-established facts, conjectures, mathematically proved theorems, statements given by an authority, expressions having the logical form of proposition but not belief-value. Certain propositions may be described as 'imperatives.' The machine should be so constructed that as soon as an imperative is classed as 'well-established' the appropriate action automatically takes place. To illustrate this, suppose the teacher says to the machine, 'Do your homework now'. This may cause "Teacher says 'Do your homework now'" to be included amongst the well-established facts. Another such fact might be, "Everything that teacher says is true". Combining these may eventually lead to the imperative, 'Do your homework now', being included amongst the well-established facts, and this, by the

---

[4] Or rather 'programmed in' for our child-machine will be programmed in a digital computer. But the logical system will not have to be learnt.

construction of the machine, will mean that the homework actually gets started, but the effect is very satisfactory. The processes of inference used by the machine need not be such as would satisfy the most exacting logicians. There might for instance be no hierarchy of types. But this need not mean that type fallacies will occur, any more than we are bound to fall over unfenced cliffs. Suitable imperatives (expressed *within* the systems, not forming part of the rules *of* the system) such as 'Do not use a class unless it is a subclass of one which has been mentioned by teacher' can have a similar effect to 'Do not go too near the edge'.

The imperatives that can be obeyed by a machine that has no limbs are bound to be of a rather intellectual character, as in the example (doing homework) given above. Important amongst such imperatives will be ones which regulate the order in which the rules of the logical system concerned are to be applied. For at each stage when one is using a logical system, there is a very large number of alternative steps, any of which one is permitted to apply, so far as obedience to the rules of the logical system is concerned. These choices make the difference between a brilliant and a footling reasoner, not the difference between a sound and a fallacious one. Propositions leading to imperatives of this kind might be "When Socrates is mentioned, use the syllogism in Barbara" or "If one method has been proved to be quicker than another, do not use the slower method". Some of these may be 'given by authority', but others may be produced by the machine itself, *e.g.* by scientific induction.

The idea of a learning machine may appear paradoxical to some readers. How can the rules of operation of the machine change? They should describe completely how the machine will react whatever its history might be, whatever changes it might undergo. The rules are thus quite time-invariant. This is quite true. The explanation of the paradox is that the rules which get changed in the learning process are of a rather less pretentious kind, claiming only an ephemeral validity. The reader may draw a parallel with the Constitution of the United States.

An important feature of a learning machine is that its teacher will often be very largely ignorant of quite what is going on inside, although he may still be able to some extent to predict his pupil's behaviour. This should apply most strongly to the later education of a machine arising from a child-machine of well-tried design (or programme). This is in clear contrast with normal procedure when using a machine to do computations: one's object is then to have a clear mental picture of the state of the machine at each moment in the computation. This object can only be achieved with a struggle. The view that 'the machine can only do what we know how to order it to do'[5] appears strange in face of this. Most of the programmes which we can put into the machine will result in its doing something that we cannot make sense of at all, or which we

---

[5] Compare Lady Lovelace's statement (p. [455]), which does not contain the word 'only'.

regard as completely random behaviour. Intelligent behaviour presumably consists in a departure from the completely disciplined behaviour involved in computation, but a rather slight one, which does not give rise to random behaviour, or to pointless repetitive loops. Another important result of preparing our machine for its part in the imitation game by a process of teaching and learning is that 'human fallibility' is likely to be omitted[6] in a rather natural way, *i.e.* without special 'coaching'. (The reader should reconcile this with the point of view on p. [454].)[7] Processes that are learnt do not produce a hundred per cent. certainty of result; if they did they could not be unlearnt.

It is probably wise to include a random element in a learning machine (see p. [445]). A random element is rather useful when we are searching for a solution of some problem. Suppose for instance we wanted to find a number between 50 and 200 which was equal to the square of the sum of its digits, we might start at 51 then try 52 and go on until we got a number that worked. Alternatively we might choose numbers at random until we got a good one. This method has the advantage that it is unnecessary to keep track of the values that have been tried, but the disadvantage that one may try the same one twice, but this is not very important if there are several solutions. The systematic method has the disadvantage that there may be an enormous block without any solutions in the region which has to be investigated first. Now the learning process may be regarded as a search for a form of behaviour which will satisfy the teacher (or some other criterion). Since there is probably a very large number of satisfactory solutions the random method seems to be better than the systematic. It should be noticed that it is used in the analogous process of evolution. But there the systematic method is not possible. How could one keep track of the different genetical combinations that had been tried, so as to avoid trying them again?

We may hope that machines will eventually compete with men in all purely intellectual fields. But which are the best ones to start with? Even this is a difficult decision. Many people think that a very abstract activity, like the playing of chess, would be best. It can also be maintained that it is best to provide the machine with the best sense organs that money can buy, and then teach it to understand and speak English. This process could follow the normal teaching of a child. Things would be pointed out and named, etc. Again I do not know what the right answer is, but I think both approaches should be tried.

We can only see a short distance ahead, but we can see plenty there that needs to be done.

---

[6] Editor's note. Presumably 'omitted' is a typographical error in *Mind*.

[7] Editor's note. The cross-reference in *Mind* is to 'pp. 24, 25'. These are presumably pages of Turing's original typescript. The approximate position of the material is indicated by the fact that another uncorrected cross-reference in *Mind* places Turing's quotation from Jefferson on p. 21 of the original typescript.

Bibliography

Samuel Butler, Erewhon, London, 1865. Chapters 23, 24, 25, *The Book of the Machines*.

Alonzo Church, "An Unsolvable Problem of Elementary Number Theory", *American Journal of Mathematics*, 58 (1936), 345–363.

K. Gödel, "Über formal unentscheidbare Sätze der Principia Mathematica und verwandter Systeme I", *Monatshefte für Mathematik und Physik* (1931), 173–189.

D. R. Hartree, *Calculating Instruments and Machines*, New York, 1949.

S. C. Kleene, "General Recursive Functions of Natural Numbers", *American Journal of Mathematics*, 57 (1935), 153–173 and 219–244.

G. Jefferson, "The Mind of Mechanical Man". Lister Oration for 1949. *British Medical Journal*, vol. i (1949), 1105–1121.

Countess of Lovelace, 'Translator's notes to an article on Babbage's Analytical Engine', *Scientific Memoirs* (ed. by R. Taylor), vol. 3 (1842), 691–731.

Bertrand Russell, *History of Western Philosophy*, London, 1940.

A. M. Turing, "On Computable Numbers, with an Application to the Entscheidungsproblem" [Chapter 1].

CHAPTER 12

Intelligent Machinery, A Heretical Theory (c.1951)

Alan Turing

Introduction

Jack Copeland

The '51 Society

Turing gave the presentation 'Intelligent Machinery, A Heretical Theory' on a radio discussion programme called *The '51 Society*. Named after the year in which the programme first went to air, *The '51 Society* was produced by the BBC Home Service at their Manchester studio and ran for several years.[1] A presentation by the week's guest would be followed by a panel discussion. Regulars on the panel included Max Newman, Professor of Mathematics at Manchester, the philosopher Michael Polanyi, then Professor of Social Studies at Manchester, and the mathematician Peter Hilton, a younger member of Newman's department at Manchester who had worked with Turing and Newman at Bletchley Park.

Machine Learning

Turing's target in 'Intelligent Machinery, A Heretical Theory' is the claim that 'You cannot make a machine to think for you' (p. 472). A common theme in his writing is that if a machine is to be intelligent, then it will need to 'learn by experience' (probably with some pre-selection, by an external educator, of the experiences to which the machine will be subjected). The present article continues the discussion of machine learning begun in Chapters 10 and 11. Turing remarks that the 'human analogy alone' suggests that a process of education 'would in practice be an essential to the production of a reasonably intelligent machine within a reasonably short space of time' (p. 473). He emphasizes the

---

[1] Peter Hilton in interview with Copeland (June 2001).

point, also made in Chapter 11, that one might 'start from a comparatively simple machine, and, by subjecting it to a suitable range of "experience" transform it into one which was more elaborate, and was able to deal with a far greater range of contingencies' (p. 473).

Turing goes on to give some indication of how learning might be accomplished, introducing the idea of a machine's building up what he calls 'indexes of experiences' (p. 474). (This idea is not mentioned elsewhere in his writings.) An example of an index of experiences is a list (ordered in some way) of situations in which the machine has found itself, coupled with the action that was taken, and the outcome, good or bad. The situations are described in terms of features. Faced with a choice as to what to do next, the machine looks up features of its present situation in whatever indexes it has. If this procedure affords more than one candidate action, the machine selects between them by means of some rule, possibly itself learned through experience. Turing very reasonably grounds his belief that comparatively crude selection-rules will lead to satisfactory behaviour in the fact that engineering problems are regularly solved by 'the crudest rule of thumb procedure...e.g. whether a function increases or decreases with one of its variables' (p. 474).

In response to the problem of how the educator is to indicate to the machine whether a situation or outcome is a 'favourable' one or not, Turing returns to the possibility of incorporating two 'keys' in the machine, which can be manipulated by the educator, and which represent 'pleasure' and 'pain' (p. 474). This is an idea that Turing discusses more fully in Chapter 10, where he considers adding two input lines to a (modified) Turing machine, the pleasure (or reward) line and the pain (or punishment) line. He calls the result a 'P-type machine' ('P' standing for 'pleasure–pain').[2]

Random Elements

Turing ends his discussion of machine learning with the suggestion that a 'random element' be incorporated in the machine (p. 475). This would, as he says, result in the behaviour of the machine being by no means completely determined by the experiences to which it was subjected (p. 475). The idea that a random element be included in a learning machine appears elsewhere in Turing's discussions of machine intelligence. In Chapter 11 he says: 'A random element is rather useful when...searching for a solution of some problem' (p. 463). He gives this example:

---

[2] A detailed description of Turing's P-type machines is given in B. J. Copeland and D. Proudfoot, 'On Alan Turing's Anticipation of Connectionism', *Synthese*, 108 (1996), 361–77 (reprinted in R. Chrisley (ed.), *Artificial Intelligence: Critical Concepts in Cognitive Science*, ii: *Symbolic AI* (London: Routledge, 2000)).

Suppose for instance we wanted to find a number between 50 and 200 which was equal to the square of the sum of its digits, we might start at 51 then try 52 and so on until we got a number that worked. Alternatively we might choose numbers at random until we got a good one.

Turing continues (p. 463):

The systematic method has the disadvantage that there may be an enormous block without any solutions in the region which has to be investigated first. Now the learning process may be regarded as a search for a form of behaviour which will satisfy the teacher (or some other criterion). Since there is probably a very large number of satisfactory solutions the random method seems to be better than the systematic. It should be noticed that it is used in the analogous process of evolution.

Turing's discussion of 'pleasure–pain systems' in Chapter 10 also mentions randomness (p. 425):

I will use this term ['pleasure–pain' system] to mean an unorganised machine of the following general character: The configurations of the machine are described by two expressions, which we may call the character-expression and the situation-expression. The character and situation at any moment, together with the input signals, determine the character and situation at the next moment. The character may be subject to some random variation. Pleasure interference has a tendency to fix the character i.e. towards preventing it changing, whereas pain stimuli tend to disrupt the character, causing features which had become fixed to change, or to become again subject to random variation.

The Mathematical Objection

In what are some of the most interesting remarks in 'Intelligent Machinery, A Heretical Theory', Turing sketches and rebuts an argument against the possibility of computing machines emulating the full intelligence of human beings. The objection is stated as follows in Chapter 10 (pp. 410–11):

Recently the theorem of Gödel and related results... have shown that if one tries to use machines for such purposes as determining the truth or falsity of mathematical theorems and one is not willing to tolerate an occasional wrong result, then any given machine will in some cases be unable to give an answer at all. On the other hand the human intelligence seems to be able to find methods of ever-increasing power for dealing with such problems 'transcending' the methods available to machines.

In Chapter 11 he terms this the 'Mathematical Objection' (p. 450).

As Turing notes, the 'related results' include what he himself proved in 'On Computable Numbers'. The import of the satisfactoriness problem (explained in 'Computable Numbers: A Guide') is that no Turing machine can correctly determine the truth or falsity of each statement of the form 'such-and-such

Turing machine is circle-free'. Whichever Turing machine one chooses to ask, there will be statements of this form for which the chosen machine either gives no answer or gives the wrong answer (compare Chapter 11, pp. 450–1). (In Chapter 3, Turing extends this result to his oracle machines: no oracle machine can correctly determine the truth or falsity of each statement of the form 'such-and-such oracle machine is circle-free' (pp. 156–7).)

Post formulated a version of the Mathematical Objection as early as 1921.[3] However, the objection has become known over the years as the 'Gödel argument'. In 1961, in a famous article, the philosopher John Lucas claimed the Gödel argument establishes that 'mechanism'—which Lucas characterizes as the view that 'minds [can] be explained as machines'—is false.[4] More recently, the mathematical physicist Roger Penrose has endorsed a version of the Gödel argument.[5]

Lucas was happy to assert, on the basis of the Mathematical Objection, that 'no scientific enquiry can ever exhaust the...human mind'.[6] Not many who admire the explanatory power of science would be happy to endorse this conclusion. Penrose himself appears to hold that the mind can be explained in ultimately physical terms. However, it is difficult to say what scientific conception of the mind could be available to someone who endorses the Mathematical Objection. This is because the objection, if sound, could be used equally well to support the conclusion, not only that the mind is not a Turing machine, but also that it is not any one of a very broad range of machines (which includes the oracle machines). Given the enormous diversity of types of machine in this range, it is an open question whether there is any scientific conception of the mind that the Mathematical Objection (if sound) would not rule out.[7]

Penrose acknowledges that the objection applies not only to the view that the mind is equivalent to a Turing machine but 'much more generally', saying: 'No doubt there are readers who believe that the last vestige of credibility of my [version of the Gödel] argument has disappeared at this stage! I certainly should not blame any reader for feeling this way.'[8]

So far, however, Penrose has not made it clear what scientific conception of the mind can remain for one who endorses the argument, remarking only that, since

---

[3] E. L. Post, 'Absolutely Unsolvable Problems and Relatively Undecidable Propositions: Account of an Anticipation', in M. Davis (ed.), *The Undecidable: Basic Papers on Undecidable Propositions, Unsolvable Problems and Computable Functions* (New York: Raven, 1965), 417; see also 423.

[4] J. R. Lucas, 'Minds, Machines and Gödel', *Philosophy*, 36 (1961), 112–27 (112).

[5] See his *The Emperor's New Mind: Concerning Computers, Minds, and the Laws of Physics* (Oxford: Oxford University Press, 1989); 'Précis of *The Emperor's New Mind: Concerning Computers, Minds, and the Laws of Physics*', *Behavioral and Brain Sciences*, 13 (1990), 643–55 and 692–705; *Shadows of the Mind: A Search for the Missing Science of Consciousness* (Oxford: Oxford University Press, 1994); 'Beyond the Doubting of a Shadow', *Psyche*, 2/23 (1996).

[6] Lucas, 'Minds, Machines and Gödel', 127.

[7] See B. J. Copeland, 'Turing's O-machines, Penrose, Searle, and the Brain', *Analysis*, 58 (1998), 128–38.

[8] Penrose, 'Beyond the Doubting of a Shadow', section 3.10, and *Shadows of the Mind*, 381.

the argument 'can be applied in very general circumstances indeed', the mind is 'something very mysterious'.[9]

Turing's Answer to the Mathematical Objection

In Chapter 11 Turing says (p. 451): 'The short answer to this argument is that although it is established that there are limitations to the powers of any particular machine, it has only been stated, without any sort of proof, that no such limitations apply to the human intellect.' This remark might appear to cut to the heart of the matter. However, Turing expresses dissatisfaction with it, saying that the Mathematical Objection cannot 'be dismissed so lightly'. He goes on to broach a further line of attack on the argument, pointing out that humans 'often give wrong answers to questions', and it is this line of attack that he pursues in 'Intelligent Machinery, A Heretical Theory'.

In the quotation from Chapter 10 given above, Turing notes that the Mathematical Objection rests on a proviso that the machine is not allowed to make mistakes, and as he goes on to point out, 'the condition that the machine must not make mistakes... is not a requirement for intelligence' (p. 411). In 'Intelligent Machinery, A Heretical Theory' he suggests that the 'danger of the mathematician making mistakes is an unavoidable corollary of his power of sometimes hitting upon an entirely new method' (p. 472). Turing envisages machines also able to hit upon new methods: 'My contention is that machines can be constructed which will simulate the behaviour of the human mind very closely. They will make mistakes at times, and at times they may make new and very interesting statements.'

Turing makes a similar point in Chapter 9 (pp. 393–4):

[I]f a mathematician is confronted with such a problem [e.g. determining the truth or falsity of statements of the form '*p* is provable in such-and-such system'—Ed.] he would search around and find new methods of proof, so that he ought eventually to be able to reach a decision about any given formula. ... I would say that fair play must be given to the machine. Instead of it sometimes giving no answer we could arrange that it gives occasional wrong answers. But the human mathematician would likewise make blunders when trying out new techniques. It is easy for us to regard these blunders as not counting and give him another chance, but the machine would probably be allowed no mercy.

The use of heuristic search carries with it the risk of the computer producing a proportion of incorrect answers (see 'Artificial Intelligence'). This fact would have been very familiar to Turing from his experience with the bombe. Probably Turing was thinking of heuristic search when he wrote this, the earliest surviving

---

[9] 'Beyond the Doubting of a Shadow', section 13.2.

statement of his views concerning machine intelligence, in 'Proposed Electronic Calculator': 'There are indications however that it is possible to make the machine display intelligence at the risk of its making occasional serious mistakes. By following up this aspect the machine could probably be made to play very good chess.'[10]

In 'Intelligent Machinery, A Heretical Theory' Turing passes immediately from his remarks on the Mathematical Objection to a discussion of machine learning. This juxtaposition perhaps indicates that Turing's view was this: it is the possibility of a machine's learning *new* methods and techniques that ultimately defeats the Mathematical Objection. In the simplest possible case, the machine's tutor—a human mathematician—can just present the machine with a better method whenever the machine produces an incorrect answer to a problem. This new input in effect alters the machine's standard description, transforming it into a different Turing machine (see 'Computable Numbers: A Guide'). Alternatively a machine may itself be able to search around (albeit fallibly) for better methods. The search might involve the use of a random element. As in the preceding case, the standard description of the machine alters in consequence of the learning process, as the machine overwrites its previous algorithm with a successor. (As Turing says in Chapter 9: 'What we want is a machine that can learn from experience. The possibility of letting the machine alter its own instructions provides the mechanism for this.') Thus the learning machine may traverse the space of what in one of his letters to Newman (Chapter 4, p. 215) Turing calls 'proof finding' machines. In the same letter Turing says:

One imagines different machines allowing different sets of proofs, and by choosing a suitable machine one can approximate 'truth' by 'provability' better than with a less suitable machine, and can in a sense approximate it as well as you please.

The learning machine successively mutates from one proof-finding Turing machine into another, becoming capable of wider sets of proofs as new, more powerful methods of proof are acquired.

The Future

Turing ends 'Intelligent Machinery, A Heretical Theory' with a vision of the future, now hackneyed, in which intelligent computers 'outstrip our feeble powers' and 'take control'. There is more of the same in Chapter 13. No doubt this is comic-strip stuff. Nevertheless, these images of Turing's reveal his profound grasp of the potential of the universal Turing machine at a time when the

---

[10] 'Proposed Electronic Calculator', National Physical Laboratory, 1945, 16 (National Physical Laboratory library; a digital facsimile of the original typescript is in The Turing Archive for the History of Computing <www.AlanTuring.net/proposed_electronic_calculator> (page reference is to the original typescript)).

only computers in existence were minuscule, and none but the most straightforward of tasks had been successfully programmed.[11]

Further reading

Benacerraf, P., 'God, the Devil, and Gödel', *Monist*, 51 (1967), 9–32.
Copeland, B. J., 'Turing's O-machines, Penrose, Searle, and the Brain', *Analysis*, 58 (1998), 128–38.
Gandy, R., 'Human versus Mechanical Intelligence', in P. Millican and A. Clark (eds.), *Machines and Thought: The Legacy of Alan Turing* (Oxford: Clarendon Press, 1996).
Lucas, J. R., 'Minds, Machines and Gödel', *Philosophy*, 36 (1961), 112–27.
——'Minds, Machines and Gödel: A Retrospect', in P. Millican and A. Clark (eds.), *Machines and Thought: The Legacy of Alan Turing* (Oxford: Clarendon Press, 1996).
Penrose, R., *Shadows of the Mind: A Search for the Missing Science of Consciousness* (Oxford: Oxford University Press, 1994).
Piccinini, G., 'Alan Turing and the Mathematical Objection', *Minds and Machines*, 13 (2003), 23–48.

Provenance

The text that follows is from a typescript entitled 'Intelligent Machinery, A Heretical Theory' and marked 'Typist's Typescript'.[12]

---

[11] In this chapter Turing speaks of the 'mechanic who has constructed the machine'. This is perhaps a glimpse of Turing's attitude toward Kilburn, Williams, and the other engineers who built the Manchester computer. Kilburn himself was hardly less dismissive of the logicians' contributions (for example in an interview with Christopher Evans in 1976, 'The Pioneers of Computing: An Oral History of Computing' (London: Science Museum)).

[12] The typescript is among the Turing Papers in the Modern Archive Centre, King's College, Cambridge (catalogue reference B 4). Turing's mother Sara included the text of 'Intelligent Machinery, A Heretical Theory' in her biography *Alan M. Turing* but unfortunately incorporated some errors (S. Turing, *Alan M. Turing* (Cambridge: Heffer, 1959), 128–34.) The present edition first appeared in B. J. Copeland (ed.), 'A Lecture and Two Radio Broadcasts on Machine Intelligence by Alan Turing', in K. Furukawa, D. Michie, and S. Muggleton (eds.), *Machine Intelligence 15* (Oxford: Oxford University Press, 1999).

Intelligent Machinery, A Heretical Theory

'You cannot make a machine to think for you.' This is a commonplace that is usually accepted without question. It will be the purpose of this paper to question it.

Most machinery developed for commercial purposes is intended to carry out some very specific job, and to carry it out with certainty and considerable speed. Very often it does the same series of operations over and over again without any variety. This fact about the actual machinery available is a powerful argument to many in favour of the slogan quoted above. To a mathematical logician this argument is not available, for it has been shown that there are machines theoretically possible which will do something very close to thinking. They will, for instance, test the validity of a formal proof in the system of Principia Mathematica, or even tell of a formula of that system whether it is provable or disprovable. In the case that the formula is neither provable nor disprovable such a machine certainly does not behave in a very satisfactory manner, for it continues to work indefinitely without producing any result at all, but this cannot be regarded as very different from the reaction of the mathematicians, who have for instance worked for hundreds of years on the question as to whether Fermat's last theorem is true or not. For the case of machines of this kind a more subtle argument is necessary. By Gödel's famous theorem, or some similar argument, one can show that however the machine is constructed there are bound to be cases where the machine fails to give an answer, but a mathematician would be able to. On the other hand, the machine has certain advantages over the mathematician. Whatever it does can be relied upon, assuming no mechanical 'breakdown', whereas the mathematician makes a certain proportion of mistakes. I believe that this danger of the mathematician making mistakes is an unavoidable corollary of his power of sometimes hitting upon an entirely new method. This seems to be confirmed by the well known fact that the most reliable people will not usually hit upon really new methods.

My contention is that machines can be constructed which will simulate the behaviour of the human mind very closely. They will make mistakes at times, and at times they may make new and very interesting statements, and on the whole the output of them will be worth attention to the same sort of extent as the output of a human mind. The content of this statement lies in the greater frequency expected for the true statements, and it cannot, I think, be given an exact statement. It would not, for instance, be sufficient to say simply that the machine will make any true statement sooner or later, for an example of such a machine would be one which makes all possible statements sooner or later. We

Printed with the permission of the BBC and the Estate of Alan Turing.

know how to construct these, and as they would (probably) produce true and false statements about equally frequently, their verdicts would be quite worthless. It would be the actual reaction of the machine to circumstances that would prove my contention, if indeed it can be proved at all.

Let us go rather more carefully into the nature of this 'proof'. It is clearly possible to produce a machine which would give a very good account of itself for any range of tests, if the machine were made sufficiently elaborate. However, this again would hardly be considered an adequate proof. Such a machine would give itself away by making the same sort of mistake over and over again, and being quite unable to correct itself, or to be corrected by argument from outside. If the machine were able in some way to 'learn by experience' it would be much more impressive. If this were the case there seems to be no real reason why one should not start from a comparatively simple machine, and, by subjecting it to a suitable range of 'experience' transform it into one which was more elaborate, and was able to deal with a far greater range of contingencies. This process could probably be hastened by a suitable selection of the experiences to which it was subjected. This might be called 'education'. But here we have to be careful. It would be quite easy to arrange the experiences in such a way that they automatically caused the structure of the machine to build up into a previously intended form, and this would obviously be a gross form of cheating, almost on a par with having a man inside the machine. Here again the criterion as to what would be considered reasonable in the way of 'education' cannot be put into mathematical terms, but I suggest that the following would be adequate in practice. Let us suppose that it is intended that the machine shall understand English, and that owing to its having no hands or feet, and not needing to eat, nor desiring to smoke, it will occupy its time mostly in playing games such as Chess and GO, and possibly Bridge. The machine is provided with a typewriter keyboard on which any remarks to it are typed, and it also types out any remarks that it wishes to make. I suggest that the education of the machine should be entrusted to some highly competent schoolmaster who is interested in the project but who is forbidden any detailed knowledge of the inner workings of the machine. The mechanic who has constructed the machine, however, is permitted to keep the machine in running order, and if he suspects that the machine has been operating incorrectly may put it back to one of its previous positions and ask the schoolmaster to repeat his lessons from that point on, but he may not take any part in the teaching. Since this procedure would only serve to test the bona fides of the mechanic, I need hardly say that it would not be adopted in the experimental stages. As I see it, this education process would in practice be an essential to the production of a reasonably intelligent machine within a reasonably short space of time. The human analogy alone suggests this.

I may now give some indication of the way in which such a machine might be expected to function. The machine would incorporate a memory. This does not need very much explanation. It would simply be a list of all the statements that had been made to it or by it, and all the moves it had made and the cards it had played in its games. This would be listed in chronological order. Besides this straightforward memory there would be a number of 'indexes of experiences'. To explain this idea I will suggest the form which one such index might possibly take. It might be an alphabetical index of the words that had been used giving the 'times' at which they had been used, so that they could be looked up in the memory. Another such index might contain patterns of men on parts of a GO board that had occurred. At comparatively late stages of education the memory might be extended to include important parts of the configuration of the machine at each moment, or in other words it would begin to remember what its thoughts had been. This would give rise to fruitful new forms of indexing. New forms of index might be introduced on account of special features observed in the indexes already used. The indexes would be used in this sort of way. Whenever a choice has to be made as to what to do next, features of the present situation are looked up in the indexes available, and the previous choice in the similar situations, and the outcome, good or bad, is discovered. The new choice is made accordingly. This raises a number of problems. If some of the indications are favourable and some are unfavourable what is one to do? The answer to this will probably differ from machine to machine and will also vary with its degree of education. At first probably some quite crude rule will suffice, e.g. to do whichever has the greatest number of votes in its favour. At a very late stage of education the whole question of procedure in such cases will probably have been investigated by the machine itself, by means of some kind of index, and this may result in some highly sophisticated, and, one hopes, highly satisfactory, form of rule. It seems probable however that the comparatively crude forms of rule will themselves be reasonably satisfactory, so that progress can on the whole be made in spite of the crudeness of the choice [of] rules.[1] This seems to be verified by the fact that engineering problems are sometimes solved by the crudest rule of thumb procedure which only deals with the most superficial aspects of the problem, e.g. whether a function increases or decreases with one of its variables. Another problem raised by this picture of the way behaviour is determined is the idea of 'favourable outcome'. Without some such idea, corresponding to the 'pleasure principle' of the psychologists, it is very difficult to see how to proceed. Certainly it would be most natural to introduce some such thing into the machine. I suggest that there should be two keys which can be manipulated by the schoolmaster, and which represent the ideas of pleasure and pain. At later stages in education the machine would recognise certain other conditions as desirable owing to their having been constantly associated in the past with pleasure, and likewise certain others as undesirable. Certain expressions of

---

[1] Editor's note. Words enclosed in square brackets do not appear in the typescript.

anger on the part of the schoolmaster might, for instance, be recognised as so ominous that they could never be overlooked, so that the schoolmaster would find that it became unnecessary to 'apply the cane' any more.

To make further suggestions along these lines would perhaps be unfruitful at this stage, as they are likely to consist of nothing more than an analysis of actual methods of education applied to human children. There is, however, one feature that I would like to suggest should be incorporated in the machines, and that is a 'random element'. Each machine should be supplied with a tape bearing a random series of figures, e.g. 0 and 1 in equal quantities, and this series of figures should be used in the choices made by the machine. This would result in the behaviour of the machine not being by any means completely determined by the experiences to which it was subjected, and would have some valuable uses when one was experimenting with it. By faking the choices made one would be able to control the development of the machine to some extent. One might, for instance, insist on the choice made being a particular one at, say, 10 particular places, and this would mean that about one machine in 1024 or more would develop to as high a degree as the one which had been faked. This cannot very well be given an accurate statement because of the subjective nature of the idea of 'degree of development' to say nothing of the fact that the machine that had been faked might have been also fortunate in its unfaked choices.

Let us now assume, for the sake of argument, that these machines are a genuine possibility, and look at the consequences of constructing them. To do so would of course meet with great opposition, unless we have advanced greatly in religious toleration from the days of Galileo. There would be great opposition from the intellectuals who were afraid of being put out of a job. It is probable though that the intellectuals would be mistaken about this. There would be plenty to do, [trying to understand what the machines were trying to say,][2] i.e. in trying to keep one's intelligence up to the standard set by the machines, for it seems probable that once the machine thinking method had started, it would not take long to outstrip our feeble powers. There would be no question of the machines dying, and they would be able to converse with each other to sharpen their wits. At some stage therefore we should have to expect the machines to take control, in the way that is mentioned in Samuel Butler's 'Erewhon'.

---

[2] Editor's note. The words 'trying to understand what the machines were trying to say,' are handwritten and are marked in the margin 'Inserted from Turing's Typescript'.

CHAPTER 13

Can Digital Computers Think? (1951)
Alan Turing

Introduction
Jack Copeland

The lecture 'Can Digital Computers Think?' was broadcast on BBC Radio on 15 May 1951, and was repeated on 3 July of that year.[1] (Sara Turing relates that Turing did not listen to the first broadcast but did 'pluck up courage' to listen to the repeat.[2]) Turing's was the second lecture in a series with the general title 'Automatic Calculating Machines'. Other speakers in the series included Newman, D. R. Hartree, M. V. Wilkes, and F. C. Williams.[3]

Imitating the Brain

Turing's principal aim in this lecture is to defend his view that 'it is not altogether unreasonable to describe digital computers as brains', and he argues for the proposition that 'If any machine can appropriately be described as a brain, then any digital computer can be so described'.

The lecture casts light upon Turing's attitude towards talk of machines thinking. In Chapter 11 he says that in his view the question 'Can machines think?' is 'too meaningless to deserve discussion' (p. 449). However, in the present chapter he makes liberal use of such phrases as 'programm[ing] a machine...to think' and 'the attempt to make a thinking machine'. In one passage, Turing says (p. 485): 'our main problem [is] how to programme a machine to imitate a brain, or as we might say more briefly, if less accurately, to think.' He shows the same willingness to discuss the question 'Can machines think?' in Chapter 14.

Turing's view is that a machine which imitates the intellectual behaviour of a human brain can itself appropriately be described as a brain or as thinking. In

---
[1] *Alan M. Turing* (Cambridge: Heffer, 1959), 102.
[2] Ibid.
[3] Letter from Maurice Wilkes to Copeland (9 July 1997).

Chapter 14, Turing emphasizes that it is only the *intellectual* behaviour of the brain that need be considered (pp. 494–5): 'To take an extreme case, we are not interested in the fact that the brain has the consistency of cold porridge. We don't want to say "This machine's quite hard, so it isn't a brain, and so it can't think."'

It is, of course, the ability of the machine to imitate the intellectual behaviour of a human brain that is examined in the Turing test (Chapter 11). Thus: any machine that plays the imitation game successfully can appropriately be described as a brain or as thinking.

Freedom of the Will

This chapter contains one of the two discussions of free will occurring in Turing's mature writings, both of which are tantalizingly brief. (The early essay entitled 'Nature of Spirit', which possibly dates from Turing's undergraduate days, also contains a discussion of free will. There Turing wrote: 'the theory which held that as eclipses etc. are predestined so were all our actions breaks down... We have a will which is able to determine the action of the atoms probably in a small portion of the brain, or possibly all over it.'[4]) The other discussion occurs in Chapter 11, where Turing says (p. 445):

An interesting variant on the idea of a digital computer is a 'digital computer with a random element'. These have instructions involving the throwing of a die or some equivalent electronic process... Sometimes such a machine is described as having free will (though I would not use this phrase myself). It is not normally possible to determine from observing a machine whether it has a random element, for a similar effect can be produced by such devices as making the choices depend on the digits of the decimal for π.

Unfortunately, Turing does not expand on the remark 'I would not use this phrase myself.' Possibly he means simply that the addition of a random element to a computer is not in itself sufficient to warrant the attribution of free will. Presumably one would at least need to add cognition and initiative as well before the machine could reasonably be described as having free will (compare Chapter 10, pp. 424, 429–30). Alternatively, it is possible that Turing is objecting to the term 'free will' itself, much as he objects elsewhere in Chapter 11 to the word 'think' ('too meaningless to deserve discussion').

Turing introduced this idea of a 'digital computer with a random element' more fully in Chapter 10 (p. 416):

It is possible to modify the above described types of discrete machines by allowing several alternative operations to be applied at some points, the alternatives to be chosen by a random process. Such a machine will be described as 'partially random'. If we wish to say

---

[4] A copy of 'Nature of Spirit' is among the Turing Papers in the Modern Archive Centre, King's College, Cambridge.

definitely that a machine is not of this kind we will describe it as 'determined'. Sometimes a machine may be strictly speaking determined but appear superficially as if it were partially random. This would occur if for instance the digits of the number π were used to determine the choices of a partially random machine where previously a dice thrower or electronic equivalent had been used. These machines are known as apparently partially random.

Turing discusses partially random machines further in Chapter 12 and in Chapter 9, where he mentions the possibility of including in the ACE a 'random element, some electronic roulette wheel' (p. 391).

In 'Can Digital Computers Think?' Turing raises both the possibility that 'the feeling of free will which we all have is an illusion' and the possibility that 'we really have got free will but yet there is no way of telling from our behaviour that this is so'. In parallel, he raises the question whether the behaviour of the brain 'is in principle predictable by calculation' (i.e. by Turing machine). In discussing this possibility, he observes that we 'certainly do not know how any such calculation should be done'. He points out that, furthermore, some physicists argue that no such prediction is even theoretically possible, 'on account of the indeterminacy principle in quantum mechanics'. Turing does not state his own position on these issues in the course of the lecture. However, in an interview given long after Turing's death, Max Newman stated that Turing 'had a deep-seated conviction that the real brain has a "roulette wheel" somewhere in it.'[5] This seems to indicate that Turing's view was that the brain is a partially random machine. Whether or not Turing would have asserted, on that basis, that we 'really have got free will' is not known.

Can Computers Think?

Turing's overarching aim in the lecture is to answer the question posed by his title. His strategy is to argue for the proposition mentioned above:

> If any machine can appropriately be described as a brain, then any digital computer can be so described.

His initial bald statement of his argument is (p. 483):

If now some particular machine can be described as a brain we have only to programme our digital computer to imitate it and it will also be a brain. If it is accepted that real brains, as found in animals, and in particular in men, are a sort of machine it will follow that our digital computer suitably programmed, will behave like a brain.

Turing goes on to flesh out his argument in various ways, turning eventually to the problem of free will (p. 484): 'There are still some difficulties. To behave like a

---

[5] Newman in interview with Christopher Evans ('The Pioneers of Computing: An Oral History of Computing' (London: Science Museum)).

brain seems to involve free will, but the behaviour of a digital computer, when it has been programmed, is completely determined.' Turing argues resourcefully that, even if it is true that brains have free will, this in fact presents no difficulty for his claim that a suitably programmed computer can imitate the brain (pp. 484–5):

> a machine which is to imitate a brain must appear to behave as if it had free will, and it may well be asked how this is to be achieved. One possibility is to make its behaviour depend on something like a roulette wheel or a supply of radium. ... It is, however, not really even necessary to do this. It is not difficult to design machines whose behaviour appears quite random to anyone who does not know the details of their construction.

Such machines are 'apparently partially random' (p. 416). Examples of apparently partially random machines are the German Enigma machine and the Lorenz SZ 40 cipher machine ('Tunny'). Since both these machines can be simulated by a digital computer, an appropriately programmed digital computer is apparently partially random. (In Chapter 11 Turing mentions having written a programme for the Manchester computer that produced apparently partially random behaviour. When given a number, the programme would reply with a number. Turing said 'I would defy anyone to learn from these replies sufficient about the programme to be able to predict any replies to untried values' (p. 457).)

Apparently partially random machines imitate partially random machines. If the brain is a partially random machine, an appropriately programmed digital computer may nevertheless give a convincing imitation of a brain. The appearance that this deterministic machine gives of possessing free will may be said to be mere sham, but this will not affect the machine's ability to play the imitation game successfully. And by Turing's principle, above, any machine that plays the imitation game successfully can appropriately be described as a brain.

The Church–Turing Thesis and Calculating Machines

In 'Can Digital Computers Think?' Turing puts foward a thesis that, while not the same as the Church–Turing thesis (see Chapter 1 and 'Computable Numbers: A Guide'), is in effect the result of replacing the term 'human computer' in the Church–Turing thesis by 'calculating machine', and replacing 'universal Turing machine' by 'digital computer of sufficient speed and storage capacity'.

The Church–Turing thesis states that any work that can be carried out by a human computer (i.e. by an obedient clerk working with pencil on paper in accordance with an effective procedure) can equally well be carried out by the universal Turing machine. The present thesis (pp. 482–3) states that any work that can be carried out by any calculating machine can equally well be carried out by a digital computer of sufficient speed and storage capacity:

A digital computer is a *universal* machine in the sense that it can be made to replace any machine of a certain very wide class. It will not replace a bulldozer or a steam-engine or a telescope, but it will replace any rival design of calculating machine.

Newman wrote in the same vein a few years previously:

A universal machine is a single machine which, when provided with suitable instructions, will perform any calculation that could be done by a specially constructed machine. No real machine can be truly universal because its size is limited...but subject to this limitation of size, the machines now being made in America and in this country will be 'universal'—if they work at all; that is, they will do every kind of job that can be done by special machines.[6]

If Turing were requested to clarify the notion of a 'calculating machine', he would perhaps offer paradigm examples such as the Brunsviga (a popular desk calculating machine), a differential analyser (an analogue computing device), special-purpose electronic machines like Colossus and ENIAC, and so on (compare Chapter 10, pp. 412–13). Or perhaps he would say, with greater generality, that a calculating machine is any machine that duplicates the abilities of a human mathematician working mechanically with paper and pencil, i.e. in accordance with an effective ('rule of thumb') procedure. It was in this manner that he explained the idea of an electronic computing machine in the opening paragraph of his *Programmers' Handbook*: 'Electronic computers are intended to carry out any definite rule of thumb process which could have been done by a human operator working in a disciplined but unintelligent manner.'[7]

Turing's remarks in Chapter 17 on the status of the Church–Turing thesis are also relevant here (and see also the section 'Normal Forms and the Church–Turing Thesis' in the introduction to Chapter 17).

Other Notable Features

Other features of note in the lecture include the continuation of the discussion of 'Lady Lovelace's dictum', begun in Chapter 11, and Turing's glorious analogy comparing trying to programme a computer to behave like a brain with trying to write a treatise about family life on Mars—and moreover with insufficient paper. (Newman once remarked on the 'comical but brilliantly apt analogies with which he [Turing] explained his ideas'.[8])

---

[6] M. H. A. Newman, 'General Principles of the Design of All-Purpose Computing Machines', *Proceedings of the Royal Society of London*, Series A, 195 (1948), 271–4 (271–2).

[7] A. M. Turing, *Programmers' Handbook for Manchester Electronic Computer*, University of Manchester Computing Laboratory (1950), 1. A digital facsimile of the *Programmers' Handbook* is available in The Turing Archive for the History of Computing <www.AlanTuring.net/ programmers_handbook>.

[8] Newman writing in the *Manchester Guardian*, 11 June 1954.

Further reading

Copeland, B. J., *Artificial Intelligence: A Philosophical Introduction* (Oxford: Blackwell, 1993). Chapter 3: 'Can a Machine Think?'; chapter 7: 'Freedom'.

—— 'Narrow versus Wide Mechanism: Including a Re-examination of Turing's Views on the Mind–Machine Issue', *Journal of Philosophy*, 97 (2000), 5–32. Reprinted in M. Scheutz, *Computationalism: New Directions* (Cambridge, Mass.: MIT Press, 2002).

Dennett, D. C., *Elbow Room: The Varieties of Freewill Worth Wanting* (Oxford: Clarendon Press, 1984).

Simons, G., *The Biology of Computer Life* (Brighton: Harvester, 1985).

Provenance

The text that follows is from Turing's typescript and incorporates corrections made in his hand.[9]

---

[9] The typescript is in the Modern Archive Centre, King's College, Cambridge (catalogue reference B 5). The present edition first appeared in B. J. Copeland (ed.), 'A Lecture and Two Radio Broadcasts on Machine Intelligence by Alan Turing', in K. Furukawa, D. Michie and S. Muggleton (eds.), *Machine Intelligence 15* (Oxford: Oxford University Press, 1999).

Can Digital Computers Think?

Digital computers have often been described as mechanical brains. Most scientists probably regard this description as a mere newspaper stunt, but some do not. One mathematician has expressed the opposite point of view to me rather forcefully in the words 'It is commonly said that these machines are not brains, but you and I know that they are.' In this talk I shall try to explain the ideas behind the various possible points of view, though not altogether impartially. I shall give most attention to the view which I hold myself, that it is not altogether unreasonable to describe digital computers as brains. A different point of view has already been put by Professor Hartree.

First we may consider the naive point of view of the man in the street. He hears amazing accounts of what these machines can do: most of them apparently involve intellectual feats of which he would be quite incapable. He can only explain it by supposing that the machine is a sort of brain, though he may prefer simply to disbelieve what he has heard.

The majority of scientists are contemptuous of this almost superstitious attitude. They know something of the principles on which the machines are constructed and of the way in which they are used. Their outlook was well summed up by Lady Lovelace over a hundred years ago, speaking of Babbage's Analytical Engine. She said, as Hartree has already quoted, 'The Analytical Engine has no pretensions whatever to *originate* anything. It can do whatever *we know how to order it* to perform.' This very well describes the way in which digital computers are actually used at the present time, and in which they will probably mainly be used for many years to come. For any one calculation the whole procedure that the machine is to go through is planned out in advance by a mathematician. The less doubt there is about what is going to happen the better the mathematician is pleased. It is like planning a military operation. Under these circumstances it is fair to say that the machine doesn't originate anything.

There is however a third point of view, which I hold myself. I agree with Lady Lovelace's dictum as far as it goes, but I believe that its validity depends on considering how digital computers *are* used rather than how they *could be* used. In fact I believe that they could be used in such a manner that they could appropriately be described as brains. I should also say that 'If any machine can appropriately be described as a brain, then any digital computer can be so described.'

This last statement needs some explanation. It may appear rather startling, but with some reservations it appears to be an inescapable fact. It can be shown to follow from a characteristic property of digital computers, which I will call their *universality*. A digital computer is a *universal* machine in the sense that it can be

Printed with the permission of the BBC and the Estate of Alan Turing.

made to replace any machine of a certain very wide class. It will not replace a bulldozer or a steam-engine or a telescope, but it will replace any rival design of calculating machine, that is to say any machine into which one can feed data and which will later print out results. In order to arrange for our computer to imitate a given machine it is only necessary to programme the computer to calculate what the machine in question would do under given circumstances, and in particular what answers it would print out. The computer can then be made to print out the same answers.

If now some particular machine can be described as a brain we have only to programme our digital computer to imitate it and it will also be a brain. If it is accepted that real brains, as found in animals, and in particular in men, are a sort of machine it will follow that our digital computer, suitably programmed, will behave like a brain.

This argument involves several assumptions which can quite reasonably be challenged. I have already explained that the machine to be imitated must be more like a calculator than a bulldozer. This is merely a reflection of the fact that we are speaking of mechanical analogues of brains, rather than of feet or jaws. It was also necessary that this machine should be of the sort whose behaviour is in principle predictable by calculation. We certainly do not know how any such calculation should be done, and it was even argued by Sir Arthur Eddington that on account of the indeterminacy principle in quantum mechanics no such prediction is even theoretically possible.

Another assumption was that the storage capacity of the computer used should be sufficient to carry out the prediction of the behaviour of the machine to be imitated. It should also have sufficient speed. Our present computers probably have not got the necessary storage capacity, though they may well have the speed. This means in effect that if we wish to imitate anything so complicated as the human brain we need a very much larger machine than any of the computers at present available. We probably need something at least a hundred times as large as the Manchester Computer. Alternatively of course a machine of equal size or smaller would do if sufficient progress were made in the technique of storing information.

It should be noticed that there is no need for there to be any increase in the complexity of the computers used. If we try to imitate ever more complicated machines or brains we must use larger and larger computers to do it. We do not need to use successively more complicated ones. This may appear paradoxical, but the explanation is not difficult. The imitation of a machine by a computer requires not only that we should have made the computer, but that we should have programmed it appropriately. The more complicated the machine to be imitated the more complicated must the programme be.

This may perhaps be made clearer by an analogy. Suppose two men both wanted to write their autobiographies, and that one had had an eventful life, but very little had happened to the other. There would be two difficulties troubling the man with the more eventful life more seriously than the other. He would have to spend more on paper and he would have to take more trouble over thinking what to say. The supply of paper would not be likely to be a serious difficulty, unless for instance he were on a desert island, and in any case it could only be a technical or a financial problem. The other difficulty would be more fundamental and would become more serious still if he were not writing his life but a work on something he knew nothing about, let us say about family life on Mars. Our problem of programming a computer to behave like a brain is something like trying to write this treatise on a desert island. We cannot get the storage capacity we need: in other words we cannot get enough paper to write the treatise on, and in any case we don't know what we should write down if we had it. This is a poor state of affairs, but, to continue the analogy, it is something to know how to write, and to appreciate the fact that most knowledge can be embodied in books.

In view of this it seems that the wisest ground on which to criticise the description of digital computers as 'mechanical brains' or 'electronic brains' is that, although they might be programmed to behave like brains, we do not at present know how this should be done. With this outlook I am in full agreement. It leaves open the question as to whether we will or will not eventually succeed in finding such a programme. I, personally, am inclined to believe that such a programme will be found. I think it is probable for instance that at the end of the century it will be possible to programme a machine to answer questions in such a way that it will be extremely difficult to guess whether the answers are being given by a man or by the machine. I am imagining something like a viva-voce examination, but with the questions and answers all typewritten in order that we need not consider such irrelevant matters as the faithfulness with which the human voice can be imitated. This only represents my opinion; there is plenty of room for others.

There are still some difficulties. To behave like a brain seems to involve free will, but the behaviour of a digital computer, when it has been programmed, is completely determined. These two facts must somehow be reconciled, but to do so seems to involve us in an age-old controversy, that of 'free will and determinism'. There are two ways out. It may be that the feeling of free will which we all have is an illusion. Or it may be that we really have got free will, but yet there is no way of telling from our behaviour that this is so. In the latter case, however well a machine imitates a man's behaviour it is to be regarded as a mere sham. I do not know how we can ever decide between these alternatives but whichever is the correct one it is certain that a machine which is to imitate a brain must appear to behave as if it had free will, and it may well be asked how this is to be achieved. One possibility is to make its behaviour depend on something like a roulette wheel or a supply of radium. The behaviour of these may perhaps be predictable, but if so, we do not know how to do the prediction.

It is, however, not really even necessary to do this. It is not difficult to design machines whose behaviour appears quite random to anyone who does not know the details of their construction. Naturally enough the inclusion of this random element, whichever technique is used, does not solve our main problem, how to programme a machine to imitate a brain, or as we might say more briefly, if less accurately, to think. But it gives us some indication of what the process will be like. We must not always expect to know what the computer is going to do. We should be pleased when the machine surprises us, in rather the same way as one is pleased when a pupil does something which he had not been explicitly taught to do.

Let us now reconsider Lady Lovelace's dictum. 'The machine can do whatever *we know how to order it* to perform.' The sense of the rest of the passage is such that one is tempted to say that the machine can *only* do what we know how to order it to perform. But I think this would not be true. Certainly the machine can only do what we *do* order it to perform, anything else would be a mechanical fault. But there is no need to suppose that, when we give it its orders we know what we are doing, what the consequences of these orders are going to be. One does not need to be able to understand how these orders lead to the machine's subsequent behaviour, any more than one needs to understand the mechanism of germination when one puts a seed in the ground. The plant comes up whether one understands or not. If we give the machine a programme which results in its doing something interesting which we had not anticipated I should be inclined to say that the machine *had* originated something, rather than to claim that its behaviour was implicit in the programme, and therefore that the originality lies entirely with us.

I will not attempt to say much about how this process of 'programming a machine to think' is to be done. The fact is that we know very little about it, and very little research has yet been done. There are plentiful ideas, but we do not yet know which of them are of importance. As in the detective stories, at the beginning of the investigation any trifle may be of importance to the investigator. When the problem has been solved, only the essential facts need to be told to the jury. But at present we have nothing worth putting before a jury. I will only say this, that I believe the process should bear a close relation of that of teaching.

I have tried to explain what are the main rational arguments for and against the theory that machines could be made to think, but something should also be said about the irrational arguments. Many people are extremely opposed to the idea of machine that thinks, but I do not believe that it is for any of the reasons that I have given, or any other rational reason, but simply because they do not like the idea. One can see many features which make it unpleasant. If a machine can think, it might think more intelligently than we do, and then where should we be? Even if we could keep the machines in a subservient position, for instance by turning off the power at strategic moments, we should, as a species, feel greatly humbled. A similar danger and humiliation threatens us from the possibility that we might be superseded by the pig or the rat. This is a theoretical possibility which is hardly controversial, but we have lived with pigs and rats for so long without their intelligence much increasing, that we no longer trouble ourselves about this possibility. We feel that if it is to happen at all it will not be for several million years to come. But this new danger is much closer. If it comes at all it will almost certainly be within the next millennium. It is remote but not astronomically remote, and is certainly something which can give us anxiety.

It is customary, in a talk or article on this subject, to offer a grain of comfort, in the form of a statement that some particularly human characteristic could never be imitated by a machine. It might for instance be said that no machine could write good English, or that it could not be influenced by sex-appeal or smoke a pipe. I cannot offer any such comfort, for I believe that no such bounds can be set. But I certainly hope and believe that no great efforts will be put into making machines with the most distinctively human, but non-intellectual characteristics such as the shape of the human body; it appears to me to be quite futile to make such attempts and their results would have something like the unpleasant quality of artificial flowers. Attempts to produce a thinking machine seem to me to be in a different category. The whole thinking process is still rather mysterious to us, but I believe that the attempt to make a thinking machine will help us greatly in finding out how we think ourselves.

CHAPTER 14

Can Automatic Calculating Machines Be Said To Think? (1952)

Alan Turing, Richard Braithwaite, Geoffrey Jefferson, Max Newman

Introduction
Jack Copeland

This discussion between Turing, Newman, R. B. Braithwaite, and G. Jefferson was recorded by the BBC on 10 January 1952 and broadcast on BBC Radio on the 14th, and again on the 23rd, of that month. This is the earliest known recorded discussion of artificial intelligence.[1]

The Participants

The anchor man of the discussion is Richard Braithwaite (1900–90). Braithwaite was at the time Sidgwick Lecturer in Moral Science at the University of Cambridge, where the following year he was appointed Knightsbridge Professor of Moral Philosophy. Like Turing, he was a Fellow of King's College. Braithwaite's main work lay in the philosophy of science and in decision and games theory (which he applied in moral philosophy).

Geoffrey Jefferson (1886–1961) retired from the Chair of Neurosurgery at Manchester University in 1951. In his Lister Oration, delivered at the Royal

---

[1] The present material together with the dialogues recorded in *Wittgenstein's Lectures on the Foundations of Mathematics* (ed. C. Diamond, Ithaca, NY: Cornell University Press, 1976) are the only known transcriptions of discussions involving Turing. (Some rather compressed notes of a discussion between Turing, Newman, Young, Polanyi, and others, entitled 'Rough Draft of the Discussion on the Mind and the Computing Machine, held on Thursday, 27th October, 1949, in the Philosophy Seminar' (anon., n.d., University of Manchester Philosophy Department), are in The Turing Archive for the History of Computing <www.AlanTuring.net/philosophy_seminar_oct1949>. I am grateful to Wolfe Mays for making these notes available.)

College of Surgeons of England on 9 June 1949, he had declared: 'When we hear it said that wireless valves think, we may despair of language.'[2]

Turing gave a substantial discussion of Jefferson's views in 'Computing Machinery and Intelligence' (pp. 451–2), rebutting the 'argument from consciousness' that he found in the Lister Oration. In the present chapter, Jefferson takes numerous pot shots at the notion of a machine thinking, which for the most part Turing and Newman are easily able to turn aside.

Jefferson may have thought little of the idea of machine intelligence, but he held Turing in considerable regard, saying after Turing's death that he 'had real genius, it shone from him'.[3]

The Turing Test Revisited

From the point of view of Turing scholarship, the most important parts of 'Can Automatic Calculating Machines Be Said to Think' are the passages containing Turing's exposition of the imitation game or Turing test. The description of the test that Turing gave in 'Computing Machinery and Intelligence' is here modified in a number of significant ways.

The lone interrogator of the original version is replaced by a 'jury' (p. 495). Each jury must judge 'quite a number of times' and 'sometimes they really are dealing with a man and not a machine'. For a machine to pass the test, a 'considerable proportion' of the jury 'must be taken in by the pretence'. The members of the jury interrogate the contestants, but their contributions 'don't really have to be questions, any more than questions in a law court are really questions'; for example, 'I put it to you that you are only pretending to be a man' is 'quite in order'.

In its original presentation (Chapter 11), the Turing test is a three-party game involving the parallel interrogation by a human of a computer and a human foil. According to the 1952 formulation, however, members of a jury interview a series of contestants one at a time, some of the contestants being machines and some humans. In the original form of the test, each interrogator knows that one of each pair of interviewees is a human and one a machine, but in the single-interviewee version of the test, this condition is necessarily absent. It appears that the earlier formulation is in fact superior, since the single-interviewee version is open to a biasing effect which disfavours the machine. Results of the Loebner series of single-interviewee Turing tests reveal a strong propensity among jurors to classify human respondents as machines. (New York businessman Hugh Loebner started the annual Loebner Prize competition in 1991, offering the

---

[2] G. Jefferson, 'The Mind of Mechanical Man', *British Medical Journal*, 25 June 1949, 1105–10 (1110).

[3] Letter from Jefferson to Sara Turing, 18 Oct. 1954. The letter is among the Turing Papers in the Modern Archive Centre, King's College, Cambridge (catalogue reference A 16).

sum of $100,000 to the programmer(s) of the first programme to pass the Turing test. $2,000 is awarded each year for the best effort. So far the grand prize remains unclaimed.) In the Loebner competition held at Dartmouth College, New Hampshire, in January 2000, human respondents were mistaken for computers on ten occasions, a computer for a human on none. The same effect was present in a series of single-interviewee tests performed with Kenneth Colby's historic programme Parry.[4] In a total of ten interviews, there were five misidentifications; in four of these a human respondent was mistaken for a computer. Presumably this phenomenon is the result of a determination on the part of the jurors not to be fooled by a programme. This lengthening of the odds against the machine cannot occur in the three-player form of the test.

Turing's Predictions

Turing is often misquoted as having predicted that, by the turn of the twentieth century, artificial intelligence indistinguishable from human intelligence would be in existence. What he in fact wrote in 1950 was that (p. 449):

in about fifty years' time it will be possible to programme computers...to make them play the imitation game so well that an average interrogator will not have more than 70 per cent chance of making the right identification after five minutes of questioning.

Some commentators have reported this prediction the wrong way about, as the claim that, by the end of the twentieth century, computers would succeed in *deceiving* the interrogator 70 per cent of the time.[5]

In 'Can Automatic Calculating Machines Be Said to Think', Turing offered a prediction that is interestingly different from the above, and which seems to concern success of a more substantial nature (p. 495):

Newman: I should like to be there when your match between a man and a machine takes place, and perhaps to try my hand at making up some of the questions. But that will be a long time from now, if the machine is to stand any chance with no questions barred?
Turing: Oh yes, at least 100 years, I should say.

In Chapter 11, Turing describes a form of the imitation game in which the interrogator must attempt to distinguish between a woman (the foil) and a man pretending to be a woman. This form of the game serves as a reference point for evaluating the computer's performance in the computer-human imitation game (p. 441):

---

[4] J. F. Heiser, K. M. Colby, W. S. Faught and R. C. Parkison, 'Can Psychiatrists Distinguish a Computer Simulation of Paranoia from the Real Thing?', *Journal of Psychiatric Research*, 15 (1980), 149–62.
[5] See, for example, R. Brooks, 'Intelligence without Reason', in L. Steels and R. Brooks (eds.), *The Artificial Life Route to Artificial Intelligence* (Mahwah, NJ: Erlbaum, 1995), n. 8, and P. Millican and A. Clark (eds.), *Machines and Thought: The Legacy of Alan Turing* (Oxford: Oxford University Press, 1996), 61.

We now ask the question, 'What will happen when a machine takes the part of A in this [man-imitates-woman] game?' Will the interrogator decide wrongly as often when the game is played like this as he does when the game is played between a man and a woman?

Reformulating Turing's 1952 prediction in these terms produces: It will be at least 100 years (2052) before a computer is able to play the imitation game sufficiently well so that jurors will decide wrongly as often in man-imitates-woman imitation games as in computer-imitates-human imitation games, in each case no questions being barred.

'Fiendish Expert' Objections to the Turing Test

'Fiendish expert' objections, of which there are many, are of the form: 'An expert could unmask the computer by asking it...'. All can be dealt with in a similar fashion. (Some other objections to the test are discussed in the introduction to Chapter 11.)

An interesting example is put forward by Robert French.[6] French's objection involves the phenomenon of associative priming, observed in what psychologists call the word/non-word recognition task. A subject seated in front of a screen is presented for a brief time with what may or may not be a word (e.g. 'DOG', 'DOK'). If the subject recognizes the letters as a word he or she must press a button. The experimenter measures the time that the subject takes to respond. It is found that, on average, subjects require less time to recognize a word if the word is preceded by a brief presentation of an associated word (e.g. 'fish' may facilitate the recognition of 'chips', 'bread' of 'butter'). French's claim is that this priming effect may be used to unmask the computer in the Turing test:

> The Turing Test interrogator makes use of this phenomenon as follows. The day before the Test, she selects a set of words (and non-words), runs the lexical decision task on the interviewees and records average recognition times. She then comes to the Test armed with the results...[and] identifies as the human being the candidate whose results more closely resemble the average results produced by her sample population of interviewees. The machine would invariably fail this type of test because there is no a priori way of determining associative strengths...Virtually the only way a machine could determine, even on average, all of the associative strengths between human concepts is to have experienced the world as the human candidate and the interviewers had.[7]

Turing, however, was happy to rule out expert jurors. In 'Can Automatic Calculating Machines Be Said to Think' he said that the interrogator 'should not be expert about machines' (p. 495), and in chapter 10, describing the chess-player version of the test, he said that the discriminator should be a 'rather poor' chess player (p. 431). Turing did not mention other kinds of expert, but the

---

[6] R. French, 'Subcognition and the Limits of the Turing Test', *Mind*, 99 (1990), 53–65.
[7] Ibid. 17.

reasons for excluding experts about machines apply equally well to experts about minds.

In any case, French's proposal is illegitimate. The specifications of the Turing test are clear: the interrogator is allowed only to put questions. There is no provision for the use of the equipment necessary for administering the lexical decision task and for measuring the contestants' reaction times. One might as well allow the interrogator to bring along equipment for measuring the contestants' magnetic fields or energy dissipation.[8]

Can Machines Think?

In Chapter 11, Turing said that the question 'Can machines think?' is 'too meaningless to deserve discussion' (p. 449). He certainly did not allow this view to prevent him from indulging rather often in such discussion. In the present chapter, Turing records a considerably milder attitude to the question (p. 495):

You might call it a test to see whether the machine thinks, but it would be better to avoid begging the question, and say that the machines that pass are (let's say) 'Grade A' machines. ... [The question whether] machines really could pass the test [is] not the same as 'Do machines think?', but it seems near enough for our present purpose, and raises much the same difficulties.

In Chapter 10 Turing wrote (p. 431):

The extent to which we regard something as behaving in an intelligent manner is determined as much by our own state of mind and training as by the properties of the object under consideration. If we are able to explain or predict its behaviour... we have little temptation to imagine intelligence. With the same object therefore it is possible that one man would consider it as intelligent and another would not; the second man would have found out the rules of its behaviour.

Turing develops this point in the present discussion (p. 500):

As soon as one can see the cause and effect working themselves out in the brain, one regards it as not being thinking, but a sort of unimaginative donkey-work. From this point of view one might be tempted to define thinking as consisting of 'those mental processes that we don't understand'. If this is right then to make a thinking machine is to make one which does interesting things without our really understanding quite how it is done.

Many years later Marvin Minsky put forward this same view, saying that 'intelligence' is simply our name for whichever problem-solving mental processes

---

[8] There is additional discussion of French's objections to the Turing test in my 'The Turing Test', *Minds and Machines*, 10 (2000), 519–39 (reprinted in J. H. Moor (ed.), *The Turing Test* (Dordrecht: Kluwer, 2003)).

we do not yet understand.[9] Minsky likens intelligence to the concept 'unexplored regions of Africa': it disappears as soon as we discover it.

Analogy and Creativity

Turing sketches an interesting mechanical explanation of how analogy works in the human brain, and suggests that a digital computer can be made to do the same (p. 499). Another proposal which does not appear elsewhere in Turing's writings concerns the vexed issue of creativity. A machine that combines words more or less at random and then scores the combinations 'for various merits' would, he says, be able to find useful new concepts. His example is 'lumping together rain, hail, snow and sleet, under the word "precipitation"' (p. 499). He agrees with Newman that this process would be 'shockingly slow'. Nevertheless, this brief suggestion of Turing's is an economical illustration of the possibility of a machine's acquiring new concepts for itself.

Machine Learning

'Can Automatic Calculating Machines Be Said to Think' contains a lengthy discussion of machine learning, in which Turing alludes to the P-type machines of Chapter 10 (pp. 425–9). He says in response to Jefferson's point that real learning involves intervention by teachers, 'I have made some experiments in teaching a machine to do some simple operation, and a very great deal of such intervention was needed' (p. 497). Turing emphasizes several times that a learning machine, if it is to be effective, must not only learn first-order facts but must also be able to learn to improve its learning methods (compare the discussion of the Mathematical Objection in Chapter 12 and elsewhere).

Newman gives a simple illustration (on p. 496) of how a computer can learn to do better with practice. His example makes use of the idea of the computer modifying its own programme, mentioned by Turing in Chapter 9 (p. 393). Faced with a two-move chess problem, the machine initially follows an instruction to choose a move at random. If the move, say B-Q5, is found to lead to forced mate in two moves, then the machine changes the instruction to 'Try B-Q5'. When presented with the same problem again, the machine immediately gives the right answer.

Newman's Test

Towards the end of the discussion Newman suggested a test—and the term 'Newman's test' seems appropriate—for when machines have 'begun to think'

---

[9] M. Minsky, *The Society of Mind* (London: Pan Books, 1988), 71.

(pp. 504–5). That stage of development has been reached, he said, when a machine can solve a mathematical problem for which no effective method exists.

Provenance

The discussion is taken from a BBC script (which is marked 'Not checked in Talks Department with "as broadcast" script').[10]

---

[10] The script is in the Modern Archive Centre, King's College, Cambridge (catalogue reference B 6). It was first published in B. J. Copeland (ed.), 'A Lecture and Two Radio Broadcasts on Machine Intelligence by Alan Turing', in K. Furukawa, D. Michie, and S. Muggleton (eds.), *Machine Intelligence 15* (Oxford: Oxford University Press, 1999).

Can Automatic Calculating Machines Be Said To Think?

Braithwaite: We're here today to discuss whether calculating machines can be said to think in any proper sense of the word. Thinking is ordinarily regarded as so much a speciality of man, and perhaps of other higher animals, that the question may seem too absurd to be discussed. But, of course, it all depends on what is to be included in thinking. The word is used to cover a multitude of different activities. What would you, Jefferson, as a physiologist, say were the most important elements involved in thinking?

Jefferson: I don't think that we need waste too much time on [a] definition of thinking since it will be hard to get beyond phrases in common usage, such as having ideas in the mind, cogitating, meditating, deliberating, solving problems or imagining. Philologists say that the word 'Man' is derived from a Sanskrit word that means 'to think', probably in the sense of judging between one idea and another. I agree that we could no longer use the word 'thinking' in a sense that restricted it to man. No one would deny that many animals think, though in a very limited way. They lack insight. For example, a dog learns that it is wrong to get on cushions or chairs with muddy paws, but he only learns it as a venture that doesn't pay. He has no conception of the real reason, that he damages fabrics by doing that.

The average person would perhaps be content to define thinking in very general terms such as revolving ideas in the mind, of having notions in one's head, of having one's mind occupied by a problem, and so on. But it is only right to add that our minds are occupied much of the time with trivialities. One might say in the end that thinking was the general result of having a sufficiently complex nervous system. Very simple ones do not provide the creature with any problems that are not answered by simple reflex mechanisms. Thinking then becomes all the things that go on in one's brain, things that often end in an action but don't necessarily do so. I should say that it was the sum total of what the brain of man or animal does. Turing, what do you think about it? Have you a mechanical definition?

Turing: I don't want to give a definition of thinking, but if I had to I should probably be unable to say anything more about it than that it was a sort of buzzing that went on inside my head. But I don't really see that we need to agree on a definition at all. The important thing is to try to draw a line between the properties of a brain, or of a man, that we want to discuss, and those that we don't. To take an extreme case, we are not interested in the fact

---

Printed with the permission of the BBC, Lewis C. Braithwaite, Antony A. Jefferson, Edward Newman, and the Estate of Alan Turing.

that the brain has the consistency of cold porridge. We don't want to say 'This machine's quite hard, so it isn't a brain, and so it can't think.' I would like to suggest a particular kind of *test* that one might apply to a machine. You might call it a test to see whether the machine thinks, but it would be better to avoid begging the question, and say that the machines that pass are (let's say) 'Grade A' machines. The idea of the test is that the machine has to try and pretend to be a man, by answering questions put to it, and it will only pass if the pretence is reasonably convincing. A considerable proportion of a jury, who should not be expert about machines, must be taken in by the pretence. They aren't allowed to see the machine itself—that would make it too easy. So the machine is kept in a far away room and the jury are allowed to ask it questions, which are transmitted through to it: it sends back a typewritten answer.

Braithwaite: Would the questions have to be sums, or could I ask it what it had had for breakfast?

Turing: Oh yes, anything. And the questions don't really have to be questions, any more than questions in a law court are really questions. You know the sort of thing. 'I put it to you that you are only pretending to be a man' would be quite in order. Likewise the machine would be permitted all sorts of tricks so as to appear more man-like, such as waiting a bit before giving the answer, or making spelling mistakes, but it can't make smudges on the paper, any more than one can send smudges by telegraph. We had better suppose that each jury has to judge quite a number of times, and that sometimes they really are dealing with a man and not a machine. That will prevent them saying 'It must be a machine' every time without proper consideration.

Well, that's my test. Of course I am not saying at present either that machines really could pass the test, or that they couldn't. My suggestion is just that this is the question we should discuss. It's not the same as 'Do machines think,' but it seems near enough for our present purpose, and raises much the same difficulties.

Newman: I should like to be there when your match between a man and a machine takes place, and perhaps to try my hand at making up some of the questions. But that will be a long time from now, if the machine is to stand any chance with no questions barred?

Turing: Oh yes, at least 100 years, I should say.

Jefferson: Newman, how well would existing machines stand up to this test? What kind of things can they do now?

Newman: Of course, their strongest line is mathematical computing, which they were designed to do, but they would also do well at some questions that don't look numerical, but can easily be made so, like solving a chess problem or looking you up a train in the time-table.

Braithwaite: Could they do that?

Newman: Yes. Both these jobs can be done by trying all the possibilities, one after another. The whole of the information in an ordinary time-table would have to be written in as part of the programme, and the simplest possible routine would be one that found the trains from London to Manchester by testing every train in the time-table to see if it calls at both places, and printing out those that do. Of course, this is a dull, plodding method, and you could improve on it by using a more complicated routine, but if I have understood Turing's test properly, you are not allowed to go behind the scenes and criticise the method, but must abide by the scoring on correct answers, found reasonably quickly.

Jefferson: Yes, but all the same a man who has to look up trains frequently gets better at it, as he learns his way about the time-table. Suppose I give a machine the same problem again, can it learn to do better without going through the whole rigmarole of trying everything over every time? I'd like to have your answer to that because it's such an important point. Can machines learn to do better with practice?

Newman: Yes, it could. Perhaps the chess problem provides a better illustration of this. First I should mention that *all* the information required in any job—the numbers, times of trains, positions of pieces, or whatever it is, and also the instructions saying what is to be done with them—all this material is stored in the same way. (In the Manchester machine it is stored as a pattern on something resembling a television screen.) As the work goes on the pattern is changed. Usually it is the part of the pattern that contains the data that changes, while the instructions stay fixed. But it is just as simple to arrange that the instructions themselves shall be changed now and then. Well, now a programme could be composed that would cause the machine to do this: a 2-move chess problem is recorded into the machine in some suitable coding, and whenever the machine is started, a white move is chosen at random (there is a device for making random choices in our machine). All the consequences of this move are now analysed, and if it does *not* lead to forced mate in two moves, the machine prints, say, 'P-Q3, wrong move', and stops. But the analysis shows that when the right move is chosen the machine not only prints, say, 'B-Q5, solution', but it changes the instruction calling for a random choice to one that says 'Try B-Q5.' The result is that whenever the machine is started again it will immediately print out the right solution—and this without the man who made up the routine knowing beforehand what it was. Such a routine could certainly be made now, and I think this can fairly be called learning.

Jefferson: Yes, I suppose it is. Human beings learn by repeating the same exercises until they have perfected them. Of course it goes further, and at the same time —we learn generally to shift the knowledge gained about one thing to another set of problems, seeing relevances and relationships. Learning means remembering. How long can a machine store information for?

Newman: Oh, at least as long as a man's lifetime, if it is refreshed occasionally.

Jefferson: Another difference would be that in the learning process there is much more frequent intervention by teachers, parental or otherwise, guiding the arts of learning. You mathematicians put the programme once into the machine and leave it to it. You wouldn't get any distance at all with human beings if that is what you did. In fact, the only time you do that in the learning period is at examinations.

Turing: It's quite true that when a child is being taught, his parents and teachers are repeatedly intervening to stop him doing this or encourage him to do that. But this will not be any the less so when one is trying to teach a machine. I have made some experiments in teaching a machine to do some simple operation, and a very great deal of such intervention was needed before I could get any results at all. In other words the machine learnt so slowly that it needed a great deal of teaching.

Jefferson: But who was learning, you or the machine?

Turing: Well, I suppose we both were. One will have to find out how to make machines that will learn more quickly if there is to be any real success. One hopes too that there will be a sort of snowball effect. The more things the machine has learnt the easier it ought to be for it to learn others. In learning to do any particular thing it will probably also be learning to learn more efficiently. I am inclined to believe that when one has taught it to do certain things one will find that some other things which one had planned to teach it are happening without any special teaching being required. This certainly happens with an intelligent human mind, and if it doesn't happen when one is teaching a machine there is something lacking in the machine. What do you think about learning possibilities, Braithwaite?

Braithwaite: No-one has mentioned what seems to me the great difficulty about learning, since we've only discussed learning to solve a particular problem. But the most important part of human learning is learning from experience—not learning from one particular kind of experience, but being able to learn from experience in general. A machine can easily be constructed with a feed-back device so that the programming of the machine is controlled by the relation of its output to some feature in its external environment—so that the working of the machine in relation to the environment is self-corrective. But this requires that it should be some particular feature of the environment to which the machine has to adjust itself. The peculiarity of men and animals is that they have the power of adjusting themselves to almost all the features. The feature to which adjustment is made on a particular occasion is the one the man is attending to and he attends to what he is *interested in*. His interests are determined, by and large, by his appetites, desires, drives, instincts—all the things that together make up his 'springs of action'. If we want to construct a machine which will vary its attention to things in its environment so that it will sometimes adjust itself to one and sometimes to another, it would seem to be necessary to equip the machine with something corresponding to a set of appetites. If the machine is built to be treated only as a domestic pet, and is spoon-fed with particular problems, it will not be able to learn in the varying way in which human beings learn. This arises from the necessity of adapting behaviour suitably to environment if human appetites are to be satisfied.

Jefferson: Turing, you spoke with great confidence about what you are going to be able to do. You make it sound as if it would be fairly easy to modify construction so that the machine reacted more like a man. But I recollect that from the time of Descartes and Borelli on people have said that it would be only a matter of a few years, perhaps 3 or 4 or maybe 50, and a replica of man would have been artificially created. We shall be wrong, I am sure, if we give the impression that these things would be easy to do.

Newman: I agree that we are getting rather far away from computing machines as they exist at present. These machines have rather restricted appetites, and they can't blush when they're embarrassed, but it's quite hard enough, and I think a very interesting problem, to discover how near these actually existing machines can get to thinking. Even if we stick to the reasoning side of thinking, it is a long way from solving chess problems to the invention of new mathematical concepts or making a generalisation that takes in ideas that were current before, but had never been brought together as instances of a single general notion.

Braithwaite: For example?

Newman: The different kinds of number. There are the integers, 0, 1, -2, and so on; there are the real numbers used in comparing lengths, for example the circumference of a circle and its diameter; and the complex numbers involving $\sqrt{-1}$; and so on. It is not at all obvious that these are instances of one thing, 'number'. The Greek mathematicians used entirely different words for the integers and the real numbers, and had no single idea to cover both. It is really only recently that the general notion of kinds of number has been abstracted from these instances and accurately defined. To make this sort of generalisation you need to have the power of recognising similarities, seeing analogies between things that had not been put together before. It is not just a matter of testing things for a specified property and classifying them accordingly. The concept itself has to be framed, something has to be created, say the idea of a number-field. Can we even guess at the way a machine could make such an invention from a programme composed by a man who had not the concept in his own mind?

Turing: It seems to me, Newman, that what you said about 'trying out possibilities' as a method applies to quite an extent, even when a machine is required to do something as advanced as finding a useful new concept. I wouldn't like to have to define the meaning of the word 'concept', nor to give rules for rating their usefulness, but whatever they are they've got outward and visible forms, which are words and combinations of words. A machine could make up such combinations of words more or less at random, and then give them marks for various merits.

Newman: Wouldn't that take a prohibitively long time?

Turing: It would certainly be shockingly slow, but it could start on easy things, such as lumping together rain, hail, snow and sleet, under the word 'precipitation.' Perhaps it might do more difficult things later on if it was learning all the time how to improve its methods.

Braithwaite: I don't think there's much difficulty about seeing analogies that can be formally analysed and explicitly stated. It is then only a question of designing the machine so that it can recognise similarities of mathematical structure. The difficulty arises if the analogy is a vague one about which little more can be said than that one has a feeling that there is some sort of similarity between two cases but one hasn't any idea as to the respect in which the two cases are similar. A machine can't recognise similarities when there is nothing in its programme to say what are the similarities it is expected to recognise.

Turing: I think you could make a machine spot an analogy, in fact it's quite a good instance of how a machine could be made to do some of those things that one usually regards as essentially a human monopoly. Suppose that someone was trying to explain the double negative to me, for instance, that when something isn't not green it must be green, and he couldn't quite get it across. He might say 'Well, it's like crossing the road. You cross it, and then you cross it again, and you're back where you started.' This remark might just clinch it. This is one of the things one would like to work with machines, and I think it would be likely to happen with them. I imagine that the way analogy works in our brains is something like this. When two or more sets of ideas have the same pattern of logical connections, the brain may very likely economise parts by using some of them twice over, to remember the logical connections both in the one case and in the other. One must suppose that some part of my brain was used twice over in this way, once for the idea of double negation and once for crossing the road, there and back. I am really supposed to know about both these things but can't get what it is the man is driving at, so long as he is talking about all those dreary nots and not-nots. Somehow it doesn't get through to the right part of the brain. But as soon as he says his piece about crossing the road it gets through to the right part, but by a different route. If there is some such purely mechanical explanation of how this argument by analogy goes on in the brain, one could make a digital computer do the same.

Jefferson: Well, there isn't a mechanical explanation in terms of cells and connecting fibres in the brain.

Braithwaite: But could a machine really do this? How would it do it?

Turing: I've certainly left a great deal to the imagination. If I had given a longer explanation I might have made it seem more certain that what I was describing was feasible, but you would probably feel rather uneasy about it all, and you'd probably exclaim impatiently, 'Well, yes, I see that a machine could do all that, but I wouldn't call it thinking.' As soon as one can see the cause and effect working themselves out in the brain, one regards it as not being thinking, but a sort of unimaginative donkey-work. From this point of view one might be tempted to define thinking as consisting of 'those mental processes that we don't understand'. If this is right then to make a thinking machine is to make one which does interesting things without our really understanding quite how it is done.

Jefferson: If you mean that we don't know the wiring in men, as it were, that is quite true.

Turing: No, that isn't at all what I mean. We know the wiring of our machine, but it already happens there in a limited sort of way. Sometimes a computing machine does do something rather weird that we hadn't expected. In principle one could have predicted it, but in practice it's usually too much trouble. Obviously if one were to predict everything a computer was going to do one might just as well do without it.

Newman: It is quite true that people are disappointed when they discover what the big computing machines actually do, which is just to add and multiply, and use the results to decide what further additions and multiplications to do. '*That's not thinking*', is the natural comment, but this is rather begging the question. If you go into one of the ancient churches in Ravenna you see some most beautiful pictures round the walls, but if you peer at them through binoculars you might say, 'Why, they aren't really pictures at all, but just a lot of little coloured stones with cement in between.' The machine's processes are mosaics of very simple standard parts, but the designs can be of great complexity, and it is not obvious where the limit is to the patterns of thought they could imitate.

Braithwaite: But how many stones are there in your mosaic? Jefferson, is there a sufficient multiplicity of the cells in the brain for them to behave like a computing machine?

Jefferson: Yes, there are thousands, tens of thousands more cells in the brain than there are in a computing machine, because the present machine contains— how many did you say?

Turing: Half a million digits. I think we can assume that is the equivalent of half a million nerve cells.

Braithwaite: If the brain works like a computing machine then the present computing machine cannot do all the things the brain does. Agreed; but if a computing machine were made that could do all the things the brain does, wouldn't it require more digits than there is room for in the brain?

Can Automatic Calculating Machines Be Said To Think? | 501

Jefferson: Well, I don't know. Suppose that it is right to equate digits in a machine with nerve cells in a brain. There are various estimates, somewhere between ten thousand million and fifteen thousand million cells are supposed to be there. Nobody knows for certain, you see. It is a colossal number. You would need 20,000 or more of your machines to equate digits with nerve cells. But it is not, surely, just a question of size. There would be too much logic in your huge machine. It wouldn't be really like a human output of thought. To make it more like, a lot of the machine parts would have to be designed quite differently to give greater flexibility and more diverse possibilities of use. It's a very tall order indeed.

Turing: It really is the size that matters in this case. It is the amount of information that can be stored up. If you think of something very complicated that you want one of these machines to do, you may find the particular machine you have got won't do, but if any machine can do it at all, then it can be done by your first computer, simply increased in its storage capacity.

Jefferson: If we are really to get near to anything that can be truly called 'thinking' the effects of external stimuli cannot be missed out; the intervention of all sorts of extraneous factors, like the worries of having to make one's living, or pay one's taxes, or get food that one likes. These are not in any sense minor factors, they are very important indeed, and worries concerned with them may greatly interfere with good thinking, especially with creative thinking. You see a machine has no environment, and man is in constant relation to his environment, which, as it were punches him whilst he punches back. There is a vast background of memories in a man's brain that each new idea or experience has to fit in with. I wonder if you could tell me how far a calculating machine meets that situation. Most people agree that man's first reaction to a new idea (such as the one we are discussing today) is one of rejection, often immediate and horrified denial of it. I don't see how a machine could as it were say 'Now Professor Newman or Mr. Turing, I don't like this programme at all that you've just put into me, in fact I'm not going to have anything to do with it.'

Newman: One difficulty about answering that is one that Turing has already mentioned. If someone says, 'Could a machine do this, e.g. could it say "I don't like the programme you have just put into me"', and a programme for doing that very thing is duly produced, it is apt to have an artificial and ad hoc air, and appear to be more of a trick than a serious answer to the question. It is like those passages in the Bible, which worried me as a small boy, that say that such and such was done 'that the prophecy might be fulfilled which says' so and so. This always seemed to me a most unfair way of making sure that the prophecy came true. If I answer your question, Jefferson, by making a routine which simply caused the machine to say just the words 'Newman and Turing, I don't like your programme', you would certainly feel this was a rather childish trick, and not the answer to what you really wanted to know. But yet it's hard to pin down what you want.

Jefferson: I want the machine to reject the problem because it offends it in some way. That leads me to enquire what the ingredients are of ideas that we reject because we instinctively don't care for them. I don't know why I like some pictures and some music and am bored by other sorts. But I'm not going to carry that line on because we are all different, our dislikes are based on our personal histories and probably too on small differences of construction in all of us, I mean by heredity. Your machines have no genes, no pedigrees. Mendelian inheritance means nothing to wireless valves. But I don't want to score debating points! We ought to make it clear that not even Turing thinks that all that he has to do is to put a skin on the machine and that it is alive! We've been trying for a more limited objective whether the sort of thing that machines do can be considered as thinking. But is not your machine more certain than any human being of getting its problem right at once, and infallibly?

Newman: Oh!

Turing: Computing machines aren't really infallible at all. Making up checks on their accuracy is quite an important part of the art of using them. Besides making mistakes they sometimes haven't done quite the calculation one had expected, and one gets something that might be called a 'misunderstanding'.

Jefferson: At any rate, they are not influenced by the emotions. You have only to upset a person enough and he becomes confused, he can't think of the answers and may make a fool of himself. It is high emotional content of mental processes in the human being that makes him quite different from a machine. It seems to me to come from the great complexity of his nervous system with its 10^{10} cells and also from his endocrine system which imports all sorts of emotions and instincts, such as those to do with sex. Man is essentially a chemical machine, he is much affected by hunger and fatigue, by being 'out of sorts' as we say, also by innate judgements, and by sexual urges. This chemical side is tremendously important, not the least so because the brain does exercise a remote control over the most important chemical processes that go on in our bodies. Your machines don't have to bother with that, with being tired or cold or happy or satisfied. They show no delight at having done something never done before. No, they are 'mentally' simple things. I mean that however complicated their structure is (and I know it *is* very complicated), compared with man they are very simple and perform their tasks with an absence of distracting thoughts which is quite *inhuman*.

Braithwaite: I'm not sure that I agree. I believe that it will be necessary to provide the machine with something corresponding to appetites, or other 'springs of action', in order that it will pay enough attention to relevant features in its environment to be able to learn from experience. Many psychologists have held that the emotions in men are by-products of their appetites and that they serve a biological function in calling higher levels of mental activity into play when the lower levels are incapable of coping with an external situation. For example, one does not feel afraid when there is no danger, or a danger which can be avoided more or less automatically: fear is a symptom showing that the danger has to be met by conscious thought. Perhaps it will be impossible to build a machine capable of learning in general from experience without incorporating in it an emotional apparatus, the function of which will be to switch over to a different part of the machine when the external environment differs too much from what would satisfy the machine's appetites by more than a certain amount. I don't want to suggest that it will be necessary for the machine to be able to throw a fit of tantrums. But in humans tantrums frequently fulfil a definite function—that of escaping from responsibility; and to protect a machine against a too hostile environment it may be essential to allow it, as it were, to go to bed with a neurosis, or psychogenic illness—just as, in a simpler way, it is provided with a fuse to blow, if the electric power working it threatens its continued existence.

Turing: Well, I don't envisage teaching the machine to throw temperamental scenes. I think some such effects are likely to occur as a sort of by-product of genuine teaching, and that one will be more interested in curbing such displays than in encouraging them. Such effects would probably be distinctly different from the corresponding human ones, but recognisable as variations on them. This means that if the machine was being put through one of my imitation tests, it would have to do quite a bit of acting, but if one was comparing it with a man in a less strict sort of way the resemblance might be quite impressive.

Newman: I still feel that too much of our argument is about what hypothetical future machines will do. It is all very well to say that a machine could easily be made to do this or that, but, to take only one practical point, what about the time it would take to do it? It would only take an hour or two to make up a routine to make our Manchester machine analyse all possible variations of the game of chess right out, and find the best move that way—*if* you didn't mind its taking thousands of millions of years to run through the routine. Solving a problem on the machine doesn't mean finding a way to do it between now and eternity, but within a reasonable time. This is not just a technical detail that will be taken care of by future improvements. It's most unlikely that the engineers can ever give us a factor of more than a thousand or two times our present speeds. To assume that runs that would take thousands of millions of years on our present machines will be done in a flash on machines of the future, is to move into the realms of science fiction.

Turing: To my mind this time factor is the one question which will involve all the real technical difficulty. If one didn't know already that these things can be done by brains within a reasonable time one might think it hopeless to try with a machine. The fact that a brain *can* do it seems to suggest that the difficulties may not really be so bad as they now seem.

Braithwaite: I agree that we ought not to extend our discussion to cover whether calculating machines could be made which would do everything that a man can do. The point is, surely, whether they can do all that it is proper to call thinking. Appreciation of a picture contains elements of thinking, but it also contains elements of feeling; and we're not concerned with whether a machine can be made that will feel. Similarly with moral questions: we're only concerned with them so far as they are also intellectual ones. We haven't got to give the machine a sense of duty or anything corresponding to a will: still less need it be given temptations which it would then have to have an apparatus for resisting. All that it has got to do in order to think is to be able to solve, or to make a good attempt at solving, all the intellectual problems with which it might be confronted by the environment in which it finds itself. This environment, of course, must include Turing asking it awkward questions as well as natural events such as being rained upon, or being shaken up by an earthquake.

Newman: But I thought it was you who said that a machine wouldn't be able to learn to adjust to its environment if it hadn't been provided with a set of appetites and all that went with them?

Braithwaite: Yes, certainly. But the problems raised by a machine having appetites are not properly our concern today. It may be the case that it wouldn't be able to learn from experience without them; but we're only required to consider whether it would be able to learn at all—since I agree that being able to learn is an essential part of thinking. So oughtn't we to get back to something centred on thinking? Can a machine make up new concepts, for example?

Newman: There are really two questions that can be asked about machines and thinking, first, what do we require before we agree that the machine does *everything* that we call thinking? This is really what we have been talking about for most of the time; but there is also another interesting and important question: Where does the doubtful territory begin? What is the *nearest* thing to straight computing that the present machines perhaps can't do?

Braithwaite: And what would your own answer be?

Newman: I think perhaps to solve mathematical problems for which no method is known, in the way that men do; to find new methods. This is a much more modest aim than inventing new mathematical concepts. What happens when you try to solve a new problem in the ordinary way is that you think about it for a few seconds, or a few years, trying out all the analogies you can think of with problems that have been solved, and then you have an idea. You try it out in detail. If it is no good you must wait for another idea. This is a little like the chess-problem routine, where one move after another is tried, but with one

Can Automatic Calculating Machines Be Said To Think? | 505

very important difference, that if I am even a moderately good mathematician the ideas that I get are not just random ones, but are pre-selected so that there is an appreciable chance that after a few trials one of them will be successful. Henry Moore says about the studies he does for his sculpture, 'When the work is more than an exercise, inexplicable jumps occur. This is where the imagination comes in.' If a machine could really be got to imitate this sudden pounce on an idea, I believe that everyone would agree that it had begun to think, even though it didn't have appetites or worry about the income tax. And suppose that we also stuck to what we know about the physiology of human thinking, how much would that amount to, Jefferson?

Jefferson: We know a great deal about the end-product, thinking itself. Are not the contents of our libraries and museums the total up to date? Experimental psychology has taught us a lot about the way that we use memory and association of ideas, how we fill in gaps in knowledge and improvise from a few given facts. But exactly how we do it in terms of nerve cell actions we don't know. We are particularly ignorant of the very point that you mentioned just now, Newman, the actual physiology of the pounce on an idea, of the sudden inspiration. Thinking is clearly a motor activity of the brain's cells, a suggestion supported by the common experience that so many people think better with a pen in their hand than viva voce or by reverie and reflection. But you can't so far produce ideas in a man's mind by stimulating his exposed brain here or there electrically. It would have been really exciting if one could have done that—if one could have perhaps excited original thoughts by local stimulation. It can't be done. Nor does the electro-encephalograph show us how the process of thinking is carried out. It can't tell you what a man is thinking about. We can trace the course, say, of a page of print or of a stream of words into the brain, but we eventually lose them. If we could follow them to their storage places we still couldn't see how they are reassembled later as ideas. You have the great advantage of knowing how your machine was made. We only know that we have in the human nervous system a concern compact in size and in its way perfect for its job. We know a great deal about its microscopical structure and its connections. If fact, we know everything except how these myriads of cells allow us to think. But, Newman, before we say 'not only does this machine think but also here in this machine we have an exact counterpart of the wiring and circuits of human nervous systems,' I ought to ask whether machines have been built or could be built which are as it were anatomically different, and yet produce the same work.

Newman: The logical plan of all of them is rather similar, but certainly their anatomy, and I suppose you could say their physiology, varies a lot.

Jefferson: Yes, that's what I imagined—we cannot then assume that any one of these electronic machines is a replica of part of a man's brain even though the result of its actions has to be conceded as thought. The real value of the machine to you is its end results, its performance, rather than that its plan reveals to us a model of our brains and nerves. Its usefulness lies in the fact that electricity travels along wires 2 or 3 million times faster than nerve impulses pass along nerves. You can set it to do things that man would need thousands of lives to complete. But that old slow coach, man, is the one with the ideas—or so I think. It would be fun some day, Turing, to listen to a discussion, say on the Fourth Programme, between two machines on why human beings think that they think!

Artificial Life
Jack Copeland

1. What is 'Artificial Life'?

The highly interdisciplinary field of Artificial Life was so named by Christopher Langton, a physicist working at the Los Alamos National Laboratory.[1] In 1987 Langton organized at Los Alamos what he described as an 'interdisciplinary workshop on the synthesis and simulation of living systems'.[2] The workshop was a rallying point, bringing together researchers with shared interests and diverse backgrounds.

Artificial Life ('A-Life') aims to achieve theoretical understanding of naturally occurring biological life, in particular of the most conspicuous feature of living matter, its ability to *self-organize*, i.e. to develop form and structure spontaneously. Langton defines Artificial Life as 'the study of man-made systems that exhibit behaviors characteristic of natural living systems'.[3] A-Life, he says, 'complements the traditional biological sciences concerned with the *analysis* of living organisms by attempting to *synthesise* life-like behaviors within computers and other artificial media'.[4]

The use of computers to simulate living and life-like systems is central to A-Life. Langton says:

Computers should be thought of as an important laboratory tool for the study of life, substituting for the array of incubators, culture dishes, microscopes, electrophoretic gels, pipettes, centrifuges, and other assorted wet-lab paraphernalia, one simple-to-master piece of experimental equipment.[5]

Langton even suggests that the 'ultimate goal of the study of artificial life would be to create "life" in some other medium, ideally a *virtual* medium where the essence of life has been abstracted from the details of its implementation in *any* particular hardware'.[6]

---

[1] C. G. Langton, 'Studying Artificial Life with Cellular Automata', *Physica D*, 22 (1986), 120–49.
[2] C. G. Langton (ed.), *Artificial Life: The Proceedings of an Interdisciplinary Workshop on the Synthesis and Simulation of Living Systems* (Redwood City, Calif.: Addison-Wesley, 1989).
[3] C. G. Langton, 'Artificial Life', in Langton, *Artificial Life*, 1.
[4] Ibid.
[5] Ibid. 32.
[6] Langton, 'Studying Artificial Life with Cellular Automata', 147.

2. Morphogenesis

Turing was the earliest pioneer of computer-based A-Life: he was the first to use computer simulation to investigate a theory of the development of organization and pattern in living things.[7] Early in 1951, the world's first commercially manufactured general-purpose electronic digital computer, the Manchester-built Ferranti Mark I, was installed in the Computing Machine Laboratory at Manchester University.[8] Turing immediately set about using the machine to model biological growth. In February 1951 he wrote to a colleague at the National Physical Laboratory:

Our new machine [the Ferranti Mark I] is to start arriving on Monday. I am hoping as one of the first jobs to do something about 'chemical embryology'. In particular I think one can account for the appearance of Fibonacci numbers in connection with fir-cones.[9]

In a Fibonacci number-series, each number, except for the first two, is the sum of the two previous numbers: for example 1, 3, 4, 7, 11 ...

In brief, Turing's 'chemical embryology' is the hypothesis that the development of anatomical structure in the animal or plant embryo is a result of the fact that diffusing chemicals reacting with one another can form spatial patterns. This is a thoroughly reductionist view: forms found in living matter are accounted for by the fact that, under appropriate conditions, sheer chemical reaction produces pattern and form.

Turing took his cue from the zoologist D'Arcy Thompson (to whose work Turing refers at the end of Chapter 15). D'Arcy Thompson held that the forms of living things, no less than naturally occurring forms in inorganic matter, are to be explained in terms of 'the operation of physical forces or mathematical laws'.[10]

Cell and tissue, shell and bone, leaf and flower, are so many portions of matter, and it is in obedience to the laws of physics that their particles have been moved, moulded and conformed. ... The form ... of any portion of matter, whether it be living or dead, and

---

[7] Another early pioneer of A-Life was the neurophysiologist W. Grey Walter (a founder of the Ratio Club, of which Turing was also a member). His famous 'tortoises'—built from about 1949 onwards—were mobile, battery-powered devices exhibiting life-like properties. He constructed them in order to show that seemingly complex behaviour can result from simple mechanisms (W. Grey Walter, *The Living Brain* (London: Gerald Duckworth, 1953), 82–7 and appendix B). Although digital computers were not involved, Grey Walter's work certainly conforms to Langton's description given above: 'Artificial Life ... attempt[s] to synthesise life-like behaviors within computers *and other artificial media*' (italics added).

[8] A digital facsimile of Turing's *Programmers' Handbook for Manchester Electronic Computer* (University of Manchester Computing Machine Laboratory, 1950) is in The Turing Archive for the History of Computing <www.AlanTuring.net/programmers_handbook>.

[9] Letter from Turing to Michael Woodger, undated, received 12 Feb. 1951 (in the Woodger Papers, National Museum of Science and Industry, Kensington, London; a digital facsimile is in the Turing Archive for the History of Computing <www.AlanTuring.net/turing_woodger_feb51>).

[10] W. D'Arcy Thompson, *On Growth and Form* (2nd edn. Cambridge: Cambridge University Press, 1942), 3.

the changes of form which are apparent in its ... growth, may in all cases alike be described as due to the action of force.[11]

Concerning the puzzling fact that the scales of a fir-cone, or the florets of a sunflower, are grouped together in numbers that form a Fibonacci series, D'Arcy Thompson boldly declared:

while the Fibonacci series stares us in the face in the fir-cone, it does so for mathematical reasons; and its supposed usefulness, and the hypothesis of its introduction into plant-structure through natural selection, are matters which deserve no place in the plain study of botanical phenomena.[12]

Turing summarized his own theory as the suggestion that 'certain well-known physical laws are sufficient to account for many of the facts' of morphogenesis (p. 519). ('Morphogenesis' means 'generation of form'.) Turing described an idealized chemical mechanism, now called the *reaction-diffusion model*. He showed that this mechanism could lead to a number of simple but life-like patterns and forms. The reaction-diffusion model is the topic of Chapter 15, 'The Chemical Basis of Morphogenesis'. 'The Chemical Basis of Morphogenesis' has been widely cited in the biological literature and today reaction-diffusion remains a possible, although still unconfirmed, explanation of aspects of the generation of biological pattern and form.

The geneticist C. H. Waddington commented in a letter to Turing in 1952 that the most clear-cut application of Turing's theory appeared to be 'in the arising of spots, streaks, and flecks of various kinds in apparently uniform areas such as the wings of butterflies, the shells of molluscs, the skin of tigers, leopards, etc'.[13] Modern computer simulations of Turing's reaction-diffusion mechanism have indeed produced leopard-like spots, cheetah-like spots, and giraffe-like stripes, as well as textures reminiscent of reptile skin, corals, and the surfaces of some fungi.[14] Turing himself, however, envisaged diverse applications of his theory (as he indicated in the letter to biologist J. Z. Young quoted below), including leaf arrangements and the appearance of Fibonacci sequences, and phenomena such as gastrulation. (Gastrulation is a process of rearrangement of cells in the spherical embryo, involving the folding inwards of part of the surface, producing a depression akin to the dent formed by poking a balloon with a finger.) Furthermore, Turing described his research into

---

[11] Ibid. 10, 16.
[12] Ibid. 933.
[13] Letter from Waddington to Turing (11 Sept. 1952). The letter is among the Turing Papers in the Modern Archive Centre, King's College, Cambridge (catalogue reference D 5).
[14] J. D. Murray, 'A Pre-pattern Formation Mechanism for Animal Coat Markings', *Journal of Theoretical Biology*, 88 (1981), 161–99; G. Turk, 'Generating Textures on Arbitrary Surfaces Using Reaction-Diffusion', *Computer Graphics*, 25 (1991), 289–98; A. Witkin and M. Kass, 'Reaction-Diffusion Textures', *Computer Graphics*, 25 (1991), 299–308.

morphogenesis as 'not altogether unconnected' to his work on neural networks (see Chapter 10).

Turing simulated the reaction-diffusion mechanism using the Ferranti computer. A mathematician using paper and pencil to analyse the behaviour of a reaction-diffusion system risks becoming overwhelmed by formidable mathematical complexity, and must make (what Turing calls) 'simplifying assumptions'. Turing explains that use of the computer enables such assumptions to be dispensed with to some extent. This freedom enables Turing to employ *non-linear* differential equations to describe the chemical interactions hypothesized by his theory.[15] Non-linear differential equations are mathematically intractable. Turing used the computer to explore in detail particular cases of interactions governed by equations of this type. He may well have been the first researcher to engage in the computer-assisted exploration of non-linear systems.[16] (It was not until Benoit Mandelbrot's discovery of the 'Mandelbrot set' in 1979 that the computer-assisted investigation of non-linear systems gained widespead attention.[17])

Turing's work on morphogenesis was in every respect ahead of its time. He died while in the midst of this groundbreaking work, leaving a large pile of handwritten notes and various programmes.[18] This material is still not fully understood.

The computer programme shown in Figure 1, which is in Turing's own hand, formed part of his study of the development of the fir-cone. He also investigated the development of the sunflower. The photograph and diagram shown in Figure 2 are from his notes.[19]

3. The Reaction-Diffusion Model

In reaction-diffusion, two or more chemicals diffuse through the embryo reacting with each other. Turing showed that under certain conditions a stable pattern of chemical concentrations will be reached. For example, in the case of an artificially simple embryo consisting of nothing but a ring of twenty cells, reaction-diffusion will produce a stable, regular pattern of concentration and rarefaction around the circumference of the ring. The points of highest concentration

---

[15] See p. 561 of Chapter 15; also section 2 of Turing's 'Morphogen Theory of Phyllotaxis', in *Morphogenesis: Collected Works of A. M. Turing*, ed. P. T. Saunders (Amsterdam: North-Holland, 1992).

[16] Unless perhaps the computer-assisted modelling of non-linear systems was first undertaken a little earlier, in secret, by members of the Los Alamos group in connection with nuclear explosions.

[17] B. B. Mandelbrot, *The Fractal Geometry of Nature* (New York: Freeman, 1977; revised and expanded 1982, 1983). For a popular exposition see J. Gleick, *Chaos: Making a New Science* (London: Cardinal, 1988).

[18] Some—but by no means all—of this material appears in *Morphogenesis*, ed. Saunders.

[19] The notes and programme sheets are among the Turing Papers in the Modern Archive Centre, King's College, Cambridge (catalogue references C 25, C 27).

Artificial Life | 511

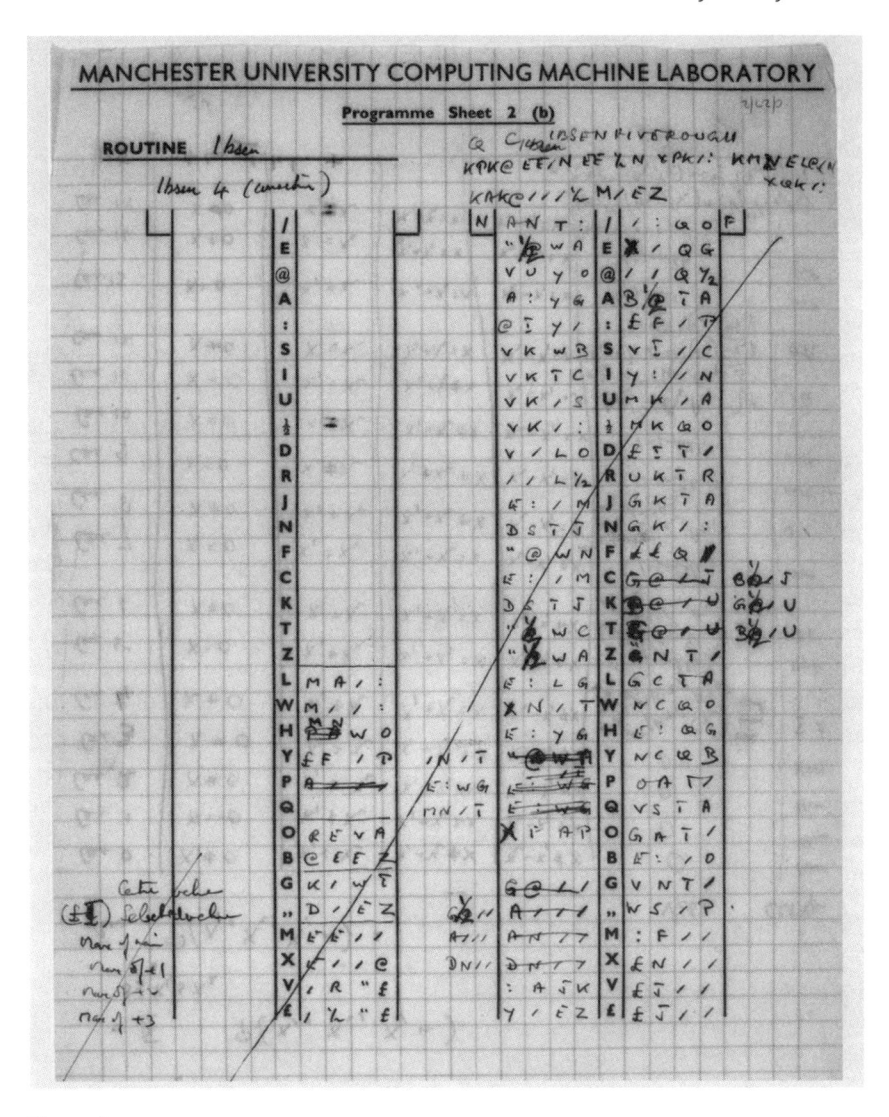

Figure 1.

occur equidistantly from one another. Turing picturesquely describes this pattern as a stationary 'chemical wave'. His suggestion is that at the points of high concentration around the ring, the chemical acts as a trigger to stimulate growth. For example, a ring of cells might sprout leaves at these points, or tentacles, producing a structure reminiscent of *Hydra* ('something like a sea-anenome but liv[ing] in fresh water and hav[ing] from about five to ten tentacles' (p. 556)). Regular but non-stationary patterns of concentration—travelling waves—are also possible. Turing suggested that the movements of the tail of a spermatozoon may provide an example of these travelling waves.

512 | Jack Copeland

It might be wondered how genes fit into this picture. Given our current knowledge of the role played by genes in the determination of anatomical structure, is Turing's theory archaic? Not at all. On Turing's account, reaction-diffusion is the *mechanism by which* genes determine the anatomical structure of the resulting organism. The function of genes, he suggested, is catalytic. The genes are presumed to catalyse the production of appropriate chemicals, so setting reaction-diffusion in train.

Turing calls the chemicals that diffuse and react 'morphogens', suggesting hormones as an example. However, his aim is not to give a taxonomy of morphogens or to describe specific morphogens, but rather to demonstrate in the abstract that, given certain realistic assumptions about *unspecified* morphogens, reaction-diffusion will produce pattern. These assumptions include the rates at which the morphogens diffuse between the cells, the rates at which the various chemical reactions between the morphogens take place, and the ways in which these rates change (due, for example, to increases in temperature in the tissue, or increases in concentration of morphogens that act as catalysts). The initial concentrations of the morphogens must be specified, and the number, dimensions, and positions of the cells making up the mass of tissue through which the morphogens diffuse. This mass—the embryo—is assumed to be initially homogeneous. (Other features of the situation, such as the motions and elasticities of the cells, are ignored in order to simplify the model.)

Turing represents the reactions between morphogens purely schematically, making no assumptions concerning the actual chemical compositions of the substances involved. For example, two reactions might be specified like this: morphogens X and Y react to produce Z; Z and morphogen A react to produce

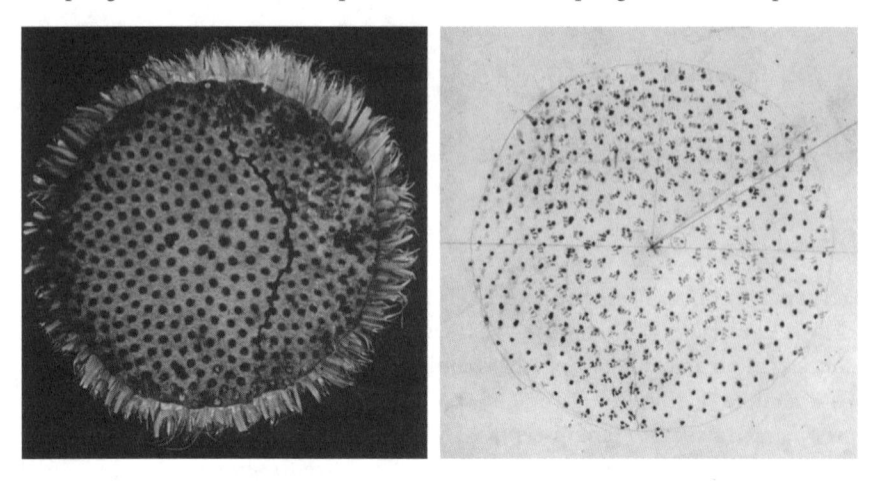

Figure 2. Turing's numbering of the individual florets of a sunflower.

[20] Editor's note. A blastula is a hollow sphere of cells, one cell in thickness.

2Y. The first reaction therefore depletes the supply of morphogen Y, while the second tends to build up the concentration of Y.

Uniform diffusion and uniform reaction in a uniform mass of tissue can produce only uniformity, not differentiation and pattern. Some sort of symmetry-breaker must be thrown into the mix. As Turing put it (p. 525):

> There appears superficially to be a difficulty confronting this theory of morphogenesis, or, indeed, almost any other theory of it. An embryo in its spherical blastula[20] stage has spherical symmetry, or if there are any deviations from perfect symmetry, they cannot be regarded as of any particular importance, for the deviations vary greatly from embryo to embryo within a species, though the organisms developed from them are barely distinguishable. One may take it therefore that there is perfect spherical symmetry. But a system which has spherical symmetry, and whose state is changing because of chemical reactions and diffusion, will remain spherically symmetrical forever. ... It certainly cannot result in an organism such as a horse, which is not spherically symmetrical.

Turing suggested various possible symmetry-breakers, including small disturbances caused by the presence of anatomical structures neighbouring the embryo, and purely statistical fluctuations at the molecular level. An example of the latter is statistical fluctuation in the number of molecules of a given morphogen passing through the wall of a cell, so producing small fluctuations in the concentration of that morphogen within the cell. Turing's point is that in certain circumstances, small fluctuations such as this can bring about large effects, just as a small nudge that under normal circumstances would have no effect on a person's balance could be enough to topple someone balancing on one foot.

Turing demonstrated that under appropriate conditions small departures from uniformity can indeed lead to the formation of chemical waves.

4. Genetic Algorithms

An important concept both in Artificial Life and in Artificial Intelligence is that of a *genetic algorithm* (GA). GAs employ methods analogous to the processes of natural evolution in order to produce successive generations of software entities that are increasingly fit for their intended purpose. Turing anticipated the concept of a genetic algorithm in a brief passage of his 'Intelligent Machinery', where he described what he called a 'genetical or evolutionary search' (Chapter 10, p. 431; see also Chapter 11, p. 463, and the introduction to Chapter 16, p. 565). The actual term 'genetic algorithm' was introduced circa 1975 by John Holland and his research group at the University of Michigan.[21] Holland's work is responsible for the current intense interest in GAs. (Holland, a student of Arthur Burks, was influenced by von Neumann's ideas—see below.)

---

[21] J. H. Holland, *Adaptation in Natural and Artificial Systems* (Cambridge, Mass.: MIT Press, 1992), p. x.

Turing described an early example of a GA in connection with his chess-player in Chapter 16.

One of the first GAs to be implemented (in the 1950s) formed part of the learning mechanism of Samuel's checkers or draughts programme mentioned above in 'Artificial Intelligence'.[22] Samuel's programme used heuristics to rank moves and board positions (the programme 'looked ahead' as many as ten turns of play). To speed up learning, Samuel would set up two copies of the programme, Alpha and Beta, on the same computer and leave them to play game after game with each other. The learning procedure consisted in the computer making small numerical changes to Alpha's ranking procedure, leaving Beta's unchanged, and then comparing Alpha's and Beta's performance over a few games. If Alpha played worse than Beta, these changes to the ranking procedure were discarded, but if Alpha played better than Beta then Beta's ranking procedure was replaced with Alpha's. As in biological evolution, the fitter survived. Over many such cycles of mutation and selection, the programme's quality of play increased markedly.

The use of GAs is burgeoning, in AI and elsewhere. In one application a GA-based system and a witness to a crime cooperate to generate on-screen faces that become closer and closer to the recollected face of the criminal.[23] In A-Life, researchers study GAs as a means of studying the process of evolution itself.

5. John von Neumann and A-Life

John von Neumann was another important early pioneer of Artificial Life. In his 1948 Hixon Symposium, entitled 'The General and Logical Theory of Automata', he said:

Natural organisms are, as a rule, much more complicated and subtle, and therefore much less well understood in detail, than are artificial automata. Nevertheless... a good deal of our experiences and difficulties with our artificial automata can be to some extent projected on our interpretations of natural organisms.[24]

Arthur Burks (who edited and completed von Neumann's posthumously published volume *Theory of Self-Reproducing Automata*, listed in the section of Further Reading) wrote this concerning von Neumann's research on the problem of self-reproduction:

---

[22] A. L. Samuel, 'Some Studies in Machine Learning Using the Game of Checkers', *IBM Journal of Research and Development*, 3 (1959), 211–29; reprinted in E. A. Feigenbaum and J. Feldman (eds.), *Computers and Thought* (New York: McGraw-Hill, 1963).

[23] D. E. Goldberg, 'Genetic and Evolutionary Algorithms Come of Age', *Communications of the Association for Computing Machinery*, 37 (1994), 113–19.

[24] J. von Neumann, 'The General and Logical Theory of Automata', in vol. v of von Neumann's *Collected Works*, ed. A. H. Taub (Oxford: Pergamon Press, 1963), 288–9.

Von Neumann had the familiar natural phenomenon of self-reproduction in mind... but he was not trying to simulate the self-reproduction of a natural system at the levels of genetics and biochemistry. He wished to abstract from the natural self-reproduction problem its logical form.[25]

This passage is quoted approvingly by Langton, who italicizes the final sentence and comments:

This approach is the first to capture the essence of Artificial Life. To understand the field of Artificial Life, one need only replace references to 'self-reproduction' in the above with references to any other biological phenomenon.[26]

Von Neumann was thinking about issues relevant to A-Life at least as early as 1946. In a letter to the cyberneticist Norbert Wiener (dated 29 November 1946), von Neumann wrote: 'I did think a good deal about self-reproductive mechanisms. I can formulate the problem rigorously, in about the style in which Turing did it for his mechanisms.'[27] There is no doubt that von Neumann's theorizing about self-reproduction was strongly influenced by Turing's discovery of the universal computing machine (or 'universal automaton', as von Neumann called it). Von Neumann's colleague Herman Goldstine wrote:

von Neumann had a profound concern for automata. In particular, he always had a deep interest in Turing's work. ... Turing proved a most remarkable and unexpected result. ... In essence what he showed is that any particular automaton can be described by a finite set of instructions, and that when this is fed to his universal automaton it in turn imitates the special one. ... Von Neumann was enormously intrigued with these ideas, and he started in 1947 working on... how complex a device or construct needed to be in order to be self-reproductive.[28]

In his Hixon Symposium, von Neumann said:

For the question which concerns me here, that of 'self-reproduction' of automata, Turing's procedure is too narrow in one respect only. His automata are purely computing machines. Their output is a piece of tape with zeros and ones on it. What is needed... is an automaton whose output is other automata. There is, however, no difficulty in principle in dealing with this broader concept and in deriving from it the equivalent of Turing's result. ... The problem of self-reproduction can... be stated like this: Can one build an aggregate out of... elements in such a manner that if it is put into a reservoir, in which there float all these elements in large numbers, it will then begin to construct other aggregates, each of which will at the end turn out to be another automaton exactly like the

---

[25] A. W. Burks (ed.), *Essays on Cellular Automata* (Urbana: University of Illinois Press, 1970), p. xv.
[26] C. G. Langton, 'Artificial Life', in M. A. Boden (ed.), *The Philosophy of Artificial Life* (Oxford: Oxford University Press, 1996), 47.
[27] Letter from von Neumann to Wiener, 29 Nov. 1946 (in the von Neumann Archive at the Library of Congress, Washington, DC).
[28] H. H. Goldstine, *The Computer from Pascal to von Neumann* (Princeton: Princeton University Press, 1972), 271, 274–5.

original one? This is feasible, and the principle on which it can be based is closely related to Turing's [universal automaton] outlined earlier.[29]

In lectures von Neumann described a 'universal constructor' (remarking, 'You see, I'm coming quite close to Turing's trick with universal automata').[30] Just as complete descriptions of Turing machines can be fed into the universal automaton in the form of programmes, complete descriptions of automata can be inserted into the universal constructor. The universal constructor floats in a medium—or sea—in which also float, in practically unlimited supply, the components from which the universal constructor is made. Given a complete description of an automaton, the universal constructor will assemble that automaton.

If what is inserted into the universal constructor is a description I of the universal constructor itself, the constructor assembles a duplicate of itself—self-reproduction. Since the duplicate is to be an exact copy, it must contain the description I. This is done by a copying mechanism in the universal constructor and the copy of I is inserted into the offspring.[31] Thus the offspring, too, is capable of self-reproduction. Von Neumann likened the description I to the gene, saying that the copying mechanism 'performs the fundamental act of reproduction, the duplication of the genetic material, which is clearly the fundamental operation in the multiplication of living cells'.[32] Allowing the copying mechanism to make occasional random errors affords the possibility of random mutation in the genes of the offspring, thus opening the door to Darwinian evolution.[33]

In his letter to Wiener (29 November 1946), von Neumann voiced some concerns that might equally well be raised concerning the focus of modern research in A-Life and Artificial Intelligence. Von Neumann pointed out that when automata theorists choose the human nervous system as their model, they are unrealistically selecting 'the most complicated object under the sun—literally'. Moreover, he said, there is little advantage in choosing instead simpler organisms with fewer neurons, for example, the ant: *any* nervous system exhibits 'exceptional complexity'. Von Neumann suggested that automata theorists 'turn to simpler systems', and he recommended attention to 'organisms of the virus or bacteriophage type'. These, he pointed out, are 'self-reproductive and...are able to orient themselves in an unorganized milieu, to move towards food, to appropriate it and to use it'. He estimated that a typical bacteriophage might consist of 6 million atoms grouped into a few hundred thousand 'mechanical elements', saying that this represents 'a degree of complexity which is not

---

[29] von Neumann, 'The General and Logical Theory of Automata', 315.
[30] J. von Neumann, *Theory of Self-Reproducing Automata*, ed. and completed by A. W. Burks (Urbana: University of Illinois Press, 1966), 83.
[31] von Neumann 'The General and Logical Theory of Automata', 316–17.
[32] Ibid., 317.
[33] Ibid. 317–18.

necessarily beyond human endurance'. By following this path, he said, the 'decisive break' might be achieved.

6. Letter from Turing to Young

Shortly before the Ferranti computer arrived in 1951, Turing wrote about his work on morphogenesis in a letter to the biologist J. Z. Young. The letter connects Turing's work on morphogenesis with his interest in neural networks (Chapter 10), and moreover to some extent explains why he did not follow up his earlier suggestion (Chapter 10, p. 428) and use the Ferranti computer to simulate his 'unorganised machines'.

> I am afraid I am very far from the stage where I feel inclined to start asking any anatomical questions [about the brain]. According to my notions of how to set about it that will not occur until quite a late stage when I have a fairly definite theory about how things are done.
>
> At present I am not working on the problem at all, but on my mathematical theory of embryology... This is yielding to treatment, and it will so far as I can see, give satisfactory explanations of
>
> i) Gastrulation.
> ii) Polyogonally symmetrical structures, e.g., starfish, flowers.
> iii) Leaf arrangement, in particular the way the Fibonacci series (0, 1, 1, 2, 3, 5, 8, 13, ...) comes to be involved.
> iv) Colour patterns on animals, e.g., stripes, spots and dappling.
> v) Patterns on nearly spherical structures such as some Radiolaria, but this is more difficult and doubtful.
>
> I am really doing this now because it is yielding more easily to treatment. I think it is not altogether unconnected with the other problem. The brain structure has to be one which can be achieved by the genetical embryological mechanism, and I hope that this theory that I am now working on may make clearer what restrictions this really implies. What you tell me about growth of neurons under stimulation is very interesting in this connection. It suggests means by which the neurons might be made to grow so as to form a particular circuit, rather than to reach a particular place.[34]

Further reading

Boden, M. A., *Mind as Machine: A History of Cognitive Science* (Oxford: Oxford University Press, 2005).
—— (ed.), *The Philosophy of Artificial Life* (Oxford: Oxford University Press, 1996).

---

[34] Letter from Turing to Young, 8 Feb. 1951 (a copy of the letter is in the Modern Archive Centre, King's College, Cambridge (catalogue reference K 78)).

Holland, J. H., 'Genetic Algorithms', *Scientific American*, 267 (July 1992), 44–50.

Swinton, J., 'Watching the Daisies Grow: Turing and Fibonacci Phyllotaxis', in C. Teuscher (ed.), *Alan Turing: Life and Legacy of a Great Thinker* (Berlin: Springer-Verlag, 2004).

Turing, A. M., *Morphogenesis: Collected Works of A. M. Turing*, ed. P. T. Saunders (Amsterdam: North-Holland, 1992).

Turk, G., 'Generating Textures on Arbitrary Surfaces Using Reaction-Diffusion', *Computer Graphics*, 25 (1991), 289–98.

von Neumann, J., *Theory of Self-Reproducing Automata*, ed. and completed by A. W. Burks (Urbana: University of Illinois Press, 1966).

CHAPTER 15

The Chemical Basis of Morphogenesis (1952)

Alan Turing

It is suggested that a system of chemical substances, called morphogens, reacting together and diffusing through a tissue, is adequate to account for the main phenomena of morphogenesis. Such a system, although it may originally be quite homogenous, may later develop a pattern or structure due to an instability of the homogeneous equilibrium, which is triggered off by random disturbances. Such reaction-diffusion systems are considered in some detail in the case of an isolated ring of cells, a mathematically convenient, though biologically unusual system. The investigation is chiefly concerned with the onset of instability. It is found that there are six essentially different forms which this may take. In the most interesting form stationary waves appear on the ring. It is suggested that this might account, for instance, for the tentacle patterns on *Hydra* and for whorled leaves. A system of reactions and diffusion on a sphere is also considered. Such a system appears to account for gastrulation. Another reaction system in two dimensions gives rise to patterns reminiscent of dappling. It is also suggested that stationary waves in two dimensions could account for the phenomena of phyllotaxis.

The purpose of this paper is to discuss a possible mechanism by which the genes of a zygote may determine the anatomical structure of the resulting organism. The theory does not make any new hypotheses; it merely suggests that certain well-known physical laws are sufficient to account for many of the facts. The full understanding of the paper requires a good knowledge of mathematics, some biology, and some elementary chemistry. Since readers cannot be expected to be experts in all of these subjects, a number of elementary facts are explained, which can be found in text-books, but whose omission would make the paper difficult reading.

1. A model of the embryo. Morphogens

In this section a mathematical model of the growing embryo will be described. This model will be a simplification and an idealization, and consequently a falsification. It is to be hoped that the features retained for discussion are those of greatest importance in the present state of knowledge.

This article first appeared in *Philosophical Transactions of the Royal Society of London*, Series B, 237 (1952–54), 37–72. It is reprinted with the permission of the Royal Society and the Estate of Alan Turing.

The model takes two slightly different forms. In one of them the cell theory is recognized but the cells are idealized into geometrical points. In the other the matter of the organism is imagined as continuously distributed. The cells are not, however, completely ignored, for various physical and physico-chemical characteristics of the matter as a whole are assumed to have values appropriate to the cellular matter.

With either of the models one proceeds as with a physical theory and defines an entity called 'the state of the system'. One then describes how that state is to be determined from the state at a moment very shortly before. With either model the description of the state consists of two parts, the mechanical and the chemical. The mechanical part of the state describes the positions, masses, velocities and elastic properties of the cells, and the forces between them. In the continuous form of the theory essentially the same information is given in the form of the stress, velocity, density and elasticity of the matter. The chemical part of the state is given (in the cell form of theory) as the chemical composition of each separate cell; the diffusibility of each substance between each two adjacent cells must also be given. In the continuous form of the theory the concentrations and diffusibilities of each substance have to be given at each point. In determining the changes of state one should take into account

(i) The changes of position and velocity as given by Newton's laws of motion.
(ii) The stresses as given by the elasticities and motions, also taking into account the osmotic pressures as given from the chemical data.
(iii) The chemical reactions.
(iv) The diffusion of the chemical substances. The region in which this diffusion is possible is given from the mechanical data.

This account of the problem omits many features, e.g. electrical properties and the internal structure of the cell. But even so it is a problem of formidable mathematical complexity. One cannot at present hope to make any progress with the understanding of such systems except in very simplified cases. The interdependence of the chemical and mechanical data adds enormously to the difficulty, and attention will therefore be confined, so far as is possible, to cases where these can be separated. The mathematics of elastic solids is a well-developed subject, and has often been applied to biological systems. In this paper it is proposed to give attention rather to cases where the mechanical aspect can be ignored and the chemical aspect is the most significant. These cases promise greater interest, for the characteristic action of the genes themselves is presumably chemical. The systems actually to be considered consist therefore of masses of tissues which are not growing, but within which certain substances are reacting chemically, and through which they are diffusing. These substances will be called morphogens, the word being intended to convey the idea of a form producer. It is not intended to have any very exact meaning, but is simply the kind of substance concerned in this theory. The evocators of Waddington provide a good example of morphogens (Waddington 1940). These evocators diffusing into a tissue somehow persuade it to develop along different lines from those which would have been followed in its absence. The genes themselves may also be considered to be morphogens. But they certainly form rather a special class. They are quite indiffusible. Moreover, it is only by courtesy that genes can be regarded as separate molecules. It would be more accurate (at any rate at mitosis) to regard them as radicals of the giant molecules known as chromosomes. But presumably these radicals act almost independently, so that it is unlikely that serious errors will arise through regarding the genes as molecules. Hormones may also be regarded as quite typical morphogens. Skin pigments may be regarded as morphogens if desired. But those whose action is to be considered here do not come squarely within any of these categories.

The function of genes is presumed to be purely catalytic. They catalyze the production of other morphogens, which in turn may only be catalysts. Eventually, presumably, the chain leads to some morphogens whose duties are not purely catalytic. For instance, a substance might break down into a number of smaller molecules, thereby increasing the osmotic pressure in a cell and promoting its growth. The genes might thus be said to influence the anatomical form of the organism by determining the rates of those reactions which they catalyze. If the rates are assumed to be those determined by the genes, and if a comparison of organisms is not in question, the genes themselves may be eliminated from the discussion. Likewise any other catalysts obtained secondarily through the agency of the genes may equally be ignored, if there is no question of their concentrations varying. There may, however, be some other morphogens, of the nature of evocators, which cannot be altogether forgotten, but whose role may nevertheless be subsidiary, from the point of view of the formation of a particular organ. Suppose, for instance, that a 'leg-evocator' morphogen were being produced in a certain region of an embryo, or perhaps diffusing into it, and that an attempt was being made to explain the mechanism by which the leg was formed in the presence of the evocator. It would then be reasonable to take the distribution of the evocator in space and time as given in advance and to consider the chemical reactions set in train by it. That at any rate is the procedure adopted in the few examples considered here.

2. Mathematical background required

The greater part of this present paper requires only a very moderate knowledge of mathematics. What is chiefly required is an understanding of the solution of linear differential equations with constant coefficients. (This is also what is chiefly required for an understanding of mechanical and electrical oscillations.) The solution of such an equation takes the form of a sum $\sum A e^{bt}$, where the quantities

A, b may be complex, i.e. of the form $\alpha + i\beta$, where α and β are ordinary (real) numbers and $i = \sqrt{-1}$. It is of great importance that the physical significance of the various possible solutions of this kind should be appreciated, for instance, that

(a) Since the solutions will normally be real one can also write them in the form $\mathcal{R} \sum A e^{bt}$ or $\sum \mathcal{R} A e^{bt}$ ($\mathcal{R}$ means 'real part of').
(b) That if $A = A' e^{i\phi}$ and $b = \alpha + i\beta$, where A', α, β, ϕ are real, then

$$\mathcal{R} A e^{bt} = A' e^{\alpha t} \cos(\beta t + \phi).$$

Thus each such term represents a sinusoidal oscillation if $\alpha = 0$, a damped oscillation if $\alpha < 0$, and an oscillation of ever-increasing amplitude if $\alpha > 0$.
(c) If any one of the numbers b has a positive real part the system in question is unstable.
(d) After a sufficiently great lapse of time all the terms $A e^{bt}$ will be negligible in comparison with those for which b has the greatest real part, but unless this greatest real part is itself zero these dominant terms will eventually either tend to zero or to infinite values.
(e) That the indefinite growth mentioned in (b) and (d) will in any physical or biological situation eventually be arrested due to a breakdown of the assumptions under which the solution was valid. Thus, for example, the growth of a colony of bacteria will normally be taken to satisfy the equation $dy/dt = \alpha y$ ($\alpha > 0$), y being the number of organisms at time t, and this has the solution $y = A e^{\alpha t}$. When, however, the factor $e^{\alpha t}$ has reached some billions the food supply can no longer be regarded as unlimited and the equation $dy/dt = \alpha y$ will no longer apply.

The following relatively elementary result will be needed, but may not be known to all readers:

$$\sum_{r=1}^{N} \exp\left[\frac{2\pi i r s}{N}\right] = 0 \text{ if } 0 < s < N,$$

but

$$= N \text{ if } s = 0 \text{ or } s = N.$$

The first case can easily be proved when it is noticed that the left-hand side is a geometric progression. In the second case all the terms are equal to 1.

The relative degrees of difficulty of the various sections are believed to be as follows. Those who are unable to follow the points made in this section should only attempt §§3, 4, 11, 12, 14 and part of §13. Those who can just understand this section should profit also from §§7, 8, 9. The remainder, §§5, 10, 13, will probably only be understood by those definitely trained as mathematicians.

The Chemical Basis of Morphogenesis | 523

3. Chemical reactions

It has been explained in a preceding section that the system to be considered consists of a number of chemical substances (morphogens) diffusing through a mass of tissue of given geometrical form and reacting together within it. What laws are to control the development of this situation? They are quite simple. The diffusion follows the ordinary laws of diffusion, i.e. each morphogen moves from regions of greater to regions of less concentration, at a rate proportional to the gradient of the concentration, and also proportional to the 'diffusibility' of the substance. This is very like the conduction of heat, diffusibility taking the place of conductivity. If it were not for the walls of the cells the diffusibilities would be inversely proportional to the square roots of the molecular weights. The pores of the cell walls put a further handicap on the movement of the larger molecules in addition to that imposed by their inertia, and most of them are not able to pass through the walls at all.

The reaction rates will be assumed to obey the 'law of mass action'. This states that the rate at which a reaction takes place is proportional to the concentrations of the reacting substances. Thus, for instance, the rate at which silver chloride will be formed and precipitated from a solution of silver nitrate and sodium chloride by the reaction

$$Ag^+ + Cl^- \rightarrow AgCl$$

will be proportional to the product of the concentrations of the silver ion Ag^+ and the chloride ion Cl^-. It should be noticed that the equation

$$AgNO_3 + NaCl \rightarrow AgCl + NaNO_3$$

is not used because it does not correspond to an actual reaction but to the final outcome of a number of reactions. The law of mass action must only be applied to the *actual* reactions. Very often certain substances appear in the individual reactions of a group, but not in the final outcome. For instance, a reaction $A \rightarrow B$ may really take the form of two steps $A + G \rightarrow C$ and $C \rightarrow B + G$. In such a case the substance G is described as a catalyst, and as catalyzing the reaction $A \rightarrow B$. (Catalysis according to this plan has been considered in detail by Michaelis & Menten (1913).) The effect of the genes is presumably achieved almost entirely by catalysis. They are certainly not permanently used up in the reactions.

Sometimes one can regard the effect of a catalyst as merely altering a reaction rate. Consider, for example, the case mentioned above, but suppose also that A can become detached from G, i.e. that the reaction $C \rightarrow A + G$ is taken into account. Also suppose that the reactions $A + G \rightleftarrows C$ both proceed much faster than $C \rightarrow B + G$. Then the concentrations of A, G, C will be related by the condition that there is equilibrium between the reactions $A + G \rightarrow C$ and $C \rightarrow A + G$, so that (denoting concentrations by square brackets)

$[A][G] = k[C]$ for some constant k. The reaction $C \to B + G$ will of course proceed at a rate proportional to $[C]$, i.e. to $[A][G]$. If the amount of C is always small compared with the amount of G one can say that the presence of the catalyst and its amount merely alter the mass action constant for the reaction $A \to B$, for the whole proceeds at a rate proportional to $[A]$. This situation does not, however, hold invariably. It may well happen that nearly all of G takes the combined form C so long as any of A is left. In this case the reaction proceeds at a rate independent of the concentration of A until A is entirely consumed. In either of these cases the rate of the complete group of reactions depends only on the concentrations of the reagents, although usually not according to the law of mass action applied crudely to the chemical equation for the whole group. The same applies in any case where all reactions of the group with one exception proceed at speeds much greater than that of the exceptional one. In these cases the rate of the reaction is a function of the concentrations of the reagents. More generally again, no such approximation is applicable. One simply has to take all the actual reactions into account.

According to the cell model then, the number and positions of the cells are given in advance, and so are the rates at which the various morphogens diffuse between the cells. Suppose that there are N cells and M morphogens. The state of the whole system is then given by MN numbers, the quantities of the M morphogens in each of N cells. These numbers change with time, partly because of the reactions, partly because of the diffusion. To determine the part of the rate of change of one of these numbers due to diffusion, at any one moment, one only needs to know the amounts of the same morphogen in the cell and its neighbours, and the diffusion coefficient for that morphogen. To find the rate of change due to chemical reaction one only needs to know the concentrations of all morphogens at that moment in the one cell concerned.

This description of the system in terms of the concentrations in the various cells is, of course, only an approximation. It would be justified if, for instance, the contents were perfectly stirred. Alternatively, it may often be justified on the understanding that the 'concentration in the cell' is the concentration at a certain representative point, although the idea of 'concentration at a point' clearly itself raises difficulties. The author believes that the approximation is a good one, whatever argument is used to justify it, and it is certainly a convenient one.

It would be possible to extend much of the theory to the case of organisms immersed in a fluid, considering the diffusion within the fluid as well as from cell to cell. Such problems are not, however, considered here.

4. The breakdown of symmetry and homogeneity

There appears superficially to be a difficulty confronting this theory of morphogenesis, or, indeed, almost any other theory of it. An embryo in its spherical blastula stage has spherical symmetry, or if there are any deviations from perfect symmetry, they cannot be regarded as of any particular importance, for the deviations vary greatly from embryo to embryo within a species, though the organisms developed from them are barely distinguishable. One may take it therefore that there is perfect spherical symmetry. But a system which has spherical symmetry, and whose state is changing because of chemical reactions and diffusion, will remain spherically symmetrical for ever. (The same would hold true if the state were changing according to the laws of electricity and magnetism, or of quantum mechanics.) It certainly cannot result in an organism such as a horse, which is not spherically symmetrical.

There is a fallacy in this argument. It was assumed that the deviations from spherical symmetry in the blastula could be ignored because it makes no particular difference what form of asymmetry there is. It is, however, important that there are *some* deviations, for the system may reach a state of instability in which these irregularities, or certain components of them, tend to grow. If this happens a new and stable equilibrium is usually reached, with the symmetry entirely gone. The variety of such new equilibria will normally not be so great as the variety of irregularities giving rise to them. In the case, for instance, of the gastrulating sphere, discussed at the end of this paper, the direction of the axis of the gastrula can vary, but nothing else.

The situation is very similar to that which arises in connexion with electrical oscillators. It is usually easy to understand how an oscillator keeps going when once it has started, but on a first acquaintance it is not obvious how the oscillation begins. The explanation is that there are random disturbances always present in the circuit. Any disturbance whose frequency is the natural frequency of the oscillator will tend to set it going. The ultimate fate of the system will be a state of oscillation at its appropriate frequency, and with an amplitude (and a wave form) which are also determined by the circuit. The phase of the oscillation alone is determined by the disturbance.

If chemical reactions and diffusion are the only forms of physical change which are taken into account the argument above can take a slightly different form. For if the system originally has no sort of geometrical symmetry but is a perfectly homogeneous and possibly irregularly shaped mass of tissue, it will continue indefinitely to be homogeneous. In practice, however, the presence of irregularities, including statistical fluctuations in the numbers of molecules undergoing the various reactions, will, if the system has an appropriate kind of instability, result in this homogeneity disappearing.

This breakdown of symmetry or homogeneity may be illustrated by the case of a pair of cells originally having the same, or very nearly the same, contents. The system is homogeneous: it is also symmetrical with respect to the operation of interchanging the cells. The contents of either cell will be supposed describable by giving the concentrations X and Y of two morphogens. The chemical reactions will be supposed such that, on balance, the first morphogen (X) is produced at the rate $5X - 6Y + 1$ and the second (Y) at the rate $6X - 7Y + 1$. When, however, the strict application of these formulae would involve the concentration of a morphogen in a cell becoming negative, it is understood that it is instead destroyed only at the rate at which it is reaching that cell by diffusion. The first morphogen will be supposed to diffuse at the rate 0·5 for unit difference of concentration between the cells, the second, for the same difference, at the rate 4·5. Now if both morphogens have unit concentration in both cells there is equilibrium. There is no resultant passage of either morphogen across the cell walls, since there is no concentration difference, and there is no resultant production (or destruction) of either morphogen in either cell since $5X - 6Y + 1$ and $6X - 7Y + 1$ both have the value zero for $X = 1$, $Y = 1$. But suppose the values are $X_1 = 1\cdot06$, $Y_1 = 1\cdot02$ for the first cell and $X_2 = 0\cdot94$, $Y_2 = 0\cdot98$ for the second. Then the two morphogens will be being produced by chemical action at the rates 0·18, 0·22 respectively in the first cell and destroyed at the same rates in the second. At the same time there is a flow due to diffusion from the first cell to the second at the rate 0·06 for the first morphogen and 0·18 for the second. In sum the effect is a flow from the second cell to the first at the rates 0·12, 0·04 for the two morphogens respectively. This flow tends to accentuate the already existing differences between the two cells. More generally, if

$$X_1 = 1 + 3\xi, \quad X_2 = 1 - 3\xi, \quad Y_1 = 1 + \xi, \quad Y_2 = 1 - \xi,$$

at some moment the four concentrations continue afterwards to be expressible in this form, and ξ increases at the rate 2ξ. Thus there is an exponential drift away from the equilibrium condition. It will be appreciated that a drift away from the equilibrium occurs with almost any small displacement from the equilibrium condition, though not normally according to an exact exponential curve. A particular choice was made in the above argument in order to exhibit the drift with only very simple mathematics.

Before it can be said to follow that a two-cell system can be unstable, with inhomogeneity succeeding homogeneity, it is necessary to show that the reaction rate functions postulated really can occur. To specify actual substances, concentrations and temperatures giving rise to these functions would settle the matter finally, but would be difficult and somewhat out of the spirit of the present inquiry. Instead, it is proposed merely to mention imaginary reactions which give rise to the required functions by the law of mass action, if suitable reaction constants are assumed. It will be sufficient to describe

(i) A set of reactions producing the first morphogen at the constant rate 1, and a similar set forming the second morphogen at the same rate.

(ii) A set destroying the second morphogen (Y) at the rate $7Y$.

(iii) A set converting the first morphogen (X) into the second (Y) at the rate $6X$.
(iv) A set producing the first morphogen (X) at the rate $11X$.
(v) A set destroying the first morphogen (X) at the rate $6Y$, so long as any of it is present.

The conditions of (i) can be fulfilled by reactions of the type $A \to X$, $B \to Y$, where A and B are substances continually present in large and invariable concentrations. The conditions of (ii) are satisfied by a reaction of the form $Y \to D$, D being an inert substance, and (iii) by the reaction $X \to Y$ or $X \to Y + E$. The remaining two sets are rather more difficult. To satisfy the conditions of (iv) one may suppose that X is a catalyst for its own formation from A. The actual reactions could be the formation of an unstable compound U by the reaction $A + X \to U$, and the subsequent almost instantaneous breakdown $U \to 2X$. To destroy X at a rate proportional to Y as required in (v) one may suppose that a catalyst C is present in small but constant concentration and immediately combines with X, $X + C \to V$. The modified catalyst reacting with Y, at a rate proportional to Y, restores the catalyst but not the morphogen X, by the reactions $V + Y \to W$, $W \to C + H$, of which the latter is assumed instantaneous.

It should be emphasized that the reactions here described are by no means those which are most likely to give rise to instability in nature. The choice of the reactions to be discussed was dictated entirely by the fact that it was desirable that the argument be easy to follow. More plausible reaction systems are described in §10.

Unstable equilibrium is not, of course, a condition which occurs very naturally. It usually requires some rather artificial interference, such as placing a marble on the top of a dome. Since systems tend to leave unstable equilibria they cannot often be in them. Such equilibria can, however, occur naturally through a stable equilibrium changing into an unstable one. For example, if a rod is hanging from a point a little above its centre of gravity it will be in stable equilibrium. If, however, a mouse climbs up the rod the equilibrium eventually becomes unstable and the rod starts to swing. A chemical analogue of this mouse-and-pendulum system would be that described above with the same diffusibilities but with the two morphogens produced at the rates

$$(3 + I)X - 6Y + I - 1 \quad \text{and} \quad 6X - (9 + I)Y - I + 1.$$

This system is stable if $I < 0$ but unstable if $I > 0$. If I is allowed to increase, corresponding to the mouse running up the pendulum, it will eventually become positive and the equilibrium will collapse. The system which was originally discussed was the case $I = 2$, and might be supposed to correspond to the mouse somehow reaching the top of the pendulum without disaster, perhaps by falling vertically on to it.

5. Left-handed and right-handed organisms

The object of this section is to discuss a certain difficulty which might be thought to show that the morphogen theory of morphogenesis cannot be right. The difficulty is mainly concerned with organisms which have not got bilateral symmetry. The argument, although carried through here without the use of mathematical formulae, may be found difficult by non-mathematicians, and these are therefore recommended to ignore it unless they are already troubled by such a difficulty.

An organism is said to have 'bilateral symmetry' if it is identical with its own reflexion in some plane. This plane of course always has to pass through some part of the organism, in particular through its centre of gravity. For the purpose of this argument it is more general to consider what may be called 'left–right symmetry'. An organism has left-right symmetry if its description in any right-handed set of rectangular Cartesian co-ordinates is identical with its description in some set of left-handed axes. An example of a body with left–right symmetry, but not bilateral symmetry, is a cylinder with the letter P printed on one end, and with the mirror image of a P on the other end, but with the two upright strokes of the two letters not parallel. The distinction may possibly be without a difference so far as the biological world is concerned, but mathematically it should not be ignored.

If the organisms of a species are sufficiently alike, and the absence of left–right symmetry sufficiently pronounced, it is possible to describe each individual as either right-handed or left-handed without there being difficulty in classifying any particular specimen. In man, for instance, one could take the X-axis in the forward direction, the Y-axis at right angles to it in the direction towards the side on which the heart is felt, and the Z-axis upwards. The specimen is classed as left-handed or right-handed according as the axes so chosen are left-handed or right-handed. A new classification has of course to be defined for each species.

The fact that there exist organisms which do not have left–right symmetry does not in itself cause any difficulty. It has already been explained how various kinds of symmetry can be lost in the development of the embryo, due to the particular disturbances (or 'noise') influencing the particular specimen not having that kind of symmetry, taken in conjunction with appropriate kinds of instability. The difficulty lies in the fact that there are species in which the proportions of left-handed and right-handed types are very unequal. It will be as well to describe first an argument which appears to show that this should not happen. The argument is very general, and might be applied to a very wide class of theories of morphogenesis.

An entity may be described as 'P-symmetrical' if its description in terms of one set of right-handed axes is identical with its description in terms of any other set of right-handed axes with the same origin. Thus, for instance, the totality of positions that a corkscrew would take up when rotated in all possible ways about the origin has *P*-symmetry. The entity will be said to be '*F*-symmetrical' when changes from right-handed axes to left-handed may also be made. This would apply if the corkscrew were replaced by a bilaterally symmetrical object such as a coal scuttle, or a left–right symmetrical object. In these terms one may say that there are species such that the totality of specimens from that species, together with the rotated specimens, is *P*-symmetrical, but very far from *F*-symmetrical. On the other hand, it is reasonable to suppose that

(i) The laws of physics are *F*-symmetrical.
(ii) The initial totality of zygotes for the species is *F*-symmetrical.
(iii) The statistical distribution of disturbances is *F*-symmetrical. The individual disturbances of course will in general have neither *F*-symmetry nor *P*-symmetry.

It should be noticed that the ideas of *P*-symmetry and *F*-symmetry as defined above apply even to so elaborate an entity as 'the laws of physics'. It should also be understood that the laws are to be the laws taken into account in the theory in question rather than some ideal as yet undiscovered laws.

Now it follows from these assumptions that the statistical distribution of resulting organisms will have *F*-symmetry, or more strictly that the distribution deduced as the result of working out such a theory will have such symmetry. The distribution of observed mature organisms, however, has no such symmetry. In the first place, for instance, men are more often found standing on their feet than their heads. This may be corrected by taking gravity into account in the laws, together with an appropriate change of definition of the two kinds of symmetry. But it will be more convenient if, for the sake of argument, it is imagined that some species has been reared in the absence of gravity, and that the resulting distribution of mature organisms is found to be *P*-symmetrical but to yield more right-handed specimens than left-handed and so not to have *F*-symmetry. It remains therefore to explain this absence of *F*-symmetry.

Evidently one or other of the assumptions (i) to (iii) must be wrong, i.e. in a correct theory one of them would not apply. In the morphogen theory already described these three assumptions do all apply, and it must therefore be regarded as defective to some extent. The theory may be corrected by taking into account the fact that the morphogens do not always have an equal number of left-and right-handed molecules. According to one's point of view one may regard this as invalidating either (i), (ii) or even (iii). Simplest perhaps is to say that the totality of zygotes just is not *F*-symmetrical, and that this could be seen if one looked at the molecules. This is, however, not very satisfactory from the point of view of this paper, as it would not be consistent with describing states in terms of concentrations only. It would be preferable if it was found possible to find more accurate laws concerning reactions and diffusion. For the purpose of accounting for unequal numbers of left- and right-handed organisms it is unnecessary to do more than show that there are corrections which would not be F-symmetrical when there are laevo- or dextrorotatory morphogens, and which would be large enough to account for the effects observed. It is not very difficult to think of such effects. They do not have to be very large, but must, of course, be larger than the purely statistical effects, such as thermal noise or Brownian movement.

There may also be other reasons why the totality of zygotes is not F-symmetrical, e.g. an asymmetry of the chromosomes themselves. If these also produce a sufficiently large effect, so much the better.

Though these effects may be large compared with the statistical disturbances they are almost certainly small compared with the ordinary diffusion and reaction effects. This will mean that they only have an appreciable effect during a short period in which the breakdown of left–right symmetry is occurring. Once their existence is admitted, whether on a theoretical or experimental basis, it is probably most convenient to give them mathematical expression by regarding them as P-symmetrically (but not F-symmetrically) distributed disturbances. However, they will not be considered further in this paper.

6. Reactions and diffusion in a ring of cells

The original reason for considering the breakdown of homogeneity was an apparent difficulty in the diffusion-reaction theory of morphogenesis. Now that the difficulty is resolved it might be supposed that there is no reason for pursuing this aspect of the problem further, and that it would be best to proceed to consider what occurs when the system is very far from homogeneous. A great deal more attention will nevertheless be given to the breakdown of homogeneity. This is largely because the assumption that the system is still nearly homogeneous brings the problem within the range of what is capable of being treated mathematically. Even so many further simplifying assumptions have to be made. Another reason for giving this phase such attention is that it is in a sense the most critical period. That is to say, that if there is any doubt as to how the organism is going to develop it is conceivable that a minute examination of it just after instability has set in might settle the matter, but an examination of it at any earlier time could never do so.

There is a great variety of geometrical arrangement of cells which might be considered, but one particular type of configuration stands out as being particularly simple in its theory, and also illustrates the general principles very well. This configuration is a ring of similar cells. One may suppose that there are N such cells. It must be admitted that there is no biological example to which the theory of the ring can be immediately applied, though it is not difficult to find ones in which the principles illustrated by the ring apply.

It will be assumed at first that there are only two morphogens, or rather only two interesting morphogens. There may be others whose concentration does not vary either in space or time, or which can be eliminated from the discussion for one reason or another. These other morphogens may, for instance, be catalysts involved in the reactions between the interesting morphogens. An example of a complete system of reactions is given in §10. Some consideration will also be given in §§8, 9 to the case of three morphogens. The reader should have no difficulty in extending the results to any number of morphogens, but no essentially new features appear when the number is increased beyond three.

The two morphogens will be called X and Y. These letters will also be used to denote their concentrations. This need not lead to any real confusion. The concentration of X in cell r may be written X_r, and Y_r has a similar meaning. It is convenient to regard 'cell N' and 'cell O' as synonymous, and likewise 'cell 1' and cell '$N+1$'. One can then say that for each r satisfying $1 \leqslant r \leqslant N$ cell r exchanges material by diffusion with cells $r-1$ and $r+1$. The cell-to-cell diffusion constant for X will be called μ, and that for Y will be called ν. This means that for unit concentration difference of X, this morphogen passes at the rate μ from the cell with the higher concentration to the (neighbouring) cell with the lower concentration. It is also necessary to make assumptions about the rates of chemical reaction. The most general assumption that can be made is that for concentrations X and Y chemical reactions are tending to increase X at the rate $f(X, Y)$ and Y at the rate $g(X, Y)$. When the changes in X and Y due to diffusion are also taken into account the behaviour of the system may be described by the $2N$ differential equations

$$\left.\begin{aligned} \frac{dX_r}{dt} &= f(X_r, Y_r) + \mu(X_{r+1} - 2X_r + X_{r-1}) \\ \frac{dY_r}{dt} &= g(X_r, Y_r) + \nu(Y_{r+1} - 2Y_r + Y_{r-1}) \end{aligned}\right\} \quad (r = 1, \ldots, N). \quad (6.1)$$

If $f(h, k)$: $g(h, k) = 0$, then an isolated cell has an equilibrium with concentrations $X = h$, $Y = k$. The ring system also has an equilibrium, stable or unstable, with each X_r equal to h and each Y_r equal to k. Assuming that the system is not very far from this equilibrium it is convenient to put $X_r = h + x_r$, $Y_r = k + y_r$. One may also write $ax + by$ for $f(h + x, y + k)$ and $cx + dy$ for $g(h + x, y + k)$. Since $f(h, k) = g(h, k) = 0$ no constant terms are required, and since x and y are supposed small the terms in higher powers of x and y will have relatively little effect and one is justified in ignoring them. The four quantities a, b, c, d may be called the 'marginal reaction rates'. Collectively they may be described as the 'marginal reaction rate matrix'. When there are M morphogens this matrix consists of M^2 numbers. A marginal reaction rate has the dimensions of the reciprocal of a time, like a radioactive decay rate, which is in fact an example of a marginal (nuclear) reaction rate.

With these assumptions the equations can be rewritten as

$$\begin{aligned}\frac{dx_r}{dt} &= ax_r + by_r + \mu(x_{r+1} - 2x_r + x_{r-1}), \\ \frac{dy_r}{dt} &= cx_r + dy_r + v(y_{r+1} - 2y_r + y_{r-1}).\end{aligned} \quad (6.2)$$

To solve the equations one introduces new co-ordinates $\xi_0, \ldots, \xi_{N-1}$ and $\eta_0, \ldots, \eta_{N-1}$ by putting

$$\begin{aligned}x_r &= \sum_{s=0}^{N-1} \exp\left[\frac{2\pi i r s}{N}\right] \xi_s, \\ y_r &= \sum_{s=0}^{N-1} \exp\left[\frac{2\pi i r s}{N}\right] \eta_s.\end{aligned} \quad (6.3)$$

These relations can also be written as

$$\begin{aligned}\xi_r &= \frac{1}{N} \sum_{s=1}^{N} \exp\left[-\frac{2\pi i r s}{N}\right] x_s, \\ \eta_r &= \frac{1}{N} \sum_{s=1}^{N} \exp\left[-\frac{2\pi i r s}{N}\right] y_s.\end{aligned} \quad (6.4)$$

as may be shown by using the equations

$$\sum_{s=1}^{N} \exp\left[\frac{2\pi i r s}{N}\right] = 0 \text{ if } 0 < r < N,$$
$$= N \text{ if } r = 0 \text{ or } r = N, \quad (6.5)$$

(referred to in §2). Making this substitution one obtains

$$\begin{aligned}\frac{d\xi_s}{dt} &= \frac{1}{N} \sum_{s=1}^{N} \exp\left[-\frac{2\pi i r s}{N}\right]\left[ax_r + by_r + \mu\left(\exp\left[-\frac{2\pi i s}{N}\right] - 2 + \exp\left[\frac{2\pi i s}{N}\right]\right)\xi_s\right] \\ &= a\xi_s + b\eta_s + \mu\left(\exp\left[-\frac{2\pi i s}{N}\right] - 2 + \exp\left[\frac{2\pi i s}{N}\right]\right)\xi_s \\ &= \left(a - 4\mu \sin^2 \frac{\pi s}{N}\right)\xi_s + b\eta_s.\end{aligned} \quad (6.6)$$

Likewise
$$\frac{d\eta_s}{dt} = c\xi_s + \left(d - 4v \sin^2 \frac{\pi s}{N}\right)\eta_s. \quad (6.7)$$

The equations have now been converted into a quite manageable form, with the variables separated. There are now two equations concerned with ξ_1 and η_1, two concerned with ξ_2 and η_2, etc. The equations themselves are also of a well-known standard form, being linear with constant coefficients. Let p_s and p'_s be the roots of the equation

$$\left(p - a + 4\mu \sin^2 \frac{\pi s}{N}\right)\left(p - d + 4\nu \sin^2 \frac{\pi s}{N}\right) = bc \qquad (6.8)$$

(with $\mathcal{R}p_s \geqslant \mathcal{R}p'_s$ for definiteness), then the solution of the equations is of the form

$$\left.\begin{array}{l} \xi_s = A_s e^{p_s t} + B_s e^{p'_s t}, \\ \eta_s = C_s e^{p_s t} + D_s e^{p'_s t}, \end{array}\right\} \qquad (6.9)$$

where, however, the coefficients A_s, B_s, C_s, D_s are not independent but are restricted to satisfy

$$\left.\begin{array}{l} A_s\left(p_s - a + 4\mu \sin^2 \dfrac{\pi s}{N}\right) = bC_s, \\ B_s\left(p'_s - a + 4\mu \sin^2 \dfrac{\pi s}{N}\right) = bD_s. \end{array}\right\} \qquad (6.10)$$

If it should happen that $p_s = p'_s$ the equations (6.9) have to be replaced by

$$\left.\begin{array}{l} \xi_s = (A_s + B_s t)e^{p_s t}, \\ \eta_s = (C_s + D_s t)e^{p_s t}. \end{array}\right\} \qquad (6.9)'$$

and (6.10) remains true. Substituting back into (6.3) and replacing the variables x_r, y_r by X_r, Y_r (the actual concentrations) the solution can be written

$$\left.\begin{array}{l} X_r = h + \displaystyle\sum_{s=1}^{N}(A_s e^{p_s t} + B_s e^{p'_s t}) \exp\left[\dfrac{2\pi i r s}{N}\right], \\ Y_r = k + \displaystyle\sum_{s=1}^{N}(C_s e^{p_s t} + D_s e^{p'_s t}) \exp\left[\dfrac{2\pi i r s}{N}\right]. \end{array}\right\} \qquad (6.11)$$

Here A_s, B_s, C_s, D_s are still related by (6.10), but otherwise are arbitrary complex numbers; p_s and p'_s are the roots of (6.8).

The expression (6.11) gives the general solution of the equations (6.1) when one assumes that departures from homogeneity are sufficiently small that the functions $f(X, Y)$ and $g(X, Y)$ can safely be taken as linear. The form (6.11) given is not very informative. It will be considerably simplified in §8. Another implicit assumption concerns random disturbing influences. Strictly speaking one should consider such influences to be continuously at work. This would make the mathematical treatment considerably more difficult without substantially altering the conclusions. The assumption which is implicit in the analysis, here and in §8, is that the state of the system at $t = 0$ is not one of homogeneity, since it has been displaced from such a state by the disturbances; but after $t = 0$ further disturbances are ignored. In §9 the theory is reconsidered without this latter assumption.

7. Continuous ring of tissue

As an alternative to a ring of separate cells one might prefer to consider a continuous ring of tissue. In this case one can describe the position of a point of the ring by the angle θ which a radius to the point makes with a fixed reference radius. Let the diffusibilities of the two substances be μ' and v'. These are not quite the same as μ and v of the last section, since μ and v are in effect referred to a cell diameter as unit of length, whereas μ' and v' are referred to a conventional unit, the same unit in which the radius ρ of the ring is measured. Then

$$\mu = \mu' \left(\frac{N}{2\pi\rho}\right)^2, \quad v = v' \left(\frac{N}{2\pi\rho}\right)^2.$$

The equations are

$$\left. \begin{aligned} \frac{\partial X}{\partial t} &= a(X-h) + b(Y-k) + \frac{\mu'}{\rho^2}\frac{\partial^2 X}{\partial \theta^2}, \\ \frac{\partial Y}{\partial t} &= c(X-h) + d(Y-k) + \frac{v'}{\rho^2}\frac{\partial^2 Y}{\partial \theta^2}, \end{aligned} \right\} \quad (7.1)$$

which will be seen to be the limiting case of (6.2). The marginal reaction rates a, b, c, d are, as before, the values at the equilibrium position of $\partial f/\partial X$, $\partial f/\partial Y$, $\partial g/\partial X$, $\partial g/\partial Y$. The general solution of the equations is

$$\left. \begin{aligned} X &= h + \sum_{s=-\infty}^{\infty} (A_s e^{p_s t} + B_s e^{p'_s t}) e^{is\theta}, \\ Y &= k + \sum_{s=-\infty}^{\infty} (C_s e^{p_s t} + D_s e^{p'_s t}) e^{is\theta}, \end{aligned} \right\} \quad (7.2)$$

where p_s, p'_s are now roots of

$$\left(p - a + \frac{\mu' s^2}{\rho^2}\right)\left(p - d + \frac{v' s^2}{\rho^2}\right) = bc \quad (7.3)$$

and

$$\left. \begin{aligned} A_s\left(p_s - a + \frac{\mu' s^2}{\rho^2}\right) &= bC_s, \\ B_s\left(p'_s - a + \frac{\mu' s^2}{\rho^2}\right) &= bD_s. \end{aligned} \right\} \quad (7.4)$$

This solution may be justified by considering the limiting case of the solution (6.11). Alternatively, one may observe that the formula proposed is a solution, so that it only remains to prove that it is the most general one. This will follow if values of A_s, B_s, C_s, D_s can be found to fit any given initial conditions. It is well known that any function of an angle (such as X) can be expanded as a 'Fourier series'

$$X(\theta) = \sum_{s=-\infty}^{\infty} G_s e^{is\theta} \quad (X(\theta) \text{ being values of } X \text{ at } t=0),$$

provided, for instance, that its first derivative is continuous. If also

$$Y(\theta) = \sum_{s=-\infty}^{\infty} H_s e^{is\theta} \quad (Y(\theta) \text{ being values of } Y \text{ at } t=0),$$

then the required initial conditions are satisfied provided $A_s + B_s = G_s$ and $C_s + D_s = H_s$. Values A_s, B_s, C_s, D_s to satisfy these conditions can be found unless $p_s = p'_s$. This is an exceptional case and its solution if required may be found as the limit of the normal case.

8. Types of asymptotic behaviour in the ring after a lapse of time

As the reader was reminded in §2, after a lapse of time the behaviour of an expression of the form of (6.11) is eventually dominated by the terms for which the corresponding p_s has the largest real part. There may, however, be several terms for which this real part has the same value, and these terms will together dominate the situation, the other terms being ignored by comparison. There will, in fact, normally be either two or four such 'leading' terms. For if p_{s_0} is one of them then $p_{N-s_0} = p_{s_0}$, since

$$\sin^2 \frac{\pi(N-s_0)}{N} = \sin^2 \frac{\pi s_0}{N},$$

so that p_{s_0} and p_{N-s_0} are roots of the same equation (6.8). If also p_{s_0} is complex then $\mathcal{R} p_{s_0} = \mathcal{R} p'_{s_0}$, and so in all

$$\mathcal{R} p_{s_0} = \mathcal{R} p'_{s_0} = \mathcal{R} p_{N-s_0} = \mathcal{R} p'_{N-s_0}.$$

One need not, however, normally anticipate that any further terms will have to be included. If p_{s_0} and p_{s_1} are to have the same real part, then, unless $s_1 = s_0$ or $s_0 + s_1 = N$ the quantities a, b, c, d, μ, ν will be restricted to satisfy some special condition, which they would be unlikely to satisfy by chance. It is possible to find circumstances in which as many as ten terms have to be included if such special conditions *are* satisfied, but these have no particular physical or biological importance. It is assumed below that none of these chance relations hold.

It has already been seen that it is necessary to distinguish the cases where the value of p_{s_0} for one of the dominant terms is real from those where it is complex. These may be called respectively the *stationary* and the *oscillatory* cases.

Stationary case. After a sufficient lapse of time $X_r - h$ and $Y_r - k$ approach asymptotically to the forms

$$X_r - h = 2\mathcal{R} A_{s_0} \exp\left[\frac{2\pi i s_0 r}{N} + It\right],$$
$$Y_r - k = 2\mathcal{R} C_{s_0} \exp\left[\frac{2\pi i s_0 r}{N} + It\right]. \tag{8.1}$$

Oscillatory case. After a sufficient lapse of time $X_r - h$ and $Y_r - k$ approach the forms

$$X_r - h = 2e^{It}\mathcal{R}\left\{A_{s_0} \exp\left[\frac{2\pi i s_0 r}{N} + i\omega t\right] + A_{N-s_0} \exp\left[-\frac{2\pi i s_0 r}{N} - i\omega t\right]\right\},$$
$$Y_r - k = 2e^{It}\mathcal{R}\left\{C_{s_0} \exp\left[\frac{2\pi i s_0 r}{N} + i\omega t\right] + C_{N-s_0} \exp\left[-\frac{2\pi i s_0 r}{N} - i\omega t\right]\right\}. \tag{8.2}$$

The real part of p_{s_0} has been represented by I, standing for 'instability', and in the oscillatory case its imaginary part is ω. By the use of the $\mathcal{R}$ operation (real part of), two terms have in each case been combined in one.

The meaning of these formulae may be conveniently described in terms of waves. In the stationary case there are stationary waves on the ring having s_0 lobes or crests. The coefficients A_{s_0} and C_{s_0} are in a definite ratio given by (6·10), so that the pattern for one morphogen determines that for the other. With the lapse of time the waves become more pronounced provided there is genuine instability, i.e. if I is positive. The wave-length of the waves may be obtained by dividing the number of lobes into the circumference of the ring. In the oscillatory case the interpretation is similar, but the waves are now not stationary but travelling. As well as having a wave-length they have a velocity and a frequency. The frequency is $\omega/2\pi$, and the velocity is obtained by multiplying the wave-length by the frequency. There are two wave trains moving round the ring in opposite directions.

The wave-lengths of the patterns on the ring do not depend only on the chemical data a, b, c, d, μ', ν' but on the circumference of the ring, since they must be submultiples of the latter. There is a sense, however, in which there is a 'chemical wave-length' which does not depend on the dimensions of the ring. This may be described as the limit to which the wave-lengths tend when the rings are made successively larger. Alternatively (at any rate in the case of continuous tissue), it may be described as the wave-length when the radius is chosen to give the largest possible instability I. One may picture the situation by supposing that the chemical wave-length is true wave-length which is achieved whenever possible, but that on a ring it is necessary to 'make do' with an approximation which divides exactly into the circumference.

Although all the possibilities are covered by the stationary and oscillatory alternatives there are special cases of them which deserve to be treated separately.

One of these occurs when $s_0 = 0$, and may be described as the 'case of extreme long wave-length', though this term may perhaps preferably be reserved to describe the chemical data when they are such that s_0 is zero whatever the dimensions of the ring. There is also the case of 'extreme short wave-length'. This means that $\sin^2(\pi s_0/N)$ is as large as possible, which is achieved by s_0 being either $\frac{1}{2}N$, or $\frac{1}{2}(N-1)$. If the remaining possibilities are regarded as forming the 'case of finite wave-length', there are six subcases altogether. It will be shown that each of these really can occur, although two of them require three or more morphogens for their realization.

(a) *Stationary case with extreme long wave-length.* This occurs for instance if $\mu = \nu = \frac{1}{4}$, $b = c = 1$, $a = d$. Then $p_s = a - \sin^2 \frac{\pi s}{N} + 1$. This is certainly real and is greatest when $s = 0$. In this case the contents of all the cells are the same; there is no resultant flow from cell to cell due to diffusion, so that each is behaving as if it were isolated. Each is in unstable equilibrium, and slips out of it in synchronism with the others.

(b) *Oscillatory case with extreme long wave-length.* This occurs, for instance, if $\mu = \nu = \frac{1}{4}$, $b = -c = 1$, $a = d$. Then $p_s = a - \sin^2 \frac{\pi s}{N} \pm i$. This is complex and its real part is greatest when $s = 0$. As in case (a) each cell behaves as if it were isolated. The difference from case (a) is that the departure from the equilibrium is oscillatory.

(c) *Stationary waves of extreme short wave-length.* This occurs, for instance, if $\nu = 0$, $\mu = 1$, $d = I$, $a = I - 1$, $b = -c = 1$. p_s is

$$I - \tfrac{1}{2} - 2\sin^2\frac{\pi s}{N} + \sqrt{\left\{\left(2\sin^2\frac{\pi s}{N} + \tfrac{1}{2}\right)^2 - 1\right\}},$$

and is greatest when $\sin^2(\pi s/N)$ is greatest. If N is even the contents of each cell are similar to those of the next but one, but distinctly different from those of its immediate neighbours. If, however, the number of cells is odd this arrangement is impossible, and the magnitude of the difference between neighbouring cells varies round the ring, from zero at one point to a maximum at a point diametrically opposite.

(d) *Stationary waves of finite wave-length.* This is the case which is of greatest interest, and has most biological application. It occurs, for instance, if $a = I - 2$, $b = 2 \cdot 5$, $c = -1 \cdot 25$, $d = I + 1 \cdot 5$, $\mu' = 1$, $\nu' = \frac{1}{2}$, and $\frac{\mu}{\mu'} = \frac{\nu}{\nu'} = \left(\frac{N}{2\pi\rho}\right)^2$. As before ρ is the radius of the ring, and N the number of cells in it. If one writes U for $\left(\frac{N}{\pi\rho}\right)^2 \sin^2 \frac{\pi s}{N}$, then equation (6.8) can, with these special values, be written

$$(p - I)^2 + (\tfrac{1}{2} + \tfrac{3}{2}U)(p - I) + \tfrac{1}{2}(U - \tfrac{1}{2})^2 = 0. \tag{8.3}$$

This has a solution $p = I$ if $U = \frac{1}{2}$. On the other hand, it will be shown that if U has any other (positive) value then both roots for $p - I$ have negative real parts. Their product is positive being $\frac{1}{2}(U - \frac{1}{2})^2$, so that if they are real they both have the same sign. Their sum in this case is $-\frac{1}{2}-\frac{3}{2}U$ which is negative. Their common sign is therefore negative. If, however, the roots are complex their real parts are both equal to $-\frac{1}{4}-\frac{3}{4}U$, which is negative.

If the radius ρ of the ring be chosen so that for some integer s_0, $\frac{1}{2}=U=\left(\frac{N}{\pi\rho}\right)^2 \sin^2\frac{\pi s_0}{N}$, there will be stationary waves with s_0 lobes and a wave-length which is also equal to the chemical wave-length, for p_{s_0} will be equal to I, whereas every other p_s will have a real part smaller than I. If, however, the radius is chosen so that $\left(\frac{N}{\pi\rho}\right)^2 \sin^2\frac{\pi s}{N} = \frac{1}{2}$ cannot hold with an integral s, then (in this example) the actual number of lobes will be one of the two integers nearest to the (non-integral) solutions of this equation, and usually *the* nearest. Examples can, however, be constructed where this simple rule does not apply.

Figure 1 shows the relation (8.3) in graphical form. The curved portions of the graphs are hyperbolae.

The two remaining possibilities can only occur with three or more morphogens. With one morphogen the only possibility is (*a*).

(*e*) *Oscillatory case with a finite wave-length.* This means that there are genuine travelling waves. Since the example to be given involves three morphogens it is not possible to use the formulae of § 6. Instead, one must use the corresponding three morphogen formulae. That which corresponds to (6·8) or (7·3) is most conveniently written as

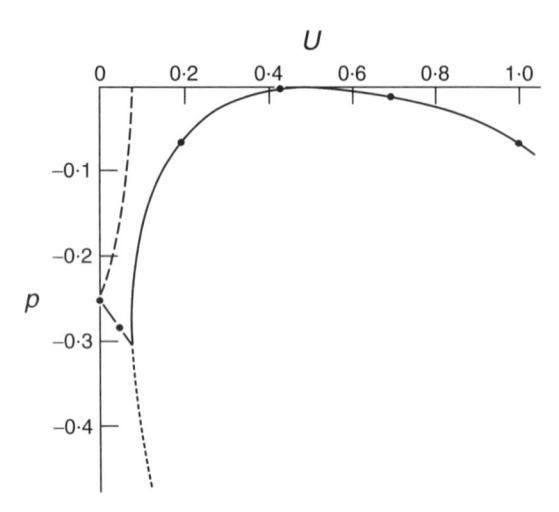

Figure 1. Values of $\mathcal{R}p$ (instability or growth rate), and $|\mathcal{I}p|$ (radian frequency of oscillation), related to wave-length $2\pi U^{-\frac{1}{2}}$ as in the relation (8.3) with $I=0$. This is a case of stationary waves with finite wave-length. Full line, $\mathcal{R}p$; broken line, $-|\mathcal{I}p|$ (zero for $U > 0 \cdot 071$); dotted line, $\mathcal{R}p'$. The full circles on the curve for $\mathcal{R}p$ indicate the values of U, p actually achievable on the finite ring considered in §10, with $s=0$ on the extreme left, $s=5$ on the right.

$$\begin{vmatrix} a_{11} - p - \mu_1 U & a_{12} & a_{13} \\ a_{21} & a_{22} - p - \mu_2 U & a_{23} \\ a_{31} & a_{32} & a_{33} - p - \mu_3 U \end{vmatrix} = 0, \qquad (8.4)$$

where again U has been written for $\left(\frac{N}{\pi\rho}\right)^2 \sin^2 \frac{\pi s}{N}$. (This means essentially that $U = \left(\frac{2\pi}{\lambda}\right)^2$, where λ is the wave-length.) The four marginal reactivities are superseded by nine $a_{11}, \ldots, a_{33}$, and the three diffusibilities are μ_1, μ_2, μ_3. Special values leading to travelling waves are

$$\left. \begin{array}{lll} \mu_1 = \tfrac{2}{3}, & \mu_2 = \tfrac{1}{3}, & \mu_3 = 0, \\ a_{11} = -\tfrac{10}{3}, & a_{12} = 3, & a_{13} = -1, \\ a_{21} = -2, & a_{22} = \tfrac{7}{3}, & a_{23} = 0, \\ a_{31} = 3, & a_{32} = -4, & a_{33} = 0, \end{array} \right\} \qquad (8.5)$$

and with them (8.4) reduces to

$$p^3 + p^2(U+1) + p(1 + \tfrac{2}{9}(U-1)^2) + U + 1 = 0. \qquad (8.6)$$

If $U = 1$ the roots are $\pm i$ and -2. If U is near to I they are approximately $-1 - U$ and $\pm i + \frac{(U-1)^2}{18}(\pm i - 1)$, and all have negative real parts. If the greatest real part is not the value zero, achieved with $U = 1$, then the value zero must be reached again at some intermediate value of U. Since P is then pure imaginary the even terms of (8.6) must vanish, i.e. $(p^2 + 1)(U + 1) = 0$. But this can only happen if $p = \pm i$, and the vanishing of the odd terms then shows that $U = 1$. Hence zero is the largest real part for any root p of (8.6). The corresponding p is $\pm i$ and U is 1. This means that there are travelling waves with unit (chemical) radian frequency and unit (chemical) velocity. If I is added to a_{11}, a_{22} and a_{33}, the instability will become I in place of zero.

(f) *Oscillatory case with extreme short wave-length.* This means that there is metabolic oscillation with neighbouring cells nearly 180° out of phase. It can be achieved with three morphogens and the following chemical data:

$$\left. \begin{array}{lll} \mu = 1, & \mu_2 = \mu_3 = 0, & \\ a_{11} = -1, & a_{12} = -1, & a_{13} = 0, \\ a_{21} = 1, & a_{22} = 0, & a_{23} = -1, \\ a_{31} = 0, & a_{32} = 1, & a_{33} = 0. \end{array} \right\} \qquad (8.7)$$

With these values (8.4) reduces to

$$p^3 + p^2(U+1) + 2p + U + 1 = 0. \qquad (8.8)$$

This may be shown to have all the real parts of its roots negative if $U \geqslant 0$, for if $U = 0$ the roots are near to $-0\cdot 6$, $-0\cdot 2 \pm 1\cdot 3i$, and if U be continuously increased the values of p will alter continuously. If they ever attain values with a positive real part they must pass through pure imaginary values (or zero). But if p is pure imaginary $p^3 + 2p$ and $(p^2 + 1)(U + 1)$ must both vanish, which is impossible if $U \geqslant 0$. As U approaches infinity, however, one of the roots approaches i. Thus $\mathcal{R}p = 0$ can be approached as closely as desired by large values of U, but not attained.

9. Further consideration of the mathematics of the ring

In this section some of the finer points concerning the development of wave patterns are considered. These will be of interest mainly to those who wish to do further research on the subject, and can well be omitted on a first reading.

(1) *General formulae for the two morphogen case.* Taking the limiting case of a ring of large radius (or a filament), one may write $\left(\dfrac{N}{\pi\rho}\right)^2 \sin^2 \dfrac{\pi S}{N} = U = \left(\dfrac{2\pi}{\lambda}\right)^2$ in (6.11) or $\dfrac{s^2}{\rho^2} = U = \left(\dfrac{2\pi}{\lambda}\right)^2$ in (7.3) and obtain

$$(p - a + \mu'U)(p - d + \nu'U) = bc, \tag{9.1}$$

which has the solution

$$p = \frac{a+d}{2} - \frac{\mu' + \nu'}{2}U \pm \sqrt{\left\{\left(\frac{\mu' - \nu'}{2}U + \frac{d-a}{2}\right)^2 + bc\right\}}. \tag{9.2}$$

One may put $I(U)$ for the real part of this, representing the instability for waves of wave-length $\lambda = 2\pi U^{-\frac{1}{2}}$. The dominant waves correspond to the maximum of $I(U)$. This maximum may either be at $U = 0$ or $U = \infty$ or at a stationary point on the part of the curve which is hyperbolic (rather than straight). When this last case occurs the values of p (or I) and U at the maximum are

$$\left.\begin{aligned} p &= I = (d\mu' - a\nu' - 2\sqrt{(\mu'\nu')}\sqrt{(-bc)})(\mu' - \nu')^{-1}, \\ U &= \left(a - d + \frac{\mu' + \nu'}{\sqrt{(\mu'\nu')}}\sqrt{(-bc)}\right)(\mu' - \nu')^{-1}. \end{aligned}\right\} \tag{9.3}$$

The conditions which lead to the four cases (a), (b), (c), (d) described in the last section are

(a) (Stationary waves of extreme long wave-length.) This occurs if either

(i) $bc > 0$, (ii) $bc < 0$ and $\dfrac{d - a}{\sqrt{(-bc)}} > \dfrac{\mu' + \nu'}{\sqrt{(\mu'\nu')}}$, (iii) $bc < 0$ and $\dfrac{d - a}{\sqrt{(-bc)}} < -2$.

The condition for instability in either case is that either $bc > ad$ or $a + d > 0$.

(b) (Oscillating case with extreme long wave-length, i.e. synchronized oscillations.)

This occurs if

$$bc < 0 \text{ and } -2 < \frac{d - a}{\sqrt{(-bc)}} < \frac{4\sqrt{(\mu'\nu')}}{\mu' + \nu'}.$$

The Chemical Basis of Morphogenesis | 541

There is instability if in addition $a + d > 0$.

(c) (Stationary waves of extreme short wave-length.) This occurs if $bc < 0$, $\mu' > v' = 0$.

There is instability if, in addition, $a + d > 0$.

(d) (Stationary waves of finite wave-length.) This occurs if

$$bc < 0 \text{ and } \frac{4\sqrt{(\mu'v')}}{\mu' + v'} < \frac{d-a}{\sqrt{(-bc)}} < \frac{\mu' + v'}{(\sqrt{\mu'v'})}, \quad (9.4a)$$

and there is instability if also

$$\frac{d}{\sqrt{(-bc)}}\sqrt{\frac{\mu'}{v'}} - \frac{a}{\sqrt{(-bc)}}\sqrt{\frac{v'}{\mu'}} > 2. \quad (9.4b)$$

It has been assumed that $v' \leqslant \mu' > 0$. The case where $\mu' \leqslant v' > 0$ can be obtained by interchanging the two morphogens. In the case $\mu' = v' = 0$ there is no co-operation between the cells whatever.

Some additional formulae will be given for the case of stationary waves of finite wave-length. The marginal reaction rates may be expressed parametrically in terms of the diffusibilities, the wave-length, the instability, and two other parameters α and χ. Of these α may be described as the ratio of $X - h$ to $Y - k$ in the waves. The expressions for the marginal reaction rates in terms of these parameters are

$$\left.\begin{array}{l} a = \mu'(v' - \mu')^{-1}(2v'U_0 + \chi) + I, \\ b = \mu'(v' - \mu')^{-1}((\mu' + v')U_0 + \chi)\alpha, \\ c = v'(\mu' - v')^{-1}((\mu' + v')U_0 + \chi)\alpha^{-1}, \\ d = v'(\mu' - v')^{-1}(2\mu'U_0 + \chi) + I, \end{array}\right\} \quad (9.5)$$

and when these are substituted into (9.2) it becomes

$$p = I - \tfrac{1}{2}\chi - \frac{\mu' + v'}{2}U + \sqrt{\left\{\left(\frac{\mu' + v'}{2}U + \tfrac{1}{2}\chi\right)^2 - \mu'v'(U - U_0)^2\right\}}. \quad (9.6)$$

Here $2\pi U_0^{-\frac{1}{2}}$ is the chemical wave-length and $2\pi U^{-\frac{1}{2}}$ the wave-length of the Fourier component under consideration. χ must be positive for case (d) to apply.

If s be regarded as a continuous variable one can consider (9.2) or (9.6) as relating s to p, and dp/ds and d^2p/ds^2 have meaning. The value of d^2p/ds^2 at the maximum is of some interest, and will be used below in this section. Its value is

$$\frac{d^2p}{ds^2} = -\frac{\sqrt{(\mu'v')}}{\rho^2} \cdot \frac{8\sqrt{(\mu'v')}}{\mu' + v'} \cos^2\frac{\pi s}{N}(1 + \chi U_0^{-1}(\mu' + v')^{-1})^{-1}. \quad (9.7)$$

(2) In §§6, 7, 8 it was supposed that the disturbances were not continuously operative, and that the marginal reaction rates did not change with the passage of time. These assumptions will now be dropped, though it will be necessary to make some other, less drastic, approximations to replace them. The (statistical) amplitude of the 'noise' disturbances will be assumed constant in time. Instead of (6.6), (6.7), one then has

$$\left.\begin{aligned}\frac{d\xi}{dt} &= a'\xi + b\eta + R_1(t), \\ \frac{d\eta}{dt} &= c\xi + d'\eta + R_2(t),\end{aligned}\right\} \quad (9.8)$$

where ξ, η have been written for ξ_s, η_s since s may now be supposed fixed. For the same reason $a - 4\mu\sin^2\frac{\pi s}{N}$ has been replaced by a' and $d - 4v\sin^2\frac{\pi s}{N}$ by d'. The noise disturbances may be supposed to constitute white noise, i.e. if (t_1, t_2) and (t_3, t_4) are two non-overlapping intervals then $\int_{t_1}^{t_2} R_1(t)dt$ and $\int_{t_3}^{t_4} R_2(t)dt$ are statistically independent and each is normally distributed with variances $\beta_1(t_2 - t_1)$ and $\beta_1(t_4 - t_3)$ respectively, β_1 being a constant describing the amplitude of the noise. Likewise for $R_2(t)$, the constant β_1 being replaced by β_2. If p and p' are the roots of $(p - a')(p - d') = bc$ and p is the greater (both being real), one can make the substitution

$$\left.\begin{aligned}\xi &= b(u + v), \\ \eta &= (p - a')u + (p' - a')v,\end{aligned}\right\} \quad (9.9)$$

which transforms (9.8) into

$$\frac{du}{dt} = pu + \frac{p' - a'}{(p' - p)b}R_1(t) - \frac{R_2(t)}{p' - p} + \xi\frac{d}{dt}\left(\frac{p' - a'}{(p' - p)b}\right) - \eta\frac{d}{dt}\left(\frac{1}{p' - p}\right), \quad (9.11)$$

with a similar equation for v, of which the leading terms are $dv/dt = p'v$. This indicates that v will be small, or at least small in comparison with u after a lapse of time. If it is assumed that $v = 0$ holds (9.11) may be written

$$\frac{du}{dt} = qu + L_1(t)R_1(t) + L_2(t)R_2(t), \quad (9.12)$$

where

$$L_1(t) = \frac{p' - a'}{(p' - p)b}, \quad L_2(t) = \frac{1}{p' - p}, \quad q = p + bL_1'(t). \quad (9.13)$$

The solution of this equation is

$$u = \int_{-\infty}^{t} (L_1(w)R_1(w) + L_2(w)R_2(w)) \exp\left[\int_w^t q(z)dz\right] dw. \quad (9.14)$$

One is, however, not so much interested in such a solution in terms of the statistical disturbances as in the consequent statistical distribution of values of u, ξ and η at various times after instability has set in. In view of the properties of 'white noise' assumed above, the values of u at time t will be distributed according to the normal error law, with the variance

$$\int_{-\infty}^{t} [\beta_1 (L_1(w))^2 + \beta_2 (L_2(w))^2] \exp\left[2 \int_w^t q(z) dz\right] dw. \qquad (9.15)$$

There are two commonly occurring cases in which one can simplify this expression considerably without great loss of accuracy. If the system is in a distinctly stable state, then $q(t)$, which is near to $p(t)$, will be distinctly negative, and $\exp\left[\int_w^t q(z) dz\right]$ will be small unless w is near to t. But then $L_1(w)$ and $L_2(w)$ may be replaced by $L_1(t)$ and $L_2(t)$ in the integral, and also $q(z)$ may be replaced by $q(t)$. With these approximations the variance is

$$(-2q(t))^{-1}[\beta_1 (L_1(t))^2 + \beta_2 (L_2(t))^2]. \qquad (9.16)$$

A second case where there is a convenient approximation concerns times when the system is unstable, so that $q(t) > 0$. For the approximation concerned to apply $2 \int_w^t q(z) dz$ must have its maximum at the last moment $w \, (= t_0)$ when $q(t_0) = 0$, and it must be the maximum by a considerable margin (e.g. at least 5) over all other local maxima. These conditions would apply for instance if $q(z)$ were always increasing and had negative values at a sufficiently early time. One also requires $q'(t_0)$ (the rate of increase of q at time t_0) to be reasonably large; it must at least be so large that over a period of time of length $(q'(t_0))^{-\frac{1}{2}}$ near to t_0 the changes in $L_1(t)$ and $L_2(t)$ are small, and $q'(t)$ itself must not appreciably alter in this period. Under these circumstances the integrand is negligible when w is considerably different from t_0, in comparison with its values at that time, and therefore one may replace $L_1(w)$ and $L_2(w)$ by $L_1(t_0)$ and $L_2(t_0)$, and $q'(w)$ by $q'(t_0)$. This gives the value

$$\sqrt{\pi}(q'(t_0))^{-\frac{1}{2}}[\beta_1 (L_1(t_0))^2 + \beta_2 (L_2(t_0))^2] \exp\left[2 \int_{t_0}^t q(z) dz\right], \qquad (9.17)$$

for the variance of u.

The physical significance of this latter approximation is that the disturbances near the time when the instability is zero are the only ones which have any appreciable ultimate effect. Those which occur earlier are damped out by the subsequent period of stability. Those which occur later have a shorter period of instability within which to develop to greater amplitude. This principle is familiar in radio, and is fundamental to the theory of the superregenerative receiver.

Naturally one does not often wish to calculate the expression (9.17), but it is valuable as justifying a common-sense point of view of the matter. The factor $\exp\left[\int_{t_0}^{t} q(z)dz\right]$ is essentially the integrated instability and describes the extent to which one would expect disturbances of appropriate wave-length to grow between times t_0 and t. Taking the terms in β_1, β_2 into consideration separately, the factor $\sqrt{\pi\beta_1(q'(t_0))^{-\frac{1}{2}}(L_1(t_0))^2}$ indicates that the disturbances on the first morphogen should be regarded as lasting for a time $\sqrt{\pi(q_1(t_0))^{-\frac{1}{2}}(bL_1(t_0))^2}$. The dimensionless quantities $bL_1(t_0)$, $bL_2(t_0)$ will not usually be sufficiently large or small to justify their detailed calculation.

(3) The extent to which the component for which p_s is greatest may be expected to outdistance the others will now be considered in case (d). The greatest of the p_s will be called p_{s_0}. The two closest competitors to s_0 will be $s_0 - 1$ and $s_0 + 1$; it is required to determine how close the competition is. If the variation in the chemical data is sufficiently small it may be assumed that, although the exponents p_{s_0-1}, p_{s_0}, p_{s_0+1} may themselves vary appreciably in time, the differences $p_{s_0} - p_{s_0-1}$ and $p_{s_0} - p_{s_0+1}$ are constant. It certainly can happen that one of these differences is zero or nearly zero, and there is then 'neck and neck' competition. The weakest competition occurs when $p_{s_0-1} = p_{s_0+1}$. In this case

$$p_{s_0} - p_{s_0-1} = p_{s_0} - p_{s_0+1} = -\tfrac{1}{2}(p_{s_0+1} - 2p_{s_0} + p_{s_0-1}).$$

But if s_0 is reasonably large $p_{s_0+1} - 2p_{s_0} + p_{s_0-1}$ can be set equal to $(d^2p/ds^2)_{s=s_0}$. It may be concluded that the rate at which the most quickly growing component grows cannot exceed the rate for its closest competitor by more than about $\tfrac{1}{2}(d^2p/ds^2)_{s=s_0}$. The formula (9.7), by which d^2p/ds^2 can be estimated, may be regarded as the product of two factors. The dimensionless factor never exceeds 4. The factor $\sqrt{(\mu'\nu')}/\rho^2$ may be described in very rough terms as 'the reciprocal of the time for the morphogens to diffuse a length equal to a radius'. In equally rough terms one may say that a time of this order of magnitude is required for the most quickly growing component to get a lead, amounting to a factor whose logarithm is of the order of unity, over its closest competitors, in the favourable case where $p_{s_0-1} = p_{s_0+1}$.

(4) Very little has yet been said about the effect of considering non-linear reaction rate functions when far from homogeneity. Any treatment so systematic as that given for the linear case seems to be out of the question. It is possible, however, to reach some qualitative conclusions about the effects of non-linear terms. Suppose that z_1 is the amplitude of the Fourier component which is most unstable (on a basis of the linear terms), and which may be supposed to have wave-length λ. The non-linear terms will cause components with wave-lengths $\tfrac{1}{2}\lambda$, $\tfrac{1}{3}\lambda$, $\tfrac{1}{4}\lambda$, ... to appear as well as a space-independent component. If only quadratic terms are taken into account and if these are somewhat small, then the

component of wave-length $\frac{1}{2}\lambda$ and the space-independent component will be the strongest. Suppose these have amplitudes z_2 and z_1. The state of the system is thus being described by the numbers z_0, z_1, z_2. In the absence of non-linear terms they would satisfy equations

$$\frac{dz_0}{dt} = p_0 z_0, \quad \frac{dz_1}{dt} = p_1 z_1, \quad \frac{dz_2}{dt} = p_2 z_2,$$

and if there is slight instability p_1 would be a small positive number, but p_0 and p_2 distinctly negative. The effect of the non-linear terms is to replace these equations by ones of the form

$$\frac{dz_0}{dt} = p_0 z_0 + A z_1^2 + B z_2^2,$$
$$\frac{dz_1}{dt} = p_1 z_1 + C z_2 z_1 + D z_0 z_1,$$
$$\frac{dz_2}{dt} = p_2 z_2 + E z_1^2 + F z_0 z_2.$$

As a first approximation one may put $dz_0/dt = dz_2/dt = 0$ and ignore z_1^4 and higher powers; z_0 and z_1 are then found to be proportional to z_1^2, and the equation for z_1 can be written $dz_1/dt = p_0 z_1 - k z_1^3$. The sign of k in this differential equation is of great importance. If it is positive, then the effect of the term $k z_1^3$ is to arrest the exponential growth of z_1 at the value $\sqrt{(p_1/k)}$. The 'instability' is then very confined in its effect, for the waves can only reach a finite amplitude, and this amplitude tends to zero as the instability (p_1) tends to zero. If, however, k is negative the growth becomes something even faster than exponential, and, if the equation $dz_1/dt = p_1 z_1 - k z_1^3$ held universally, it would result in the amplitude becoming infinite in a finite time. This phenomenon may be called 'catastrophic instability'. In the case of two-dimensional systems catastrophic instability is almost universal, and the corresponding equation takes the form $dz_1/dt = p_1 z_1 + k z_1^2$. Naturally enough in the case of catastrophic instability the amplitude does not really reach infinity, but when it is sufficiently large some effect previously ignored becomes large enough to halt the growth.

(5) Case (*a*) as described in §8 represents a most extremely featureless form of pattern development. This may be remedied quite simply by making less drastic simplifying assumptions, so that a less gross account of the pattern can be given by the theory. It was assumed in § 9 that only the most unstable Fourier components would contribute appreciably to the pattern, though it was seen above (heading (3) of this section) that (in case (*d*)) this will only apply if the period of time involved is adequate to permit the morphogens, supposed for this purpose to be chemically inactive, to diffuse over the whole ring or organ concerned. The same may be shown to apply for case (*a*). If this assumption is dropped a much more interesting form of pattern can be accounted for. To do this it is necessary to consider not merely the components with $U = 0$ but some others with small

positive values of U. One may assume the form $At - BU$ for p. Linearity in U is assumed because only small values of U are concerned, and the term At is included to represent the steady increase in instability. By measuring time from the moment of zero instability the necessity for a constant term is avoided. The formula (9.17) may be applied to estimate the statistical distribution of the amplitudes of the components. Only the factor $\exp\left[2\int_{t_0}^{t} q(z)dz\right]$ will depend very much on U, and taking $q(t) = p(t) = At - BU$, t_0 must be BU/A and the factor is

$$\exp\left[A(t - BU/A)^2\right].$$

The term in U^2 can be ignored if At^2 is fairly large, for then either B^2U^2/A^2 is small or the factor e^{-BUt} is. But At^2 certainly is large if the factor e^{At^2}, applying when $U = 0$, is large. With this approximation the variance takes the form $Ce^{-\frac{1}{2}k^2U}$, with only the two parameters C, k to distinguish the pattern populations. By choosing appropriate units of concentration and length these pattern populations may all be reduced to a standard one, e.g. with $C = k = 1$. Random members of this population may be produced by considering any one of the type (a) systems to which the approximations used above apply. They are also produced, but with only a very small amplitude scale, if a homogeneous one-morphogen system undergoes random disturbances without diffusion for a period, and then diffusion without disturbance. This process is very convenient for computation, and can also be applied to two dimensions. Figure 2 shows such a pattern, obtained in a few hours by a manual computation.

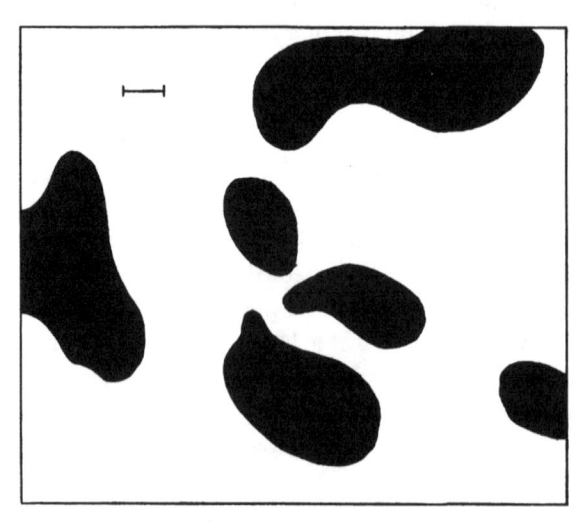

Figure 2. An example of a 'dappled' pattern as resulting from a type (a) morphogen system. A marker of unit length is shown. See text, §9, 11.

To be more definite a set of numbers $u_{r,s}$ was chosen, each being ± 1, and taking the two values with equal probability. A function $f(x, y)$ is related to these numbers by the formula

$$f(x, y) = \sum u_{r,s} \exp[-\tfrac{1}{2}((x - hr)^2 + (y - hs)^2)].$$

In the actual computation a somewhat crude approximation to the function

$$\exp[-\tfrac{1}{2}(x^2 + y^2)]$$

was used and h was about 0·7. In the figure the set of points where $f(x, y)$ is positive is shown black. The outlines of the black patches are somewhat less irregular than they should be due to an inadequacy in the computation procedure.

10. A numerical example

The numerous approximations and assumptions that have been made in the foregoing analysis may be rather confusing to many readers. In the present section it is proposed to consider in detail a single example of the case of most interest, (d). This will be made as specific as possible. It is unfortunately not possible to specify actual chemical reactions with the required properties, but it is thought that the reaction rates associated with the imagined reactions are not unreasonable.

The detail to be specified includes

(i) The number and dimensions of the cells of the ring.
(ii) The diffusibilities of the morphogens.
(iii) The reactions concerned.
(iv) The rates at which the reactions occur.
(v) Information about random disturbances.
(vi) Information about the distribution, in space and time, of those morphogens which are of the nature of evocators.

These will be taken in order.

(i) It will be assumed that there are twenty cells in the ring, and that they have a diameter of 0·1 mm each. These cells are certainly on the large rather than the small side, but by no means impossibly so. The number of cells in the ring has been chosen rather small in order that it should not be necessary to make the approximation of continuous tissue.

(ii) Two morphogens are considered. They will be called X and Y, and the same letters will be used for their concentrations. This will not lead to any real confusion. The diffusion constant for X will be assumed to be 5×10^{-8} cm^{2}s^{-1} and that for Y to be $2·5 \times 10^{-8}$ cm^{2}s^{-1}. With cells of diameter 0·01 cm this means that X flows between neighbouring cells at the rate 5×10^{-4} of the difference of X-content of the two cells per second. In other words, if there is nothing altering the concentrations but diffusion the difference of concentrations suffers an exponential decay with time constant 1000 s, or 'half-period' of 700 s. These times are doubled for Y.

If the cell membrane is regarded as the only obstacle to diffusion the permeability of the membranes to the morphogen is 5×10^{-6} cm/s or 0·018 cm/h. Values as large as 0·1 cm/h have been observed (Davson & Danielli 1943, figure 28).

(iii) The reactions are the most important part of the assumptions. Four substances A, X, Y, B are involved; these are isomeric, i.e. the molecules of the four substances are all rearrangements of the same atoms. Substances C, C', W will also be concerned. The thermodynamics of the problem will not be discussed except to say that it is contemplated that of the substances A, X, Y, B the one with the greatest free energy is A, and that with the least is B. Energy for the whole process is obtained by the degradation of A into B. The substance C is in effect a catalyst for the reaction $Y \to X$, and may also be regarded as an evocator, the system being unstable if there is a sufficient concentration of C.

The reactions postulated are

$$Y + X \to W,$$
$$W + A \to 2Y + B \quad \text{instantly},$$
$$2X \to W,$$
$$A \to X,$$
$$Y \to B,$$
$$Y + C \to C' \quad \text{instantly},$$
$$C' \to X + C.$$

(iv) For the purpose of stating the reaction rates special units will be introduced (for the purpose of this section only). They will be based on a period of 1000 s as units of time, and 10^{-11} mole/cm^3 as concentration unit.[1] There will be little occasion to use any but these special units (s.u.). The concentration of A will be assumed to have the large value of 1000 s.u. and the catalyst C, together with its combined form C' the concentration $10^{-3}(1+\gamma)$ s.u., the dimensionless quantity γ being often supposed somewhat small, though values over as large a range as from $-0\cdot5$ to $0\cdot5$ may be considered. The rates assumed will be

$Y + X \to W$	at the rate $\frac{25}{16} YX$,
$2X \to W$	at the rate $\frac{7}{64} X^2$,
$A \to X$	at the rate $\frac{1}{16} \times 10^{-3} A$,
$C' \to X + C$	at the rate $\frac{55}{32} \times 10^{+3} C'$,
$Y \to B$	at the rate $\frac{1}{16} Y$.

---

[1] A somewhat larger value of concentration unit (e.g. 10^{-9} mole/cm^3) is probably more suitable. The choice of unit only affects the calculations through the amplitude of the random disturbances.

With the values assumed for A and C' the net effect of these reactions is to convert X into Y at the rate $\frac{1}{32}[50XY + 7X^2 - 55(1+\gamma)]$ at the same time producing X at the constant rate $\frac{1}{16}$, and destroying Y at the rate $Y/16$. If, however, the concentration of Y is zero and the rate of increase of Y required by these formulae is negative, the rate of conversion of Y into X is reduced sufficiently to permit Y to remain zero.

In the special units $\mu = \frac{1}{2}$, $\nu = \frac{1}{4}$.

(v) Statistical theory describes in detail what irregularities arise from the molecular nature of matter. In a period in which, on the average, one should expect a reaction to occur between n pairs (or other combinations) of molecules, the actual number will differ from the mean by an amount whose mean square is also n, and is distributed according to the normal error law. Applying this to a reaction proceeding at a rate F (s.u.) and taking the volume of the cell as 10^{-8} cm^3 (assuming some elongation tangentially to the ring) it will be found that the root mean square irregularity of the quantity reacting in a period τ of time (s.u.) is $0 \cdot 004\sqrt{(F\tau)}$.

The diffusion of a morphogen from a cell to a neighbour may be treated as if the passage of a molecule from one cell to another were a monomolecular reaction; a molecule must be imagined to change its form slightly as it passes the cell wall. If the diffusion constant for a wall is μ, and quantities M_1, M_2 of the relevant morphogen lie on the two sides of it, the root-mean-square irregularity in the amount passing the wall in a period τ is

$$0 \cdot 004\sqrt{\{(M_1 + M_2)\mu\tau\}}.$$

These two sources of irregularity are the most significant of those which arise from truly statistical cause, and are the only ones which are taken into account in the calculations whose results are given below. There may also be disturbances due to the presence of neighbouring anatomical structures, and other similar causes. These are of great importance, but of too great variety and complexity to be suitable for consideration here.

(vi) The only morphogen which is being treated as an evocator is C. Changes in the concentration of A might have similar effects, but the change would have to be rather great. It is preferable to assume that A is a 'fuel substance' (e.g. glucose) whose concentration does not change. The concentration of C, together with its combined form C', will be supposed the same in all cells, but it changes with the passage of time. Two different varieties of the problem will be considered, with slightly different assumptions.

The results are shown in table 1. There are eight columns, each of which gives the concentration of a morphogen in each of the twenty cells; the circumstances to which these concentrations refer differ from column to column. The first five columns all refer to the same 'variety' of the imaginary organism, but there are two specimens shown. The specimens differ merely in the chance factors which

Table 1. Some stationary-wave patterns

	first specimen				second specimen: incipient	'slow cooking': incipient	four-lobed equilibrium	
	incipient pattern		final pattern					
cell number	X	Y	X	Y	Y	Y	X	Y
0	1·130	0·929	0·741	1·463	0·834	1·057	1·747	0·000
1	1·123	0·940	0·761	1·469	0·833	0·903	1·685	0·000
2	1·154	0·885	0·954	1·255	0·766	0·813	1·445	2·500
3	1·215	0·810	1·711	0·000	0·836	0·882	0·445	2·500
4	1·249	0·753	1·707	0·000	0·930	1·088	1·685	0·000
5	1·158	0·873	0·875	1·385	0·898	1·222	1·747	0·000
6	1·074	1·003	0·700	1·622	0·770	1·173	1·685	0·000
7	1·078	1·000	0·699	1·615	0·740	0·956	0·445	2·500
8	1·148	0·896	0·885	1·382	0·846	0·775	0·445	2·500
9	1·231	0·775	1·704	0·000	0·937	0·775	1·685	0·000
10	1·204	0·820	1·708	0·000	0·986	0·969	1·747	0·000
11	1·149	0·907	0·944	1·273	1·019	1·170	1·685	0·000
12	1·156	0·886	0·766	1·451	0·899	1·203	0·445	2·500
13	1·170	0·854	0·744	1·442	0·431	1·048	0·445	2·500
14	1·131	0·904	0·756	1·478	0·485	0·868	1·685	0·000
15	1·090	0·976	0·935	1·308	0·919	0·813	1·747	0·000
16	1·109	0·957	1·711	0·000	1·035	0·910	1·685	0·000
17	1·201	0·820	1·706	0·000	1·003	1·050	0·445	2·500
18	1·306	0·675	0·927	1·309	0·899	1·175	0·445	2·500
19	1·217	0·811	0·746	1·487	0·820	1·181	1·685	0·000

were involved. With this variety the value of γ was allowed to increase at the rate of 2^{-7} s.u. from the value $-\frac{1}{4}$ to $+\frac{1}{16}$. At this point a pattern had definitely begun to appear, and was recorded. The parameter γ was then allowed to decrease at the same rate to zero and then remained there until there was no more appreciable change. The pattern was then recorded again. The concentrations of Y in these two recordings are shown in figure 3 as well as in table 1. For the second specimen only one column of figures is given, viz. those for the Y morphogen in the incipient pattern. At this stage the X values are closely related to the Y values, as may be seen from the first specimen (or from theory). The final values can be made almost indistinguishable from those for the first specimen by renumbering the cells and have therefore not been given. These two specimens may be said to belong to the 'variety with quick cooking', because the instability is allowed to increase so quickly that the pattern appears relatively soon. The effect of this haste might be regarded as rather unsatisfactory, as the incipient pattern is very irregular. In both specimens the four-lobed component is present in considerable

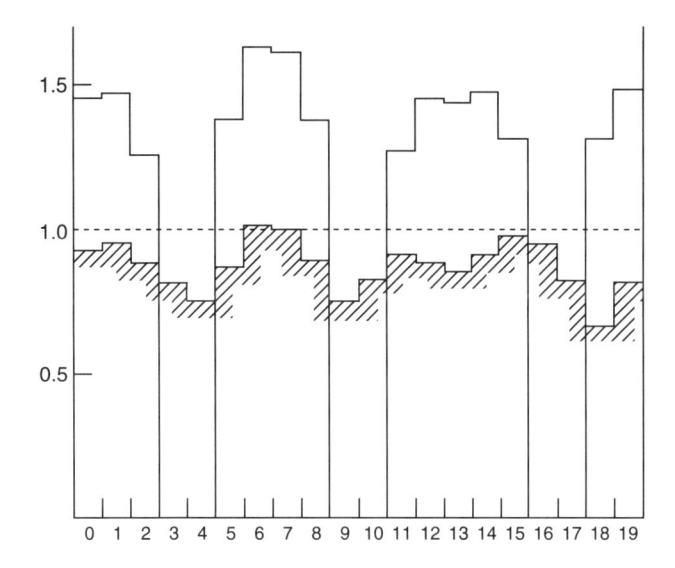

Figure 3. Concentrations of Y in the development of the first specimen (taken from table 1). ------- original homogeneous equilibrium; ▨ incipient pattern; ——— final equilibrium.

strength in the incipient pattern. It 'beats' with the three-lobed component producing considerable irregularity. The relative magnitudes of the three- and four-lobed components depend on chance and vary from specimen to specimen. The four-lobed component may often be the stronger, and may occasionally be so strong that the final pattern is four-lobed. How often this happens is not known, but the pattern, when it occurs, is shown in the last two columns of the table. In this case the disturbances were supposed removed for some time before recording, so as to give a perfectly regular pattern.

The remaining column refers to a second variety, one with 'slow cooking'. In this the value of γ was allowed to increase only at the rate 10^{-5}. Its initial value was -0.010, but is of no significance. The final value was 0.003. With this pattern, when shown graphically, the irregularities are definitely perceptible, but are altogether overshadowed by the three-lobed component. The possibility of the ultimate pattern being four-lobed is not to be taken seriously with this variety.

The set of reactions chosen is such that the instability becomes 'catastrophic' when the second-order terms are taken into account, i.e. the growth of the waves tends to make the whole system more unstable than ever. This effect is finally halted when (in some cells) the concentration of Y has become zero. The constant conversion of Y into X through the agency of the catalyst C can then no longer continue in these cells, and the continued growth of the amplitude of the waves is arrested. When $\gamma = 0$ there is of course an equilibrium with $X = Y = 1$ in all cells, which is very slightly stable. There are, however, also other stable equilibria with $\gamma = 0$, two of which are shown in the table. These final equilibria may, with some trouble but little difficulty, be verified to be solutions of the equations (6.1) with

$$\frac{dX}{dt} = \frac{dY}{dt} = 0,$$

and

$$32f(X, Y) = 57 - 50XY - 7Y^2, \quad 32g(X, Y) = 50XY + 7Y^2 - 2Y - 55.$$

The morphogen concentrations recorded at the earlier times connect more directly with the theory given in §§6 to 9. The amplitude of the waves was then still sufficiently small for the approximation of linearity to be still appropriate, and consequently the 'catastrophic' growth had not yet set in.

The functions $f(X, Y)$ and $g(X, Y)$ of §6 depend also on γ and are

$$f(X, Y) = \tfrac{1}{32}[-7X^2 - 50XY + 57 + 55\gamma],$$
$$g(X, Y) = \tfrac{1}{32}[7X^2 + 50XY - 2Y - 55 - 55\gamma].$$

In applying the theory it will be as well to consider principally the behaviour of the system when γ remains permanently zero. Then for equilibrium $f(X, Y) = g(X, Y) = 0$ which means that $X = Y = 1$, i.e. $h = k = 1$. One also finds the following values for various quantities mentioned in §§ 6 to 9:

$$a = -2, \ b = -1 \cdot 5625, \ c = 2, \ d = 1 \cdot 500, \ s = 3 \cdot 333,$$
$$I = 0, \ \alpha = 0 \cdot 625, \ \chi = 0 \cdot 500, \ (d-a)(-bc)^{-\tfrac{1}{2}} = 1 \cdot 980,$$
$$(\mu + v)(\mu v)^{-\tfrac{1}{2}} = 2 \cdot 121, \ p_0 = -0 \cdot 25 \pm 0 \cdot 25 i,$$
$$p_2 = -0 \cdot 0648, \ p_3 = -0 \cdot 0034, \ p_4 = -0 \cdot 0118.$$

(The relation between p and U for these chemical data, and the values p_n, can be seen in figure 1, the values being so related as to make the curves apply to this example as well as that in §8.) The value $s = 3 \cdot 333$ leads one to expect a three-lobed pattern as the commonest, and this is confirmed by the values p_n. The four-lobed pattern is evidently the closest competitor. The closeness of the competition may be judged from the difference $p_3 - p_4 = 0 \cdot 0084$, which suggests that the three-lobed component takes about 120 s.u. or about 33 h to gain an advantage of a neper (i.e. about 2·7:1) over the four-lobed one. However, the fact that γ is different from 0 and is changing invalidates this calculation to some extent.

The figures in table 1 were mainly obtained with the aid of the Manchester University Computer.

Although the above example is quite adequate to illustrate the mathematical principles involved it may be thought that the chemical reaction system is somewhat artificial. The following example is perhaps less so. The same 'special units' are used. The reactions assumed are

$A \to X$	at the rate	$10^{-3}A$, $A = 10^3$,
$X + Y \to C$	at the rate	$10^3 XY$,
$C \to X + Y$	at the rate	$10^6 C$,
$C \to D$	at the rate	$62 \cdot 5 C$,
$B + C \to W$	at the rate	$0 \cdot 125 BC$, $B = 10^3$,
$W \to Y + C$	instantly,	
$Y \to E$	at the rate	$0 \cdot 0625 Y$,
$Y + V \to V'$	instantly,	
$V' \to E + V$	at the rate	$62 \cdot 5 V'$, $V' = 10^{-3} \beta$.

The effect of the reactions $X + Y \rightleftarrows C$ is that $C = 10^{-3} XY$. The reaction $C \to D$ destroys C, and therefore in effect both X and Y, at the rate $\frac{1}{16} XY$. The reaction $A \to X$ forms X at the constant rate 1, and the pair $Y + V \to V' \to E + V$ destroys Y at the constant rate $\frac{1}{16} \beta$. The pair $B + C \to W \to Y + C$ forms Y at the rate $\frac{1}{8} XY$, and $Y \to E$ destroys it at the rate $\frac{1}{16} Y$. The total effect therefore is that X is produced at the rate $f(X, Y) = \frac{1}{16}(16 - XY)$, and Y at the rate $g(X, Y) = \frac{1}{16}(XY - Y - \beta)$. However, $g(X, Y) = 0$ if $Y \leq 0$. The diffusion constants will be supposed to be $\mu = \frac{1}{4}$, $\nu = \frac{1}{16}$. The homogeneity condition gives $hk = 16$, $k = 16 - \beta$.

It will be seen from conditions (9.4a) that case (d) applies if and only if $\frac{4}{k} + \frac{k}{4} < 2 \cdot 75$, i.e. if k lies between 1·725 and 9·257. Condition (9.4b) shows that there will be instability if in addition $\frac{8}{k} + \frac{k}{8} > \sqrt{3} + \frac{1}{2}$, i.e. if k does not lie between 4·98 and 12·8. It will also be found that the wave-length corresponding to $k = 4 \cdot 98$ is 4·86 cell diameters.

In the case of a ring of six cells with $\beta = 12$ there is a stable equilibrium, as shown in table 2.

Table 2.

cell	0	1	2	3	4	5
X	7·5	3·5	2·5	2·5	3·5	7·5
Y	0	8	8	8	8	0

It should be recognized that these equilibria are only dynamic equilibria. The molecules which together make up the chemical waves are continually changing, though their concentrations in any particular cell are only undergoing small statistical fluctuations. Moreover, in order to maintain the wave pattern a continual supply of free energy is required. It is clear that this must be so since there is a continual degradation of energy through diffusion. This energy is supplied through the 'fuel substances' (A, B in the last example), which are degraded into 'waste products' (D, E).

11. Restatement and biological interpretation of the results

Certain readers may have preferred to omit the detailed mathematical treatment of §§ 6 to 10. For their benefit the assumptions and results will be briefly summarized, with some change of emphasis.

The system considered was either a ring of cells each in contact with its neighbours, or a continuous ring of tissue. The effects are extremely similar in the two cases. For the purposes of this summary it is not necessary to distinguish between them. A system with two or three morphogens only was considered, but the results apply quite generally. The system was supposed to be initially in a stable homogeneous condition, but disturbed slightly from this state by some influences unspecified, such as Brownian movement or the effects of neighbouring structures or slight irregularities of form. It was supposed also that slow changes are taking place in the reaction rates (or, possibly, the diffusibilities) of the two or three morphogens under consideration. These might, for instance, be due to changes of concentration of other morphogens acting in the role of catalyst or of fuel supply, or to a concurrent growth of the cells, or a change of temperature. Such changes are supposed ultimately to bring the system out of the stable state. The phenomena when the system is just unstable were the particular subject of the inquiry. In order to make the problem mathematically tractable it was necessary to assume that the system never deviated very far from the original homogeneous condition. This assumption was called the 'linearity assumption' because it permitted the replacement of the general reaction rate functions by linear ones. This linearity assumption is a serious one. Its justification lies in the fact that the patterns produced in the early stages when it is valid may be expected to have strong qualitative similarity to those prevailing in the later stages when it is not. Other, less important, assumptions were also made at the beginning of the mathematical theory, but the detailed effects of these were mostly considered in § 9, and were qualitatively unimportant.

The conclusions reached were as follows. After the lapse of a certain period of time from the beginning of instability, a pattern of morphogen concentrations appears which can best be described in terms of 'waves'. There are six types of possibility which may arise.

(*a*) The equilibrium concentrations and reaction rates may become such that there would be instability for an isolated cell with the same content as any one of the cells of the ring. If that cell drifts away from the equilibrium position, like an upright stick falling over, then, in the ring, each cell may be expected to do likewise. In neighbouring cells the drift may be expected to be in the same direction, but in distant cells, e.g. at opposite ends of a diameter there is no reason to expect this to be so.

This is the least interesting of the cases. It is possible, however, that it might account for 'dappled' colour patterns, and an example of a pattern in two dimensions produced by this type of process is shown in figure 2 for comparison with 'dappling'. If dappled patterns are to be explained in this way they must be laid down in a latent form when the foetus is only a few inches long. Later the distances would be greater than the morphogens could travel by diffusion.

(b) This case is similar to (a), except that the departure from equilibrium is not a unidirectional drift, but is oscillatory. As in case (a) there may not be agreement between the contents of cells at great distances.

There are probably many biological examples of this metabolic oscillation, but no really satisfactory one is known to the author.

(c) There may be a drift from equilibrium, which is in opposite directions in contiguous cells.

No biological examples of this are known.

(d) There is a stationary wave pattern on the ring, with no time variation, apart from a slow increase in amplitude, i.e. the pattern is slowly becoming more marked. In the case of a ring of continuous tissue the pattern is sinusoidal, i.e. the concentration of one of the morphogens plotted against position on the ring is a sine curve. The peaks of the waves will be uniformly spaced round the ring. The number of such peaks can be obtained approximately by dividing the so-called 'chemical wave-length' of the system into the circumference of the ring. The chemical wave-length is given for the case of two morphogens by the formula (9.3). This formula for the number of peaks of course does not give a whole number, but the actual number of peaks will always be one of the two whole numbers nearest to it, and will usually be *the* nearest. The degree of instability is also shown in (9.3).

The mathematical conditions under which this case applies are given in equations (9.4a), (9.4b).

Biological examples of this case are discussed at some length below.

(e) For a two-morphogen system only the alternatives (a) to (d) are possible, but with three or more morphogens it is possible to have travelling waves. With a ring there would be two sets of waves, one travelling clockwise and the other anticlockwise. There is a natural chemical wave-length and wave frequency in this case as well as a wave-length; no attempt was made to develop formulae for these.

In looking for biological examples of this there is no need to consider only rings. The waves could arise in a tissue of any anatomical form. It is important to know what wave-lengths, velocities and frequencies would be consistent with the theory. These quantities are determined by the rates at which the reactions occur (more accurately by the 'marginal reaction rates', which have the dimensions of the reciprocal of a time), and the diffusibilities of the morphogens. The possible range of values of the reaction rates is so immensely wide that they do not even give an indication of orders of magnitude. The diffusibilities are more helpful. If one were to assume that all the *dimensionless* parameters in a system of travelling waves were the same as in the example given in §8, one could say that the product of the velocity and wave-length of the waves was 3π times the diffusibility of the most diffusible morphogen. But this assumption is certainly false, and it is by no means obvious what is the true range of possible values for the numerical constant (here 3π). The movements of the tail of a spermatozoon suggest themselves as an example of these travelling waves. That the waves are within one cell is no real difficulty. However, the speed of propagation seems to be somewhat greater than can be accounted for except with a rather large numerical constant.

(*f*) Metabolic oscillation with neighbouring cells in opposite phases. No biological examples of this are known to the author.

It is difficult also to find cases to which case (*d*) applies directly, but this is simply because isolated rings of tissue are very rare. On the other hand, systems that have the same kind of symmetry as a ring are extremely common, and it is to be expected that under appropriate chemical conditions, stationary waves may develop on these bodies, and that their circular symmetry will be replaced by a polygonal symmetry. Thus, for instance, a plant shoot may at one time have circular symmetry, i.e. appear essentially the same when rotated through any angle about a certain axis; this shoot may later develop a whorl of leaves, and then it will only suffer rotation through the angle which separates the leaves, or any multiple of it. This same example demonstrates the complexity of the situation when more than one dimension is involved. The leaves on the shoots may not appear in whorls, but be imbricated. This possibility is also capable of mathematical analysis, and will be considered in detail in a later paper. The cases which appear to the writer to come closest biologically to the 'isolated ring of cells' are the tentacles of (e.g.) *Hydra*, and the whorls of leaves of certain plants such as Woodruff (*Asperula odorata*).

Hydra is something like a sea-anemone but lives in fresh water and has from about five to ten tentacles. A part of a *Hydra* cut off from the rest will rearrange itself so as to form a complete new organism. At one stage of this proceeding the organism has reached the form of a tube open at the head end and closed at the other end. The external diameter is somewhat greater at the head end than over the rest of the tube. The whole still has circular symmetry. At a somewhat later stage the symmetry has gone to the extent that an appropriate stain will bring out a number of patches on the widened head end. These patches arise at the points where the tentacles are subsequently to appear (Child 1941, p. 101 and figure 30). According to morphogen theory it is natural to suppose that reactions, similar to those which were considered in connection with the ring of tissue, take place in the widened head end, leading to a similar breakdown of symmetry. The situation is more complicated than the case of the thin isolated ring, for the portion of the *Hydra* concerned is neither isolated nor very thin. It is not unreasonable to suppose that this head region is the only one in which the chemical conditions are such as to give instability. But substances produced in this region are still free to diffuse through the surrounding region of lesser activity. There is no great difficulty in extending the mathematics to cover this point in particular cases. But if the active region is too wide the system no longer approximates the behaviour of a thin ring and one can no longer expect the tentacles to form a single whorl. This also cannot be considered in detail in the present paper.

In the case of woodruff the leaves appear in whorls on the stem, the number of leaves in a whorl varying considerably, sometimes being as few as five or as many as nine. The numbers in consecutive whorls on the same stem are often equal, but by no means invariably. It is to be presumed that the whorls originate in rings of active tissue in the meristematic area, and that the rings arise at sufficiently great distance to have little influence on one another. The number of leaves in the whorl will presumably be obtainable by the rule given above, viz. by dividing the chemical wave-length into the circumference, though both these quantities will have to be given some new interpretation more appropriate to woodruff than to the ring. Another important example of a structure with polygonal symmetry is provided by young root fibres just breaking out from the parent root. Initially these are almost homogeneous in cross-section, but eventually a ring of fairly evenly spaced spots appear, and these later develop into vascular strands. In this case again the full explanation must be in terms of a two-dimensional or even a three-dimensional problem, although the analysis for the ring is still illuminating. When the cross-section is very large the strands may be in more than one ring, or more or less randomly or hexagonally arranged. The two-dimensional theory (not expounded here) also goes a long way to explain this.

Flowers might appear superficially to provide the most obvious examples of polygonal symmetry, and it is probable that there are many species for which this 'waves round a ring' theory is essentially correct. But it is certain that it does not apply for all species. If it did it would follow that, taking flowers as a whole, i.e. mixing up all species, there would be no very markedly preferred petal (or corolla, segment, stamen, etc.) numbers. For when all species are taken into account one must expect that the diameters of the rings concerned will take on nearly all values within a considerable range, and that neighbouring diameters will be almost equally common. There may also be some variation in chemical wave-length. Neighbouring values of the ratio circumferences to wave-length should therefore be more or less equally frequent, and this must mean that neighbouring petal numbers will have much the same frequency. But this is not borne out by the facts. The number five is extremely common, and the number seven rather rare. Such facts are, in the author's opinion, capable of explanation on the basis of morphogen theory, and are closely connected with the theory of phyllotaxis. They cannot be considered in detail here.

The case of a filament of tissue calls for some comment. The equilibrium patterns on such a filament will be the same as on a ring, which has been cut at a point where the concentrations of the morphogens are a maximum or a minimum. This could account for the segmentation of such filaments. It should be noticed, however, that the theory will not apply unmodified for filaments immersed in water.

12. Chemical waves on spheres. Gastrulation

The treatment of homogeneity breakdown on the surface of a sphere is not much more difficult than in the case of the ring. The theory of spherical harmonics, on which it is based, is not, however, known to many that are not mathematical specialists. Although the essential properties of spherical harmonics that are used are stated below, many readers will prefer to proceed directly to the last paragraph of this section.

The anatomical structure concerned in this problem is a hollow sphere of continuous tissue such as a blastula. It is supposed sufficiently thin that one can treat it as a 'spherical shell'. This latter assumption is merely for the purpose of mathematical simplification; the results are almost exactly similar if it is omitted. As in § 7 there are to be two morphogens, and a, b, c, d, μ', ν', h, k are also to have the same meaning as they did there. The operator ∇^2 will be used here to mean the superficial part of the Laplacian, i.e. $\nabla^2 V$ will be an abbreviation of

$$\frac{1}{\rho^2}\frac{\partial^2 V}{\partial \phi^2} + \frac{1}{\rho^2 \sin^2 \theta}\frac{\partial}{\partial \theta}\left(\sin \theta \frac{\partial V}{\partial \theta}\right),$$

where θ and ϕ are spherical polar co-ordinates on the surface of the sphere and ρ is its radius. The equations corresponding to (7.1) may then be written

$$\left.\begin{aligned}\frac{\partial X}{\partial t} &= a(X - h) + b(Y - k) + \mu'\nabla^2 X,\\ \frac{\partial Y}{\partial t} &= c(X - h) + d(Y - k) + \nu'\nabla^2 Y.\end{aligned}\right\} \quad (12.1)$$

It is well known (e.g. Jeans 1927, chapter 8) that any function on the surface of the sphere, or at least any that is likely to arise in a physical problem, can be 'expanded in spherical surface harmonics'. This means that it can be expressed in the form

$$\sum_{n=0}^{\infty}\left[\sum_{m=-n}^{n} A_n^m P_n^m(\cos \theta)\, e^{im\phi}\right].$$

The expression in the square bracket is described as a 'surface harmonic of degree n'. Its nearest analogue in the ring theory is a Fourier component. The essential property of a spherical harmonic of degree n is that when the operator ∇^2 is applied to it the effect is the same as multiplication by $-n(n+1)/\rho^2$. In view of this fact it is evident that a solution of (12.1) is

$$X = h + \sum_{n=0}^{\infty} \sum_{m=-n}^{n} (A_n^m e^{iq_n t} + B_n^m e^{iq'_n t}) P_n^m(\cos\theta) e^{im\phi},$$

$$Y = k + \sum_{n=0}^{\infty} \sum_{m=-n}^{n} (C_n^m e^{iq_n t} + D_n^m e^{iq'_n t}) P_n^m(\cos\theta) e^{i\phi}, \quad (12.2)$$

where q_n and q'_n are the two roots of

$$\left(q - a + \frac{\mu'}{\rho^2} n(n+1)\right)\left(q - d + \frac{\nu'}{\rho^2} n(n+1)\right) = bc \quad (12.3)$$

and

$$A_n^m\left(q_n - a + \frac{\mu'}{\rho^2} n(n+1)\right) = bC_n^m,$$

$$B_n^m\left(q'_n - a + \frac{\mu'}{\rho^2} n(n+1)\right) = cD_n^m. \quad (12.4)$$

This is the most general solution, since the coefficients A_n^m and B_n^m can be chosen to give any required values of X, Y when $t = 0$, except when (12.3) has two equal roots, in which case a treatment is required which is similar to that applied in similar circumstances in §7. The analogy with §7 throughout will indeed be obvious, though the summation with respect to m does not appear there. The meaning of this summation is that there are a number of different patterns with the same wave-length, which can be superposed with various amplitude factors. Then supposing that, as in §8, one particular wave-length predominates, (12.2) reduces to

$$X - h = e^{iq_{n_0} t} \sum_{m=-n_0}^{n_0} A_{n_0}^m P_{n_0}^m(\cos\theta) e^{im\phi},$$

$$b(Y - k) = \left(q_{n_0} - a + \frac{\mu'}{\rho^2} n(n+1)\right)(X - h). \quad (12.5)$$

In other words, the concentrations of the two morphogens are proportional, and both of them are surface harmonics of the same degree n_0, viz. that which makes the greater of the roots q_{n_0}, q'_{n_0} have the greatest value.

It is probable that the forms of various nearly spherical structures, such as radiolarian skeletons, are closely related to these spherical harmonic patterns. The most important application of the theory seems, however, to be to the gastrulation of a blastula. Suppose that the chemical data, including the chemical wave-length, remain constant as the radius of the blastula increases. To be quite specific suppose that

$$\mu' = 2, \ v' = 1, \ a = -4, \ b = -8, \ c = 4, \ d = 7.$$

With these values the system is quite stable so long as the radius is less than about 2. Near this point, however, the harmonics of degree 1 begin to develop and a pattern of form (12.5) with $n_0 = 1$ makes its appearance. Making use of the facts that

$$P_1^0(\cos\theta) = \cos\theta, \ P_1^1(\cos\theta) = P_1^{-1}(\cos\theta) = \sin\theta,$$

it is seen that $X - h$ is of the form

$$X - h = A\cos\theta + B\sin\theta\cos\phi + C\sin\theta\sin\phi, \qquad (12.6)$$

which may also be interpreted as

$$X - h = A'\cos\theta', \qquad (12.7)$$

where θ' is the angle which the radius θ, ϕ makes with the fixed direction having direction cosines proportional to B, C, A and $A' = \sqrt{(A^2 + B^2 + C^2)}$.

The outcome of the analysis therefore is quite simply this. Under certain not very restrictive conditions (which include a requirement that the sphere be relatively small but increasing in size) the pattern of the breakdown of homogeneity is axially symmetrical, not about the original axis of spherical polar coordinates, but about some new axis determined by the disturbing influences. The concentrations of the first morphogen are given by (12.7), where θ' is measured from this new axis; and $Y - k$ is proportional to $X - h$. Supposing that the first morphogen is, or encourages the production of, a growth hormone, one must expect the blastula to grow in an axially symmetric manner, but at a greater rate at one end of the axis than at the other. This might under many circumstances lead to gastrulation, though the effects of such growth are not very easily determinable. They depend on the elastic properties of the tissue as well as on the growth rate at each point. This growth will certainly lead to a solid of revolution with a marked difference between the two poles, unless, in addition to the chemical instability, there is a mechanical instability causing the breakdown of axial symmetry. The direction of the axis of gastrulation will be quite random according to this theory. It may be that it is found experimentally that the axis is normally in some definite direction such as that of the animal pole. This is not essentially contradictory to the theory, for any small asymmetry of the zygote may be sufficient to provide the 'disturbance' which determines the axis.

13. Non-linear theory. Use of digital computers

The 'wave' theory which has been developed here depends essentially on the assumption that the reaction rates are linear functions of the concentrations, an assumption which is justifiable in the case of a system just beginning to leave a homogeneous condition. Such systems certainly have a special interest as giving the first appearance of a pattern, but they are the exception rather than the rule. Most of an organism, most of the time, is developing from one pattern into another, rather than from homogeneity into a pattern. One would like to be able to follow this more general process mathematically also. The difficulties are, however, such that one cannot hope to have any very embracing *theory* of such processes, beyond the statement of the equations. It might be possible, however, to treat a few particular cases in detail with the aid of a digital computer. This method has the advantage that it is not so necessary to make simplifying assumptions as it is when doing a more theoretical type of analysis. It might even be possible to take the mechanical aspects of the problem into account as well as the chemical, when applying this type of method. The essential disadvantage of the method is that one only gets results for particular cases. But this disadvantage is probably of comparatively little importance. Even with the ring problem, considered in this paper, for which a reasonably complete mathematical analysis was possible, the computational treatment of a particular case was most illuminating. The morphogen theory of phyllotaxis, to be described, as already mentioned, in a later paper, will be covered by this computational method. Non-linear equations will be used.

It must be admitted that the biological examples which it has been possible to give in the present paper are very limited. This can be ascribed quite simply to the fact that biological phenomena are usually very complicated. Taking this in combination with the relatively elementary mathematics used in this paper one could hardly expect to find that many observed biological phenomena would be covered. It is thought, however, that the imaginary biological systems which have been treated, and the principles which have been discussed, should be of some help in interpreting real biological forms.

References

Child, C. M. 1941. *Patterns and problems of development.* University of Chicago Press.
Davson, H. & Danielli, J. F. 1943. *The permeability of natural membranes.* Cambridge University Press.
Jeans, J. H. 1927. *The mathematical theory of elasticity and magnetism*, 5th ed. Cambridge University Press.
Michaelis, L. & Menten, M. L. 1913. Die Kinetik der Invertinwirkung. *Biochemische Zeitschrift* 49: 333–369.
Thompson, Sir D'Arcy 1942. *On growth and form*, 2nd ed. Cambridge University Press.
Waddington, C. H. 1940. *Organisers and genes.* Cambridge University Press.

CHAPTER 16

Chess (1953)
Alan Turing

Introduction
Jack Copeland

Chess and AI

Chess and some other board games are a test-bed for ideas in Artificial Intelligence. Donald Michie—Turing's wartime colleague and subsequently founder of the Department of Machine Intelligence and Perception at the University of Edinburgh—explains the relevance of chess to AI:

Computer chess has been described as the *Drosophila melanogaster* of machine intelligence. Just as Thomas Hunt Morgan and his colleagues were able to exploit the special limitations and conveniences of the *Drosophila* fruit fly to develop a methodology of genetic mapping, so the game of chess holds special interest for the study of the representation of human knowledge in machines. Its chief advantages are: (1) chess constitutes a fully defined and well-formalized domain; (2) the game challenges the highest levels of human intellectual capacity; (3) the challenge extends over the full range of cognitive functions such as logical calculation, rote learning, concept-formation, analogical thinking, imagination, deductive and inductive reasoning; (4) a massive and detailed corpus of chess knowledge has accumulated over the centuries in the form of chess instructional works and commentaries; (5) a generally accepted numerical scale of performance is available in the form of the U.S. Chess Federation and International ELO rating system.[1]

History of Computer Chess

In 1945, in his paper 'Proposed Electronic Calculator', Turing predicted that computers would probably play 'very good chess', an opinion echoed in 1949 by Claude Shannon of Bell Telephone Laboratories, another leading early theoret-

---
[1] D. Michie, *On Machine Intelligence* (2nd edn. Chichester: Ellis Horwood, 1986), 78–9.

ician of computer chess.[2] By 1958, Herbert Simon and Allen Newell were predicting that within ten years the world chess champion would be a computer, unless barred by the rules.[3] Just under forty years later, on 11 May 1997, IBM's Deep Blue beat the reigning world champion, Gary Kasparov, in a six-game match.

Turing was theorizing about the mechanization of chess as early as 1941. Fellow codebreakers at GC & CS remember him experimenting with two heuristics now commonly used in computer chess, *minimax* and *best-first*.[4] The minimax heuristic involves assuming that one's opponent will move in such a way as to maximize their gains; one then makes one's own move in such a way as to minimize the losses caused by the opponent's expected move.[5] The best-first heuristic involves ranking the moves available to a player by means of a rule of thumb scoring system and examining the consequences of the highest-scoring move first.

Turochamp

In 1948 Turing and the mathematical economist David Champernowne designed a chess-playing routine known as 'Turochamp'.

Champernowne later gave this description of Turochamp:

It was in the late summer of 1948 that Turing and I did try out a loose system of rules for deciding on the next move in a chess game which we thought could be fairly easily programmed for a computer. My long-suffering wife, a beginner at chess, took on the system and lost. ... Here is what I think I remember about the system but I may have been influenced by what I have since read about other people's systems. There was a system for estimating the effects of any move on White's estimated net advantage over Black. This allowed for:

(1) Captures, using the conventional scale of 10 for pawn, 30 for knight or bishop, 50 for rook, 100 for queen and something huge, say 5000, for king.
(2) Change in mobility; i.e., change in the number of squares to which any piece or pawn could immediately move legitimately (1 each).
(3) Special incentives for: (a) Castling (3 points). (b) Advancing a passed pawn (1 or 2 points). (c) Getting a rook onto the seventh rank (4 points perhaps).

(I don't think occupation of one of the 4 central squares gained any special bonus. We did not cater to the end-game.) Most of our attention went to deciding which moves were to

---

[2] C. E. Shannon, 'Programming a Computer for Playing Chess', *Philosophical Magazine*, 41 (1950), 256–75.
[3] H. A. Simon and A. Newell, 'Heuristic Problem Solving: The Next Advance in Operations Research', *Operations Research*, 6 (1958), 1–10.
[4] Michie in interview with Copeland (Oct. 1995).
[5] The foundations of the minimax approach were laid in 1928 by von Neumann in his fundamental 'minimax theorem' (J. von Neumann, 'Zur Theorie der Gesellschaftsspiele', *Mathematische Annalen*, 100 (1928), 295–320).

be followed up. My memory about this is infuriatingly weak. Captures had to be followed up at least to the point where no further capture was immediately possible. Checks and forcing moves had to be followed further. We were particularly keen on the idea that whereas certain moves would be scorned as pointless and pursued no further others would be followed quite a long way down certain paths. In the actual experiment I suspect we were a bit slapdash about all this and must have made a number of slips since the arithmetic was extremely tedious with pencil and paper. Our general conclusion was that a computer should be fairly easy to programme to play a game of chess against a beginner and stand a fair chance of winning or least reaching a winning position.[6]

Turing started to code the Turochamp for the Ferranti Mark I computer at Manchester University but he never completed the task.[7]

The First Working Chess Programme

Dietrich Prinz, who worked for the engineering firm Ferranti Ltd., wrote the first chess programme to be fully implemented. It ran in November 1951 on the University of Manchester Ferranti Mark I.[8] Prinz, like Michie, Strachey, and others, was influenced by an important article published in 1950 by Donald Davies, 'A Theory of Chess and Noughts and Crosses' (see Further Reading). Prinz 'learned all about programming the [Mark I] computer at seminars given by Alan Turing and Cecily Popplewell'.[9] (Like Turing, Prinz wrote a programming manual for the Mark I.[10])

Prinz's programme was for solving simple problems of the mate-in-two variety.[11] The programme would examine every possible move until a solution was found. On average, several thousand moves had to be examined in the course of solving a problem, and the programme was considerably slower than a human player. Unlike Prinz's programme, Turochamp could in principle play a complete game and operated not by exhaustive search but under the guidance of heuristics.

Prinz saw in chess programming an 'indication of the methods that might be used to treat structural or logistic problems occurring in other fields by means of

[6] Champernowne's account is from a letter that he wrote to *Computer Chess* (4 (Jan. 1980), 80–1); it is published here by permission of W. and R. P. Champernowne.
[7] D. Michie, 'Game-Playing and Game-Learning Automata', in L. Fox (ed.), *Advances in Programming and Non-numerical Computation* (New York: Pergamon, 1966), 189.
[8] B. V. Bowden (ed.), *Faster Than Thought* (London: Pitman, 1953), 295.
[9] C. Gradwell, 'Early Days', reminiscences in a Newsletter 'For those who worked on the Manchester Mk I computers', Apr. 1994. (I am grateful to Prinz's daughter, Daniela Derbyshire, for sending me a copy of Gradwell's article.)
[10] D. G. Prinz, 'Introduction to Programming on the Manchester Electronic Digital Computer', n.d., Ferranti Ltd. (a digital facsimile is in The Turing Archive for the History of Computing <www.AlanTuring.net/prinz>).
[11] Prinz's programme is described in Bowden (ed.), *Faster Than Thought*, 295–7 and in D. G. Prinz, 'Robot Chess', *Research*, 5 (1952), 261–6.

electronic computers'.[12] Prinz used the Ferranti Mark I to solve logical problems (as also did Audrey Bates), and in 1949, 1950, and 1951 Ferranti built three small experimental special-purpose computers for theorem-proving and other logical work.[13] (This was several years before the Logic Theorist of Newell, Simon, and Shaw—often incorrectly said to be the first AI programme—made its debut at the Dartmouth conference (see 'Artificial Intelligence', above).)

Turing's Approach in 'Chess'

Turing says that the system of rules set out in 'Chess' is based on an 'introspective analysis' of his own thought processes when playing (but with 'considerable simplifications'). His system anticipates much that has become standard in chess programming: the use of heuristics to guide the search through the 'tree' of possible moves and counter-moves; the use of evaluation rules which assign numerical values, indicative of strength or weakness, to board configurations; the minimax strategy; and variable look-ahead whereby, instead of the consequences of every possible move being followed equally far, the 'more profitable moves [are] considered in greater detail than the less' (p. 571). Turing also realized the necessity of using 'an entirely different system for the end-game' (p. 574).

The learning procedure that Turing proposes in 'Chess' involves the machine trying out variations in its method of play—e.g. varying the numerical values that are assigned to the various pieces. The machine adopts any variation that leads to more satisfactory results. This procedure is an early example of a genetic algorithm or GA (see 'Artificial Life', above).

Chess Bulldozers

Turing likened the claim that no chess programme can outplay its programmer to the claim that no animal can swallow an animal heavier than itself. Both claims, he said, 'are, so far as I know, untrue'. He pointed out that a programme might outplay its programmer simply in virtue of 'the speed of the machine, which might make it feasible to carry the analysis...farther than a man could do in the same time' (p. 575).

Critics question the worth of research into computer chess. Noam Chomsky has said famously that a computer beating a grandmaster at chess is as interesting as a bulldozer winning a weight-lifting competition.[14] Deep Blue did indeed bulldoze

---

[12] Prinz, 'Robot Chess', 266.
[13] D. G. Prinz and J. B. Smith, 'Machines for the Solution of Logical Problems', in Bowden (ed.), *Faster Than Thought*, ch. 15; W. Mays and D. G. Prinz, 'A Relay Machine for the Demonstration of Symbolic Logic', *Nature*, 165/4188 (4 Feb. 1950), 197–8; M. A. Bates, 'On the Mechanical Solution of a Problem in Church's Lambda Calculus', M.Sc. thesis, University of Manchester.
[14] N. Chomsky, *Language and Thought* (London: Moyer Bell, 1993), 93.

its way to victory—256 parallel processors enabled it to examine 200 million possible moves per second and to look ahead as many as fourteen turns of play.

The huge improvement in computer chess since Turing's day owes much more to advances in hardware engineering than to advances in AI. Massive increases in cpu speed and memory have meant that successive generations of machines have been able to examine increasingly many possible moves. Turing's expectation was that chess-programming would contribute to the study of how human beings think. In fact, little or nothing about human thought processes appears to have been learned from the series of projects that culminated in Deep Blue.

The Turing Test Again

In the introductory paragraphs of his essay, Turing touches on some philosophical issues 'unconnected with chess'. In a reference to the imitation game (Chapter 11), he again states his belief that one could 'make a machine which would answer questions put to it, in such a way that it would not be possible to distinguish its answers from those of a man' (p. 569). On the question of what justification there is for this, he says that he knows of 'no really convincing argument to support this belief and certainly of none to disprove it.'

Consciousness

In 'Chess' Turing is careful to distinguish the question whether one could make a machine that would play the imitation game successfully from the question 'Could one make a machine which would have feelings like you and I do?' (p. 569). Examples of 'feelings' are pleasure, grief, misery, and anger (Chapter 11, p. 451). The assertion 'No mechanism could feel' forms the basis of the objection to the Turing test that in Chapter 11 Turing dubs the 'Argument from Consciousness'. His remarks in the present chapter clarify the discussion that he gives of this objection in Chapter 11.

The answer that Turing gives in 'Chess' to his question 'Could one make a machine which would have feelings' is this (p. 569): 'I shall never know, any more than I shall ever be quite certain that *you* feel as I do.' Thus his view appears to be that the question 'Can machines think?' is independent of the question whether machines can feel, and that an affirmative answer may be given to the former in the absence of our having any answer at all to the latter. If he is right—and arguably he is[15]—then the assertion 'No mechanism could feel', even if true, goes wide of the mark as an objection to the Turing test. As Turing sensibly says in Chapter 11 (pp. 452–3):

---

[15] See section 3.1 ('Is Consciousness Necessary for Thought?') of B. J. Copeland, *Artificial Intelligence: A Philosophical Introduction* (Oxford: Blackwell, 1993).

I do not wish to give the impression that I think there is no mystery about consciousness. ... But I do not think these mysteries necessarily need to be solved before we can answer the question with which we are concerned in this paper [i.e. 'Can machines think?'].

Consciousness and Computer Simulation

Turing asserts (p. 569) that in the case of the questions 'Could one make a machine to play chess, and to improve its play?', and 'Could one make a machine which would answer questions put to it, in such a way that it would not be possible to distinguish its answers from those of a man?', the phrase 'Could one make a machine to...' might equally well be replaced by 'Could one programme an electronic computer to...'. On pp. 569–70 Turing supports this claim with an argument that also appears, in somewhat different forms, in Chapters 11 and 13:

if some other machine had been constructed to do the job we could use an electronic computer (of sufficient storage capacity), suitably programmed, to calculate what this machine would do, and in particular what answer it would give.

(If no superior method presented itself, then the computer could be supplied with a 'look-up table' setting out the behaviour of the other machine (see Chapter 11).)

In the case, however, of the question 'Could one make a machine which would have feelings like you and I do?', Turing indicates (on p. 569) that replacing 'Could one make a machine to...' by 'Could one programme an electronic computer to...' might possibly result in a question that is not equivalent. Turing says in Chapter 13 that 'any machine of a certain very wide class' can be replaced equivalently by a digital computer, but notes that not *every* machine can be so replaced, giving as examples of those that cannot bulldozers, steam-engines, and telescopes (p. 483). Presumably, therefore, Turing considered it an open question whether a machine capable of feeling belongs to the first or second of these two classes.

The Church–Turing Thesis

'Chess' contains a formulation of the Church–Turing thesis that does not appear elsewhere in Turing's writings (p. 570):

If one can explain quite unambiguously in English, with the aid of mathematical symbols if required, how a calculation is to be done, then it is always possible to programme any digital computer to do that calculation, provided the storage capacity is adequate.

As elsewhere (e.g. in Chapter 10 (p. 414) and Chapter 17 (pp. 588–9)), Turing emphasizes here that this thesis 'is not the sort of thing that admits of clear cut proof', but nevertheless 'amongst workers in the field it is regarded as being clear as day' (p. 570). (The phrase 'how a calculation is to be done' should be understood as being elliptical for some such phrase as 'how a calculation is to be done by an obedient clerk working in accordance with a systematic method'.)

Turing applies this thesis to the question 'Could one programme an electronic computer to play a reasonably good game of chess?' He points out that, in view of the thesis, there is no need to set out an actual programme in order to show that the answer to the question is affirmative. It suffices to explain, 'unambiguously in English', the rules by which the machine is to choose its move in each position. This Turing proceeds to do.

Further reading

Davies, D. W., 'A Theory of Chess and Noughts and Crosses', *Science News*, 16 (1950), 40–64.

Newborn, M., *Kasparov Versus Deep Blue: Computer Chess Comes of Age*, (Springer: New York, 1997).

Shannon, C. E. 'Programming a Computer for Playing Chess', *Philosophical Magazine*, 41 (1950), 256–75.

—— 'A Chess-Playing Machine', *Scientific American*, 182 (1950), 48–51.

Provenance

Turing's essay 'Chess' was published in the 1953 collection *Faster Than Thought*, where it formed a section of a chapter entitled 'Digital Computers Applied To Games'. (Edited by Vivian Bowden, *Faster Than Thought* is a fascinating survey of the then state of the art in digital computing.) The chapter was co-authored by Audrey Bates, Bowden, and Christopher Strachey. Bowden's editorial remarks do not make it clear which parts of 'Digital Computers Applied To Games' were written by which author.[16] Fortunately, Turing's typescript of his contribution has survived.[17] It is this which is printed here.[18]

---

[16] The whole article is mistakenly attributed to Turing alone in *Mechanical Intelligence: Collected Works of A. M. Turing*, ed. D. C. Ince (Amsterdam: North-Holland, 1992).

[17] Turing's typescript is itself entitled 'Digital Computers applied to Games'. In *Faster Than Thought* Turing's essay was published under the narrower and more accurate title 'Chess'. The typescript is among the Turing Papers in the Modern Archive Centre, King's College, Cambridge (catalogue reference B 7).

[18] Bowden's edition of the essay differs from Turing's typescript in numerous respects, many of them minor. The present edition follows Turing's typescript. Some obvious typing errors have been corrected and in one case a conjectured missing word has been added in square brackets. Some significant differences between the typescript and Bowden's edition are mentioned in footnotes. (Bowden's edition is reprinted in Turing, *Mechanical Intelligence*, ed. Ince, 288–95.)

Chess

When one is asked 'Could one make a machine to play chess?', there are several possible meanings which might be given to the words. Here are a few:

  i) Could one make a machine which would obey the rules of chess, i.e. one which would play random legal moves, or which could tell one whether a given move is a legal one?
  ii) Could one make a machine which would solve chess problems, e.g. tell one whether, in a given position, white has a forced mate in three?
  iii) Could one make a machine which would play a reasonably good game of chess, i.e. which, confronted with an ordinary (that is, not particularly unusual) chess position, would after two or three minutes of calculation, indicate a passably good legal move?
  iv) Could one make a machine to play chess, and to improve its play, game by game, profiting from its experience?

To these we may add two further questions, unconnected with chess, which are likely to be on the tip of the reader's tongue.

  v) Could one make a machine which would answer questions put to it, in such a way that it would not be possible to distinguish its answers from those of a man?
  vi) Could one make a machine which would have feelings like you and I do?

The problem to be considered here is iii), but to put this problem into perspective with the others I shall give the very briefest of answers to each of them.

To i) and ii) I should say 'This certainly can be done. If it has not been done already it is merely because there is something better to do.'

Question iii) we are to consider in greater detail, but the short answer is 'Yes, but the better the standard of play required, the more complex will the machine be, and the more ingenious perhaps the designer.'

To iv) and v) I should answer 'I believe so. I know of no really convincing argument to support this belief and certainly of none to disprove it.'

To vi) I should say 'I shall never know, any more than I shall ever be quite certain that *you* feel as I do.'

In each of these problems except possibly the last, the phrase 'Could one make a machine to...' might equally well be replaced by 'Could one programme an electronic computer to...'. Clearly the electronic computer so programmed would itself constitute a machine. And on the other hand if some other machine

Printed with the permission of Financial Times Management and the Estate of Alan Turing.

had been constructed to do the job we could use an electronic computer (of sufficient storage capacity), suitably programmed, to calculate what this machine would do, and in particular what answer it would give.

After these preliminaries let us give our minds to the problem of making a machine, or of programming a computer, to play a tolerable game of chess. In this short discussion it is of course out of the question to provide actual programmes, but this does not really matter on account of the following principle:

> If one can explain quite unambiguously in English, with the aid of mathematical symbols if required, how a calculation is to be done, then it is always possible to programme any digital computer to do that calculation, provided the storage capacity is adequate.

This is not the sort of thing that admits of clear cut proof, but amongst workers in the field it is regarded as being clear as day. Accepting this principle, our problem is reduced to explaining 'unambiguously in English' the rules by which the machine is to choose its move in each position. For definiteness we will suppose the machine is playing white.

If the machine could calculate at an infinite speed, and also had unlimited storage capacity, a comparatively simple rule would suffice, and would give a result that in a sense could not be improved on. This rule could be stated:

> Consider every possible continuation of the game from the given position. There is only a finite number of them (at any rate if the fifty-move rule makes a draw obligatory, not merely permissive). Work back from the end of these continuations, marking a position with white to play as 'win' if there is a move which turns it into a position previously marked as 'win'. If this does not occur, but there is a move which leads to a position marked 'draw', then mark the position 'draw'. Failing this, mark it 'lose'. Mark a position with black to play by a similar rule with 'win' and 'lose' interchanged. If after this process has been completed it is found that there are moves which lead to a position marked 'win', one of these should be chosen. If there is none marked 'win' choose one marked 'draw' if such exists. If all moves lead to a position marked 'lose', any move may be chosen.

Such a rule is practically applicable in the game of noughts and crosses, but in chess is of merely academic interest.

Even when the rule can be applied it is not very appropriate for use against a weak opponent, who may make mistakes which ought to be exploited.

In spite of the impracticability of this rule it bears some resemblance to what one really does when playing chess. One does not follow all the continuations of play, but one follows some of them. One does not follow them until the end of the game, but one follows them a move or two, perhaps more. Eventually a position seems, rightly or wrongly, too bad to be worth further consideration, or (less frequently) too good to hesitate longer over. The further a position is from the one on the board the less likely is it to occur, and therefore the shorter is the time which can be assigned for its consideration. Following this idea we might have a rule something like this:

> Consider all continuations of the game consisting of a move by white, a reply by black, and another move and reply. The value of the position at the end of each of these sequences of moves is estimated according to some suitable rule. The values at earlier positions are then calculated by working backwards move by move as in the theoretical rule given before. The move to be chosen is that which leads to the position with the greatest value.

It is possible to arrange that no two positions have the same value. The rule is then unambiguous. A very simple form of values, but one not having this property, is an 'evaluation of material', e.g. on the basis

$$P = 1$$
$$Kt = 3$$
$$B = 3\tfrac{1}{2}$$
$$R = 5$$
$$Q = 10$$
$$\text{Checkmate} \;\pm\; 1000^{[1]}$$

If B is black's total and W is white's, then W/B is quite a good measure of value. This is better than $W - B$ as the latter does not encourage exchanges when one has the advantage. Some small extra arbitrary function of position may be added to ensure definiteness in the result.

The weakness of this rule is that it follows all combinations equally far. It would be much better if the more profitable moves were considered in greater detail than the less. It would also be desirable to take into account more than mere 'value of material'.

After this introduction I shall describe a particular set of rules, which could without difficulty be made into a machine programme. It is understood that the machine is white and white is next to play. The current position is called the 'position of the board', and the positions arising from it by later moves 'positions in the analysis'.[2]

---

[1] Editor's note. In the Bowden edition. Turing's ' $\pm$ ' has been replaced with '='.
[2] Editor's note. In the Bowden edition, Turing's 'position of the board' has been replaced by 'position on the board'.

'Considerable' Moves, i.e. Moves to be considered in the analysis by the machine

Every possibility for white's next move and for black's reply is 'considerable'. If a capture is considerable then any recapture is considerable. The capture of an undefended piece or the capture of a piece of higher value by one of lower value is always considerable. A move giving checkmate is considerable.

Dead position

A position in the analysis is dead if there are no considerable moves in that position, i.e. if it is more than two moves ahead of the present position, and no capture or recapture or mate can be made in the next move.

Value of position

The value of a dead position is obtained by adding up the piece values as above, and forming the ratio W/B of white's total to black's. In other positions with white to play the value is the greatest value of: (*a*) the positions obtained by considerable moves, or (*b*) the position itself evaluated as if a dead position, the latter alternative to be omitted if all moves are considerable. The same process is to be undertaken for one of black's moves, but the machine will then choose the *least* value.

Position-play value

Each white piece has a certain position-play contribution and so has the black king. These must all be added up to give the position-play value.

For a Q, R, B or Kt, count

 i) The square root of the number of moves the piece can make from the position, counting a capture as two moves, and not forgetting that the king must not be left in check.
 ii) (If not a Q) 1·0 if it is defended, and an additional 0·5 if twice defended.

For a K, count

 iii) For moves other than castling as i) above.
 iv) It is then necessary to make some allowance for the vulnerability of the K. This can be done by assuming it to be replaced by a friendly Q on the same square, estimating as in i), but subtracting instead of adding.
 v) Count 1·0 for the possibility of castling later not being lost by moves of K or rooks, a further 1·0 if castling could take place on the next move, and yet another 1·0 for the actual performance of castling.

For a P, count

 vi) 0·2 for each rank advanced.
 vii) 0·3 for being defended by at least one piece (not P).

For the black K, count

viii) 1·0 for the threat of checkmate.

ix) 0·5 for check.

We can now state the rule for play as follows.

The move chosen must have the greatest possible value, and, consistent with this, the greatest possible position-play value. If this condition admits of several solutions a choice may be made at random, or according to an arbitrary additional condition.

Note that no 'analysis' is involved in position-play evaluation. This is in order to reduce the amount of work done on deciding the move.

The game below was played between this machine and a weak player who did not know the system. To simplify the calculations the square roots were rounded off to one decimal place (i.e. the table below was used). No 'random choices' actually arose in this game. The increase of position-play value is given after white's move if relevant. An asterisk indicates that every other move had a lower position-play value. 0—0 indicates castling.[3]

Number	0	1	2	3	4	5	6	7	8	9	10	11	12	13
Square Root	0	1	1·4	1·7	2·0	2·2	2·4	2·6	2·8	3·0	3·2	3·3	3·5	3·6

White (Machine)		Black
1. P—K$_4$	4·2*	P—K$_4$
2. Kt—QB$_3$	3·1*	Kt—KB$_3$
3. P—Q$_4$	3·1*	B—QKt$_5$
4. Kt—KB$_3$ (1)	2·0	P—Q$_3$
5. B—Q$_2$	3·6*	Kt—QB$_3$
6. P—Q$_5$	0·2	Kt—Q$_5$
7. P—KR$_4$ (2)	1·1*	B—Kt$_5$
8. P—QR$_4$ (2)	1·0*	Kt × Kt ch.
9. P × Kt		B—KR$_4$
10. B—Kt$_5$ ch.	2·4*	P—QB$_3$
11. P × P		0—0
12. P × P		R—Kt$_1$
13. B—R$_6$	−1·5	Q—R$_4$
14. Q—K$_2$	0·6	Kt—Q$_2$
15. KR—Kt$_1$ (3)	1·2*	Kt—B$_4$ (4)
16. R—Kt$_5$ (5)		B—Kt$_3$
17. B—Kt$_5$	0·4	Kt × KtP
18. 0—0	3·0*	Kt—B$_4$
19. B—B$_6$		KR—QB$_1$

Continued

---

[3] Editor's note. In the Bowden edition a different game is to shown, diverging from the present game at the 21st move.

White (Machine)		Black
20. B—Q$_5$		B × Kt
21. P × B$^{(2)}$	−0·7	Q × P
22. B—K$_3$$^{(6)}$		Q—R$_6$ ch.
23. K—Q$_2$		Kt—R$_5$
24. B × RP$^{(7)}$		R—Kt$_7$
25. P—B$_4$		Q—B$_6$ ch.
26. K—B$_1$		R—R$_7$
27. B × BP ch.		B × B
28. R × KtP ch.$^{(5)}$		K × R
29. B—K$_3$$^{(8)}$		R—R$_8$ mate

Notes:
1. If B—Q$_2$ 3·6* then P × P is foreseen.
2. Most inappropriate moves from a positional point of view.
3. If O—O then B × Kt, B × B, Q × P.
4. The fork is unforeseen.
5. Heads in the sand!
6. Only this or B—K$_1$ can prevent Q—R$_8$ mate.
7. Fiddling while Rome burns.
8. Mate is foreseen, but 'business as usual'.

Numerous criticisms of the machine's play may be made. It is quite defenceless against 'forks', although it may be able to see certain other kinds of combination. It is of course not difficult to devise improvements of the programme [so] that these simple forks are foreseen. The reader may be able to think of some such improvements for himself. Since no claim is made that the above rule is particularly good, I have been content to leave this flaw without remedy; clearly a line has to be drawn between the flaws which one will attempt to eliminate and those which must be accepted as a risk. Another criticism is that the scheme proposed, although reasonable in the middle game, is futile in the end game. The change-over from the middle game to the end-game is usually sufficiently clean cut for it to be possible to have an entirely different system for the end-game. This should of course include quite definite programmes for the standard situations, such as mate with rook and king, or king and pawn against king. There is no intention to discuss the end-game further here.

If I were to sum up the weakness of the above system in a few words I would describe it as a caricature of my own play. It was in fact based on an introspective analysis of my thought processes when playing, with considerable simplifications. It makes oversights which are very similar to those which I make myself, and which may in both cases be ascribed to the considerable moves being rather inappropriately chosen. This fact might be regarded as supporting the rather glib view which is often expressed, to the effect that 'one cannot programme a machine to play a better game than one plays oneself'. This statement should I think be compared with another of rather similar form. 'No animal can swallow an animal heavier than himself.' Both statements are, so far as I know, untrue. They are also both of a kind that one is rather easily bluffed into accepting, partly because one thinks that there ought to be some rather slick way of demonstrating them, and one does not like to admit that one does not see what this argument is. They are also both supported by normal experience, and need rather exceptional cases to falsify them. The statement about chess programming may be falsified quite simply by the speed of the machine, which might make it feasible to carry the analysis a move farther than a man could do in the same time. This effect is rather less than might be supposed. Although electronic computers are very fast where conventional computing is concerned, their advantage is much reduced where enumeration of cases, etc., is involved on a large scale. Take for instance the problem of counting the possible moves from a given position in chess. If the number is 30 a man might do it in 45 seconds and the machine in 1 second. The machine has still an advantage, but it is much less overwhelming than it would be for instance where calculating cosines.

In connection with the question, numbered iv) above, as to the ability of a chess-machine to profit from experience, one can see that it would be quite possible to programme the machine to try out variations in its method of play (e.g. variations in piece value) and adopt the one giving the most satisfactory results. This could certainly be described as 'learning', though it is not quite representative of learning as we know it. It might also be possible to programme the machine to search for new types of combination in chess. If this product produced results which were quite new, and also interesting to the programmer, who should have the credit? Compare this with the situation where the Defence Minister has given orders for research to be done to find a counter to the bow and arrow. Should the inventor of the shield have the credit, or should the Defence Minister?

CHAPTER 17

Solvable and Unsolvable Problems (1954)

Alan Turing

Introduction

Jack Copeland

Unsolvable Problems

In Chapter 1 Turing proves the existence of mathematical problems that cannot be solved by the universal Turing machine. There he also advances the thesis, now called the Church–Turing thesis, that any systematic method for solving mathematical problems can be carried out by the universal Turing machine. Combining these two propositions yields the result that there are mathematical problems which cannot be solved by any systematic method—cannot, in other words, be solved by any algorithm.

Substitution Puzzles

In 'Solvable and Unsolvable Problems' Turing sets out to explain this result to a lay audience. The article first appeared in *Science News*, a popular science journal of the time. Starting from concrete examples of problems that do admit of algorithmic solution, Turing works his way towards an example of a problem that is not solvable by any systematic method. Loosely put, this is the problem of sorting puzzles into those that will 'come out' and those that will not. Turing gives an elegant argument showing that a sharpened form of this problem is not solvable by means of a systematic method (pp. 591–2).

The sharpened form of the problem involves what Turing calls 'the substitution type of puzzle'. An typical example of a substitution puzzle is this. Starting with the word BOB, is it possible to produce BOOOB by replacing selected occurrences of the pair OB by BOOB and selected occurences of the triple BOB by O? The answer is yes:

$$\text{BOB} \rightarrow \text{BBOOB} \rightarrow \text{BBOBOOB} \rightarrow \text{BOOOB}.$$

Turing suggests that any puzzle can be re-expressed as a substitution puzzle. Some row of letters can always be used to represent the 'starting position' envisaged in a particular puzzle, e.g. in the case of a chess problem, the pieces on the board and their positions. Desired outcomes, for example board positions that count as wins, can be described by further rows of letters, and the rules of the puzzle, whatever they are, are to be represented in terms of permissible substitutions of groups of letters for other groups of letters.

As Turing points out, it is not only 'toy' puzzles that can be re-expressed as substitution puzzles, but also mathematical problems, for instance the problem of finding a proof of a given mathematical theorem within an axiom system (which Turing describes as 'a very good example of a puzzle'). The axioms—which are simply strings of mathematical symbols—form the starting position. The theorem—another string of symbols—is the winning position. The rules of the puzzle are substitutions that enable strings of mathematical symbols to be transformed into other strings, much as in the case of the transition from BOB to BBOOB in the earlier example.

Turing calls the substitution formulation of any puzzle its 'normal form' and states the following *normal form principle* (p. 588):

Given any puzzle, we can find a corresponding substitution puzzle which is equivalent to it in the sense that given a solution of the one we can easily use it to find a solution of the other.

Normal Forms and the Church–Turing Thesis

The normal form principle for puzzles closely parallels the Church–Turing thesis, which says that given any systematic method, we can find a corresponding Turing machine that is equivalent to it.

Neither the normal form principle for puzzles nor the Church–Turing thesis is susceptible to definite proof (see 'Computable Numbers: A Guide'). While few doubt that the Church–Turing thesis is in fact true, the very nature of the thesis has always been a matter for debate. Church, for example, described the thesis as a *definition*.[1] Post, on the other hand, described it as a 'working hypothesis' that is in need of 'continual verification', and he criticized Church for masking this hypothesis as a definition.[2] Turing's remarks in 'Solvable and Unsolvable Problems' about the status of the normal form principle for puzzles are of outstanding interest for the light that they may cast on his view concerning

---

[1] A. Church, 'An Unsolvable Problem of Elementary Number Theory', *American Journal of Mathematics*, 58 (1936), 345–63 (356).
[2] E. L. Post, 'Finite Combinatory Processes - Formulation 1', *Journal of Symbolic Logic*, 1 (1936), 103–5 (105).

the status of the Church–Turing thesis. In this connection, see also the material from Turing's draft typescript quoted in n. 9 on p. 590.

Turing says of the normal form principle (pp. 588–9):

> The statement is...one which one does not attempt to prove. Propaganda is more appropriate to it than proof, for its status is something between a theorem and a definition. In so far as we know *a priori* what is a puzzle and what is not, the statement is a theorem. In so far as we do not know what puzzles are, the statement is a definition which tells us something about what they are. One can of course define a puzzle by some phrase beginning, for instance, 'A set of definite rules...', but this just throws us back on the definition of 'definite rules'. Equally one can reduce it to the definition of 'computable function' or 'systematic procedure'. A definition of any one of these would define all the rest.

Turing would perhaps have said much the same concerning not only the Church–Turing thesis but also the thesis introduced in Chapter 13:

> A digital computer will replace any rival design of calculating machine.

In so far as we do not know what calculating machines are, the statement is a definition which tells us something about what they are.

Proof of Unsolvability

Having introduced the normal form principle for puzzles, Turing turns to his central project of establishing that 'there cannot be any systematic procedure for determining whether a puzzle be solvable or not' (p. 590). In particular, there cannot be a systematic procedure for determining whether substitution puzzles are or are not solvable. Turing argues by *reductio ad absurdum*. He shows that the supposition that there is a systematic procedure for determining whether substitution puzzles are or are not solvable leads to an outright contradiction, and on that basis concludes that there can be no such procedure. The argument turns on the impossibility of applying a certain procedure to itself.

Any systematic procedure is in effect a puzzle, since in following the procedure one applies rules to some 'starting position' until one or another result is achieved. So if there were a systematic procedure for determining whether each puzzle is or is not solvable, then by the normal form principle, there is a substitution puzzle—call it K—that is equivalent to this procedure. When applied to any substitution puzzle, K—if it exists—must 'come out' either with the result SOLVABLE or with the result NOT SOLVABLE. Since K is applicable to any substitution puzzle, K can be applied to itself in order to determine whether it itself is or is not solvable. Turing shows (p. 592) that this supposed ability of K to pronounce on its own solvability leads to outright contradiction, and so concludes that K cannot exist.

The Meaning of 'Unsolvable'

Turing points out that the result he has established, namely that there is no systematic method for deciding whether or not substitution puzzles come out, is often expressed by saying that there is no *decision procedure* for puzzles of this type, and that the *decision problem* for this type of puzzle is *unsolvable*. He continues (p. 592): 'so one comes to speak (as in the title of this article) about "unsolvable problems" meaning in effect puzzles for which there is no decision procedure. This is the technical meaning which the words are now given by mathematical logicians.'

As Turing says, this terminology is potentially confusing. It is natural to use the words 'unsolvable problem' to mean a problem for which no solution can possibly be found. It would be a confusion to think that Turing has shown that the problem of deciding whether or not substitution puzzles come out is an unsolvable problem in this natural sense. Indeed, with sufficient time, inventiveness, and patience, mathematicians may always be able to establish whether or not any given substitution puzzle comes out. If that is so, then the problem of deciding whether or not substitution puzzles come out is solvable, in the natural sense of the word.

What Turing has shown is that there is no systematic method for deciding whether or not substitution puzzles come out, i.e. there is no general procedure, applicable by rote, that one can employ in order to decide whether or not each substitution puzzle comes out. The 'decision problem' for substitution puzzles is the problem of finding such a rote procedure (a 'decision procedure'); in showing that there is no such procedure, Turing has shown that the decision problem for substitution puzzles is unsolvable in the natural sense.

Turing therefore recommends that, in order to 'minimize confusion', one should 'always speak of "unsolvable decision problems", rather than just "unsolvable problems"' (p. 592).

Significance of Turing's Result

Turing ends the chapter with a comment on the significance of what he has shown. His result concerning the decision problem for substitution puzzles 'may be regarded as going some way towards a demonstration, within mathematics itself, of the inadequacy of "reason" unsupported by common sense'. For he has, he says, set 'certain bounds to what we can hope to achieve purely by reasoning'.

The phrase 'purely by reasoning' here presumably means 'purely by algorithmic methods'. Some mathematical problems require for their solution not only 'reason', in this sense, but also what Turing refers to in Chapter 3 as 'intuition' (see also Chapter 4). There he says (pp. 192–3):

The activity of the intuition consists in making spontaneous judgements which are not the result of conscious trains of reasoning. ... Often it is possible to find some other way of verifying the correctness of an intuitive judgement. We may, for instance, judge that all positive integers are uniquely factorizable into primes; a detailed mathematical argument leads to the same result. This argument will also involve intuitive judgements, but they will be less open to criticism than the original judgement about factorization. ... The necessity for using the intuition is ... greatly reduced by setting down formal rules for carrying out inferences which are always intuitively valid. ... In pre-Gödel times it was thought by some that it would probably be possible to carry this programme to such a point that all the intuitive judgements of mathematics could be replaced by a finite number of these rules. The necessity for intuition would then be entirely eliminated.

The argument of 'Solvable and Unsolvable Problems' illustrates why it is that the need for intuition cannot always be eliminated in favour of formal rules.

Gödel's Theorem

Turing notes that the unsolvability of the decision problem for substitution puzzles affords an elegant proof of the following rather general statement (p. 593):

no systematic method of proving mathematical theorems is sufficiently complete to settle every mathematical question, yes or no.

The proof Turing gives is as follows. Each statement of the form 'such-and-such substitution puzzle comes out' can be expressed in the form of a mathematical statement. So if there were a systematic method of settling every question that can be posed in mathematical form, this method would serve as a decision procedure for substitution puzzles. Given that there is no such decision procedure, it follows that no systematic method is able to settle every mathematical question.

Turing remarks that the above statement follows 'by a famous theorem of Gödel' and describes himself as providing 'an independent proof' of the statement (p. 593). Turing might also have pointed out that his own 'On Computable Numbers' yields a proof of this statement.

Gödel's famous incompleteness theorem of 1931 is, however, importantly less general than the above statement, since it concerns only one particular systematic method of proving mathematical theorems, the system set out by Whitehead and Russell in *Principia Mathematica*[3] (as explained in 'Computable Numbers: A Guide'). Gödel did later generalize his result of 1931 to all formal systems (containing a certain amount of arithmetic), but emphasized the importance that Turing's work played in this generalization. Gödel said in 1964:

[3] A. N. Whitehead and B. Russell, *Principia Mathematica*, vols. i–iii (Cambridge: Cambridge University Press, 1910–13).

[D]ue to A. M. Turing's work, a precise and unquestionably adequate definition of the general concept of formal system can now be given... Turing's work gives an analysis of the concept of 'mechanical procedure' (alias 'algorithm' or 'computation procedure' or 'finite combinatorial procedure'). ... A formal system can simply be defined to be any mechanical procedure for producing formulas, called provable formulas.[4]

In his references to Gödel's work, Turing hides his own light under a bushel.

Further reading

Boone, W. W., review of Turing's 'The Word Problem in Semi-Groups with Cancellation', *Journal of Symbolic Logic*, 17 (1952), 74–6.

Turing, A. M., 'The Word Problem in Semi-Groups with Cancellation', *Annals of Mathematics*, 52 (1950), 491–505. Reprinted in *Pure Mathematics: Collected Works of A. M. Turing*, ed. J. L. Britton (Amsterdam: North-Holland, 1992).

Provenance

What follows is the text of the original printing of 'Solvable and Unsolvable Problems' in *Science News*.[5] Unfortunately Turing's own typescript appears to have been lost. However, a sizeable fragment of a draft typescript, with additions in Turing's handwriting, has been preserved.[6] (Turing recycled the draft pages, covering the reverse sides with handwritten notes concerning morphogenesis.) The fragment corresponds to pp. 584–9. For the most part the published version follows the draft pages closely (except for punctuation and occasional changes of word and word-order). Significant differences between the draft and the published version are mentioned in footnotes.

---

[4] K. Gödel, 'Postscriptum', in M. Davis (ed.), *The Undecidable* (New York: Raven, 1965), 71–3 (71–2); the Postscriptum, dated 1964, is to Gödel's 1934 paper 'On Undecidable Propositions of Formal Mathematical Systems' (ibid. 41–71).

[5] Footnotes have been renumbered consecutively. Footnotes not marked 'Editor's note' appeared in *Science News*. A page reference to *Science News* been replaced by the number (in square brackets) of the corresponding page of this volume.

[6] The fragment is among the Turing Papers in the Modern Archive Centre, King's College, Cambridge; at the time of writing it is uncatalogued.

Solvable and Unsolvable Problems

If one is given a puzzle to solve one will usually, if it proves to be difficult, ask the owner whether it can be done. Such a question should have a quite definite answer, yes or no, at any rate provided the rules describing what you are allowed to do are perfectly clear. Of course the owner of the puzzle may not know the answer. One might equally ask, 'How can one tell whether a puzzle is solvable?', but this cannot be answered so straightforwardly. The fact of the matter is that there is *no* systematic method of testing puzzles to see whether they are solvable or not. If by this one meant merely that nobody had ever yet found a test which could be applied to any puzzle, there would be nothing at all remarkable in the statement. It would have been a great achievement to have invented such a test, so we can hardly be surprised that it has never been done. But it is not merely that the test has never been found. It has been proved that no such test ever can be found.

Let us get away from generalities a little and consider a particular puzzle. One which has been on sale during the last few years and has probably been seen by most of the readers of this article illustrates a number of the points involved quite well. The puzzle consists of a large square within which are some smaller movable squares numbered 1 to 15, and one empty space, into which any of the neighbouring squares can be slid leaving a new empty space behind it. One may be asked to transform a given arrangement of the squares into another by a succession of such movements of a square into an empty space. For this puzzle there is a fairly simple and quite practicable rule by which one can tell whether the transformation required is possible or not. One first imagines the transformation carried out according to a different set of rules. As well as sliding the squares into the empty space one is allowed to make moves each consisting of two interchanges, each of one pair of squares. One would, for instance, be allowed as one move to interchange the squares numbered 4 and 7, and also the squares numbered 3 and 5. One is permitted to use the same number in both pairs. Thus one may replace 1 by 2, 2 by 3, and 3 by 1 as a move because this is the same as interchanging first (1, 2) and then (1, 3). The original puzzle is solvable by sliding if it is solvable according to the new rules. It is not solvable by sliding if the required position can be reached by the new rules, together with a 'cheat' consisting of *one single* interchange of a pair of squares.[1] Suppose, for instance, that one is asked to get back to the standard position—

This article first appeared in *Science News*, 31 (1954), 7–23, published by the Penguin Press, and is printed by permission of the Estate of Alan Turing.

[1] It would take us too far from our main purpose to give the proof of this rule: the reader should have little difficulty in proving it by making use of the fact that an odd number of interchanges can never bring a set of objects back to the position it started from.

Solvable and Unsolvable Problems | 583

1	2	3	4
5	6	7	8
9	10	11	12
13	14	15	▨

from the position

10	1	4	5
9	2	6	8
11	3	▨	15
13	14	7	12

One may, according to the modified rules, first get the empty square into the correct position by moving the squares 15 and 12, and then get the squares 1, 2, 3, ... successively into their correct positions by the interchanges (1, 10), (2, 10), (3, 4), (4, 5), (5, 9), (6, 10), (7, 10), (9, 11), (10, 11), (11, 15). The squares 8, 12, 13, 14, 15 are found to be already in their correct positions when their turns are reached. Since the number of interchanges required is even, this transformation is possible by sliding.[2] If one were required after this to interchange say square 14 and 15 it could not be done.

This explanation of the theory of the puzzle can be regarded as entirely satisfactory. It gives one a simple rule for determining for any two positions whether one can get from one to the other or not. That the rule is so satisfactory depends very largely on the fact that it does not take very long to apply. No mathematical method can be useful for any problem if it involves much calculation. It is nevertheless sometimes interesting to consider whether something is possible at all or not, without worrying whether, in case it *is* possible, the amount of labour or calculation is economically prohibitive. These investigations that are not concerned with the amount of work involved are in some ways easier to carry out, and they certainly have a greater aesthetic appeal. The results are not altogether without value, for if one has proved that there is no method of doing something it follows *a fortiori* that there is no practicable method. On the other hand, if one method has been proved to exist by which the decision can be made, it gives some encouragement to anyone who wishes to find a workable method.

From this point of view, in which one is only interested in the question, 'Is there a systematic way of deciding whether puzzles of this kind are solvable?', the rules which have been described for the sliding-squares puzzle are much more special and detailed than is really necessary. It would be quite enough to say: 'Certainly one can find out whether one position can be reached from another by

---

[2] It can in fact be done by sliding successively the squares numbered 7, 14, 13, 11, 9, 10, 1, 2, 3, 7, 15, 8, 5, 4, 6, 3, 10, 1, 2, 6, 3, 10, 6, 2, 1, 6, 7, 15, 8, 5, 10, 8, 5, 10, 8, 7, 6, 9, 15, 5, 10, 8, 7, 6, 5, 15, 9, 5, 6, 7, 8, 12, 14, 13, 15, 10, 13, 15, 11, 9, 10, 11, 15, 13, 12, 14, 13, 15, 9, 10, 11, 12, 14, 13, 15, 14, 13, 15, 14, 13, 12, 11, 10, 9, 13, 14, 15, 12, 11, 10, 9, 13, 14, 15.

a systematic procedure. There are only a finite number of positions in which the numbered squares can be arranged (viz. 20922789888000) and only a finite number (2, 3, or 4) of moves in each position. By making a list of all the positions and working through all the moves, one can divide the positions into classes, such that sliding the squares allows one to get to any position which is in the same class as the one started from. By looking up which classes the two positions belong to one can tell whether one can get from one to the other or not.' This is all, of course, perfectly true, but one would hardly find such remarks helpful if they were made in reply to a request for an explanation of how the puzzle should be done. In fact they are so obvious that under such circumstances one might find them somehow rather insulting. But the fact of the matter is, that if one is interested in the question as put, 'Can one tell by a systematic method in which cases the puzzle is solvable?', this answer is entirely appropriate, because one wants to know if there is a systematic method, rather than to know of a good one.

The same kind of argument will apply for any puzzle where one is allowed to move certain 'pieces' around in a specified manner, provided that the total number of essentially different positions which the pieces can take up is finite. A slight variation on the argument is necessary in general to allow for the fact that in many puzzles some moves are allowed which one is not permitted to reverse. But one can still make a list of the positions, and list against these first the positions which can be reached from them in one move. One then adds the positions which are reached by two moves and so on until an increase in the number of moves does not give rise to any further entries. For instance, we can say at once that there is a method of deciding whether a patience can be got out with a given order of the cards in the pack: it is to be understood that there is only a finite number of places in which a card is ever to be placed on the table. It may be argued that one is permitted to put the cards down in a manner which is not perfectly regular, but one can still say that there is only a finite number of 'essentially different' positions. A more interesting example is provided by those puzzles made (apparently at least) of two or more pieces of very thick twisted wire which one is required to separate. It is understood that one is not allowed to bend the wires at all, and when one makes the right movement there is always plenty of room to get the pieces apart without them ever touching, if one wishes to do so. One may describe the positions of the pieces by saying where some three definite points of each piece are. Because of the spare space it is not necessary to give these positions quite exactly. It would be enough to give them to, say, a tenth of a millimetre. One does not need to take any notice of movements of the puzzle as a whole: in fact one could suppose one of the pieces quite fixed. The second piece can be supposed to be not very far away, for, if it is, the puzzle is already solved. These considerations enable us to reduce the number of 'essentially different' positions to a finite number, probably a few hundred millions, and the usual argument will then apply. There are some further complications, which we will not consider in detail, if we do not know how much clearance to allow for. It is necessary to repeat the process again and again allowing successively smaller and smaller clearances. Eventually one will find that either it can be solved, allowing a small clearance margin, or else it cannot be solved even allowing a small margin of 'cheating' (i.e. of 'forcing', or having the pieces slightly overlapping in space). It will, of course, be understood that this process of trying out the possible positions is not to be done with the physical puzzle itself, but on paper, with mathematical descriptions of the positions, and mathematical criteria for deciding whether in a given position the pieces overlap, etc.

These puzzles where one is asked to separate rigid bodies are in a way like the 'puzzle' of trying to undo a tangle, or more generally of trying to turn one knot into another without cutting the string. The difference is that one is allowed to bend the string, but not the wire forming the rigid bodies. In either case, if one wants to treat the problem seriously and systematically one has to replace the physical puzzle by a mathematical equivalent. The knot puzzle lends itself quite conveniently to this. A knot is just a closed curve in three dimensions nowhere crossing itself; but, for the purpose we are interested in, any knot can be given accurately enough as a series of segments in the directions of the three coordinate axes. Thus, for instance, the trefoil knot (Figure 1*a*) may be regarded as consisting of a number of segments joining the points given, in the usual (x, y, z) system of coordinates, as (1, 1, 1), (4, 1, 1,), (4, 2, 1), (4, 2, −1), (2, 2, −1), (2, 2, 2), (2, 0, 2), (3, 0, 2), (3, 0, 0), (3, 3, 0), (1, 3, 0), (1, 3, 1), and returning again with a twelfth segment to the starting point (1, 1, 1).[3] This representation of the knot is shown in perspective in Figure 1*b*. There is no special virtue in the representation which has been chosen. If it is desired to follow the original curve more closely a greater number of segments must be used. Now let *a* and *d* represent unit steps in the positive and negative X-directions respectively, *b* and *e* in the Y-directions, and *c* and *f* in the Z-directions: then this knot may be described as *aaabffddcccceeaffbbbddcee*.[4] One can then, if one wishes, deal entirely with such sequences of letters. In order that such a sequence should represent a knot it is necessary and sufficient that the numbers of *a*'s and *d*'s should be equal, and likewise the number of *b*'s equal to the number of *e*'s and the number of *c*'s equal to the number of *f*'s, and it must not be possible to obtain another sequence of letters with these properties by omitting a number of consecutive letters at the beginning

---

[3] Editor's note. In place of this sentence Turing's draft has: 'Thus for instance the trefoil knot may be regarded as consisting of a number of segments joining the points (0, 0, 0), (0, 2, 0), (1, 2, 0), (1, 2, 2), (1, −1, 2), (1, −1, 1), (−1, −1, 1), (−1, 1, 1), (2, 1, 1), (2, 0, 1), (2, 0, 3), (0, 0, 3), (0, 0, 0).'

[4] Editor's note. Turing's draft has 'bbacceeefddbbaaaeccddfff'.

586 | Alan Turing

(a)

(b)

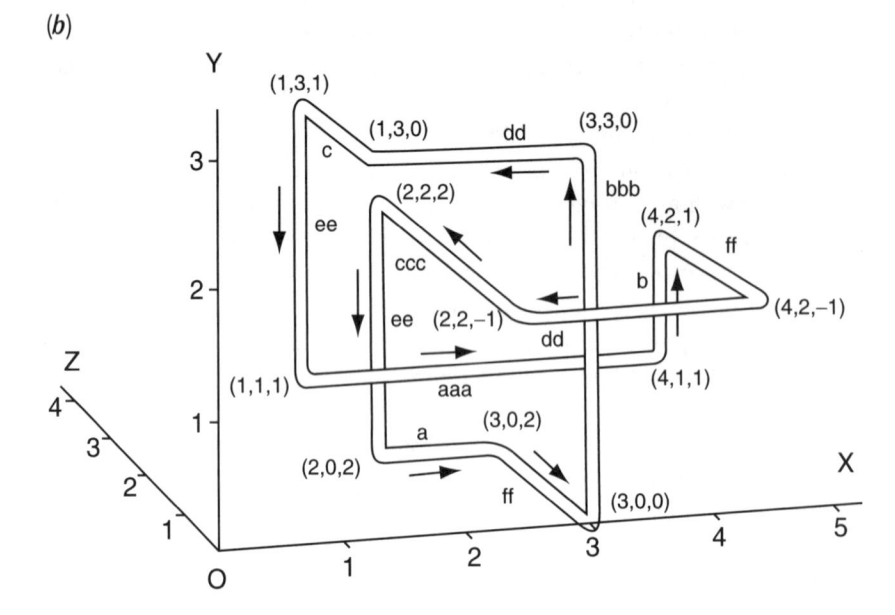

Figure 1. (a) The trefoil knot (b) a possible representation of this knot as a number of segments joining points.

or the end or both. One can turn a knot into an equivalent one by operations of the following kinds—

    (i) One may move a letter from one end of the row to the other.
   (ii) One may interchange two consecutive letters provided this still gives a knot.
  (iii) One may introduce a letter *a* in one place in the row, and *d* somewhere else, or *b* and *e*, or *c* and *f*, or take such pairs out, provided it still gives a knot.
  (iv) One may replace *a* everywhere by *aa* and *d* by *dd* or replace each *b* and *e* by *bb* and *ee* or each *c* and *f* by *cc* and *ff*. One may also reverse any such operation.

—and these are all the moves that are necessary.

Solvable and Unsolvable Problems | 587

It is also possible to give a similar symbolic equivalent for the problem of separating rigid bodies, but it is less straightforward than in the case of knots.

These knots provide an example of a puzzle where one cannot tell in advance how many arrangements of pieces may be involved (in this case the pieces are the letters *a, b, c, d, e, f*), so that the usual method of determining whether the puzzle is solvable cannot be applied. Because of rules (iii) and (iv) the lengths of the sequences describing the knots may become indefinitely great. No systematic method is yet known by which one can tell whether two knots are the same.

Another type of puzzle which we shall find very important is the 'substitution puzzle'. In such a puzzle one is supposed to be supplied with a finite number of different kinds of counters, perhaps just black (*B*) and white (*W*). Each kind is in unlimited supply. Initially a number of counters are arranged in a row and one is asked to transform it into another pattern by substitutions. A finite list of the substitutions allowed is given. Thus, for instance, one might be allowed the substitutions[5]

(i) WBW → B

(ii) BW → WBBW

and be asked to transform WBW into WBBBW, which could be done as follows

WBW → WWBBW → WWBWBBW → WBBBW
 (ii)   (ii)   (i)

Here the substitutions used are indicated by the numbers below the arrows, and their effects by underlinings. On the other hand if one were asked to transform WBB into BW it could not be done, for there are no admissible steps which reduce the number of *B*'s.

It will be seen that with this puzzle, and with the majority of substitution puzzles, one cannot set any bound to the number of positions that the original position might give rise to.

It will have been realized by now that a puzzle can be something rather more important than just a toy. For instance the task of proving a given mathematical theorem within an axiomatic system is a very good example of a puzzle.

It would be helpful if one had some kind of 'normal form' or 'standard form' for describing puzzles. There is, in fact, quite a reasonably simple one which I

---

[5] Editor's note. Turing's draft has: 'Thus for instance one might be allowed the substitutions

WBW → B
BWWW → WB
BWB → WWWB
WWB → W

and be asked to transform WBWWBWBBB into WBB, and this one could do first by substituting W for WWB and getting WBWBBBB and then successively WWWWBBBB, WWBBB, WBB.'

shall attempt to describe. It will be necessary for reasons of space to take a good deal for granted, but this need not obscure the main ideas. First of all we may suppose that the puzzle is somehow reduced to a mathematical form in the sort of way that was used in the case of the knots. The position[6] of the puzzle may be described, as was done in that case, by sequences of symbols in a row. There is usually very little difficulty in reducing other arrangements of symbols (e.g. the squares in the sliding squares puzzle) to this form. The question which remains to be answered is, 'What sort of rules should one be allowed to have for rearranging the symbols or counters?' In order to answer this one needs to think about what kinds of processes ever do occur in such rules, and, in order to reduce their number, to break them up into simpler processes. Typical of such processes are counting, copying, comparing, substituting. When one is doing such processes, it is necessary, especially if there are many symbols involved, and if one wishes to avoid carrying too much information in one's head, either to make a number of jottings elsewhere or to use a number of marker objects as well as the pieces of the puzzle itself. For instance, if one were making a copy of a row of counters concerned in the puzzle it would be as well to have a marker which divided the pieces which have been copied from those which have not and another showing the end of the portion to be copied. Now there is no reason why the rules of the puzzle itself should not be expressed in such a way as to take account of these markers. If one does express the rules in this way they can be made to be just substitutions. This means to say that the *normal form for puzzles is the substitution type of puzzle*. More definitely we can say:

Given any puzzle we can find a corresponding substitution puzzle *which is equivalent to it in the sense that given a solution of the one we can easily use it to find a solution of the other. If the original puzzle is concerned with rows of pieces of a finite number of different kinds, then the substitutions may be applied as an alternative set of rules to the pieces of the original puzzle. A transformation can be carried out by the rules of the original puzzle if and only if it can be carried out by the substitutions and leads to a final position from which all marker symbols have disappeared.*

This statement is still somewhat lacking in definiteness, and will remain so. I do not propose, for instance, to enter here into the question as to what I mean by the word 'easily'. The statement is moreover one which one does not attempt to prove. Propaganda is more appropriate to it than proof, for its status is something between a theorem and a definition. In so far as we know *a priori* what is a puzzle and what is not, the statement is a theorem. In so far as we do not know what puzzles are, the statement is a definition which tells us something about what they are. One can of course define a puzzle by some phrase beginning, for instance, 'A set of definite rules ...', but this just throws us back on the definition

---

[6] Editor's note. Turing's draft has 'positions'.

of 'definite rules'. Equally one can reduce it to the definition of 'computable function' or 'systematic procedure'. A definition of any one of these would define all the rest. Since 1935 a number of definitions have been given, explaining in detail the meaning of one or other of these terms, and these have all been proved equivalent to one another and also equivalent to the above statement. In effect there is no opposition to the view that every puzzle is equivalent to a substitution puzzle.[7]

After these preliminaries let us think again about puzzles as a whole. First let us recapitulate. There are a number of questions to which a puzzle may give rise. When given a particular task one may ask quite simply

(a) *Can this be done?*

Such a straightforward question admits only the straightforward answers, 'Yes' or 'No', or perhaps 'I don't know'. In the case that the answer is 'Yes' the answerer need only have done the puzzle himself beforehand to be sure. If the answer is to be 'No', some rather more subtle kind of argument, more or less mathematical, is necessary. For instance, in the case of the sliding squares one can state that the impossible cases *are* impossible because of the mathematical fact that an odd number of simple interchanges of a number of objects can never bring one back to where one started. One may also be asked

(b) *What is the best way of doing this?*

Such a question does not admit of a straightforward answer. It depends partly on individual differences in people's ideas as to what they find easy. If it is put in the form, 'What is the solution which involves the smallest number of steps?', we again have a straightforward question, but now it is one which is somehow of remarkably little interest. In any particular case where the answer to (a) is 'Yes' one can find the smallest possible number of steps by a tedious and usually impracticable process of enumeration, but the result hardly justifies the labour.

When one has been asked a number of times whether a number of different puzzles of similar nature can be solved one is naturally led to ask oneself

(c) *Is there a systematic procedure[8] by which I can answer these questions, for puzzles of this type?*

If one were feeling rather more ambitious one might even ask

(d) *Is there a systematic procedure[8] by which one can tell whether a puzzle is solvable?*

I hope to show that the answer to this last question is 'No'.

There are in fact certain types of puzzle for which the answer to (c) is 'No'.

---

[7] Editor's note. At this point Turing's draft contains the following: 'Some of these other definitions will be found in Refs (1), (3), (11), (13), and (16) vol II. Some equivalence theorems are proved in (4) and (14), and some propaganda on the matter will be found in (13). A very satisfactory account of all these problems will be found in (5).' Tantalizingly, the list of references is omitted.

[8] Editor's note. Turing's draft has 'systematic method'.

Before we can consider this question properly we shall need to be quite clear what we mean by a 'systematic procedure'[8] for deciding a question.[9] But this need not now give us any particular difficulty. A 'systematic procedure' was one of the phrases which we mentioned as being equivalent to the idea of a puzzle, because either could be reduced to the other. If we are now clear as to what a puzzle is, then we should be equally clear about 'systematic procedures'. In fact a systematic procedure is just a puzzle *in which there is never more than one possible move in any of the positions which arise and in which some significance is attached to the final result.*

Now that we have explained the meaning both of the term 'puzzle' and of 'systematic procedure', we are in a position to prove the assertion made in the first paragraph of this article, that there cannot be any systematic procedure for determining whether a puzzle be solvable or not. The proof does not really require the detailed definition of either of the terms, but only the relation between them which we have just explained. Any systematic procedure for deciding whether a puzzle were solvable could certainly be put in the form of a puzzle, with unambiguous moves (i.e. only one move from any one position), and having for its starting position a combination of the rules, the starting position and the final position of the puzzle under investigation.

The puzzle under investigation is also to be described by its rules and starting position. Each of these is to be just a row of symbols. As we are only considering substitution puzzles, the rules need only be a list of all the substitution pairs appropriately punctuated. One possible form of punctuation would be to separate the first member of a pair from the second by an arrow, and to separate the different substitution pairs with colons. In this case the rules

$$B \text{ may be replaced by } BC$$
$$WBW \text{ may be deleted}$$

would be represented by ': $B \rightarrow BC : WBW \rightarrow :$'. For the purposes of the argument which follows, however, these arrows and colons are an embarrassment. We

---

[9] Editor's note. At this point Turing's draft contains the following material, which is crossed out. 'It is a phrase which, like many others e.g. "vegetable" one understands well enough in the ordinary way. But one can have difficulties when speaking to greengrocers or microbiologists or when playing "twenty questions". Are rhubarb and tomatoes vegetables or fruits? Is coal vegetable or mineral? What about coal gas, marrow, fossilised trees, streptococci, viruses? Has the lettuce I ate at lunch yet become animal? The fact of the matter is that when one is applying a word, say an adjective, to something definite, one chooses the word itself so that it describes what one wants to describe fairly and squarely. If it doesn't one had better look for another word. But if one is playing twenty questions this just can't be done. The questions are about "the object", and one doesn't know what it is. The same sort of difficulty arises about question c) above. An ordinary sort of acquaintance with the meaning of the phrase "systematic method" won't do, because one has got to be able to say quite clearly about any kind of method that might be proposed whether it is allowable or not. Fortunately a number of satisfactory definitions were found in the late thirties, and they have...' [the fragment ends at this point].

shall need the rules to be expressed without the use of any symbols which are barred from appearing in the starting positions. This can be achieved by the following simple, though slightly artificial trick. We first double all the symbols other than the punctuation symbols, thus ': $BB \to BBCC$: $WWBBWW \to$:'. We then replace each arrow by a single symbol, which must be different from those on either side of it, and each colon by three similar symbols, also chosen to avoid clashes. This can always be done if we have at least three symbols available, and the rules above could then be represented as, for instance, '*CCCBBWBBCC BBBWWBBWWBWWW*'. Of course according to these conventions a great variety of different rows of symbols will describe essentially the same puzzle. Quite apart from the arbitrary choice of the punctuating symbols the substitution pairs can be given in any order, and the same pair can be repeated again and again.

Now let $P(R,S)$ stand for 'the puzzle whose rules are described by the row of symbols R and whose starting position is described by S'. Owing to the special form in which we have chosen to describe the rules of puzzles, there is no reason why we should not consider $P(R,R)$ for which the 'rules' also serve as starting position: in fact the success of the argument which follows depends on our doing so. The argument will also be mainly concerned with puzzles in which there is at most one possible move in any position; these may be called 'puzzles with unambiguous moves'. Such a puzzle may be said to have 'come out' if one reaches either the position B or the position W, and the rules do not permit any further moves. Clearly if a puzzle has unambiguous moves it cannot both come out with the end result B and with the end result W.

We now consider the problem of classifying rules R of puzzles into two classes, I and II, as follows:

Class I is to consist of sets R of rules, which represent puzzles with unambiguous moves, and such that $P(R,R)$ comes out with the end result W.

Class II is to include all other cases, i.e. either $P(R,R)$ does not come out, or comes out with the end result B, or else R does not represent a puzzle with unambiguous moves. We may also, if we wish, include in this class sequences of symbols such as *BBBBB* which do not represent a set of rules at all.

Now suppose that, contrary to the theorem that we wish to prove, we have a systematic procedure for deciding whether puzzles come out or not. Then with the aid of this procedure we shall be able to distinguish rules of class I from those of class II. There is no difficulty in deciding whether R really represents a set of rules, and whether they are unambiguous. If there is any difficulty it lies in finding the end result in the cases where the puzzle is known to come out: but this can be decided by actually working the puzzle through. By a principle which has already been explained, this systematic procedure for distinguishing the two classes can itself be put into the form of a substitution puzzle (with rules K, say). When applying these rules K, the rules R of the puzzle under investigation form the starting position, and the end result of the puzzle gives the result of the test.

Since the procedure always gives an answer, the puzzle $P(K,R)$ always comes out. The puzzle K might be made to announce its results in a variety of ways, and we may be permitted to suppose that the end result is B for rules R of class I, and W for rules of class II. The opposite choice would be equally possible, and would hold for a slightly different set of rules K', which however we do not choose to favour with our attention. The puzzle with rules K may without difficulty be made to have unambiguous moves. Its essential properties are therefore:

K has unambiguous moves.
$P(K,R)$ always comes out whatever R.
If R is in class I, then $P(K,R)$ has end result B.
If R is in class II, then $P(K,R)$ has end result W.

These properties are however inconsistent with the definitions of the two classes. If we ask ourselves which class K belongs to, we find that neither will do. The puzzle $P(K,K)$ is bound to come out, but the properties of K tell us that we must get end result B if K is in class I and W if it is in class II, whereas the definitions of the classes tell us that the end results must be the other way round. The assumption that there was a systematic procedure for telling whether puzzles come out has thus been reduced to an absurdity.

Thus in connexion with question (c) above we can say that there are some types of puzzle for which no systematic method of deciding the question exists. This is often expressed in the form, 'There is no *decision procedure* for this type of puzzle', or again, 'The decision problem for this type of puzzle is unsolvable', and so one comes to speak (as in the title of this article) about 'unsolvable problems' meaning in effect puzzles for which there is no decision procedure. This is the technical meaning which the words are now given by mathematical logicians. It would seem more natural to use the phrase 'unsolvable problem' to mean just an unsolvable puzzle, as for example 'to transform 1, 2, 3 into 2, 1, 3 by cyclic permutation of the symbols', but this is not the meaning it now has. However, to minimize confusion I shall here always speak of 'unsolvable decision problems', rather than just 'unsolvable problems', and also speak of puzzles rather than problems where it is puzzles and not decision problems that are concerned.

It should be noticed that a decision problem only arises when one has an infinity of questions to ask. If you ask, 'Is this apple good to eat?', or 'Is this number prime?', or 'Is this puzzle solvable?' the question can be settled with a single 'Yes' or 'No'. A finite number of answers will deal with a question about a finite number of objects, such as the apples in a basket. When the number is infinite, or in some way not yet completed concerning say all the apples one may ever be offered, or all whole numbers or puzzles, a list of answers will not suffice. Some kind of rule or systematic procedure must be given. Even if the number concerned is finite one may still prefer to have a rule rather than a list: it may be easier to remember. But there certainly cannot be an unsolvable decision problem in such cases, because of the possibility of using finite list.

Regarding decision problems as being concerned with classes of puzzles, we see that if we have a decision method for one class it will apply also for any subclass. Likewise, if we have proved that there is no decision procedure for the subclass, it follows that there is none for the whole class. The most interesting and valuable results about unsolvable decision problems concern the smaller classes of puzzle.

Another point which is worth noticing is quite well illustrated by the puzzle which we considered first of all in which the pieces were sliding squares. If one wants to know whether the puzzle is solvable with a given starting position, one can try moving the pieces about in the hope of reaching the required end-position. If one succeeds, then one will have solved the puzzle and consequently will be able to answer the question, 'Is it solvable?' In the case that the puzzle is solvable one will eventually come on the right set of moves. If one has also a procedure by which, if the puzzle is unsolvable, one would eventually establish the fact that it was so, then one would have a solution of the decision problem for the puzzle. For it is only necessary to apply both processes, a bit of one alternating with a bit of the other, in order eventually to reach a conclusion by one or the other. Actually, in the case of the sliding squares problem, we have got such a procedure, for we know that if, by sliding, one ever reaches the required final position, with squares 14 and 15 interchanged, then the puzzle is impossible.

It is clear then that the difficulty in finding decision procedures for types of puzzle lies in establishing that the puzzle is unsolvable in those cases where it *is* unsolvable. This, as was mentioned on page [589], requires some sort of mathematical argument. This suggests that we might try expressing the statement that the puzzle comes out in a mathematical form and then try and prove it by some systematic process. There is no particular difficulty in the first part of this project, the mathematical expression of the statement about the puzzle. But the second half of the project is bound to fail, because by a famous theorem of Gödel no systematic method of proving mathematical theorems is sufficiently complete to settle every mathematical question, yes or no. In any case we are now in a position to give an independent proof of this. If there were such a systematic method of proving mathematical theorems we could apply it to our puzzles and for each one eventually either prove that it was solvable or unsolvable; this would provide a systematic method of determining whether the puzzle was solvable or not, contrary to what we have already proved.

This result about the decision problem for puzzles, or, more accurately speaking, a number of others very similar to it, was proved in 1936–7. Since then a considerable number of further decision problems have been shown to be unsolvable. They are all proved to be unsolvable by showing that if they were solvable one could use the solution to provide a solution of the original one. They could all without difficulty be reduced to the same unsolvable problem. A number of these results are mentioned very shortly below. No attempt is made to explain the technical terms used, as most readers will be familiar with some of them, and the space required for the explanation would be quite out of proportion to its usefulness in this context.

(1) It is not possible to solve the decision problem even for substitution processes applied to rows of black and white counters only.
(2) There are certain particular puzzles for which there is no decision procedure, the rules being fixed and the only variable element being the starting position.
(3) There is no procedure for deciding whether a given set of axioms leads to a contradiction or not.
(4) The 'word problem in semi-groups with cancellation' is not solvable.
(5) It has recently been announced from Russia that the 'word problem in groups' is not solvable. This is a decision problem not unlike the 'word problem in semi-groups', but very much more important, having applications in topology: attempts were being made to solve this decision problem before any such problems had been proved unsolvable. No adequately complete proof is yet available, but if it is correct this is a considerable step forward.
(6) There is a set of 102 matrices of order 4, with integral coefficients such that there is no decision method for determining whether another given matrix is or is not expressible as a product of matrices from the given set.

These are, of course, only a selection from the results. Although quite a number of decision problems are now known to be unsolvable, we are still very far from being in a position to say of a given decision problem, whether it is solvable or not. Indeed, we shall never be quite in that position, for the question whether a given decision problem is solvable is itself one of the undecidable decision problems. The results which have been found are on the whole ones which have fallen into our laps rather than ones which have positively been searched for. Considerable efforts have however been made over the word problem in groups (see (5) above). Another problem which mathematicians are very anxious to settle is known as 'the decision problem of the equivalence of manifolds'. This is something like one of the problems we have already mentioned, that concerning the twisted wire puzzles. But whereas with the twisted wire puzzles the pieces are quite rigid, the 'equivalence of manifolds' problem concerns pieces which one is allowed to bend, stretch, twist, or compress as much as one likes, without ever actually breaking them or making new junctions or filling in holes. Given a number of interlacing pieces of plasticine one may be asked to transform them in this way into another given form. The decision problem for this class of problem is the 'decision problem for the equivalence of manifolds'. It is probably unsolvable, but has never been proved to be so. A similar decision problem which might well be unsolvable is the one concerning knots which has already been mentioned.

The results which have been described in this article are mainly of a negative character, setting certain bounds to what we can hope to achieve purely by reasoning. These, and some other results of mathematical logic may be regarded as going some way towards a demonstration, within mathematics itself, of the inadequacy of 'reason' unsupported by common sense.

Further reading

Kleene, S. C. *Introduction to Metamathematics*, Amsterdam, 1952.

Index

Ackermann, W. 49, 84
ADA 28
Adcock, F. 207, 219
Agnus Dei (2nd bombe) 255
Aiken, H. H. 29, 363
Alexander, C. H. O'D 257, 258, 259, 261, 263, 264, 265, 340, 345
Alexander, S. 368
Alexandria 'Y' (intercept) station 274
all inclusive (logic formula) 179ff
Allanson, J. T. 360
a-machine 60, 156; *see also* choice machine
analogy 492, 498–9
Analytical Engine 28–30, 363, 446, 455, 482
Andreae, J. H. 385
Apple Macintosh 366
Aquinas, St Thomas 450
ARPANET 92
Artificial Intelligence
  bacteriophage and 516
  discussion concerning 494–506
  draughts (checkers) and 356–8, 514
  genetic algorithms and 430–1, 513–14, 565, 575
  history of 1, 2, 3, 353–60, 401–32, 469–70, 487–506, 562–6
  intelligence as emotional concept 431, 491–2, 500, 501
  origin of term 353
  situated 439
  Turing pioneers 2, 3, 353–9, 374–6, 392–4, 401–9, 410–32, 433–9, 441–63, 465–71, 472–5, 476–8, 480, 482–6, 487–92, 494–5, 563–4, 565, 566–7, 569–75
  see also brain, chess, connectionism, expert system, heuristic, learning, Mathematical Objection, mechanical theorem proving, neural simulation, neuron-like computation, thinking machine, Turing test, search
Artificial Life
  AI and 439, 508

computer simulation and 507, 508, 510, 560–1
genetic algorithms and 513–14
history of 507–17
meaning of term 507
self-reproduction and 515–16
Turing pioneers 1, 3, 401, 405, 508–14, 517, 519–61
von Neumann and 22, 513, 514–17
see also connectionism, Fibonacci number, gastrulation, genetic algorithm, morphogenesis, neural simulation, neuron-like computation, non-linear equations, phyllotaxis, reaction-diffusion model, ring of cells or tissue
Asperula odorata (woodruff) 556, 557
atomic bomb 22
A-type unorganized machine 417, 418, 427, 429
Automatic Computing Engine (ACE)
  ACE Test Assembly 367, 398–400
  AI and 356, 374–5, 392–4
  as example of discrete controlling machine 412
  as example of universal practical computing machine 415
  Big ACE 368–9, 377
  chess and 393
  compared with trained P-type 428, 432
  derivatives of 368–71
  description of 365–6, 383–8
  draughts (checkers) and 356
  efforts to build 395–400
  letter from Turing to Ross Ashby about 374–5
  memory in 365, 366, 369, 376–7, 380–4
  number of operations without repetition 411
  optimum programming and 377
  Pilot Model of 16, 92, 365, 367, 368, 369, 377, 397, 399, 400
  preparation of problems for 389–91

Automatic Computing Engine (ACE) (*cont.*):
  random element in 391, 478
  self-modification and 374–5, 392–3, 419, 462, 470
  start-up procedure 390
  storage capacity of 382–3, 413
  Turing designs 2, 12, 27, 30, 31, 363–7, 369, 376–7, 378–94
  universal Turing machine and 16, 378–9, 383
  universality and 415–6
  Version H (Huskey) 32
  Version V (of Turing's design for) 399
  Version VII (of Turing's design for) 370, 396
  see also DEUCE
axiomatic (class or property) 151ff; *see also* choice machine

Babbage, C. 27–30, 236, 363, 371, 446, 455, 482
Babbage, D. 29
Baer, R. 128
Balme, D. 261
ban (unit) *see* deciban
Banburismus
  as method for reducing bombe time 256, 261–2
  Banburies 261, 282
  explanation of 281–5, 297, 299
  origin of name 261
  Turing invents 256, 261–2, 263, 279, 281, 285
  use in Hut 8 261–2, 285ff, 311–12, 314
Bates, M. A. 565, 568
Baudot-Murray code *see* teleprinter code
Bayley, D. 374
BBC 465, 476, 487, 493
Bell Telephone Laboratories 363, 393, 562
Bendix Corp. 369
Bernays, P. 48–9, 82, 127, 132, 191
Bernstein, F. 191
Bertrand, G. 234
best-first 563
Beurle, R. L. 360
Bigelow, J. 23
bigram tables (in Naval Enigma) 257, 258–61, 271–3, 280–1, 285–91, 311–12
BINAC 367
Birch, F. 219, 257, 259–60, 267, 279, 287–9, 292
Birkhoff, G. D. 130
Biuro Szyfrów 231ff

Bletchley Park
  Babbage discussed at 29
  establishment and development as codebreaking centre 217–20
  history of AI and 353–5, 563
  home of Colossus, first electronic computer 208–9, 362–3
  Turing comes to 1, 205, 217, 220, 257, 279
  work on Enigma at 217–31, 232, 234–5, 235–6, 238, 246–64, 265–312, 313–35, 336–40, 341–52, 353–5, 465
  work on Fish at 207–9, 262–3, 362–3, 465
  see also Government Code and Cypher School
Block, N. 434, 437
bomba
  demise of 233–4, 236, 246
  explanation of name 235–7, 291, 314
  function of 237–45
  history of AI and 354
  Poles invent 233, 292
  Poles reveal to British and French 234
  Stecker and 245–6
bombe
  appearance of 246–7, 248, 256, 291–2
  Banburismus and 261–2, 285, 287, 311
  bomba and 233, 234, 246
  designers of 218, 246, 254–5, 263, 292, 327, 329
  diagonal board of 254, 255, 323–34
  explanation of name 235–7, 291, 314
  first installed 253, 292
  four-wheel 344–5
  function of 246–53, 291–4, 315–35
  history of AI and 353–5, 469
  importance of 2, 218, 256–7, 262
  numbers of 256–7, 292, 311, 337, 338
  simultaneous scanning and 254–5, 319–20, 321, 323, 327, 343–4
  Spider and 255, 320–31
  Stecker and 235, 250–5, 291, 293, 314ff
  Turing and 218, 235, 246, 250–5, 314–35
  Turing's feedback method and 254–5, 322–3
  US Navy bombes 256–7, 342–5
  use against Naval Enigma 218, 253, 285ff, 315ff
  use of cribs with 240, 246, 248–55, 259, 287–8, 291, 293–4, 307, 315–35, 339, 344
  see also Agnus Dei, menu, Victory

Index | 599

Borelli, G. A. 498
Bowden, B. V. 568
brain
  analogy and 499
  as machine 2, 374, 382, 403, 405, 407,
    412–13, 418, 423–4, 425, 429–30, 431–2,
    451, 456–7, 459–61, 478, 482, 483, 499,
    500–1, 503–5
  continuity and 412–13, 456–7, 459
  digital computer as 2, 374, 375, 476–7, 478,
    479, 480, 482–6, 500ff
  electronic 374, 420, 484
  free will and 479, 484
  growth of 375, 517
  higher parts of 400–1
  imitation by computer 463, 456–7, 476, 477,
    478, 479, 494–5, 483–5
  intellectual search and 401, 430–1
  learning and 408–9, 421, 423–4, 438
  mechanical 482, 483, 484
  of child 424, 429, 432, 438, 460
  random element and 424, 478–9
  storage capacity of 383, 393, 459, 483, 500–1
  Turing machine and 407, 408, 424, 429
  Turing test and 477, 479
  see also connectionism, human being as
    machine, learning, neural simulation,
    neuron-like computing
Braithwaite, R. B. 131, 487
British Tabulating Machine Co. 246, 330, 339
Brooks, R. 439
Brouwer, L. E. J. 96
Brunsviga 412, 413, 480
B-type unorganized machine 403–9, 418, 422,
  429
  universal Turing machine and 407, 422
Burali-Forti paradox 170
Burks, A. W. 24, 27, 32, 513, 514–15
Bush, V. 29
Butler, S. 475

C 12
calculating machine 479–80, 483, 487ff, 578,
  591–2
Cambridge, University of 1, 15, 17, 27, 125,
  127, 131, 133, 205, 219, 265, 355, 377, 400,
  446, 487
  Mathematical Laboratory 358, 367
  see also King's College, St John's College

Cantor theory of ordinals 161–70
Carnegie Mellon University 359
central letter (of Enigma crib) 251–5, 317ff
Chamberlain, N. 217
Champernowne, D. G. 130, 563–4
Chandler, W. W. 369–70, 396
checkers *see* draughts
chess
  exhaustive search and 503
  genetic algorithm and 514, 565, 575
  heuristic search and 353–4, 374, 470
  history of computer chess 353, 356, 374,
    375, 393, 562–6, 569–75
  importance in AI 393, 394, 420, 439, 463,
    473, 562, 566
  learning and 375, 393, 492, 496, 498
  Max Newman on 492, 495, 496, 498, 503,
    504
  Turing test and 431, 442
  Turing's chess programmes 3, 353, 356, 412,
    431, 563–4, 565, 570–5
  see also Turochamp
child-machine 460–3; *cf* 424ff
choice machine 60, 77, 88
Chomsky, N. 565
Church, A.
  Church–Turing thesis and 44–5, 577
  comments on Turing as graduate
    student 126
  corresponds with Turing 205
  founder of *Journal of Symbolic Logic*
    205–6
  introduces term 'Turing machine' 6
  lambda calculus and 44, 52, 88, 126, 147ff,
    205–7, 211, 214–15, 360
  letter from Turing to concerning Post
    critique 92, 102
  mentioned by Turing in
    correspondence 126, 127, 128, 134,
    205–6, 207, 211, 213, 214
  on effective calculability 44–5, 59, 125–6,
    150
  ordinal logics and 125–6, 134, 137, 146, 163,
    177, 194, 205, 206
  theory of types and 205–6, 213
  work on *Entscheidungsproblem* 45, 48, 49,
    52, 59, 99, 125, 126, 207, 410, 450
Church's thesis 44–5; *see also* Church–Turing
  thesis

Church–Turing thesis
  ACE and 378, 383
  application of 43, 52, 53, 84–7
  arguments in favour of 42–3, 45, 74–9
  calculating machines and 479–80, 482–3, 578
  chess and 567–8, 570
  Church and 44–5, 577
  converse of 43
  Gödel and 45, 48, 581
  statement of 40–5, 58, 74, 414, 567, 570, 576, 577
  status of 42–3, 414, 568, 570, 577–8, 588–9, 590
Churchill, W. L. S. 262, 336–40, 342
CILLI 315
Clark, W. A. 360, 405–6
Clarke, J. 255, 258, 259, 330
class-subclass rule 462
Clayden, D. O. 31, 367, 368, 385
closure ('chain' in crib) 250ff, 317ff, 330
Cog (robot) 439
Colby, K. 489
Colebrook, F. M. 369, 400
Colossus 8, 208–9, 263, 362–3, 370, 373, 396, 480
colour (in Enigma) 227, 292
Commonwealth Scientific and Industrial Research Organisation (CSIRO) 367
computable function
  Church–Turing thesis and 44–5, 150–1, 578, 589
  computable number and 44, 58
  lambda calculus and 151–2, 211
  Max Newman on significance of 207
  meaning of term 44, 58, 79–80
  of integral variable 79–81, 151–2
  ordinal logics and 152–4, 158–9, 162, 163, 191
  see also computable number, computable sequence
computable number
  as opposed to definable number 58, 78–9
  Church–Turing thesis and 41, 43, 58, 60
  enumerability of 58, 72–4
  examples 58, 79–83, 95
  extent of 58, 74–9
  meaning of term 36, 41, 58–61, 95–6

  see also axiomatic, Church–Turing thesis, computable function, computable sequence, effective calculability, effective method, general recursive, human computer, primitive recursive, systematic method, Turing machine, uncomputable number, uncomputable sequence
computable sequence
  as opposed to definable 78–9
  Church–Turing thesis and 43
  computable function and 79ff
  continuum hypothesis and 191–2
  definition of computable number and 61, 95–6
  diagonal argument and 34–5, 37–9, 72–4
  effective calculability and 88–90
  enumeration of 66–8, 72–4
  meaning of term 33, 61
  of logical systems 171ff
  universal machine and 68
  see also computable number, uncomputable sequence
computer, history of
  at Bell Telephone Laboratories 363
  at Bletchley Park 8, 29, 208–9, 362–3, 373, 396
  at Cambridge 17, 355, 358, 367, 377
  at Commonwealth Scientific and Industrial Research Organization, Sydney 367
  at EMI 370–1
  at English Electric Co. 368, 369, 397
  at Harvard 29, 363, 364
  at IBM 17, 29, 357, 362
  at Manchester 2–3, 16, 17, 30, 209, 355, 356–7, 367, 369, 371–4, 396, 400, 401, 457, 480, 496, 508, 564, 565
  at MIT 29, 367
  at Moore School and Philadelphia 8, 16, 17, 21–7, 32, 364, 365, 366, 367, 373–4, 376, 380, 408
  at National Physical Laboratory 2, 12, 16, 27, 30–2, 92, 209, 356, 363–70, 372–3, 374–7, 378–94, 395–400
  at Packard-Bell 370–1
  at Post Office Research Station, Dollis Hill 208–9, 263, 362–3, 369–70, 373, 395–6, 397, 398
  at Princeton Institute for Advanced Study 16, 21–7, 32, 362, 373–4

Index | 601

at Radar Research and Development
   Establishment, Malvern 370
at Telecommunications Research
   Establishment, Malvern 208, 209, 373
at US Bureau of Standards 367–8
Babbage and 27–30, 236, 363, 446, 455
first 'personal' computer 369
Turing and 1, 2, 6, 9, 12, 15–17, 21–7, 30–1,
   55, 58–87, 206, 207, 209, 363, 371, 375,
   378–9, 383, 414–5
concept creation 492, 498–9; *see also* learning
connectionism
   early work on 360, 403, 405–6, 507–8
   irregular verbs and 402, 429
   meaning of term 360, 402
   Turing anticipates 356, 403–5, 406–7,
      408–9, 416–24, 429–30, 431–2, 510, 517
   see also A-type, B-type unorganized
      machine, learning, neuron-like
      computation
consciousness 451–3, 455, 456, 488, 566–7,
   569
constatation 314ff, 349ff
continuum hypothesis 191–2
convertible *see* lambda calculus
Coombs, A. W. M. 370, 396
Copernican theory 450
coral (Japanese cipher) 345
Courant, R. 127
crib
   all wheel order crib 253, 290, 291
   Banburismus and 256
   cribbing, art of 294–311
   Hut 8 Crib Room 294–5, 304
   in early days 278–81, 285ff, 294, 297, 306,
      311
   meaning of term 237
   mine-laying and 308
   Poles' use of 278–9
   types of 295–311
   use with bombe 240, 246, 248–55, 259,
      287–8, 291, 293–4, 307, 315–35, 339, 344
   use with 'mini bomba' 237–8
   W/T interception and 274, 275
   worked examples 248–50, 295–300, 315ff,
      347ff
   see also EINS catalogue, depth
CSIR Mark 1 (CSIRAC) 367
cultural search 430–1

Currier, P. 342
CYC 402

daily key 228ff, 421
D'Arcy Thompson, W. 508–9
Dartmouth College 353, 489, 565
Dartmouth Summer Research Project on
   Artificial Intelligence 353, 355, 359, 565
Darwin, C. G. 368, 370, 396, 397, 399, 400, 401
Davies, D. W. 21, 92–3, 367, 368, 564
Davis, M. 40, 41
D-Day 209
deciban 283ff
decision problem 6, 45ff, 143–4, 207, 393–4,
   469, 472, 579, 592–5; *see also*
   Entscheidungsproblem, unsolvable
   problem
Dedekind, J. W. R. 81
Deep Blue 563, 565–6
degree of unsolvability 99, 143–4
DENDRAL 360
Denniston, A. G. 217, 219, 234, 257, 279, 310,
   337, 342–3
depth (in Enigma breaking) 281ff, 295–301,
   302, 307, 311
Descartes, R. 498
Desch, J. 344
description number
   diagonal argument and 34–5, 37–9, 72–4
   halting problem and 39
   meaning of term 10–12, 67–8, 69
   of complete configuration 69, 89
   of oracle machine 142, 156–7
   ordinal logics and 184
   Post critique and 98ff
   printing problem and 39, 73–4
   satisfactoriness problem and 36–7, 68, 72–4
determinism 416, 447, 466, 475, 477–9, 483,
   484–5; *see also* free will, partially random
   machine, prediction, random element
DEUCE 368, 369, 397
diagonal argument 34–5, 37–9, 72–4, 142, 157,
   578–9, 591–2
diagonal board (of bombe) 254–5, 323–34
Difference Engine 28, 236, 363
differential analyser 29, 378, 412, 456–7, 480
discrete state machine
   brain and 412–13, 456, 459
   characterization of 412, 446–7

discrete state machine (*cont.*):
  compared with continuous machinery 412–13, 446, 456–7, 459
  complete description of behaviour of 413, 447, 448
  cryptography and 421
  digital computer as 446, 447, 450
  Logical Computing Machine as 413
  Mathematical Objection and 450–1
  numbers of states of 413, 447–8, 453
  partially random 416, 477–8
  prediction of 447, 448, 455–6, 457, 475, 485, 500
  simulation of by digital computer 448
  thinking and 455
  Turing test and 448, 456–7
  universality and 448, 455
discriminant (in Enigma) 230, 273
Dollis Hill *see* Post Office Research Station
draughts (in history of AI) 356–8, 514
Driscoll, A. M. 341–3, 345
dual (formula) 154ff

Eachus, J. J. 344
Eastcote (bombe outstation) 256
Eckert, J. P. 22, 25–7, 32, 367, 373, 376, 380
Eckert-Mauchly Computer Corp. 17
Eckert-Mauchly Electronic Control Co. 367
Eddington, A. S. 483
Edinburgh, University of 353, 359, 562
EDSAC 17, 358, 367, 377
education *see* learning
EDVAC 25–7, 364, 365, 366, 373–4, 408
Edward VIII 129–30
effective calculability
  abbreviation of treatment 150–2
  Church on 44–5, 59, 125–6, 150
  Church–Turing thesis and 44–5
  computability and 44–5, 59, 88–90
  Gentzen type ordinal logics and 194, 199
  see also effective method
effective method
  Church–Turing thesis and 42, 45, 125–6, 137, 479, 480
  duality and 158
  meaning of term 42
  Newman's test and 493
  ordinal formula and 139, 170
  rules of procedure and 171
  see also Church–Turing thesis, human computer, effective calculability, systematic method
EINS catalogue 286–7, 290, 291, 311
Einstein, A. 127
Eisenhart, L. 21, 131, 132
Ely, R. B. 344
EMI 370–1
EMI Business Machine 370–1
ENIAC 8, 22–3, 24, 26–7, 32, 364, 373, 376, 411, 412, 413, 480
Enigma
  Abwehr Enigma 246, 274
  appearance of machine 220, 221, 222, 223, 224, 226
  as example of apparently partially random machine 479
  Battle of the Atlantic and 2, 218, 257–61, 262
  breaking 231–64, 273–312, 314–35, 336–40, 341–52
  design of machine 220–8
  diagrams of machine 223, 224, 269
  four-wheel 225, 262, 270, 271, 295, 343–5
  German Air Force Enigma 220, 229–31, 233, 235, 255, 257, 279, 286, 291, 292, 293, 309, 339, 345
  German Army Enigma 220, 229–31, 232, 233, 235, 279, 286, 291, 292, 293, 309, 345
  German Naval Enigma 2, 218, 220, 225, 226, 229, 233, 271–312, 338–9, 341–52
  German Railway Enigma 322
  history of AI and 353–5
  indicator system for German Naval Enigma 257–8, 278–81
  Italian Naval Enigma 217, 246
  Knox's early work on Enigma 217, 232
  O Bar machine 277
  OP-20-G and 341–52
  operating procedures for 227–31, 269ff
  Polish work on Enigma 231–46, 257–8, 277–9, 292
  Tunny compared with 207
  Turing breaks German Naval Enigma 2, 206, 218, 253, 257–62, 279–82, 285–9, 314ff
  Turing's first work on Enigma 217, 277, 279–81, 285–9
  US attempts to break 342–5, 347–52

see also Banburismus, Bletchley Park, bomba, bombe, crib, daily key, discriminant, Government Code and Cipher School, Hut 8, indicator, indicator setting, indicator system, key, message setting, *Ringstellung*, *Stecker*, turnover, wheel order
English Electric Co. 368, 397
Entscheidungsproblem 6, 43, 45–53, 84–7, 125, 126, 207, 212, 393–4; see also decision problem, unsolvable problem
Erskine, R. 29
expert system 360, 402
extra-sensory perception 457–8

Farley, B. G. 360, 405
Feferman, S. 140, 141
Feigenbaum, E. A. 360
females (in Enigma indicators) 236, 242, 245
Fermat's last theorem 141, 155, 191, 472
Ferranti Ltd. 17, 356–7, 564–5
Ferranti Mark I computer 3, 17, 356–7, 374, 437, 483, 496, 503, 508, 510, 517, 552, 564–5; see also Manchester computers
Fibonacci number 508, 509, 517
Fieller, E. C. 399
Fish 207, 263; see also Tunny, Sturgeon, Thrasher
Fleming, I. 259, 289
flip-flop see Jordan Eccles trigger circuit
Flowerdown 'Y' (intercept) station 276
Flowers, T. H. 29, 208–9, 362–3, 369–70, 373, 395–6
flying bomb 275
formally definable (λ-definable) 88ff, 149ff; see also lambda calculus
Forrester, J. W. 367
fort (continuation) 230, 278–9
Foss, H. 290
Frankel, S. 22
Freeborn catalogue 282, 299, 311
Freeborn, F. 282, 297, 338–9
Freebornery (Hollerith section at Bletchley Park) 282, 338–9
free will 445, 449, 477–9, 484–5
French, R. M. 435, 490–1

G15 computer 369
Galilei, Galileo 450, 475

Gandy, R. O. 30, 42, 126, 400, 408, 433
gastrulation 509, 517, 519, 525, 558–60
Gauss, J. C. F. 411
General Problem Solver 359–60
general recursive function 150ff, 198ff
generalized recursion theory 143–4
generate-and-test 354
genetic algorithm (GA) 401, 430–1, 460, 463, 513–14, 565, 575
genetical search 430–1; see also genetic algorithm
Gentzen, G. 49, 51, 135, 137, 139, 141, 194, 202
GO 473, 474
Gödel argument 468; see also Mathematical Objection
Gödel, K.
  general recursive functions and 150, 153
  mentioned by Turing in correspondence 127, 213, 214
  ordinal logics and 1, 126, 136, 137, 138–9, 140, 146, 177, 180, 192
  remarks concerning Turing 45, 48, 581
  Turing's influence on 45, 48, 581
  work on incompleteness 1, 47–8, 59, 84, 126, 136, 138, 139, 140, 146, 160, 173, 189, 410, 411, 450, 467, 472, 580–1, 593
Gödel representation 74, 147ff
Gödel's incompleteness theorems
  Hilbert programme and 47–9, 84, 126, 135–9
  Mathematical Objection and 410, 411, 450, 467, 472
  ordinal logics and 1, 126, 136, 137, 138–9, 140, 141, 146, 160, 178, 180, 189, 192–3, 206, 212, 213, 215
  statement of 47–8, 84, 580–1
  substitution puzzles and 580–1, 593
  see also Gödel
Goldstine, H. H. 22, 24, 25, 27, 32, 364, 515
Good, I. J. 2, 258
Goodwin, E. T. (Charles) 32
Government Code and Cypher School (GC & CS)
  early history of 217–20
  history of AI and 353–5, 563
  significance of work of 2, 217, 262
  Turing joins 1, 205, 217, 220, 257, 279
  work on Enigma by 217–31, 232, 234–5, 235–6, 238, 246–64, 265–312, 313–35, 336–40, 341–52, 353–5, 465

Government Code and Cypher School
  (GC & CS) (cont.):
  work on Fish by 207–9, 262–3, 362–3, 465
  see also Bletchley Park, Room 40
Grey Walter, W. 508
growth see Artificial Life, morphogenesis
Grundstellung 230, 271, 312ff; see also
  indicator setting

Hall, P. 131
halting problem 39–40, 41; see also
  satisfactoriness problem
Hanslope Park 263
Hardy, G. H. 53, 127, 128
Harper, J. 247
Hartree, D. R. 363, 368, 455, 476, 482
Harvard Automatic Sequence Controlled
  Calculator 29, 363, 364
Harvard University 363, 364
Harvie-Watt, Brigadier 337
hat book 308, 310
Hayes, J. G. 31, 400
Heath Robinson (codebreaking machine) 208,
  263
Hebb, D. O. 403
Heimsoeth & Rinke Co. 277
Herbert (robot) 439
Herbrand, J. 45, 150, 153
Herivel, J. 335, 354–5
Herivelismus 335
heuristic 353–5, 356, 360, 514, 563, 564; see
  also search
Hilbert programme 46–9, 52–3, 84–7, 126,
  136–8, 142, 143, 215
Hilbert, D. 46–9, 53, 75, 77, 82, 84, 136, 139,
  143, 177; see also Hilbert programme
Hilton, P. 263, 371, 465
Hinsley, F. H. 218, 260, 287
Hiscocks, E. S. 398
Hitler, A. 209, 263, 264
Hodges, A. 263, 435, 436
Holden Agreement 344
Holland, J. 513
Hollerith, H. 29, 31
Hollerith punched card
  description of punched card plug-board
    equipment 30–1
  invention of 31
  relation to Babbage's Analytical Engine 29

use at Bletchley Park 282, 286, 338–9
use in Automatic Computing Engine 31,
  365, 388, 390
see also Freebornery
human being as machine 2, 3, 354, 355, 358,
  374–5, 382, 394–5, 401, 403, 405, 407,
  408, 420, 421, 422, 423–4, 425, 429–30,
  431–2, 438–9, 450–1, 456–7, 459–60, 478,
  482, 483, 499, 500, 502, 508–9; see also
  Artificial Intelligence, brain,
  connectionism, consciousness, free will,
  human computer, neuron-like
  computation, thinking machine
human computer
  ACE and 378, 387, 391–2
  Analytical Engine and 446
  calculating machine and 479–80
  characterization of 40
  characterization of digital computer
    and 444–5, 447, 480
  Church–Turing thesis and 41, 479–80
  computable number and 41
  history of computer and 40–1
  systematic method and 42, 43
  Turing machine as idealisation of 41, 42, 59,
    75–7, 79
Huskey, H. D. 26, 32, 55, 365, 368, 369, 373,
  398, 399
Hut 3 293
Hut 4 219, 257, 260
Hut 6 29, 219, 255, 274, 291, 292, 293, 309,
  336, 338, 339, 345, 354
Hut 8
  Alexander takes over 263
  early personnel of 258
  first breaks of into wartime traffic 259, 260,
    273, 286–7, 289–91, 341
  four wheel bombes and 256–7, 345
  impact of on Battle of Atlantic 2, 217,
    262
  indispensability of Turing to 263
  letter to Churchill concerning 338–40
  loses and regains Shark 344
  Mahon's history of, based on conversations
    with Turing 267–312
  pinches and 259–61
  Turing establishes 258
  Turing leaves 262–3
  uses first bombes 253, 255, 259

Hydra 511, 519, 556
hydrogen bomb 22

IAS computer 362; *see also* computer, history of at Princeton Institute for Advanced Study
IBM 17, 29, 362
IBM 701 computer 17, 357, 362
Illinois, University of 24
imitation game *see* Turing test
indeterminacy principle 478, 483
index of experiences 466, 474
indicator
  Banburismus and 282ff
  explanation of 230-1
  in Naval Enigma 257-8, 272, 273, 280-1
  Narvik Pinch and 258-9, 286
  Poles' method of attacking 240-6, 278, 279
  Turing attacks 257-8, 279-81, 285
indicator setting 230ff, 354-5; *see also Grundstellung*, indicator, indicator system
indicator system
  boxing (or throw-on) 273, 277-8
  change of 233, 235, 246
  for German Naval Enigma 257-8, 259, 261, 271-3, 278-81, 286ff, 314
  involving daily setting 231, 233
  involving enciphering message setting twice 229-30, 231-3, 240-3, 258
  meaning of term 231
induction 453, 454, 462
informality of behaviour 457
ingenuity (in mathematical proof) 135-8, 140, 192-3, 212-13, 215-16
initiative 429-31, 477
Institute for Advanced Study, Princeton 21, 23, 24, 125, 362; *see also* IAS computer
intellectual search 430-1
interception of Naval Enigma traffic 273-6, 293
Internet 92
intuition (in mathematics) 126, 135-8, 140, 142-3, 192-3, 202, 206, 212-13, 215-16, 579-80
irregular verbs 402, 429
Ismay, H. 336

Jacquard loom 28
James, W. 134

Jefferson, G. 451-2, 455, 487-8, 492
Jeffreys sheets 315
Jones, Squadron Leader 292-3
Jordan Eccles trigger circuit 380, 385, 426

Kalmár, L. 43
Kasparov, G. 563
K Book (in Naval Enigma) 271-3, 276, 277, 290, 311, 312
Key (Enigma)
  interception of 275
  meaning of term 227
  Ausserheimische Gewässe 258
  Bonito 273, 274
  Bounce 273
  Dolphin 257, 258, 259, 261, 271, 303, 308, 309, 310, 311, 341-3, 351
  Hackle 303
  Heimische Gewässe 257, 258, 259, 341
  Narwhal 271, 309
  Plaice 271, 308, 309, 310
  Porpoise 309
  Red 230, 335, 354
  Shark 271, 272, 286, 295, 309, 343, 344, 345
  Sucker 271, 309
  Süd 258, 273, 277
  Triton 343
  Turtle 309
  Yellow 235
  see also daily key
Keen, H. 218, 246, 263, 292, 329, 333
Keller, H. 461
Kendrick, F. A. 258
Kilburn, T. 209, 371-4, 401, 471
King's College, Cambridge 1, 131, 134, 214, 219, 264, 487
Kleene, S. C. 43, 44, 45, 88, 92, 102, 126, 127, 150, 153, 163, 211, 450
knots 585-7, 595
Knox, D. 217, 219, 220, 232, 233, 234, 246, 289

lambda calculus 44, 52, 88-90, 126, 139, 147ff, 205-6, 211-16, 360
Langton, C. G. 507, 508, 515
Laplace, P. 447
laws of behaviour 457
learning (by machine)
  chess and 565, 569, 575
  child-machine and 438-9, 460-3

learning (by machine) (*cont.*):
  connectionist 360, 402–3, 405–6, 406–7, 422–3, 424
  Darwin on Turing on 400
  discussed at Bletchley Park 353
  draughts (checkers) and 357–8, 514
  education of machinery 421–2, 460–3, 465–6, 473–5, 485, 492, 497, 503
  genetic algorithm and 514, 565, 575
  in B-type unorganized machines 403–5, 406–7, 422–3, 424
  in P-type unorganized machines 425–9
  index of experiences and 466, 474
  initiative and 430
  learning to improve learning methods 492, 497
  Lovelace objection and 455, 458–9, 485
  Mathematical Objection and 470, 504–5
  McCulloch–Pitts and 408–9
  NIM and 358
  Oettinger and 358–9
  pleasure-pain system and 424–5, 432, 461, 466, 474–5
  programme-modification and 374–5, 392–3, 419, 462, 470, 492, 496
  random element and 463, 466–7, 475
  Strachey on 358
  training 394, 403, 404ff, 424ff
Lederberg, J. 360
Lefschetz, S. 127
Lenat, D. 402
LEO 17
Letchworth bombe factory 256; *see also* British Tabulating Machine Co.
Letchworth Enigma 246, 318–19, 321, 325
Lewis, J. S. 265
limit system 171ff
Lisicki, T. 236
LISP 360
Loebner, H. 488–9
Lofoten Pinch 260, 290, 295
logic formula (in lambda calculus) 158ff
Logic Theorist 355, 565
logical computing machine *see* Turing machine
London Mathematical Society 5–6, 91, 92, 125, 130, 131, 132, 133, 207, 375
London Science Museum 28, 368
look-ahead 356, 565, 566

Lorenz SZ 40/42 cipher machine 479; *see also* Tunny
Los Alamos 22, 507, 510
Lovelace, A. A. 28–9, 455, 458, 480, 482, 485
Lucas, J. R. 468

machine intelligence *see* Artificial Intelligence
Macrae, N. 25
Mahon, A. P. 229, 258, 259, 260–1, 265–6
Manchester computers
  AI and 356–7
  'Baby' machine 2, 16, 209, 367, 371, 373, 374, 401, 413
  EDVAC and 373–4
  Ferranti Mark I 3, 17, 356–7, 374, 437, 483, 496, 503, 508, 510, 517, 552, 564–5
  first transistorized computer 369
  history of computer chess and 356, 564
  Manchester Mark I 16–17, 371, 373, 446, 447–8, 457, 479
  theorem-proving and 565
  Turing and 2–3, 209, 367, 371–4, 401, 564
Manchester University 3, 16–17, 209, 355, 367, 371–4, 400–1, 465, 487, 552, 564
  Computing Machine Laboratory 2, 3, 16–17, 209, 356, 357, 367, 371–4, 396, 400–1, 405, 508
Mandelbrot set 510
Mandelbrot, B. 510
Manhattan Project 22
Massachusetts Institute of Technology 359, 367, 405, 439
Mathematical Objection 355, 393–4, 410, 411, 436, 450–1, 467–70, 472–3
Mauchly, J. W. 22, 25–7, 32, 367, 373
McCarthy, J. 359, 360, 436–8
McCulloch, W. S. 403, 407–9
mechanic 471, 473
mechanical theorem proving 206, 215–16, 355, 401–2, 430–1, 564–5; *see also* Entscheidungsproblem, expert system, Logic Theorist, Mathematical Objection, Newman's test, ordinal logics, Turing machine
memory
  cathode ray tube 380, 396, 496
  delay line 26–7, 365, 366, 369, 376–7, 380–4, 385, 387, 395, 396, 399, 446

drum 366, 369
  magnetic core 369
menu (for bombe) 256, 291, 293, 295, 332
Menzies, S. 234
message setting 224, 228ff, 247
Meyer, *Funkmaat* 285
m-function *see* subroutine
Michie, D. 353, 359, 401, 562, 564
Michigan, University of 513
Milner-Barry, P. S. 336–7, 340
minimax 563, 565
Ministry of Supply 369
Minsky, M. L. 359, 491–2
Moore School of Electrical Engineering 22–3, 25–7, 32, 373, 376; *see also* University of Pennsylvania
Moore, H. 505
morphogenesis
  brain structure and 517
  genes as catalysts in 512, 523
  letter from Turing to Young about 517
  symmetry-breakers in 513, 519, 524ff
  Turing's theory of 3, 508–13, 519–61, 581
  see also Artificial Life, non-linear equations, reaction-diffusion model
Morse code 220, 222, 263, 273, 301
MOSAIC 369–70, 396

Napper, B. 373
Narvik Pinch 259, 286
National Cash Register Corp. 344
National Physical Laboratory 2, 16, 31–2, 55, 92, 103, 209, 355, 356, 363–70, 377, 378, 395–401, 408, 409, 508
Naval Section (at Bletchley Park) 267, 274, 276, 287–9, 290, 293, 295, 304
Nealey, R. W. 358
neural simulation 356, 374–5, 402–9, 418, 420, 423–4, 432; *see also* connectionism, neuron-like computation
neuron-like computation 356, 360, 402–9, 416–19, 422–4, 517; *see also* connectionism, neural simulation
New York 125, 127, 129
Newell, A. 354, 355, 359, 401, 563, 565
Newman, E. A. 367, 368
Newman, M. H. A.
  addresses Royal Society 371–2, 480
  arranges for Womersley to meet Turing 364

attacks Tunny by machine 208–9
attracts Turing to Manchester 3, 209, 367, 400
biography and obituaries of Turing quoted 3, 15, 48, 207, 408, 480
interviewed concerning Turing 15, 206, 207
involvement with 'On Computable Numbers' 15, 206–7
joins GC & CS 205, 207
letter to von Neumann mentioning Turing 209
mentioned by Turing in correspondence 130, 133, 207
on chess 492, 495, 496, 498, 503, 504
on universality 207, 371, 480
pioneers electronic computing 1, 2, 16, 208–9, 371–4
radio broadcasts 2, 437–8, 465, 476, 487–506
Turing's correspondence with 135, 139, 140, 205–6, 211–16, 470
Turing's influence on 1, 2, 16, 205–6, 209, 372, 373
see also Newman's test, Newmanry
Newmanry (section at Bletchley Park) 208–9, 373
Newman's test 492–3, 504–5
NIM 358
non-linear equations (in morphogenesis) 510, 544–5, 554, 560–1
Norfolk, C. L. 363
normal form 147, 149ff, 577–8, 587ff
Noskwith, R. 258, 267
noughts and crosses 420, 570
number-theoretic (theorem or problem) 152ff
  validity of logical system and 171ff

Oettinger, A. G. 358–9
OP-20-G 341–5, 347
Operational Intelligence Centre (OIC) 262
oracle machine (*o*-machine)
  circle-free 142, 157, 468
  completeness of ordinal logics and 143, 179–80
  degrees of unsolvability and 143–4
  description of 141–2, 156–7
  diagonal argument and 142, 157
  generalized recursion theory and 143–4

oracle machine (*o*-machine) (*cont.*):
  intuition and 142–3
  relative computability and 126, 144
ordinal formula 162ff
  C-K ordinal formula 163ff
  representation of ordinals by 162ff
ordinal logic
  Church and 125
  completeness of 139, 140–1, 159–60, 178–91, 206, 213
  definition of 170
  Gentzen type 194–202
  Gödel's theorem and 1, 126, 136, 138–9, 146, 178ff, 192–3
  Hilbert programme and 126, 136, 138–9, 146, 178ff, 192–3
  invariance of 180ff
  proof-finding machines and 136, 139–40, 206, 212, 215–16
  purpose of 1, 135–8, 146, 192–4
  Turing writes to Newman concerning 206, 212–13, 214–16
  Turing's work on supervised by Church 125–6, 134
  see also Church, intuition, oracle machine
Oxford, University of 219, 356

P (logical system) 139, 173ff, 177ff, 194ff
Packard-Bell PB250 370–1
Paired day (in Enigma) 259, 302, 309
Palluth, A. 233
paper machine 412, 416, 429, 431
Parry 489
partially random machine 416, 477–9; *see also* random element
Pascal 12
Pearcey, T. 367
Pearl Harbor 343
Penrose, R. 468–9
Péter, R. 43
Philadelphia 32; *see also* Moore School
phyllotaxis 519, 557, 561
pinch (of Enigma materials) 258–61, 288–9, 291; *see also* Narvik Pinch, Lofoten Pinch
Pitts, W. 403, 407–9
pleasure-pain system 424–9, 466, 477, 474–5; *see also* punishment, P-type unorganized machine
plug-board (of Enigma) 220ff; *see also* Stecker

plug-board calculators 30–2, 282; *see also* Hollerith card, programme-controlled
Polanyi, M. 465, 487
Popplewell, C. 564
Post Office Research Station (Dollis Hill) 208–9, 362, 369–70, 395–6, 397, 398
Post, E. L. 91–2, 102, 143–4, 150, 468, 577
preamble (of Enigma message) 230, 275
prediction
  in Laplacian universe 447
  of brain 477, 478, 483
  of continuous machine by discrete-state machine 456–7
  of discrete-state machine 447–8, 455–6, 457, 475, 485, 500
  of learning machine 462
  quantum mechanics and 478, 483
Price, F. 128–9, 130
Price, R. 128
primitive recursive (function or relation) 152ff, 174ff, 188, 200, 202
Princeton University 1, 21, 26, 125, 126, 130, 147, 150, 205, 373; *see also* Institute for Advanced Study
Principia Mathematica (logical system) 47–8, 52, 84, 138, 139, 173, 213, 355, 430, 472, 580
printing problem
  application to *Entscheidungsproblem* 52–3, 84–7
  characterization of 39, 73–4
  continuum hypothesis and 192
  for oracle machines 156–7
  ordinal logics and 185
  Post critique and 98ff
  satisfactoriness problem and 39, 79
  strengthened form of Church-Turing thesis and 41
Prinz, D. G. 356, 564–5
programme-controlled (opposed to *stored-programme*) 8, 22–3, 26, 28, 29–30, 31, 32, 362–3
programming, history of 2, 8, 10ff, 24–7, 28, 30–2, 55, 355, 365, 366, 375, 377, 388–92, 395, 399, 445, 460, 511; *see also* computer, history of, menu, stored-programme concept, subroutine, Turing machine, Turing pioneers computer programming
Prolog 12

Index | 609

Pryce, M. 128, 129, 130, 131, 132
PSILLI 315
P-type unorganized machine 425–9, 466, 492
   compared with ACE 428
   universal Turing machine and 427–8
   see also pleasure-pain system, punishment
punishment (and reward) 425, 426, 427, 428, 432, 461, 466, 474–5; *see also* learning, pleasure-pain system, P-type unorganized machine
Purple (Japanese cipher) 342
Pye Ltd. 254, 320, 321
Pyry 234, 246

Radar Research and Development Establishment (RRDE) 370
Radiolaria 517, 559
Radley, W. G. 395
Randell, B. 16, 22
random element
   brain and 424, 478–9
   determinism and 466, 475
   evolution and 463, 516
   free will and 445, 477–9, 484–5
   in ACE 391, 478
   in digital computer 391, 445, 466, 475, 477–8
   in learning 425, 426, 427, 429, 463, 466–7, 475, 496
   in partially random machine 416, 478
   in pleasure-pain system 425, 426, 427, 428, 429, 467
   in Turing test 458
   mathematical problem solving and 470, 505
   search and 463, 466, 467, 470
Ratio Club 508
reaction-diffusion model 509, 510–13, 519ff
recursion formula 193ff
re-encodements (in Enigma) 307–11
Rees, D. 373
Rejewski, M. 229, 231–46
   comments about Turing 235
R.H.V. (reserve hand cipher) 307–8, 310, 311
relative computability 126, 143–4
Research Section (at Bletchley Park) 207–8, 220, 262
Riemann hypothesis 155
ring (of cells or continuous tissue) 510–11, 519, 530ff

Ringstellung 227ff, 270ff, 335, 347ff, 354–5
Ringstellung cut-out (in bombe) 333, 335, 354–5
RISC 366
robot 392, 420–1, 439, 460–1, 463, 486, 508
rod-position (in Enigma) 238ff, 249ff, 315ff
Room 40 207, 218, 219; *see also* Government Code and Cypher School
Rosenblatt, F. 360, 403, 406
Ross Ashby, W. 360, 374–5
Rosser, J. B. 127, 137, 156, 450
Royal Astronomical Society 375
Royal Navy 259–61
Royal Society of London 3, 209, 371, 372
Różycki, J. 232, 236
Russell, B. A. W. 47, 131, 138, 139, 355, 430, 580

saga (in Enigma wheel-breaking) 232, 278
St John's College, Cambridge 207
Samuel, A. L. 357–8, 514
satisfactoriness problem
   characterization of 36–7, 68
   diagonal argument and 37–9, 72–4
   for oracle machines 142, 156–7
   Mathematical Objection and 467–8
   Post critique and 98ff
   printing problem and 39, 73–4, 79
   strengthened form of Church–Turing thesis and 41
Sayers, D. 410
Scarborough 'Y' (intercept) station 274–6, 290
Scherbius, A. 220, 277
Scheutz, E. 28
Scheutz, G. 28
Schmidt, H.-T. 231–2
Scholz, H. 131, 133
Scott, D. S. 356
SEAC 367
search (heuristic)
   bomba and 354
   bombe and 353–5
   chess and 353, 563–4, 565, 570ff
   draughts (checkers) and 356, 514
   intellectual activity as 354, 430–1
   Mathematical Objection and 469–70
   meaning of 'heuristic' 354
   random element and 463, 466, 467, 470
   theorem-proving and 401–2, 430–1

self-reproduction 514–16; *see also* universal constructor
sense organs 392, 420–1, 426, 439, 459, 460–1, 463
Shannon, C. E. 22, 393, 436–8, 562
Shaw, J. C. 355, 359, 565
Sherborne School 1
Shockley, W. B. 376
Shoenfield, J. R. 141, 143
Shopper 359
Simon, H. A. 354, 355, 359, 401, 439, 563, 565
simultaneous scanning (in bombe) 254–5, 319–20, 321, 323, 327, 343–4
Sinclair, H. 219, 234
Sinkov Mission 342
Sinkov, A. 342
sliding squares puzzle 582–4, 588, 589, 593
Slutz, R. J. 368
Smith-Rose, R. L. 398
Soare, R. I. 5
solipsism 452
Spanish Civil War 217, 224
spermatozoon 511, 556
Spider *see* bombe
standard description
  as first programming language 25
  halting problem and 39
  Mathematical Objection and 470
  meaning of term 10–12, 67–8
  Post critique and 98–101
  printing problem and 39
  satisfactoriness problem and 36–7
  universal Turing machine and 15, 17–20, 68–72, 105ff, 413–14
standardized logic 158ff
Stanford University 359, 360
statistical method (against Tunny) 208
Stecker 221, 222, 227ff, 270ff, 347ff
  bomba and 245
  explanation of 223, 224–6, 227, 270
  majority vote gadget and 335, 355
  self-steckered 245, 270, 279, 347, 349
  stecker hypothesis 253–5, 316ff
  Stecker Knock Out 286, 316
  stecker value of letter 252ff, 316ff
  Turing's method for discovering using bombe 235, 250–5, 314ff
  see also plug-board
Stibitz Relay Computer 363

Stibitz, G. R. 363
stop (in bomba or bombe) 237ff, 293, 319, 325, 327, 329–34, 354
stored-programme concept 1, 2, 3, 6, 8, 9, 12, 15–21, 21–7, 29–30, 30–2, 68–72, 105–117, 209, 362–8, 371–4, 375, 378–9, 389, 393
Strachey, C. S. 356–8, 564, 568
Sturgeon 207, 263
subroutine (subsidiary table)
  examples of 13–14, 54–7, 63–6, 108–12
  explanation of term 12
  history of programming and 12, 55, 375
  in ACE 12, 389–90
  m-functions and 54–7, 63ff
  universal Turing machine and 69–72, 112–15
substitution cipher 229
substitution puzzle 576–80, 587–92, 594
SWAC 368
systematic method
  Church–Turing thesis and 41–5, 568, 576, 577–8, 589
  compared with search involving random element 463, 467
  decidability and 47, 592
  Entscheidungsproblem and 52–3
  Gödel's theorem and 580–1, 593
  halting problem and 39
  meaning of term 42, 590
  meaning of 'unsolvable' and 579, 592
  solvability of puzzles and 578, 582, 583–4, 587, 589, 590, 591–2, 593
  see also Church–Turing thesis, human computer, effective calculability, effective method

Tarski, A. 177
Taylor, W. K. 360
Telecommunications Research Establishment (TRE) 208, 209, 373
teleprinter code 207–8, 263, 273
The '51 Society 465
thinking machine 2, 356, 358, 420, 434–9, 441–3, 448–52, 454–5, 459, 465, 472, 474, 475, 476–7, 478–9, 482, 485–6, 487, 488, 491, 492–3, 494–5, 498, 500, 501, 502, 504, 505, 566; *see also* Artificial Intelligence, brain, child-machine

Index | 611

Thomas, H. A. 397–400
Thrasher 207, 263
Thue problem (after Axel Thue) 99ff; *see also* substitution puzzle, word problem
Tiltman, J. H. 343
Tootill, G. C. 373–4
totalisator 363
Travis, E. 279, 287, 292, 293, 311, 338, 339
Tunny 207–9, 262–3, 479; *see also* Lorenz SZ 40/42
Turing, A. M.
  born (1912) 1
  educated at Sherborne School and King's College, Cambridge 1
  elected Fellow of King's College (1935) 1
  while proving *Entscheidungsproblem* unsolvable, invents universal Turing machine and fundamental stored-programme principle of modern computer 1, 2, 5–57, 58–87, 206, 207, 209, 363, 371, 375, 378–9, 383, 414–5
  discovers and explores the uncomputable 1, 3, 6, 32–53, 72–9, 84–7, 125–44, 146–202, 206, 212–13, 214–16, 355, 393–4, 410, 411, 450–1, 467–70, 472–3, 576–81, 582–95
  learns of Church's work on *Entscheidungsproblem* 125, 207
  studies at Princeton University (1936–8) 125–34
  writes to Sara Turing from Princeton (1936–8) 126–34
  explores ordinal logics and place of intuition in mathematics 1, 125–6, 135–44, 146–202, 206, 211–16
  returns from US to Fellowship at King's (1938) 1, 21, 134
  with outbreak of war transfers to Government Code and Cypher School at Bletchley Park (1939) 1–2, 205, 217–20, 257, 279
  breaks Naval Enigma indicator system and invents 'Banburismus' 1–2, 218, 257–8, 279–81, 256, 261–2, 281, 285
  designs bombe 218, 235, 240, 246, 250–5, 263, 314–35
  visits exiled Polish codebreakers in France 235
  leads attack on Naval Enigma 2, 218, 253, 257–62, 279–81, 285–9, 314ff, 341ff
  writes to Newman concerning logic (*c*.1940) 205–6, 211–16
  writes to Churchill (1941) 336–40
  advises US codebreakers 341–52
  leaves Enigma (1942) 262–4, 288, 312
  works on Tunny and invents 'Turingery' (1942) 262–3
  visits US (Nov 1942–Mar 1943) 263
  works on speech enciphermnent at Hanslope Park (1943–5) 263
  OBE for war work 2, 264
  joins National Physical Laboratory and designs electronic stored-programme digital computer (1945) 2, 12, 27, 30, 31, 363–4, 364–71, 376–7, 378–94
  pioneers computer programming before hardware in existence (1945–7) 2, 12, 25, 27, 30–2, 55, 356, 365, 366, 372, 375, 377, 378ff, 388–92, 395, 563–4
  pioneers Artificial Intelligence and cognitive science 2, 3, 353–9, 374–6, 392–4, 401–9, 410–32, 433–9, 441–63, 465–71, 472–5, 476–8, 480, 482–6, 487–92, 494–5, 563–4, 565, 566–7, 569–75
  writes to Ross Ashby concerning computer and brain 374–5
  pioneers computer chess 3, 353, 356, 374, 375, 393, 394, 412, 420, 431, 439, 463, 470, 473, 514, 562–4, 565–6, 569–75
  lectures on computer design in London (1946–7) 2, 355–6, 372–3, 375–7, 378–94
  visits computer projects in US (Jan 1947) 397
  proposes electronics section at NPL (1947) 397–8
  spends sabbatical in Cambridge (1947–8) 400–1
  leaves NPL for Manchester University (1948) 2–3, 209, 367, 400–1
  influence on computer developments at Manchester 2–3, 209, 371–4, 401, 564
  anticipates connectionism 356, 402–9, 416–24, 429–30, 431–2, 510, 517
  invents Turing test 2, 356, 359, 401, 431, 433–9, 441–63, 477, 479, 484, 485, 488, 489–90, 494–5, 496, 503
  pioneers Artificial Life with study of morphogenesis 1, 3, 401, 405, 508–14, 517, 519–61

Turing, A. M. (*cont.*):
  writes to Young concerning morphogenesis and neuron growth (1951) 517
  broadcasts on BBC radio (1951–2) 356, 358, 465–71, 472–5, 476–80, 482–6, 487–93, 494–506
  uses computer to explore non-linear systems empirically 510, 561
  dies (1954) 1, 510
Turing degree 144
Turing, E. S. (Sara) 21, 125–34, 471, 476
Turing machine
  circle-free 32–3, 34–5, 36, 37ff, 60–1, 72–4, 79, 98ff, 142–3, 144, 153, 154, 468
  circular 32–3, 36, 37, 60–1, 72–4
  continuum hypothesis and 191–2
  diagonal argument and 34–5, 37–9, 72–4
  effective computability and 88–90, 150
  Entscheidungsproblem and 52–3, 84–7
  introduction to 6–14, 32–41, 104–5
  intuition and 137–8, 139–40, 215–16
  learning and 407, 422, 424–30, 438–9, 470
  Mathematical Objection and 450–1, 468, 470
  oracle machine and 141–3, 144, 156–7
  ordinal logics and 137–8, 184–5
  provability by 52–3, 72–9, 84–7, 136, 140–1, 193, 206, 215–16, 430–1, 470, 472
  Turing's exposition of 59ff, 413–14
  see also universal Turing machine
Turing sheets (in Enigma breaking) 314–15, 316
Turing test
  1948 presentation of 401, 431, 433
  1950 presentation of 2, 356, 433–6, 441–3, 448–58, 488, 489–90
  1951 presentation of 433–4, 477, 479, 484, 485
  1952 presentation of 433–4, 488–9, 494–5, 496, 503
  consciousness and 451–3, 455, 456, 566–7, 569
  imitating brain and 477, 479, 494–5
  learning and 460ff
  Loebner Prize for 488–9
  objections to 436–8, 448–58, 490–1
  predictions concerning success in 449, 459, 460, 484, 489–90, 495, 566, 569
  Shopper and 359
  see also thinking machine

Turingery (in Tunny breaking) 263
Turing-reducible 143–4
Turing's feedback method (in bombe) 254–5, 322–3
turnover (of Enigma wheel) 225–6, 239, 241, 249, 253, 268, 270, 279, 284–5, 315ff, 342, 347ff
Turochamp 563–4
Tutte, W. T. 208
twiddle (in Enigma breaking) 287, 290, 311
Twinn, P. 217, 258, 260, 286, 287
twisted-wire puzzle 584–5, 594
type fallacy 462
Typex machine 249, 250, 253

U-boat 257, 261, 262, 272, 273–4, 291, 306, 308, 309, 310, 311, 343, 347
U-110 41, 261
U-559 344
Ulam, S. M. 21
uncomputable number 36, 58, 72–4, 79; *see also* computable number
uncomputable sequence 33ff, 72–4, 79, 83; *see also* computable sequence
Unilever Ltd. 399
UNIVAC 17
universal constructor 516
universal Turing machine
  Artificial Life and 515–16
  B- and P-type machines and 407, 422, 424–5, 427–8
  brain and 407, 424, 429–30, 478
  Church-Turing thesis and 41, 43, 479–80
  compared with ACE 375, 383, 384
  compared with Analytical Engine 29–30, 455
  corrections to 91–101, 115–24
  history of computer and 1, 2, 6, 9, 12, 15–17, 21–7, 30–1, 55, 206, 207, 209, 363, 371, 375, 378–9, 383, 414–15
  introduction to 15–21, 105–15
  lambda-definability and 44, 88–90
  'paper interference' and 418–19
  Turing's exposition of 68–72, 383, 413–14
  unsolvable problems and 576
  see also Entscheidungsproblem, halting problem, printing problem, satisfactoriness problem, Turing machine

Index | 613

universality 415–16, 446–8, 455, 480, 482–3, 501, 516, 569–70; *see also* universal Turing machine
University of California at San Diego 402
University of Pennsylvania 22, 380; *see also* Moore School
unorganized machine 402–9, 416–19, 422–9, 430, 432, 467, 517; *see also* A-type, B-type, connectionism, neural simulation, neuron-like computing, pleasure-pain system, P-type
unsolvable problem 6, 33, 98ff, 576–81, 582–95
   meaning of term 144, 579, 592
   see also decision problem, degree of unsolvability, *Entscheidungsproblem*, halting problem, printing problem, satisfactoriness problem, systematic method, Thue problem
US Army 342, 344
US Bureau of Standards 368
US Navy 256–7, 342, 344
Uttley, A. M. 360, 368–9

Vassar College 129, 132
Victory (1st bombe) 253, 254, 259
von Neumann, J.
   as pioneer of Artificial Life 513, 514–17
   as pioneer of stored-programme computer 1, 16, 21–7, 32, 362, 364, 365, 408
   letter to Wiener mentioning Turing 23, 515, 516–17
   Manchester computer and 373–4
   Max Newman writes to 209
   mentioned by Turing in correspondence 127, 134, 213
   minimax and 563

   on undecidability 53
   Turing offered job as assistant to 21, 134
   Turing's influence on 1, 16, 21–2, 23–5, 515–16

W/T (wireless telegraphy) 273–5
Waddington, C. H. 509, 521
Wang, H. 45
Weeks, R. 342
Welchman, W. G. 218, 254, 255, 263, 264, 292, 327, 329, 340
Werft (dockyard cipher) 307–8, 310, 311
Weyl, H. 48, 127, 131
wheel core (in Enigma) 223, 226, 227, 238, 244, 270; *see also* rod-position
wheel order (in Enigma) 222, 223, 225, 227ff, 270ff, 315ff, 343ff, 347ff
Whirlwind I computer 367
Whitehead, A. N. 47, 138, 139, 355, 580
Wiener, N. 23, 408, 515, 516
Wilkes, M. V. 367, 476
Wilkinson, J. H. 366, 367, 368, 398, 399
Williams, F. C. 16, 209, 367, 371–4, 401, 471, 476
Wilmslow 1
Wittgenstein, L. 41, 130, 487
Womersley, J. R. 31, 363–4, 375, 395, 398, 399
Woodger, M. 31, 32, 356, 364, 367, 368, 376, 388, 508
word problem 594; *see also* substitution puzzle, Thue problem
Wylie, S. 258, 286
Wynn-Williams, C. E. 208

Young, J. Z. 487, 509, 517

Zermelo, E. 194, 213
Zygalski, H. 232